Inorganic Rings and Polymers of the p-Block Elements
From Fundamentals to Applications

Inorganic Rings and Polymers of the p-Block Elements
From Fundamentals to Applications

Tristram Chivers

Department of Chemistry, University of Calgary, Calgary, Alberta, Canada

Ian Manners

School of Chemistry, University of Bristol, Bristol, UK

RSCPublishing

ISBN: 978-1-84755-906-7

A catalogue record for this book is available from the British Library

© Tristram Chivers and Ian Manners 2009

All rights reserved

Apart from fair dealing for the purposes of research for non-commercial purposes or for private study, criticism or review, as permitted under the Copyright, Designs and Patents Act 1988 and the Copyright and Related Rights Regulations 2003, this publication may not be reproduced, stored or transmitted, in any form or by any means, without the prior permission in writing of The Royal Society of Chemistry or the copyright owner, or in the case of reproduction in accordance with the terms of licences issued by the Copyright Licensing Agency in the UK, or in accordance with the terms of the licences issued by the appropriate Reproduction Rights Organization outside the UK. Enquiries concerning reproduction outside the terms stated here should be sent to The Royal Society of Chemistry at the address printed on this page.

Published by The Royal Society of Chemistry,
Thomas Graham House, Science Park, Milton Road,
Cambridge CB4 0WF, UK

Registered Charity Number 207890

For further information see our website at www.rsc.org

Preface

The chemistry of inorganic ring systems of the p-block elements has a long and venerable history that dates back to the early 19th century. In addition to an assortment of homocyclic systems, many of which can be related to well-known organic rings, there is an extraordinary variety of heterocyclic systems that result from various combinations of p-block elements. Early studies focused on the development of appropriate methods for the synthesis of inorganic rings and, subsequently, on structural determinations. In the case of unsaturated systems, the unusual features of inorganic heterocycles have been addressed by theoretical chemists and their insights have led to the advancement of our understanding of chemical bonding. The development of polymers with p-block element backbones (other than carbon) with useful applications is more recent. It received a major impetus at the time of the Second World War with the evolution of the silicone industry in response to the need for polymeric materials with higher thermal stability than that of organic polymers. Today, while the fundamental interest in the synthesis, structures and bonding in novel inorganic rings continues, the emphasis has shifted to their applications as precursors for useful materials. The synergic relationship between rings and polymers through the process of ring-opening polymerisation has been a major driving force in this progression. In addition to polymers, inorganic heterocycles have been shown to provide a convenient source for the generation of a large variety of semiconductors or ceramic materials.

Although standard inorganic textbooks cover some aspects of the better known inorganic ring systems and polymers (see Chapter 1), there is no monograph that is dedicated to this topic. This book attempts to fill that gap. In particular, we aim to address both fundamental and applied aspects of these inorganic materials. The first eight chapters are intended as a general introduction to the subject that would be suitable as a supplement to a senior undergraduate course or graduate course in p-block element chemistry.

Inorganic Rings and Polymers of the p-Block Elements: From Fundamentals to Applications
By Tristram Chivers and Ian Manners
© Tristram Chivers and Ian Manners 2009
Published by the Royal Society of Chemistry, www.rsc.org

Chapters 9–12 provide a more detailed treatment of the chemistry of inorganic rings and polymers involving the elements of groups 13–16 that is addressed to a more specialised readership. In these chapters, our intention is to combine the early seminal work with the most exciting recent developments. Throughout the book, extensive cross-referencing has been included in order to facilitate access to additional information about topics that appear in the introductory Chapters 2–8 and to refer back to fundamental concepts during the more advanced exposition presented in Chapters 9–12.

We acknowledge, with gratitude, helpful (and encouraging) input from the following individual scientists in their various areas of expertise: Professor Neil Burford (Dalhousie University, Canada), Dr Paul Kelly (University of Loughborough, England), Professor Jack Passmore (University of New Brunswick, Canada), Professor Roland Roesler (University of Calgary, Canada), Professor Akira Sekiguchi (University of Tsukuba, Japan), Professor Bob West (University of Wisconsin, USA) and Professor J. Derek Woollins (University of St Andrews, Scotland). Their perceptive suggestions have enhanced the quality and accuracy of the final version of this book substantially. Nevertheless, there are undoubtedly shortcomings in the form of errors or omissions for which the authors are entirely responsible.

A special acknowledgment is accorded to Dr Jari Konu (University of Calgary), who prepared all the structural drawings, figures and schemes. Without his diligence and expertise, this book would not have come to fruition. I.M. thanks Vivienne Blackstone, Martin Bendle and Deborah O'Hanlon Manners in Bristol for helpful comments, corrections and suggestions. T.C. acknowledges Andrea Corrente, Jamie Ritch and Stuart Robertson in Calgary for their suggestions and assistance in proof-reading at various stages of manuscript preparation.

The timely and professional contributions of the production and editorial staff of RSC Publishing are also gratefully acknowledged.

Finally, we thank our wives, Sue and Deborah, for their continuing support and patience.

Tris Chivers
Calgary

Ian Manners
Bristol

Contents

Chapter 1 Introduction to Inorganic Rings and Polymers

    1.1 Classification 1
    1.2 Historical Background 2
    1.3 Industrial Applications 2
    1.4 Scope and Limitations 3
    1.5 Organisation 3
    1.6 Nomenclature 4
    1.7 Bibliography 4
    References 5

Chapter 2 Synthetic Methods

    2.1 Cyclocondensation Reactions 7
    2.2 Reductive Coupling 11
    2.3 Dehydrocoupling 13
    2.4 Alkane Elimination Reactions 15
    2.5 Cycloaddition Reactions 16
    References 18

Chapter 3 Characterisation Methods

    3.1 X-ray and Electron Diffraction 21
    3.2 Chromatography 23
    3.3 Mass Spectrometry 24
    3.4 Nuclear Magnetic Resonance (NMR) Spectroscopy 25
        3.4.1 Introduction 25
        3.4.2 Applications 27
        3.4.3 Spin System Notation 30

Inorganic Rings and Polymers of the p-Block Elements: From Fundamentals to Applications
By Tristram Chivers and Ian Manners
© Tristram Chivers and Ian Manners 2009
Published by the Royal Society of Chemistry, www.rsc.org

	3.5	Infrared and Raman Spectroscopy	31
	3.6	Electron Paramagnetic Resonance (EPR) Spectroscopy	33
	3.7	UV–Visible Spectroscopy	36
	References		37

Chapter 4 Electronic Structure and Bonding

4.1	Delocalisation (Aromaticity) in Inorganic Rings		39
	4.1.1	Introduction	39
	4.1.2	π-Electron Delocalisation	40
	4.1.3	σ-Electron Delocalisation	45
4.2	Frontier Orbital Considerations		45
	4.2.1	Introduction	45
	4.2.2	π^*–π^* Interactions	46
	4.2.3	Formation of Inorganic Ring Systems	48
	4.2.4	Reactivity of Inorganic Ring Systems	50
References			50

Chapter 5 Paramagnetic Inorganic Rings

5.1	Neutral Radicals		53
	5.1.1	Homocyclic Systems	53
	5.1.2	Heterocyclic Systems	54
5.2	Cation Radicals		55
	5.2.1	Homocyclic Systems	55
	5.2.2	Heterocyclic Systems	56
5.3	Anion Radicals		56
5.4	Biradicals		58
	5.4.1	Stable Biradicals	58
	5.4.2	Biradicaloids	59
	5.4.3	Biradicals as Reaction Intermediates	62
References			64

Chapter 6 Inorganic Macrocycles

6.1	Homocyclic Systems		67
6.2	Heterocyclic Systems		68
	6.2.1	Saturated Rings	68
	6.2.2	Unsaturated Rings	71
6.3	Host–Guest Chemistry		74
6.4	Supramolecular Assemblies		77
References			80

Chapter 7 Ligand Chemistry

7.1 σ-Complexes ... 82
   7.1.1 Homocyclic Ligands ... 82
   7.1.2 Heterocyclic Ligands ... 86
   7.1.3 Macrocyclic Ligands ... 88
7.2 π-Complexes ... 90
   7.2.1 Homocyclic Systems ... 91
   7.2.2 Heterocyclic Systems ... 93
References ... 94

Chapter 8 Synthesis and Characterisation of Inorganic Polymers

8.1 Synthesis of Inorganic Polymers ... 98
   8.1.1 Challenges ... 98
   8.1.2 Ring-opening Polymerisation (ROP) ... 100
8.2 Characterisation of Inorganic Polymeric Materials ... 102
   8.2.1 Structural Characterisation ... 102
   8.2.2 Molecular Weights and Molecular Weight Distributions ... 103
   8.2.3 Thermal Transitions: Amorphous and Crystalline Polymers ... 104
   8.2.4 Chain Conformations ... 107
   8.2.5 Thermal Stability: Thermogravimetric Analysis ... 107
References ... 108

Chapter 9 Group 13: Rings and Polymers

9.1 Homoatomic Systems ... 110
   9.1.1 Neutral Rings ... 110
   9.1.2 Anionic Rings ... 113
9.2 Boron–Nitrogen Rings ... 116
   9.2.1 Borazines ... 116
   9.2.2 Other Unsaturated Boron–Nitrogen Rings ... 120
   9.2.3 Saturated Boron–Nitrogen Rings (Cycloborazanes) ... 124
   9.2.4 Applications of Boron–Nitrogen Rings ... 125
   9.2.5 Boron–Nitrogen Polymers ... 126
9.3 Boron–Phosphorus Rings ... 128
   9.3.1 Unsaturated Systems ... 128
   9.3.2 Cyclic Biradicals ... 130
   9.3.3 Saturated Systems ... 133
   9.3.4 Boron–Phosphorus Chains and Polymers ... 134

9.4	Aluminium–, Gallium– and Indium–Nitrogen Rings		135
	9.4.1	Unsaturated Systems	135
	9.4.2	Saturated Systems	138
9.5	Aluminium–, Gallium– and Indium–Pnictogen Rings		138
	9.5.1	Unsaturated Systems	138
	9.5.2	Saturated Systems	140
	9.5.3	Single-source Precursors for III–V Semiconductors	142
9.6	Boron–Chalcogen Rings		143
	9.6.1	Boron–Oxygen Rings	143
	9.6.2	Boron–Sulfur and –Selenium Rings	144
9.7	Aluminium–, Gallium– and Indium–Oxygen Rings		147
9.8	Aluminium–, Gallium– and Indium–Chalcogen Rings		148
References			150

Chapter 10 Group 14: Rings and Polymers

10.1	Homoatomic Rings and Polymers: Saturated Systems		159
	10.1.1	Cyclopolysilanes	159
	10.1.2	Cyclopolygermanes, -stannanes and -plumbanes	162
	10.1.3	Cyclic Polyanions of Silicon, Germanium, Tin and Lead	165
	10.1.4	Polymers with Backbones of Heavier Group 14 Elements	167
10.2	Homoatomic Rings: Unsaturated Systems		170
	10.2.1	Three-membered Rings	170
	10.2.2	Four-membered and Larger Rings	174
10.3	Silicon–, Germanium– and Tin–Nitrogen Rings		179
	10.3.1	Saturated Systems	179
	10.3.2	Unsaturated Systems and Biradicals	181
	10.3.3	Silicon–Nitrogen Polymers: Ceramic Precursors	184
10.4	Silicon–, Germanium– and Tin–Phosphorus Rings		184
10.5	Silicon–, Germanium– and Tin–Oxygen Rings		187
10.6	Silicon–Oxygen Polymers: Polysiloxanes (Silicones)		193
10.7	Silicon–, Germanium–, Tin– and Lead–Chalcogen Rings		196
	10.7.1	$(R_2ME)_n$ (M=Si, Ge, Sn, Pb; E=S, Se, Te) and Metal-rich Rings	196
	10.7.2	Chalcogen-rich Rings	201
References			202

Chapter 11 Group 15: Rings and Polymers

- 11.1 Nitrogen-rich Rings — 212
  - 11.1.1 Nitrogen Homocycles: Energetic Materials — 212
  - 11.1.2 Tetrazoles — 213
- 11.2 Cyclic Anions of Phosphorus, Arsenic, Antimony and Bismuth — 213
- 11.3 Homoatomic Rings and Polymers — 217
  - 11.3.1 Cyclophosphines — 217
  - 11.3.2 Cycloarsines, -stibines and -bismuthines — 226
  - 11.3.3 Ladder Polymers $(RE)_n$ (E = As, Sb) — 230
- 11.4 Phosphorus–Nitrogen Rings and Polymers — 231
  - 11.4.1 Cyclophosphazanes — 231
  - 11.4.2 Cyclophosphazenes — 236
  - 11.4.3 Polyphosphazenes — 245
- 11.5 Arsenic–, Antimony– and Bismuth–Nitrogen Rings — 250
  - 11.5.1 Cyclopnictazanes — 250
  - 11.5.2 Cycloarsazenes — 252
- 11.6 Pnictogen–Oxygen Rings — 252
  - 11.6.1 Phosphorus–Oxygen Rings — 252
  - 11.6.2 Arsenic–, Antimony– and Bismuth–Oxygen Rings — 254
- 11.7 Pnictogen–Chalcogen Rings — 255
  - 11.7.1 Binary Phosphorus–Sulfur and –Selenium Rings — 255
  - 11.7.2 Organophosphorus–Sulfur Rings — 256
  - 11.7.3 Organophosphorus–Selenium Rings — 258
  - 11.7.4 Organophosphorus–Tellurium Rings — 259
  - 11.7.5 Arsenic–, Antimony– and Bismuth–Chalcogen Rings — 260
- References — 264

Chapter 12 Group 16: Rings and Polymers

- 12.1 Neutral Sulfur, Selenium and Tellurium Rings — 276
  - 12.1.1 Homoatomic Rings — 276
  - 12.1.2 Heteroatomic Rings — 279
  - 12.1.3 Ring Transformations — 280
- 12.2 Oxidised Chalcogen Homocycles — 282
  - 12.2.1 With Exocyclic Oxygen Substituents — 282
  - 12.2.2 With Exocyclic Halogen Substituents — 282
- 12.3 Cationic Chalcogen Rings — 284
  - 12.3.1 Homocyclic Sulfur and Selenium Cations — 285
  - 12.3.2 Homocyclic Tellurium Cations — 290
  - 12.3.3 Heteroatomic Chalcogen Cations — 293

12.4	Anionic Chalcogen Rings and Chains	294
	12.4.1 *Catena*-Sulfur Anions	294
	12.4.2 *Catena*- and Spirocyclic Selenium Anions	296
	12.4.3 Homoatomic Tellurium Anions	297
	12.4.4 Heteroatomic Selenium–Tellurium Anions	299
12.5	Polymeric Sulfur, Selenium and Tellurium	300
12.6	Sulfur–Nitrogen Heterocycles	301
	12.6.1 Saturated Sulfur–Nitrogen Rings: Cyclic Sulfur Imides	301
	12.6.2 Unsaturated Sulfur–Nitrogen Rings	304
12.7	Sulfur–Nitrogen Chains and Polymers	309
	12.7.1 Poly(Sulfur Nitride)	309
	12.7.2 Sulfur–Nitrogen Chains	311
	12.7.3 Sulfanuric Polymers	311
12.8	Selenium– and Tellurium–Nitrogen Rings	312
	12.8.1 Cyclic Selenium and Tellurium Imides	312
	12.8.2 Unsaturated Systems	314
12.9	Sulfur–, Selenium– and Tellurium–Oxygen Rings and Polymers	317
References		319

Subject Index   328

CHAPTER 1
Introduction to Inorganic Rings and Polymers

1.1 CLASSIFICATION

Ring systems represent a very important branch of organic chemistry. Benzene is perhaps the pre-eminent example and provides the benchmark for the so-called aromatic character of cyclic systems. Experimental and theoretical studies of this unsaturated compound have provided important insights into delocalised chemical bonding, and also an understanding of the substitution patterns. Cycloalkanes are another prominent class of organic compounds. These saturated ring systems form a homologous series, $(CH_2)_n$ ($n = 3$ to > 30), known as alicyclics.

Materials that are constructed from organic polymers such as polyethylene, polystyrene, polyisoprene (natural rubber and a synthetic elastomer) and poly(vinyl chloride) are common features of our daily lives. Most of these and related organic polymers are generated from acyclic precursors by free radical, anionic, cationic or organometallic polymerisation processes or by condensation reactions. Cyclic precursors are rarely used for the production of organic polymers.

The replacement of one or more carbon atoms in a homocyclic organic system by another p-block element gives rise to heterocycles. Common examples include the unsaturated rings thiophene, C_4H_4S, and pyridine, C_5H_5N, and the saturated heterocycle tetrahydrofuran, C_4H_8O. This process of heteroatom substitution can be continued with the replacement of two or more carbons by the same (or different) p-block elements. The study of this class of compounds comprises the field of heterocyclic chemistry. Since approximately half of known organic compounds contain at least one heterocyclic component, this branch of organic chemistry is vast. The extension of this heteroatom

---

Inorganic Rings and Polymers of the p-Block Elements: From Fundamentals to Applications
By Tristram Chivers and Ian Manners
© Tristram Chivers and Ian Manners 2009
Published by the Royal Society of Chemistry, www.rsc.org

substitution process to the complete replacement of all carbon atoms in a ring system by other p-block elements produces inorganic heterocycles which, together with the polymers derived from them, are the subject of this book. A useful classification of inorganic ring systems can be found in the article by Haiduc in the *Encyclopedia of Inorganic Chemistry*.[1]

A quintessential inorganic ring system is elemental sulfur, which, in its thermodynamically stable elemental form, exists as an eight-membered ring, cyclo-S_8. In common with cycloalkanes, homocyclic sulfur rings cyclo-S_n can range in size from $n=6$ to *ca* 24 (Sections 6.1 and 12.1.1). Another type of saturated inorganic ring system that forms an extensive homologous series are the cyclosilanes, $(R_2Si)_n$ ($n=3$–35) (Sections 6.1 and 10.1.1). An intriguing example of an inorganic homocycle is the arsenic-based drug Salvarsan, which, as the forerunner of chemotherapy, was used in the early part of the 20th century for the treatment of syphilis. Recently, Salvarsan was shown to consist primarily of a mixture of three- and five-membered homocyclic arsenic rings, $(RAs)_n$ ($n=3, 5$; $R=3\text{-}NH_2\text{-}4\text{-}OHC_6H_3$) (Sections 3.2 and 11.3.2).

1.2 HISTORICAL BACKGROUND

The first inorganic heterocycle, the cyclotriphosphazene $(NPCl_2)_3$, was described in 1834 and the second example, tetrasulfur tetranitride, S_4N_4, was reported a year later, but it was over century before the structures of these fascinating molecules were shown to be comprised of a six-membered ring and a folded cage structure, respectively. These structural determinations raised fundamental questions about the nature of the bonding in inorganic heterocycles, which provided a focus for much of the early work on inorganic ring systems. Borazine, $B_3N_3H_6$, was discovered in 1926. Because of its isoelectronic relationship to C_6H_6, borazine is sometimes referred to as 'inorganic benzene', but the debate over the aromaticity of this six-membered ring continues even today (Section 4.1.2.1). Interestingly, an extensive homologous series exists for the cyclophosphazenes, $(NPX_2)_n$ ($X=Cl$, F; $n=3$–40) (Sections 6.2.2 and 11.4.2), whereas ring systems other than the six-membered borazines are the exception for boron–nitrogen systems. The reasons for this different behaviour will be discussed in later chapters.

1.3 INDUSTRIAL APPLICATIONS

Industrial interest in inorganic ring systems was stimulated in the 1940s by the discovery that cyclosiloxanes, $(R_2SiO)_n$ ($n=3, 4$), are important intermediates in the manufacture of silicone (siloxane) polymers, $(R_2SiO)_n$ (Section 10.6) Today, these inorganic polymers are made on a massive scale annually because of their multifarious uses in modern society as oils, greases, rubbers, polishes, coatings and insulating materials. Polysilanes, $(R_2Si)_n$, are of interest as ceramic precursors and in the application of microlithography in the electronics industry (Section 10.1.4.1). Phosphazene polymers, $(NPR_2)_n$ ($R=HNMe$, OCH_2CF_3, OC_6H_5), have many desirable properties that have led to a variety of significant applications, *e.g.* as water repellents, non-flammable fibres, foams, fuel pipes and metal ion conductors in batteries (Section 11.4.3). However, the widespread use of phosphazene polymers has been hampered by their high cost relative to that of silicones. In addition to their importance as precursors to inorganic polymers, more recent applications of inorganic ring systems have focused on their use as sources of functional inorganic materials such as semiconductors and ceramics.

1.4 SCOPE AND LIMITATIONS

In this book, the focus is on *monocyclic* inorganic ring systems and the inorganic polymers that, in many cases, are derived from them. Catenated species and related linear oligomers are discussed when their formation or structures are connected with either the corresponding monocyclic systems or the polymers derived from them. Bicyclic or polycyclic arrangements will be considered when they are closely related to those of monocyclic systems as a result of a transannular interaction or structural isomerism, *e.g.* in a discussion of the structural trends of polychalcogen cations (Section 12.3) and anions (Section 12.4). Inorganic heterocycles that are more accurately described as coordination complexes of chelating inorganic ligands are included only when they are directly related to an inorganic homocycle or heterocycle by the replacement of one p-block element by a more metallic p-block element, *e.g.* in a discussion of the applications of p-block metal complexes of the inorganic ligand $[EPR_2NPR_2E]^-$ (E = S, Se, Te) as single-source precursors for the production of thin films of semiconducting metal chalcogenides.

1.5 ORGANISATION

The first six chapters (Chapters 2–7) are intended to provide an introduction to the field of inorganic ring systems. We begin in Chapter 2 by discussing the various methods that are available for the synthesis of inorganic heterocycles. This is followed in Chapter 3 by consideration of the techniques that are most commonly used for their characterisation. Chapter 4 deals with concepts related to the structures and bonding of inorganic ring systems in the context of electron delocalisation (aromaticity). The role of frontier orbitals in the formation and reactions of inorganic ring systems is also stressed in that chapter. The subsequent three chapters (5–7) consider (a) paramagnetic systems with unique magnetic or conducting properties, (b) inorganic macrocycles and their uses in host–guest chemistry and (c) inorganic ring systems as ligands in metal complexes. This discussion leads naturally in Chapter 8 to an account of the methods used to synthesise inorganic polymers, since the most commonly used approach to these macromolecules is ring-opening polymerisation. This final introductory chapter also gives details of the most common techniques used to characterise the structures and properties of inorganic polymers.

After this introductory background to general concepts related to the chemistry of inorganic rings and polymers of the p-block elements, the second half of the book is comprised of four chapters that discuss specific ring systems and polymers involving the elements of groups 13–16 in considerable detail. The material chosen for inclusion in these chapters is intended to reinforce, with examples, the most important concepts introduced in Chapters 2–8. Although they are not intended to be comprehensive, these chapters do include early seminal contributions to the field and also the most important recent advances up to the middle of 2008. The applications of inorganic ring systems as precursors to functional inorganic materials such as semiconducting thin films, nanoparticles, quantum dots and ceramics are included in these chapters. Group 17 is not included because the halogens generally do not form ring systems involving two-electron, two-centre bonds. The fascinating structural chemistry of polyiodides, which includes a multifarious array of linear and polycyclic architectures involving weak I–I interactions, is covered in a recent comprehensive review.[2]

1.6 NOMENCLATURE

The nomenclature of inorganic ring systems presents a number of significant challenges that have been discussed at length in various publications.[3] Interested readers are directed to the most recent IUPAC recommendations published in 2005,[4] which are intended to be the definitive guide to the topic. Since the recommended names for many inorganic ring systems are long and unwieldy, in this book we refer to inorganic ring systems by the names that are commonly used in the literature. It is hoped that the copious use of structural drawings and Figures to depict their structures will minimise any ambiguities that may arise from this pragmatic approach.

1.7 BIBLIOGRAPHY

A selection of inorganic and organometallic textbooks that give a good background discussion of various aspects of inorganic ring systems is given at the end of this introduction. This book is intended to serve as a supplement to those general inorganic chemistry texts in senior undergraduate and graduate courses. In order to minimise unnecessary duplication of background material in Chapters 9–12, references are made to the appropriate sections of these textbooks when the topics are covered in some depth in those sources. More specialised reviews and extensive leading references to the primary literature are included at the end of each individual chapter.

This bibliography also includes a list of recent monographs that deal specifically with 'Inorganic Ring Systems', 'Inorganic Polymers' or 'Cluster Molecules of the p-Block Elements'. There are several recent books that deal specifically with inorganic polymers, but the latest comprehensive treatment of inorganic ring systems is the 1992 two-volume set edited by Steudel. There is no text that makes the connection between the fundamental aspects of the chemistry of inorganic homo- and heterocycles and their applications as functional inorganic materials such as polymers, ceramics and semiconductors for the electronics industries.

REFERENCES

1. I. Haiduc, Inorganic ring systems, in: *Encyclopedia of Inorganic Chemistry*, ed. R. B. King, 2nd edn., Wiley, Chichester, 2005, pp. 2028–2055.
2. P. H. Svensson and L. Kloo, Synthesis, structure and bonding in polyiodide and metal iodide–iodine systems, *Chem. Rev.*, 2003, **103**, 1649.
3. I. Haiduc, Comments on the nomenclature of inorganic ring systems, in *The Chemistry of Inorganic Ring Systems*, ed. R. Steudel, Elsevier, Amsterdam, 1992, pp. 451–477, and references cited therein.
4. N. G. Connelly, T. Damhus, R. M. Hartshorn and A. T. Hutton (eds), *Nomenclature of Inorganic Chemistry; IUPAC Recommendations 2005*, Royal Society of Chemistry, Cambridge, 2005.

General Texts in Inorganic and Organometallic Chemistry

N. N. Greenwood and A. Earnshaw, *Chemistry of the Elements*, 2nd edn., Butterworth-Heinemann, Oxford, 1998.

N. Wiberg and B. J. Aylett (eds), *Holleman–Wiberg Inorganic Chemistry*, Academic Press, London, 2001.

C. E. Housecroft and A. G. Sharpe, *Inorganic Chemistry*, 2nd edn., Pearson Education, Harlow, 2005.

C. Elschenbroich, *Organometallics*, 3rd edn., Wiley-VCH, Weinheim, 2005.

Inorganic Rings and Chains

I. Haiduc, *The Chemistry of Inorganic Ring Systems*, Wiley-Interscience, London, 1970.

D. A. Armitage, *Inorganic Rings and Chains*, Edward Arnold, London, 1972.

A. L. Rheingold (ed.), *Homoatomic Rings Chains and Macromolecules of Main-group Elements*, Elsevier, Amsterdam, 1977.

H. G. Heal, *The Inorganic Heterocyclic Chemistry of Sulfur, Nitrogen and Phosphorus*, Academic Press, London, 1980.

F. L. Boschke (ed.), Inorganic Ring Systems, Topics in Current Chem., Vol. **102**, 1982.

A. H. Cowley (ed.), *Rings, Clusters and Polymers of the Main Group Elements, ACS Symposium Series*, Vol. **232**, 1983.

I. Haiduc and D. B. Sowerby (eds), *The Chemistry of Inorganic Homo- and Heterocycles*, Vols 1 and 2, Academic Press, London, 1987.

J. D. Woollins, *Non-metal Rings, Cages and Clusters*, Wiley, Chichester, 1988.

H. W. Roesky, *Rings, Clusters and Polymers of Main Group and Transition Elements*, Elsevier, Amsterdam, 1989.

R. Steudel (ed.), *The Chemistry of Inorganic Ring Systems*, Vols 1 and 2, Elsevier, Amsterdam, 1992.

T. Chivers, *A Guide to Chalcogen–Nitrogen Chemistry*, World Scientific, Singapore, 2005.

Inorganic Polymers

R. D. Archer, *Inorganic and Organometallic Polymers*, Wiley, New York, 2001.

H. R. Allcock, *Chemistry and Applications of Polyphosphazenes*, Wiley-Interscience, New York, 2003.

M. Gleria and R. De Jaeger, *Phosphazenes: a Worldwide Insight*, Nova Publishers, Hauppauge, NY, 2004.

V. Chandrasekhar, *Inorganic and Organometallic Polymers*, Springer-Verlag, Berlin, 2005.

J. E. Mark, H. R. Allcock and R. West, *Inorganic Polymers*, 2nd edn., Oxford University Press, Toronto, 2005.

Inorganic Clusters

C. E. Housecroft, *Cluster Molecules of the p-Block Elements*, Oxford University Press, Oxford, 1994.

M. Driess and H. Nöth (eds), *Molecular Clusters of the Main Group Elements*, Wiley-VCH, Weinheim, 2004.

CHAPTER 2
Synthetic Methods

The synthesis of ring systems usually occurs via the formation of linear intermediates which may either cyclise or undergo chain propagation to form a polymer. Intramolecular cyclisation is a first-order reaction, whereas intermolecular condensation to give a polymer is a second-order process. Consequently, the competition between ring and polymer formation is influenced strongly by the concentration of the reagents. High dilution methods favour ring formation, although this approach requires the use of undesirably large volumes of solvent. In this chapter, the different ways of constructing inorganic ring systems are compared, with emphasis on the advantages and limitations of each method. Leading examples of inorganic ring systems will be introduced in order to illustrate each synthetic approach. The synthesis of inorganic polymers of the p-block elements is discussed in Chapter 8.

2.1 CYCLOCONDENSATION REACTIONS

Cyclocondensation is a very common procedure for generating both homocyclic and heterocyclic inorganic ring systems. Typically, this method involves the combination of two reagents followed by the thermodynamically favourable elimination of a salt, *e.g.* an alkali metal halide. Alternatively, the production of a volatile trimethylsilyl halide or a hydrogen halide, which may be trapped as an ammonium salt by the addition of an amine base, provides a strong driving force for ring formation.

An extensive series of sulfur allotropes cyclo-S_n ($n = 6-24$) (Section 12.1) is obtained as an inseparable mixture from the reaction of SCl_2 with potassium iodide. The challenge of making an individual allotrope can be surmounted by employing a cyclocondensation reaction between a cyclic metallopolysulfide and a dichlorosulfane, *e.g.* in the formation of cycloheptasulfur [eqn (2.1)].[1] This reaction is also used to prepare other cyclic sulfur allotropes by using

---

Inorganic Rings and Polymers of the p-Block Elements: From Fundamentals to Applications
By Tristram Chivers and Ian Manners
© Tristram Chivers and Ian Manners 2009
Published by the Royal Society of Chemistry, www.rsc.org

dichlorosulfanes ClS_xCl with longer sulfur chains.

$$Cp_2TiS_5 + ClSSCl \rightarrow Cp_2TiCl_2 + cyclo\text{-}S_7 \qquad (2.1)$$

Inorganic heterocycles containing nitrogen are very common. Cyclocondensation processes are used widely for their synthesis. Typically this involves the reaction of a p-block element halide with ammonia or an organic amine (or their respective ammonium salts, which are easier to handle, especially in the case of ammonia or $MeNH_2$). For example, the reaction of BCl_3 with a primary alkylammonium chloride in a high boiling solvent provides a convenient route to *B*-trichloroborazines $(ClBNR)_3$ [eqn (2.2)].[2] This versatile method is also used to prepare derivatives with an organic substituent attached to boron by using $RBCl_2$ instead of BCl_3 [see Section 9.2.1 and eqn (9.3)].

$$3BCl_3 + 3[RNH_3]Cl \rightarrow (ClBNR)_3 + 9HCl \qquad (2.2)$$

The complexity of cyclocondensation processes is illustrated by the production of cyclophosphazenes from the reaction of phosphorus pentachloride with ammonium chloride in a high-boiling solvent, *e.g.* 1,1,2,2-tetrachloroethane. This synthesis is conveniently monitored by a combination of ^{31}P NMR spectroscopy and conductivity measurements, which show that it occurs in three distinct steps: (a) formation of the salt $[Cl_3P=N=PCl_3][PCl_6]$, (b) chain growth and (c) intramolecular cyclisation, as illustrated in Scheme 2.1. The key intermediate in this process is the phosphoranimine $Cl_3P=NH$, generated *in situ* by combination of $[NH_4]^+$ and $[PCl_6]^-$ [step (a)]. As a source of $NPCl_2$

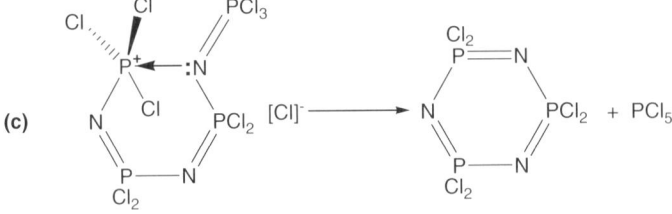

Scheme 2.1 Formation of cyclophosphazenes by cyclocondensation of phosphorus pentachloride and ammonium chloride.

units, this unstable species also acts as a chain propagator in step (b), which provides the acyclic precursors for a homologous series of ring systems $(NPCl_2)_n$ ($n = 3$–40) (see Section 11.4.2). The major product of this cyclocondensation process is the six-membered ring $(NPCl_2)_3$ formed by an intramolecular ring closure [step (c)].[3a] The use of trimethylsilylphosphoranimine $Cl_3P=NSiMe_3$ as a precursor for the phosphazene polymers $(NPCl_2)_n$ is discussed in Section 11.4.3 [see eqn (11.35)].

The acyclic cation $[Cl_3P=N=PCl_3]^+$ formed in the initial stage of the reaction of ammonium chloride with phosphorus pentachloride [Scheme 2.1(a)] is a useful building block for the synthesis of hybrid phosphazene ring systems. As illustrated in Scheme 2.2, cyclocondensation reactions with appropriate substrates can be used to generate cyclophosphazenes containing boron, carbon or sulfur in various oxidation states.[3b] These ring systems can be used as precursors for the corresponding hybrid inorganic polymers via thermal ring-opening polymerisation [see eqn (8.2)].

The cyclocondensation reaction of ammonia (or ammonium salts) with sulfur halides provides a convenient preparation of sulfur–nitrogen halides, which are versatile reagents for the synthesis of other sulfur–nitrogen heterocycles (see Section 12.6.2.2 and Scheme 12.8). The polarity of the solvent has a marked influence on the outcome of this reaction. The five-membered $[S_3N_2Cl]^+$ and six-membered $(NSCl)_3$ rings are formed in solvents such as CCl_4 or CH_2Cl_2 (Scheme 2.3).[4]

When the reaction of S_2Cl_2 is carried out in a polar solvent, e.g. DMF, however, the acidic hydrolysis of the reaction mixture produces a mixture of cyclic sulfur imides. The major product is S_7NH (**2.1**), but significant quantities of the isomers 1,3-, 1,4- and 1,5-$S_6(NH)_2$ (**2.2–2.4**), which can be separated by chromatography, are also formed (see Sections 3.2 and 12.6.1).[5]

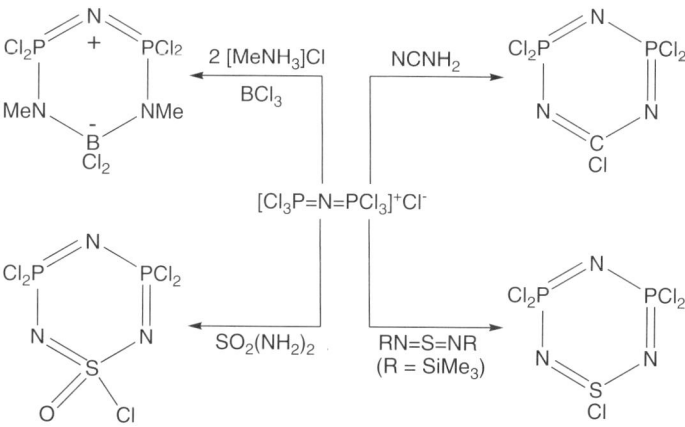

Scheme 2.2 Synthesis of hybrid cyclophosphazenes from $[Cl_3P=N=PCl_3]^+$.

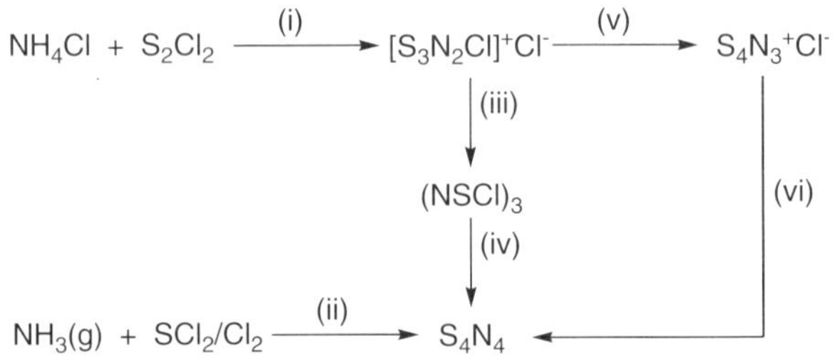

$$NH_4Cl + S_2Cl_2 \xrightarrow{(i)} [S_3N_2Cl]^+Cl^- \xrightarrow{(v)} S_4N_3^+Cl^-$$

$$\downarrow (iii)$$

$$(NSCl)_3$$

$$\downarrow (iv)$$

$$NH_3(g) + SCl_2/Cl_2 \xrightarrow{(ii)} S_4N_4 \xleftarrow{(vi)}$$

Scheme 2.3 Synthesis of S–N rings from ammonia or ammonium salts and sulfur halides: (i) reflux; (ii) CCl$_4$ or CH$_2$Cl$_2$; (iii) Cl$_2$; (iv) Fe, Hg or SbPh$_3$; (v) S$_2$Cl$_2$; (vi) KI.

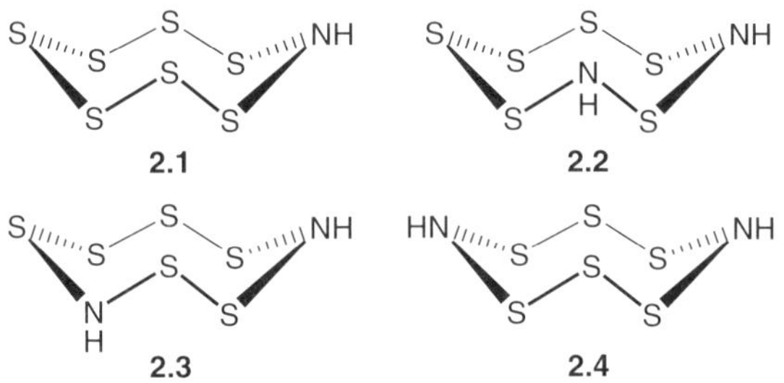

2.1 **2.2**

2.3 **2.4**

Silicon–nitrogen reagents provide an alternative to the use of amines or ammonia for the formation of inorganic rings via reactions with main group element halides. The elimination of volatile trimethylsilyl halides provides the driving force for these cyclocondensation processes. For example, trisilylated benzamidines ArCN$_2$(SiMe$_3$)$_3$ (Ar = aryl) are versatile reagents for the synthesis of chalcogen–nitrogen ring systems containing a single carbon atom [eqn (2.3)]. The five-membered cyclic cations formed in this way are important precursors for the corresponding radicals [see eqn (5.4)], which are used as building blocks for the construction of molecular conductors (see Section 3.1).

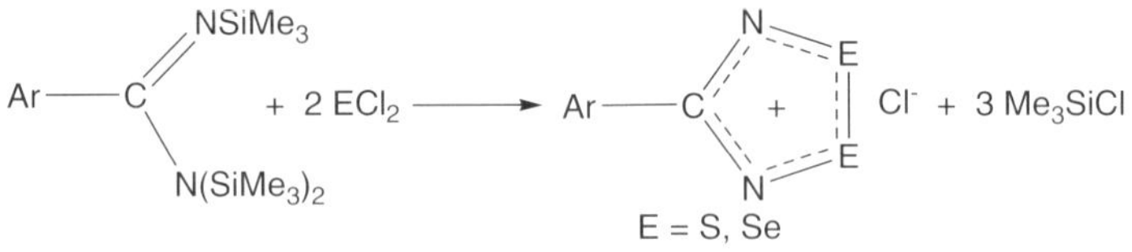

$$\text{Ar–C(NSiMe}_3\text{)(N(SiMe}_3\text{)}_2\text{)} + 2\ ECl_2 \longrightarrow \text{Ar–C}[N=E–N=E]^+ + Cl^- + 3\ Me_3SiCl$$

E = S, Se

(2.3)

The salt-elimination method involves the reaction of an organoelement halide with an alkali metal derivative of an organoelement fragment. Since the p-block element in these two reagents may differ, this synthetic approach is particularly useful for the synthesis of ring systems that contain two different elements from the same or a different group. As illustrations of this approach, the synthesis of specific examples of group 13–15 and group 14–16 heterocycles

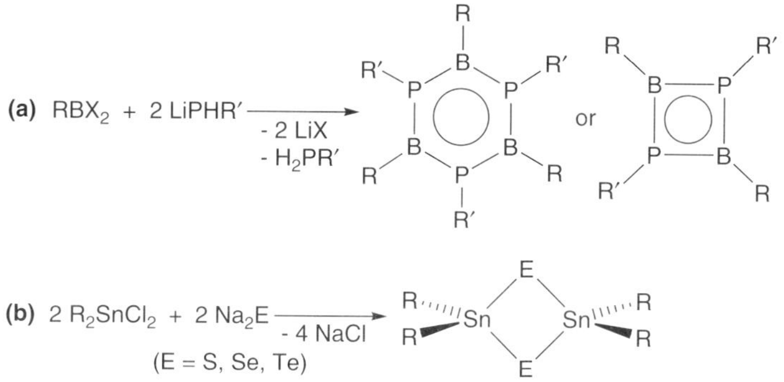

Synthetic Methods

Scheme 2.4 Synthesis of (a) boron–phosphorus and (b) tin–chalcogen rings by salt elimination.

(a) $RBX_2 + 2\ LiPHR'$ reaction scheme

(b) $2\ R_2SnCl_2 + 2\ Na_2E$ reaction scheme $(E = S, Se, Te)$

are shown in Scheme 2.4. Although six-membered B_3P_3 rings are usually obtained by method (a), the presence of extremely bulky groups on both the boron and phosphorus atoms favours the formation of four-membered rings [see Section 9.3.1 and eqn (9.10)]. The widespread use of this method is evident from the numerous other examples that are discussed in Chapters 9–11.

2.2 REDUCTIVE COUPLING

The process of reductive coupling is used widely for the synthesis of homocyclic systems. In general, this method involves the reaction of an organoelement dihalide with an electropositive metal, most commonly an alkali metal, in an ether or hydrocarbon solvent. For example, the slow addition of Me_2SiCl_2 to sodium–potassium alloy in THF produces an equilibrium mixture composed primarily of the hexamer $(Me_2Si)_6$ with smaller amounts of the pentamer $(Me_2Si)_5$ and a trace of the heptamer $(Me_2Si)_7$. Under kinetic control, *i.e.* the very slow addition of Me_2SiCl_2, a mixture of the homologous series $(Me_2Si)_n$ ($n = 5$–35) can be detected by high-performance liquid chromatography, illustrating the sensitivity of ring formation to reaction conditions (see Section 10.1.1).[6]

Cyclostannanes $(R_2Sn)_n$ are obtained, primarily as the hexamer ($n = 6$), by treatment of the diorganotin dichloride R_2SnCl_2 with sodium naphthalenide [eqn (2.4)].[7]

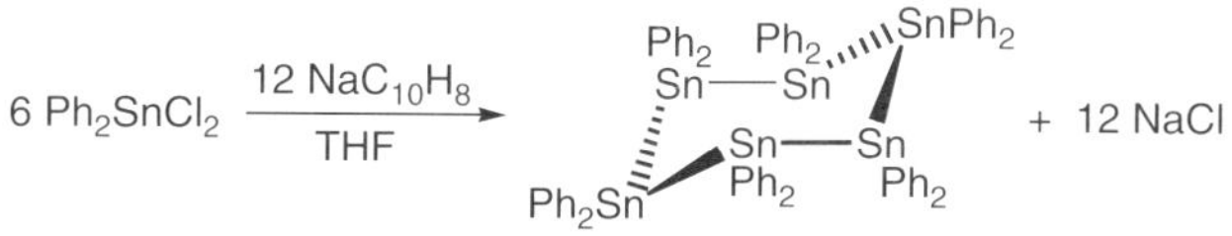

$$6\ Ph_2SnCl_2 \xrightarrow[THF]{12\ NaC_{10}H_8} (Ph_2Sn)_6 + 12\ NaCl \quad (2.4)$$

When the groups attached to tin are especially bulky, it is possible to isolate the kinetically unstable cyclotristannanes $(Ar_2Sn)_3$ ($Ar = 2,4,6\text{-}R_3H_2C_6$, $R = Me$,

^{i}Pr, ^{t}Bu) in this cyclooligomerisation process [eqn (2.5)].[8a] The corresponding cyclotrisilanes and cyclogermanes are prepared by a similar strategy.[8b,c] The photolysis of these three-membered rings in cyclohexane produces the doubly bonded species, *e.g.* Ar$_2$E=EAr$_2$ (Ar = 2,6-Me$_2$C$_6$H$_3$). The oligomer formed in the reductive coupling of Ar$_2$SiCl$_2$ is sensitive to the reaction conditions. The disilene Mes$_2$Si=SiMes$_2$ is obtained upon treatment of Mes$_2$SiCl$_2$ with metallic Li in THF under the influence of ultrasound.[8d]

$$3\ Ar_2SnCl_2 \xrightarrow[-78\ °C]{6\ LiC_{10}H_8} \begin{array}{c} Ar_2 \\ Sn \\ \triangle \\ Ar_2Sn\text{——}SnAr_2 \end{array} + 6\ LiCl \qquad (2.5)$$

The reductive coupling process can be adapted for the synthesis of molecules constructed from two different fragments. For example, the unsaturated cyclotrisilene ring R$_4$Si$_3$ (**2.5**, R = ^{t}BuMe$_2$Si) is prepared by coupling of an R$_2$Si unit with two RSi units as illustrated in eqn (2.6). This method can also be used for the synthesis of heteroatomic cyclopropene analogues (see Scheme 10.8).[9]

$$\begin{array}{c} R \\ | \\ R\text{····}Si \\ / \quad \backslash \\ Br \quad Br \end{array} + 2\ \begin{array}{c} Br \\ | \\ R\text{——}Si\text{····}Br \\ | \\ Br \end{array} \xrightarrow[RT,\ 3h]{Na/PhMe} \begin{array}{c} R \quad R \\ \backslash / \\ Si \\ / \backslash \\ Si = Si \\ / \quad \backslash \\ R \quad R \\ \mathbf{2.5} \end{array} \qquad (2.6)$$

The cyclophosphines (RP)$_n$ constitute another homologous series of homocyclic rings that are prepared by the reductive coupling method [eqn (2.7)]. For these systems, either magnesium or mercury is normally used as the reducing agent. Ring sizes ranging from $n = 3$ to 5 are most common for R = alkyl, phenyl (Section 11.3.1.1).[10a] If R is a bulky aryl group, *e.g.* 2,4,6-^{t}Bu$_3$H$_2$C$_6$), a diphosphene R$_2$P=PR$_2$ is obtained.[10b] Tetrameric and pentameric cycloarsines (CF$_3$As)$_n$ ($n = 4$, 5) are also prepared by the route depicted in eqn (2.7) (see Section 11.3.2).[11]

$$RPX_2 + 2M \rightarrow 1/n(RP)_n + 2MX \qquad (2.7)$$

(X = Cl, Br, I; M = Mg, Hg; $n = 3 - 5$)

An interesting variation of the reductive-coupling approach for the synthesis of cyclophosphines involves the addition of a stoichiometric amount of PCl$_3$ to a mixture of ^{t}BuPCl$_2$ and sodium, which produces the sodium salt of tetra-*tert*-butylcyclopentaphosphanide [^{t}Bu$_4$P$_5$]$^-$ [**2.6**, eqn (2.8); see Figure 11.10 for the structure].[12] Examples of the fascinating coordination chemistry of this monoanionic homocyclic ring are illustrated in Scheme 7.3; additional examples, and also the redox behaviour of this anion, are discussed in Section 11.3.1.3 (see structures **11.7**–**11.9**).

Synthetic Methods

$$4\,^{t}BuPCl_2 + PCl_3 + 12\,Na \xrightarrow[\text{THF}]{-11\,NaCl} \mathbf{2.6} \quad \text{[cyclic}\,(^{t}Bu)_5P_5\text{-Na(THF)}_4\text{]} \tag{2.8}$$

For the heavier group 15 elements, antimony and bismuth, the bulky alkyl group bis(trimethylsilyl)methyl, $CH(SiMe_3)_2$, is particularly effective in stabilising homocyclic rings (see Figure 11.13a). Thus, the reaction of $RSbCl_2$ with magnesium in THF produces primarily the four-membered ring $(RSb)_4$ together with smaller amounts of the trimer $(RSb)_3$ and some bicyclic products. The analogous reaction with $RBiCl_2$ produces only monocyclic ring systems [eqn (2.9)].[13]

$$RBiCl_2 + Mg \rightarrow (RBi)_3 + (RBi)_4 + MgCl_2 \tag{2.9}$$

An unusual, but historically interesting, example of the synthesis of homocyclic rings by reductive coupling involves the reaction of an arsonic acid $ArAs(O)(OH)_2$ ($Ar = 3\text{-}NO_2\text{-}4\text{-}HOC_6H_3$) with dithionite. This synthesis was first reported in 1910 and the product $(ArAs)_n$ ($Ar = 3\text{-}NH_2\text{-}4\text{-}HOC_6H_3$) was known as Salvarsan, a chemotherapeutic agent that was used for the treatment of syphilis. It was not until 2005 that Salvarsan was shown by electrospray ionisation mass spectrometry (see Figure 3.5) to be a complex mixture of homocyclic arsenic rings, with the trimer ($n = 3$) and pentamer ($n = 5$) as the predominant components [see eqn (11.16)].[14]

2.3 DEHYDROCOUPLING

An alternative to the use of cyclocondensation or reductive coupling for generating homocyclic ring systems is dehydrocoupling, which involves the loss (reductive elimination) of H_2 from organoelement dihydrides. This process is promoted thermally, photochemically or by using a catalyst.[15] For example, the decomposition of Ar_2SnH_2 ($Ar = aryl$) in DMF or pyridine (as a catalyst) generates pentameric and hexameric ring systems [eqn (2.10)].[7]

$$Ar_2SnH_2 \rightarrow (Ar_2Sn)_5 + (Ar_2Sn)_6 + H_2 \tag{2.10}$$

In recent years, dehydrocoupling reactions catalysed by early transition metal complexes have become an increasingly important method for generating catenated species of the p-block elements. In addition to producing cyclic oligomers, this approach is used to prepare linear oligomers and polymers such as polysilanes and polystannanes of the type $H(MR_2)_nH$ ($M = Si, Sn$) (see Section 10.1.4).[16]

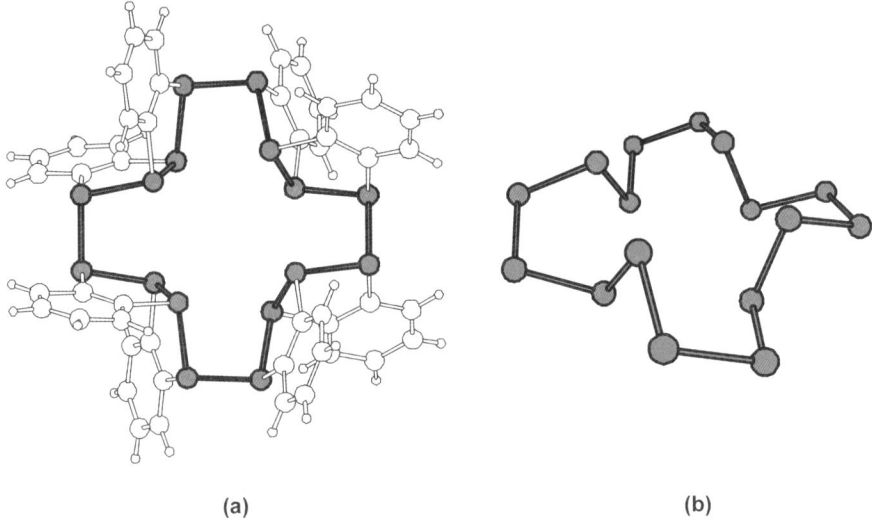

Figure 2.1 (a) Structure of $(1,2-C_6H_4P_2)_8$; (b) conformation of the P_{16} ring. Reproduced with permission from M. C. Fermin and D. W. Stephan, *J. Am. Chem. Soc.*, 1995, **117**, 12645. Copyright (2005) American Chemical Society.[17]

For group 15 elements, metal-catalysed dehydrocoupling is a versatile route for the synthesis of cyclic oligomers. For example, the zirconium hydride species $[Cp^*_2ZrH_3]^-$ catalyses the oligomerisation of primary phosphines RPH_2 (R = Ph, Cy, 2,4,6-$Me_3C_6H_2$) to the corresponding cyclic pentamers $(RP)_5$.[17] An intriguing extension of this approach involves the synthesis of the octamer $(1,2-C_6H_4P_2)_8$ by dehydrooligomerisation of $1,2-C_6H_4(PH_2)_2$.[18] The central core of 16 catenated phosphorus atoms in this macrocycle is reminiscent of a crown ether (Figure 2.1).

The zirconium hydride catalyst $[Cp_2Zr(H)Cl]$ is very effective for the high-yield synthesis of cyclotetrastibines $(ArSb)_4$ from arylstibines $ArSbH_2$ [eqn (2.11)].[19]

$$4\ ArSbH_2 \xrightarrow[C_6D_6,\ 25\ ^\circ C]{5\ mol\%\ cat.} \begin{array}{c} Ar \quad\quad Ar \\ Sb—Sb \\ | \quad\quad | \\ Sb—Sb \\ Ar \quad\quad Ar \end{array} + 4\ H_2$$

(2.11)

Catalytic dehydrocoupling can also be used for the efficient generation of heterocyclic systems by use of a suitable precursor incorporating a heteronuclear bond between two p-block elements. For example, the thermal dehydrocoupling of $Ph_2PH \cdot BH_3$ at 170 °C yields a mixture of the cyclic trimer $(Ph_2PBH_2)_3$ (**2.7**) and tetramer $(Ph_2PBH_2)_4$ (**2.8**) in an 8:1 ratio. The same two heterocycles are formed in a 2:1 ratio at a lower temperature by using an Rh(I) catalyst [eqn (2.12)].[20] The use of a primary phosphine rather than a secondary phosphine, *i.e.* the adduct $PhPH_2 \cdot BH_3$, produces the polymer $(PhPH–BH_2)_n$ in preference to cyclic systems (see Section 9.3.4).

$$Ph_2PH \cdot BH_3 \xrightarrow[\substack{120\,°C \\ -H_2}]{1\ mol\%\ Rh^I} \mathbf{2.7} + \mathbf{2.8}$$

(2.12)

Amine–boranes, notably the ammonia adduct $H_3B \cdot NH_3$, have received considerable attention as potential hydrogen storage materials (see Section 9.2.4.2),[21a] although volatile products such as borazine may act as a poison in fuel cell applications.[21b] At ambient temperature, this adduct undergoes metal-catalysed dehydrocoupling in the presence of the iridium(III) catalyst $[\eta^3\text{-}1,3\text{-}(OP^tBu_2)_2C_6H_3]IrH_2$ to give the cyclic pentamer $(H_2BNH_2)_5$ [**2.9**, eqn (2.13)].[22]

$$H_3N \cdot BH_3 \xrightarrow[\substack{20\,°C \\ -H_2}]{0.5\ mol\%\ Ir^{III}} \mathbf{2.9}$$

(2.13)

Primary and secondary amine–borane adducts are thermally dehydrocoupled at temperatures above 100 °C to give cyclic aminoboranes (borazanes) $(R_2BNR'_2)_n$ ($n=2$ or 3) and borazine $(RBNR')_3$ derivatives (see Section 9.2).[15] In the presence of a late transition metal catalyst, *e.g.* $[Rh(\mu\text{-}Cl)(1,5\text{-}cod)]_2$, this process proceeds at ambient (or slightly elevated) temperatures to give four-membered rings or, in the case of primary amine–boranes, borazines (Scheme 2.5).[23] The four-membered ring $(Me_2NBH_2)_2$ can also be generated at room temperature by using titanocene $[Cp_2Ti]$ as a catalyst.[24]

2.4 ALKANE ELIMINATION REACTIONS

The method of alkane elimination is a versatile approach for the synthesis of inorganic heterocycles comprised of alternating electropositive and electronegative p-block elements. Typically, this method involves the reaction of a homoleptic alkyl derivative of the more electropositive element with a hydride of the more electronegative element with the elimination of a volatile alkane.

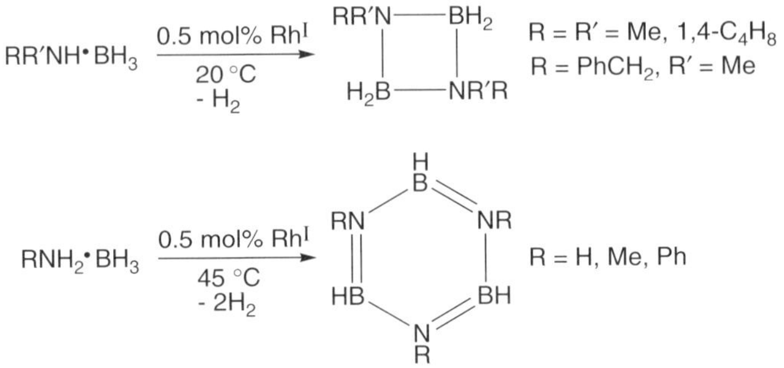

Scheme 2.5 Formation of B–N ring systems by dehydrocoupling of amine–boranes.

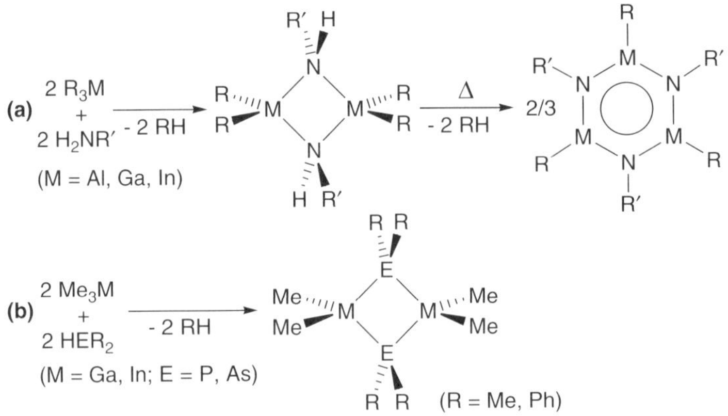

Scheme 2.6 Synthesis of saturated and unsaturated group 13–group 15 heterocycles by alkane elimination.

Specific examples of the application alkane elimination for the synthesis of either saturated or unsaturated group 13–group 15 ring systems are shown in Scheme 2.6 (for a more detailed discussion, see Sections 9.4.1 and 9.4.2).

2.5 CYCLOADDITION REACTIONS

Cycloaddition reactions involving two unsaturated reagents, *e.g.* 1,3-dipolar cycloadditions, hetero-Diels–Alder reactions and [2 + 2] cycloaddition reactions, provide useful synthetic routes to a wide range of organic heterocycles.[25] This method of ring synthesis is less common for inorganic heterocycles owing to the limited number of unsaturated reagents that are available as stable, monomeric species. However, some examples can be given to illustrate the potential scope of this method.

Synthetic Methods

Multiply bonded compounds of the heavier p-block elements, *e.g.* silicon and phosphorus, are only stable when bulky groups are attached to the main group element. In a historical context, the classic examples are the disilene Mes$_2$Si=SiMes$_2$ (Mes = mesityl) and the diphosphene Mes*P=PMes* (Mes* = 2,4,6-tri-*tert*-butylphenyl), which were reported in 1981.[26] Disilenes R$_2$Si=SiR$_2$ undergo a variety of [2 + 1], [2 + 2] and [2 + 3] cycloaddition reactions with unsaturated substrates to give a wide range of novel inorganic rings systems, as indicated by the examples depicted in Scheme 2.7.[27,28]

Salts of the linear [S=N=S]$^+$ cation (isoelectronic with CS$_2$) are readily prepared.[29] This cation undergoes quantitative, 1,3-dipolar cycloaddition reactions with unsaturated molecules such as alkenes, alkynes and nitriles, and also with NS$^+$, to give a variety of ring compounds (Scheme 2.8).[30] The

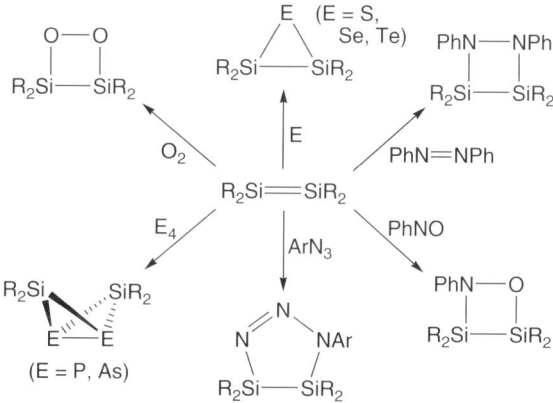

Scheme 2.7 Formation of heterocycles from cycloaddition reactions of R$_2$Si=SiR$_2$.

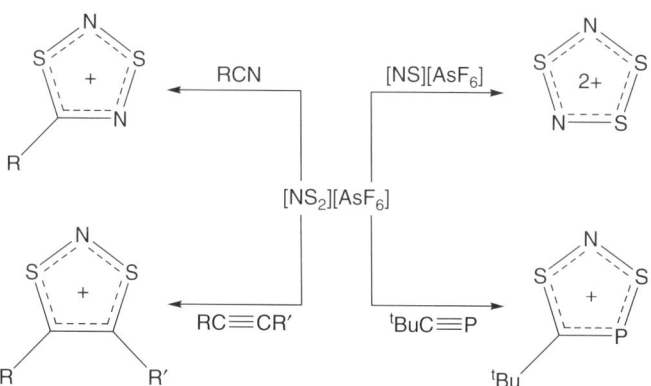

Scheme 2.8 Formation of heterocycles from cycloaddition reactions of [NS$_2$]$^+$.

cycloaddition with nitriles is especially important (see Figure 4.7), since (a) it can be applied to polynitriles, e.g. 1,2-, 1,3- or 1,4-$(CN)_2C_6H_4$ or 1,3,5-$(CN)_3C_6H_3$, and (b) the six π-electron cationic rings $[RCN_2S_2]^+$ are readily reduced to the corresponding neutral, seven π-electron radicals $[RCN_2S_2]^•$.

REFERENCES

1. J. Albertsen and R. Steudel, in *Inorganic Experiments*, ed. J. D. Woollins, 2nd edn., Wiley-VCH, Weinheim, 2003, p. 291.
2. W. Maringgele, in *The Chemistry of Inorganic Homo- and Heterocycles*, ed. I. Haiduc and D. B. Sowerby, Vol. 1, Academic Press, London, 1987, pp. 17–101.
3. (a) J. Novosad and M. Alberti, in *Inorganic Experiments*, ed. J. D. Woollins, 2nd edn., Wiley-VCH, Weinheim, 2003, p. 176; (b) D. P. Gates and I. Manners, *J. Chem. Soc., Dalton Trans.*, 1997, 2525.
4. T. Chivers, in *The Chemistry of Inorganic Homo- and Heterocycles*, ed. R. Steudel, Academic Press, London, 1987, pp. 793–870.
5. H. G. Heal and J. Kane, *Inorg. Synth.*, 1968, **11**, 184.
6. R. West and E. Carberry, *Science*, 1975, **189**, 179.
7. M. Dräger, B. Mathiasch, L. Ross and M. Ross, *Z. Anorg. Allg. Chem.*, 1983, **506**, 99.
8. (a) S. Masamune, L. Sita and D. J. Williams, *J. Am. Chem. Soc.*, 1983, **105**, 630; (b) S. Masamune, Y. Hanzawa, S. Murakami, T. Bally and J. F. Blount, *J. Am. Chem. Soc.*, 1982, **104**, 1150; (c) L. F. Brough and R. West, *J. Am. Chem. Soc.*, 1981, **103**, 3049; (d) P. Boudjouk, B. H. Han and K. R. Anderson, *J. Am. Chem. Soc.*, 1982, **104**, 4992.
9. M. Ichinohe, T. Matsuno and A. Sekiguchi, *Angew. Chem. Int. Ed.*, 1999, **38**, 2194.
10. (a) M. Baudler and K. Glinka, *Chem. Rev.*, 1993, **93**, 1623; (b) M. Yoshifuji, I. Shima, N. Inamoto, K. Hirotso and T. Higuchi, *J. Am. Chem. Soc.*, 1981, **103**, 4587.
11. A. H. Cowley, A. B. Burg and W. R. Cullen, *J. Am. Chem. Soc.*, 1966, **88**, 3178.
12. R. Wolf, M. Finger, C. Limburg, A. C. Willis, S. B. Wild and E. Hey-Hawkins, *Dalton Trans.*, 2006, 831.
13. H. J. Breunig and R. Roesler, *Chem. Soc. Rev.*, 2000, **29**, 403.
14. N. C. Lloyd, H. W. Morgan, B. K. Nicholson and R. S. Ronimus, *Angew. Chem. Int. Ed.*, 2005, **44**, 941.
15. T. J. Clark, K. Lee and I. Manners, *Chem. Eur. J.*, 2006, **12**, 8634.
16. T. Imori, V. Lu, H. Cai and T. D. Tilley, *J. Am. Chem. Soc.*, 1995, **117**, 9931.
17. M. C. Fermin and D. W. Stephan, *J. Am. Chem. Soc.*, 1995, **117**, 12645.
18. N. Etkin, M. C. Fermin and D. W. Stephan, *J. Am. Chem. Soc.*, 1997, **119**, 2954.
19. R. Waterman and T. D. Tilley, *Angew. Chem. Int. Ed.*, 2006, **44**, 2926.

20. H. Dorn, R. A. Singh, J. A. Massey, J. M. Nelson, C. A. Jaska, A. J. Lough and I. Manners, *J. Am. Chem. Soc.*, 2000, **122**, 6669.
21. (a) F. H. Stephens, R. T. Baker, M. H. Matus, D. J. Grant and D. A. Dixon, *Angew. Chem. Int. Ed.*, 2007, **46**, 726; (b) M. E. Bluhm, M. G. Bradley, R. Butterick III, U. Kusari and L. G. Sneddon, *J. Am. Chem. Soc.*, 2006, **128**, 7748.
22. M. C. Denney, V. Pons, T. J. Hebden, D. M. Heinekey and K. I. Goldberg, *J. Am. Chem. Soc.*, 2006, **128**, 12048.
23. C. A. Jaska, K. Temple, A. J. Lough and I. Manners, *J. Am. Chem. Soc.*, 2003, **125**, 9424.
24. T. J. Clark, C. A. Russell and I. Manners, *J. Am. Chem. Soc.*, 2006, **128**, 9582.
25. T. L. Gilchrist, *Heterocyclic Chemistry*, Pitman Publishing, London, 1985, pp. 79–100.
26. (a) R. West and M. J. Fink, *Science*, 1981, **214**, 1343; (b) M. Yoshifuji, I. Shima, N. Inamoto, K. Hirotsu and T. Higuchi, *J. Am. Chem. Soc.*, 1981, **103**, 4587.
27. R. West, *Polyhedron*, 2002, **21**, 467.
28. R. Okazaki and R. West, *Adv. Organomet. Chem.*, 1996, **39**, 232.
29. T. S. Cameron, A. Mailman, J. Passmore and K. V. Shuvaev, *Inorg. Chem.*, 2005, **44**, 6524.
30. S. Parsons and J. Passmore, *Acc. Chem. Res.*, 1994, **27**, 101.

CHAPTER 3
Characterisation Methods

A variety of characterisation techniques are available for the determination of the molecular structures of inorganic ring systems. The most important of these is X-ray diffraction (XRD), which reveals a picture of the atomic arrangements of the molecule in the solid state. XRD also provides accurate values of structural parameters, which are important in the discussion of bonding in heterocyclic systems. The advent of charge-coupled device (CCD) instruments in recent years has facilitated the determination of solid-state structures so that, in favourable cases, detailed information can be obtained in a few hours if suitable crystals can be grown. Several spectroscopic techniques are also used for the identification of ring systems and the most informative of these methods are discussed with examples in this chapter. When the spectroscopic data are obtained on samples in solution, the knowledge gained is complementary to that provided by an X-ray structural determination. Thus, techniques such as NMR or EPR spectroscopy may be used to determine whether the solid-state structure is maintained in solution. In addition, spectroscopic methods may furnish key structural insights in cases where X-ray structures cannot be obtained. The following discussion will focus on the application of these techniques to inorganic ring systems rather than the instrumental details. A simplified description of the theoretical background and instrumental considerations can be found in the appropriate chapters of various textbooks.[1-3] This chapter also includes a section on chromatographic methods, which are sometimes necessary for the analysis and separation of the individual members of a homologous series of inorganic ring systems prior to their characterisation by X-ray diffraction or spectroscopic methods.

3.1 X-RAY AND ELECTRON DIFFRACTION

The ability to establish atomic connectivities and ring conformations in the solid state by X-ray crystallography has been of paramount importance in the development of the chemistry of inorganic ring systems. The accurate determination of intramolecular bond lengths, bond angles and torsional (dihedral) angles is an essential prerequisite for a discussion of bonding in new inorganic ring systems. X-ray structural determinations also reveal the details of intermolecular interactions, *e.g.* chalcogen–chalcogen or chalcogen–nitrogen interactions, that have a major influence on the solid-state properties of inorganic heterocycles. The 1,2,3,5-dichalcogenodiazolyl radical $[RCN_2E_2]^{\bullet}$ (E = S, Se) provides a prominent example of the influence of intermolecular interactions on properties.[4] When R = aryl these radicals form dimers in the solid state *via* two chalcogen···chalcogen interactions, which are stronger for E = Se than E = S (Figure 3.1). The columnar stacking of these radical dimers produces materials with remarkably small band gaps for non-metallic systems. Several of the selenium derivatives are small band gap semiconductors. When R = Me, NMe_2 or CF_3, only one chalcogen···chalcogen contact is involved in dimer formation (Figure 3.1).

Table 3.1 gives the values of covalent single and double bond radii for p-block elements that form ring systems. The sum of these radii for directly bonded atoms can be used to estimate the extent of multiple bonding in inorganic rings. Table 3.1 also lists van der Waals radii, which are used to identify the presence of significant intermolecular interactions between ring systems. If the distance between atoms of neighbouring ring systems is significantly less than the sum of the van der Waals radii for those two atoms, a secondary bonding interaction can be inferred.[6]

The knowledge of torsional (dihedral) bond angles, *i.e.* the angle between bonds on adjacent atoms, defines the conformation of an individual ring system. This structural information (obtained from diffraction methods) provides important insights into properties such as ring strain that may indicate whether a particular ring system is predisposed to ring-opening polymerisation (see Section 8.1.2). The sulfur allotropes cyclo-S_8 and cyclo-S_7 furnish an interesting

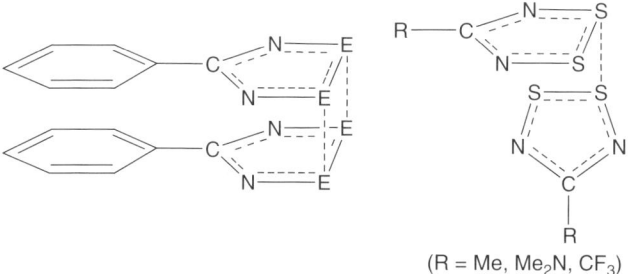

Figure 3.1 Intermolecular chalcogen···chalcogen contacts in $[RCN_2E_2]_2$ dimers.

Table 3.1 Van der Waals and covalent single and double bond radii.[a,b]

	B	C	N	O
r_w	2.08	1.85	1.54	1.40
r_{cs}	0.88	0.77	0.75	0.73
r_{cd}	–	0.67	–	–
	Al	Si	P	S
r_w	–	2.10	1.90	1.85
r_{cs}	1.30	1.18	1.10	1.03
r_{cd}	–	1.07	1.00	–
	Ga	Ge	As	Se
r_w	–	–	2.00	2.00
r_{cs}	1.22	1.22	1.22	1.17
r_{cd}	–	1.12	1.11	1.07
	In	Sn	Sb	Te
r_w	–	–	2.20	2.20
r_{cs}	1.50	1.40	1.43	1.35
r_{cd}	–	1.30	1.31	1.27
	Tl	Pb	Bi	
r_w	–	–	2.40	
r_{cs}	1.55	1.54	1.52	

[a] r_w = Van der Waals radii; r_{cs} = covalent single bond radius, r_{cd} = covalent double bond radius.
[b] Values taken from ref. 5.

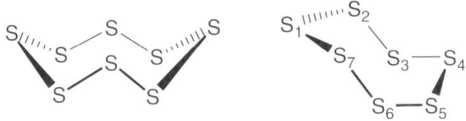

Figure 3.2 Ring conformations of cyclo-S_8 and cyclo-S_7.

illustration of the influence of torsional angles on the reactivity of inorganic ring systems (Figure 3.2). The mean torsional angle in the crown-shaped eight-membered ring of cyclo-S_8 is 98.3°, leading to a conformation that minimises interactions between non-bonding electron pairs on adjacent sulfur atoms; the mean S–S bond length is 2.04 Å. In cyclo-S_7, four of the sulfur atoms are in a planar arrangement in which the central torsional angle (S3–S4–S5–S6) is 0.3°. Consequently, the repulsions between lone pairs on S4 and S5 are maximised and the S4–S5 bond length is elongated to 2.18 Å. Cyclo-S_7 is significantly more reactive towards thermally or photochemically initiated ring-opening than cyclo-S_8 as a result of this weak sulfur–sulfur bond (see Section 12.1.1).

Electron diffraction experiments provide valuable information about structures in the gas phase. Consequently, this method of structural determination is important for inorganic ring systems that are volatile liquids or gases at room temperature. For example, the essentially planar structures of borazine (**3.1**, E = NH)[7a] and the isoelectronic boroxin ring (**3.1**, E = O),[7b] the monomeric structure of the radical [CF$_3$CNSSN]˙ (**3.2**),[8] and the all-*cis* arrangement of the

Characterisation Methods 23

three fluorine substituents in the six-membered ring (NSF)$_3$ (**3.3**) have all been established by electron diffraction.[9]

3.1   **3.2**   **3.3**

For a more detailed discussion of crystal structure determination by X-ray diffraction techniques, the reader is referred to books by Clegg[10a] and Massa.[10b]

3.2 CHROMATOGRAPHY

Some of the methods for the synthesis of inorganic ring systems described in Chapter 2 give rise to a mixture of cyclic oligomers. In cases where there are only two or three components in these mixtures, the isolation of an individual oligomer can often be achieved by fractional crystallisation. For more complex mixtures, however, it is necessary to use chromatography as a method of analysis and/or separation. For example, the separation of individual members of the series of cyclosilanes (Me$_2$Si)$_n$ ($n = 5$–19) (see Section 2.2), is achieved by using high-performance liquid chromatography (HPLC) with methanol–THF as the eluent.[11] The individual oligomers can then be characterised by NMR, UV and vibrational spectra. HPLC is also a powerful technique for the analysis of mixtures of cyclic sulfur allotropes that are produced from the reaction of SCl$_2$ with potassium iodide or upon melting elemental sulfur; the presence of all sulfur rings cyclo-S$_n$ in the series $n = 6$–22 can be detected by this method (Figure 3.3).[12] Although many of the individual sulfur allotropes have been prepared by designed synthesis (see Section 2.1), the use of chromatography provides evidence for the existence of members of this homologous series that have not been isolated, since the retention times are size-dependent.

Chromatography is also a useful technique for the separation of isomers of a given ring system, *e.g.* the eight-membered cyclic sulfur imides, in which one or more of the sulfur atoms in cyclo-S$_8$ are replaced by an imido (NH) group. The synthesis of these heterocycles from the reaction of S$_2$Cl$_2$ with gaseous ammonia in DMF produces a mainly cyclo-S$_7$NH together with a mixture of the three diimide isomers, cyclo-1,3-, -1,4- and -1,5-S$_6$(NH)$_2$, which may be separated by chromatography of CS$_2$ solutions on silica gel (see **2.1**–**2.4** in Section 2.1 and Section 12.6.1).[13]

Chromatography has also been used to separate the unsaturated phosphorus(V)–nitrogen–sulfur ring systems **3.4**–**3.6**.[14] The six-membered ring

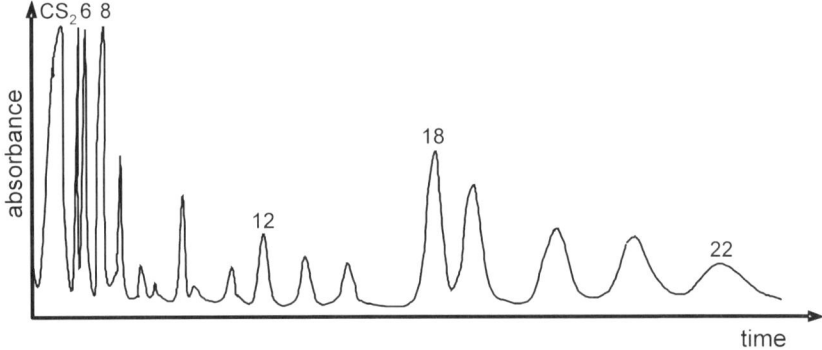

Figure 3.3 Chromatogram of the mixture of sulfur allotropes obtained from the reaction of SCl_2 with KI in CS_2. Reproduced from Figure 10 in R. Steudel, *Top. Curr. Chem.*, 1982, **102**, 149, with permission of Springer Science.[12]

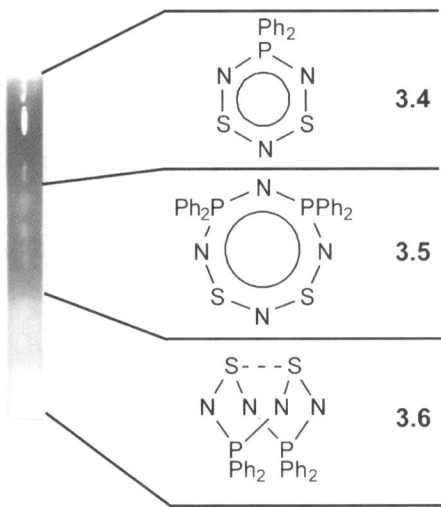

Figure 3.4 Column chromatography of a mixture of unsaturated P–N–S rings.

$Ph_2PN_3S_2$ (**3.4**) is purple, whereas the isomeric eight-membered rings 1,3- and 1,5-$Ph_4P_2N_4S_2$ (**3.5** and **3.6**) are orange and pale yellow, respectively. An efficient separation of these inorganic heterocycles is achieved by size-exclusion column chromatography, as is readily evident from the colours of the bands on the chromatographic column (shown in shades of grey–black in Figure 3.4).

3.3 MASS SPECTROMETRY

In a mass spectrometer, ions are produced from a sample and separated according to their mass-to-charge ratio (m/z) and then recorded in terms of intensity (number of ions).[1-3] Thus a mass spectrum can, in principle, provide the molecular mass of an inorganic ring system. There are a variety of different ways of producing ions, the most common of which is electron ionization (EI). This method involves bombardment of the sample with electrons whose energy (typically 70 eV) greatly exceeds the ionisation energy of the molecule under investigation. Consequently, the EI technique leads to the formation of fragment ions. The identification of these ions may provide information about the atomic arrangements in the parent molecule. In order to minimise the occurrence of fragmentation processes, soft ionisation techniques such as fast atom bombardment (FAB), chemical ionisation (CI) and electrospray ionisation (ESI) are employed. Normally, EI mass spectra are determined at low resolution, i.e. the m/z values are obtained as integers. In a high-resolution experiment, molecular masses can be determined with a high level of accuracy and this information may be used to provide identification of the molecular composition of an unknown compound. The determination of molecular masses is especially important for the identification of individual members of a homologous series, since other spectroscopic methods, e.g. NMR and UV–visible spectroscopy, often show only small changes in chemical shifts or absorption maxima for individual members of the series. For example, the sulfur allotrope cyclo-S_{12} was initially identified by the observation of the molecular ion in the EI mass spectrum.[15] In a more recent example, ESI mass spectrometry was used to solve the long-standing enigma of the composition of the drug Salvarsan [see Section 11.3.2 and eqn (11.16)]. The major components of this cycloarsine $(ArAs)_n$ (Ar = 2-NH_2-3-HOC_6H_3) are the trimer ($n=3$) and pentamer ($n=5$), although there is evidence for the presence of all members of the series ($n = 3$–8) (Figure 3.5).[16] Since Salvarsan is comprised of a mixture of cyclic oligomers with a tendency to form hydrogen bonds, it has not been possible to grow single crystals of the major component for an X-ray structural determination.

The occurrence of rearrangement processes or ion–molecule reactions in the mass spectrometer prior to ion detection is a potential source of misleading structural inferences if mass spectral data are considered in the absence of other structural information. For example, the reaction of trimethylphosphine with B_2H_6 gives rise to the six-membered ring $(Me_2PBH_2)_3$. The chair-shaped structure of this six-membered ring in the solid state has been established by X-ray crystallography.[17] However, the mass spectrum shows the fragment ions $(CH_3)_2{}^{10}B^+$ and $(CH_3)_2{}^{11}B^+$ rather than $(CH_3)_2P^+$ as a result of a rearrangement process that occurs in the mass spectrometer.

3.4 NUCLEAR MAGNETIC RESONANCE (NMR) SPECTROSCOPY

3.4.1 Introduction

NMR spectroscopy is used to provide information about inorganic ring systems through acquisition of the spectra of either the p-block element(s) within the ring or the exocyclic substituents. These substituents are usually organic

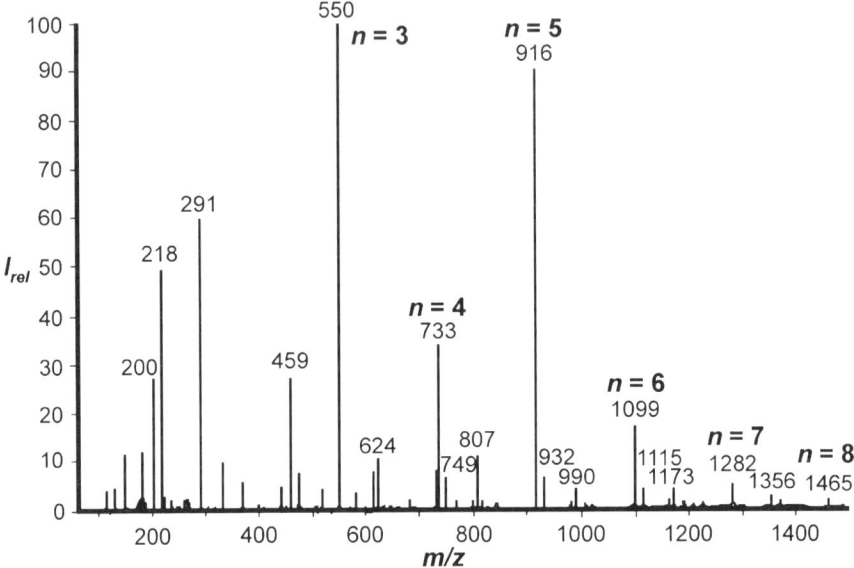

Figure 3.5 ESI mass spectrum of Salvarsan in water. Reproduced with permission from N. C. Lloyd et al., Angew. Chem. Int. Ed., 2005, **44**, 941.[16]

Table 3.2 NMR properties of p-block elements.

Nucleus	Spin	Abundance (%)	Nucleus	Spin	Abundance (%)
^{10}B	3	19.58	^{77}Se	1/2	7.58
^{11}B	3/2	80.42	^{113}In	9/2	4.28
^{13}C	1/2	1.108	^{115}In	9/2	95.72
^{14}N	1	99.63	^{117}Sn	−1/2	7.61
^{15}N	−1/2	0.37	^{119}Sn	−1/2	8.58
^{27}Al	5/2	100	^{121}Sb	5/2	57.25
^{29}Si	−1/2	4.70	^{123}Sb	7/2	42.75
^{31}P	1/2	100	^{123}Te	−1/2	0.87
^{33}S	3/2	0.76	^{125}Te	−1/2	6.99
^{69}Ga	3/2	60.4	^{203}Tl	1/2	29.50
^{71}Ga	3/2	39.6	^{205}Tl	1/2	70.50
^{73}Ge	−9/2	7.76	^{207}Pb	1/2	22.60
^{75}As	3/2	100	^{209}Bi	9/2	100

(alkyl or aryl) groups and ^{1}H or ^{13}C NMR spectra are readily obtained. Table 3.2 shows the NMR properties of the p-block elements that are most commonly encountered in inorganic ring systems. The NMR spectra of ring systems with spin $I = 1/2$ nuclei, e.g. phosphorus (^{31}P, 100%), silicon

(^{29}Si, 4.7%), selenium (^{77}Se, 7.6%), tin (^{119}Sn, 8.6%) and tellurium (^{125}Te, 7.0%), are readily accessible. However, low solubility combined with the relatively low sensitivity of the spin-active nuclei may result in long acquisition times for heterocycles involving Se or Te. Nitrogen also has a spin $I = 1/2$ nucleus (^{15}N); however, the preparation of ^{15}N-enriched samples is often necessary because of the low natural abundance (0.37%). In general, nuclei with $I > 1/2$ are less suitable for NMR studies because the large nuclear electric quadrupole moments shorten the T_2 relaxation times greatly and so broaden the NMR resonances. Nevertheless, nuclei such as ^{11}B ($I = 3/2$, 80.4%), ^{14}N ($I = 1$, 99.6%) and ^{27}Al ($I = 5/2$, 100%) are widely and routinely observed. The most abundant nuclei of oxygen and sulfur, ^{16}O (99.8%) and ^{32}S (95.0%), have zero nuclear spin ($I = 0$) and are not, therefore, amenable to NMR studies. The spin-active nuclei of these elements, ^{17}O ($I = 5/2$, 0.037%) and ^{33}S ($I = 3/2$, 0.76%), are not very informative NMR probes because of the combination of quadrupolar broadening and low natural abundances.

3.4.2 Applications

The most useful applications of NMR spectroscopy include (a) identifying individual members of a homologous series, (b) distinguishing between structural isomers, (c) monitoring ring transformations and (c) the analysis of complex reaction mixtures. Examples of each of these applications are given in this section and numerous other illustrations are discussed in Chapters 9–12.

For example, the ^{31}P NMR chemical shifts of cyclophosphazenes (NPCl$_2$)$_n$ ($n = 3, 4, 5, etc.$) show a marked dependence on ring size that allows for a facile differentiation between six- and eight-membered rings to be made.[18a] The individual members of the homologous series of permethylcyclosilanes (Me$_2$Si)$_n$ ($n = 5$–19) have been identified by ^{1}H and ^{13}C NMR spectra. The chemical shifts are in the range δ 0.135–0.197 (^{1}H) and δ −6.23 to −3.41 (^{13}C); however, only very small differences are observed for the larger cyclic oligomers ($n > 9$).[18b]

The importance of symmetry in determining the number of resonances that are expected in an NMR spectrum of different isomers of an inorganic heterocyle is illustrated by two examples from cyclophosphazene chemistry. The six-membered ring N$_3$P$_3$Cl$_3$(NMe$_2$)$_3$ may exist as *cis*, *trans* or *geminal* isomers, **3.7**, **3.8** or **3.9**, respectively. These isomers are predicted to exhibit one, two or three dimethylamino doublets (coupling to ^{31}P) in their ^{1}H NMR spectra on the basis of symmetry considerations. In **3.7**, the three Me$_2$N groups are related by a C_3 axis perpendicular to the centre of the approximately planar P$_3$N$_3$ ring. Consequently, they should exhibit chemical shift equivalence, *i.e.* only one resonance will be observed. In isomer **3.8**, the two Me$_2$N groups located above the P$_3$N$_3$ ring are related by a plane of symmetry, whereas the third Me$_2$N group is unique. Thus, two ^{1}H NMR resonances with relative intensities of 2:1 will be observed. Finally, in the third isomer **3.9**, there is no symmetry element that relates the Me$_2$N groups; consequently, three equally intense resonances are predicted for the ^{1}H NMR spectrum.

3.7   **3.8**   **3.9**

In the second example, the non-geminally disubstituted cyclotriphosphazene $N_3P_3F_4(NMe_2)_2$ exist as two isomers, *cis* (**3.10a**) and *trans* (**3.10b**), that can be distinguished by ^{19}F NMR spectroscopy. For the *cis* isomer **3.10a**, the F_C atoms are related by a plane of symmetry and, therefore, will have the same chemical shifts, whereas the fluorine atoms F_A and F_B are not symmetry related. Therefore, three resonances with relative intensities of 2:1:1 are predicted in the ^{19}F NMR spectrum. By contrast, the *trans* isomer **3.10b** has a C_2 axis that relates the fluorine atoms F_A and F_B; the same C_2 axis also gives rise to symmetry equivalence for the two F_C atoms. Consequently, two equally intense ^{19}F NMR resonances will be observed for the *trans* isomer.

3.10a   **3.10b**

A classic example of the use of isotopic enrichment is the ^{15}N NMR spectrum of the cation $[S_4{}^*N_3]^+$ (*N = 99% ^{15}N-enriched).[19] As illustrated in Figure 3.6, the spectrum consists of doublet (N_B) and triplet (1:2:1) (N_A) patterns resulting from two-bond couplings between the unique nitrogen atom N_A and the pair of equivalent nitrogen nuclei N_B in the cyclic structure.

Several saturated cyclic Se–N compounds are known, including five-, six-, eight- and 15-membered rings (see **12.17 12.20** in Section 12.8.1). The ^{77}Se NMR chemical shifts of Se–N compounds cover a range of more than 1500 ppm and the value of the shift is characteristic of the local environment of the selenium atom. Consequently, ^{77}Se NMR spectroscopy can be used to analyse a complex mixture of Se–N heterocycles as shown in Figure 3.7.[20]

NMR spectroscopy is also a powerful technique for monitoring ring transformation processes in solution. For example, the ^{14}N NMR spectrum of a solution of (NSCl)$_3$ in carbon tetrachloride has been used to determine the thermodynamic parameters for the equilibrium between this six-membered ring and the monomer NSCl in solution [eqn (3.1)].[21]

$$(NSCl)_3 \rightleftharpoons 3NSCl \qquad (3.1)$$

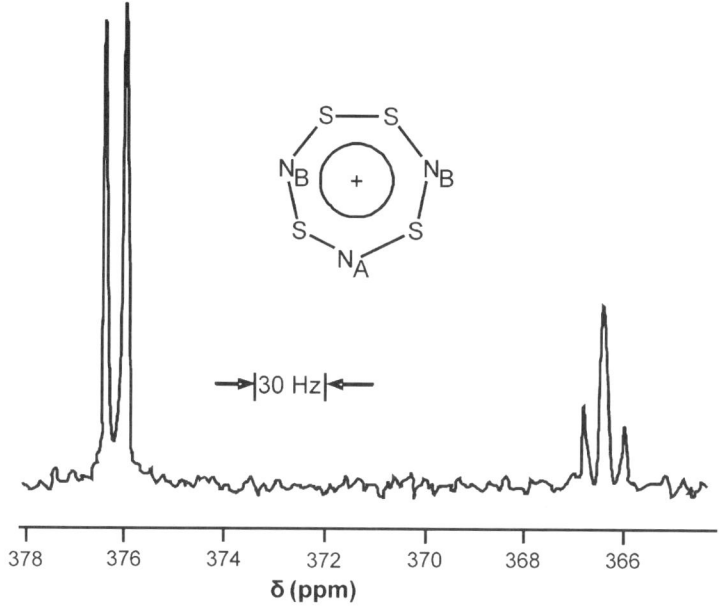

Figure 3.6 ^{15}N NMR spectrum of 99% ^{15}N-enriched [S$_4$N$_3$]$^+$ in 70% HNO$_3$. Reproduced with permission from T. Chivers et al., Inorg. Chem., 1981, **20**, 914.[19b] Copyright (1981) American Chemical Society.

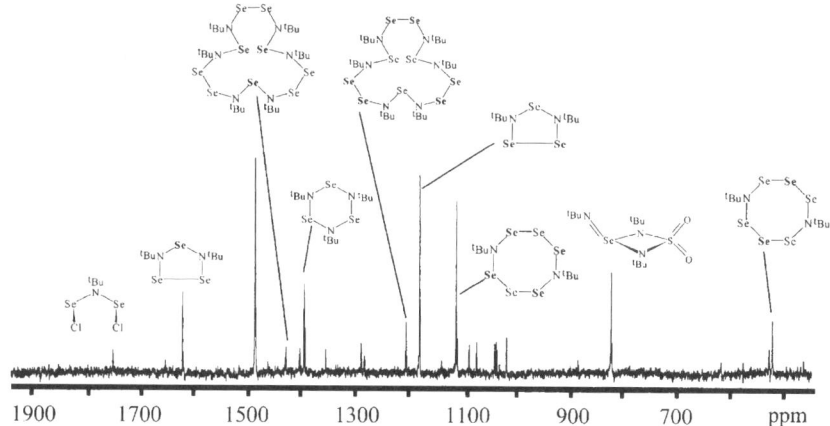

Figure 3.7 ^{77}Se NMR spectrum of a mixture of Se–N rings. Assignments are indicated by bold-face **Se** atoms. Reproduced with permission from T. Maaninen et al., Inorg. Chem., 2000, **39**, 5341.[20] Copyright (2000) American Chemical Society.

The ring-opening process for the generation of phosphazene polymers from cyclic precursors (see Scheme 11.10) provides another cogent example of the use of NMR spectroscopy for monitoring the structural reorganisation of ring systems. In this case, the transformation of the inorganic heterocycle (NPX$_2$)$_3$

to the corresponding linear polymer $(NPX_2)_n$ is accompanied by upfield shifts of 30–40 ppm in the ^{31}P NMR spectra.

3.4.3 Spin System Notation

Although the NMR spectra of some inorganic ring systems are first order and give rise to easily interpreted binomial coupling patterns, *e.g.* ^{15}N-enriched $[S_4N_3]^+$ (Figure 3.6), the spectra of many inorganic heterocycles are much more complex owing to second-order effects. In these cases, computer simulations are necessary to extract the NMR coupling constants and chemical shifts. A spin system notation is used to distinguish between first- and second-order NMR spectra for molecules with more than one nucleus with $I = 1/2$. Nuclei with $I > 1/2$ are not included in this spin system nomenclature. Familiarity with this nomenclature is desirable, since it is used in Chapters 9–12 to discuss the NMR spectra of various inorganic ring systems.

Each chemically distinct nucleus is assigned a letter and a numerical subscript is used to indicate the number of such nuclei. If the chemical shift difference between two sets of nuclei is large compared with the coupling constant between them ($\delta_1 - \delta_2 >> J_{12}$), letters that are well apart in the alphabet are used: A, X, M. Such systems are first order and give rise to simple multiplets in the NMR spectra. On the other hand, if the chemical shift difference is of the same order of magnitude as the coupling constant between the two nuclei ($\delta_1 - \delta_2 \approx J_{12}$), then consecutive letters are used: A, B, C, . . . , X, Y, Z. The latter systems give rise to second-order spectra with complex multiplet patterns.

A further refinement of this nomenclature that is often encountered for inorganic ring systems identifies magnetically inequivalent nuclei in a molecule. Nuclei which have the same chemical shift, *i.e.* are related by a symmetry element, and which couple to the same extent with all other magnetic nuclei in the molecule are referred to as magnetically equivalent. If the term 'equivalent nuclei' is used without further qualification, it is implied that such nuclei are magnetically equivalent. Nuclei which show chemical shift equivalence, but are not magnetically equivalent, are referred to by the same letter using a superscript prime to distinguish them.

Some examples to illustrate the use of this spin system notation to distinguish between first- and second-order systems and to explain the concept of magnetic inequivalence will now be discussed. Because ^{31}P is the only spin-1/2 nucleus with 100% natural abundance that forms a wide variety of inorganic ring systems, most of the examples are taken from phosphorus chemistry (for other examples, see Chapter 11).

According to the spin system notation, the ^{15}N NMR spectrum of ^{15}N-enriched $[S_4N_3]^+$ (Figure 3.6) is described as an AX_2 spin system. By contrast, the proton-decoupled $\{^1H\}$ ^{31}P NMR spectrum of the compound $(PhP)_3S_3$ has been analysed as an ABX system, yielding the following parameters: $J_{AX} = 7.5$, $J_{BX} = 13.0$ Hz; δ (A) 76.1, δ (B) 88.0, $\delta(X)$ 157.6. Two five-membered rings, **3.11a** and **3.11b**, have been considered for this structure; the original suggestion of **3.11a** was replaced by **3.11b** as a result of an X-ray structural determination.

Both isomers lack symmetry elements and, consequently, the two three-coordinate phosphorus atoms are inequivalent. The chemical environments of the two P(III) centres are similar in both isomers, but the large value of $J_{AB}=263$ Hz indicates a P–P bond as found in **3.11b**.[22]

3.11a   **3.11b**

The cyclophosphinophosphonium ions Me(PR)$_n^+$ ($n = 3, 4, 5$) (see Section 11.3.1.2 and Scheme 11.2) provide an informative example of magnetic inequivalence.[23] Whereas the ^{31}P {^{1}H} NMR spectra for the three-membered ring **3.12a** and the four-membered ring **3.12b** are essentially first order (A$_2$X and A$_2$MX, respectively), the five-membered architecture of **3.12c** results in magnetic inequivalence of the two pairs of phosphorus atoms that, in solution, are related by a C_2 axis. Consequently, complex second-order spectra resulting from an AA'BB'X spin system are observed.

3.12a   **3.12b**   **3.12c**

3.5 INFRARED AND RAMAN SPECTROSCOPY

Infrared (IR) spectra are easily obtained for any molecule and the sample may be in the gas, liquid or solid phase. Raman spectra are less frequently determined, but this technique provides information that is complementary to IR spectroscopy because of the different selection rules for the two techniques. Infrared activity requires a change in dipole, whereas Raman activity is determined by the polarisability of a vibration. Vibrational spectroscopy can be applied to inorganic ring systems at various levels of sophistication. The most common application of IR spectroscopy for acyclic inorganic (and organic) compounds involves group frequency correlations in which certain functional groups, *e.g.* E–H, E–F, E–Cl (E = p-block element), are known to occur within a certain characteristically narrow range of frequencies. The observation of an IR band in that region is taken as evidence for the presence of that functional group in a molecule. This approach is less useful for cyclic inorganic systems, since the functional groups are incorporated into a larger molecular system.

The number of IR- and Raman-active fundamentals for a particular molecule is dependent on its structure. In the case of an E_3N_3 ring system, for example, a planar structure with D_{3h} symmetry is predicted to have fewer bands than a non-planar alternative with a C_s or C_{2v} point group. Therefore, the most probable structure can be deduced by comparing the observed number of IR- and Raman-active fundamentals with that predicted by group theory.[24] For example, the six-membered ring $[S_3N_3]^-$ [see eqn (12.14)] was assigned a planar D_{3h} structure in the K^+ and $[R_4N]^+$ salts on the basis of the number of observed IR and Raman bands and their coincidences.[25] The monocyclic P_6^{4-} ion (isovalent with $[S_3N_3]^-$) has also been assigned a planar structure (in this case D_{6h}) on the basis of vibrational spectra.[26] In a similar manner, a planar (D_{5h}) structure has been attributed to the five-membered ring P_5^-.[27]

An illustrative example of the influence of symmetry on the number of vibrations is provided by the homologous series of cyclic sulfur allotropes S_n ($n = 6-12$). The IR absorptions for these ring systems are weak owing to the low polarity of S–S bonds. However, sulfur is a good Raman scatterer because S–S bonds are readily polarised. The various sulfur allotropes have different symmetries in addition to a different number of sulfur atoms and, consequently, each allotrope exhibits a characteristic Raman spectrum (Figure 3.8).[12]

Since vibrational frequencies depend on the masses of moving atoms, the substitution of an atom in a molecule by an isotope of different mass will alter the frequencies of those modes in which the substituted atom moves significantly. This technique may be useful in distinguishing between two possibilities for the assignment of a vibrational band. For example, the vibrational

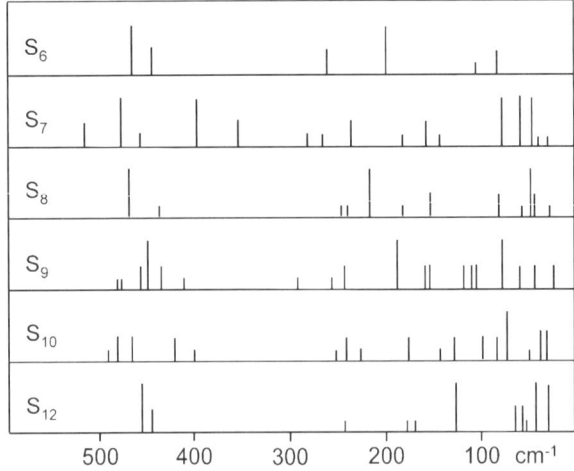

Figure 3.8 Schematic representation of the Raman vibrations for cyclic sulfur allotropes. Reproduced from Figure 3 in R. Steudel, *Top. Curr. Chem.*, 1982, **102**, 149, with permission Springer Science.[12]

assignments and force constants for the crown-shaped eight-membered ring $S_4N_4H_4$ (see Scheme 12.3b) have been verified from the IR and Raman spectra of the isotopically labelled compounds $S_4{}^{15}N_4H_4$ and $S_4{}^{15}N_4D_4$.[28]

Vibrational spectra are particularly useful in providing structural information for species that are not amenable to X-ray analysis because of thermal instability, e.g. the matrix-isolated species Si_3O_3[29] and P_3N_3,[30] both of which have been assigned planar D_{3h} structures on the basis of their IR spectra. Suitable single crystals of the $[Se_3N_2Cl]Cl$ could not be obtained for an X-ray structural analysis because of the insolubility of this explosive salt. Consequently, the structure of the five-membered ring $[Se_3N_2Cl]^+$ (see **12.26** in Section 12.8.2) was deduced by comparing the experimental spectra with the calculated fundamental vibrations based on an assumed structure.[31]

For a more detailed discussion of the theoretical aspects of infrared and Raman spectroscopy, and additional applications of this technique to inorganic ring systems, the reader is referred to the book by Nakamoto.[24]

3.6 ELECTRON PARAMAGNETIC RESONANCE (EPR) SPECTROSCOPY

The technique of EPR or ESR (electron spin resonance) spectroscopy can only be applied to species having one (or more) unpaired electrons. Thus, EPR spectroscopy can be considered as complementary to NMR spectroscopy, since paramagnetic systems normally give broad NMR resonances. Interactions between the unpaired electron and nuclear spin magnetic moments give rise to fine structure in the EPR spectra. The appearance of the splitting pattern is governed by the same rules that apply to NMR spectra. Coupling to n equivalent nuclei of spin $I = 1/2$ gives $n + 1$ lines and the intensities follow a binomial distribution.

As a simple example, the EPR spectrum of the hypothetical anion radical of a cyclotriphosphine $[(RP)_3]^{-\bullet}$ (all-cis R groups) would consist of a 1:3:3:1 quartet as a result of the coupling of the unpaired electron with three equivalent P atoms (Figure 3.9); the tetramer $[(RP)_4]^{-\bullet}$ with four equivalent phosphorus nuclei would give rise to a binomial quintet.

The schematic EPR spectra in Figure 3.9 are shown in both the absorption mode (like NMR spectra) and as a first derivative. In practice, EPR spectra are

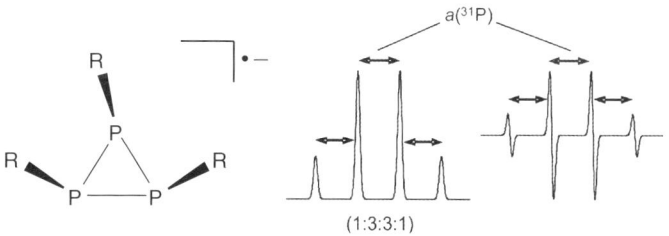

Figure 3.9 Schematic EPR spectra of the radical anion $[(RP)_3]^{-\bullet}$ ($n = 3, 4$).

usually recorded as a first derivative in order to make the lines narrower and, hence, maximise the separation of a large number of lines. This is especially important when the unpaired electron is coupled to several different sets of spin-active nuclei, some of which may have $I > 1$.

Similar to NMR spectra, two parameters may be obtained from an EPR spectrum. The first is the Landé splitting factor g, which is more commonly referred to as the 'g value'. The g value for a free electron is 2.0023. The numerical value of g is influenced by the molecular environment of the unpaired electron in the radical under investigation. For inorganic ring systems involving p-block elements, the changes in g values are small and usually involve the third decimal place. If transition metals are incorporated in the ring system, however, g values of 4–5 may be observed as a result of spin–orbit coupling. The second EPR parameter is the hyperfine coupling constant (hfcc). The hfcc can be measured from the separation between the adjacent peaks of a multiplet in an EPR spectrum. The hfcc is represented by the symbol a followed, in parentheses, by the isotope that is involved in the coupling, *e.g.* $a(^{31}P)$ in Figure 3.9. The units of a hfcc are gauss (G) or millitesla (mT) (10 G = 1 mT). As is the case for NMR coupling constants, the magnitude of the hfcc is determined primarily by the electron density at the nucleus with which the unpaired electron interacts. Hence radicals in which the unpaired electron resides primarily in an orbital with high s character will have large values of a. On the other hand, π-type radicals will exhibit relatively small a values because the unpaired electron is located in p-type orbitals.

For quadrupolar nuclei ($I = m/2$, where $m \geq 2$), coupling to a single nucleus gives $m + 1$ lines, which are of equal intensity and equally spaced. A simple example (although it is not a pure inorganic ring system) is provided by 1,3,2-dithiazolyl radicals (Figure 3.10a), which exhibit 1:1:1 triplets due to coupling to a single ^{14}N nucleus ($I = 1$); the natural abundance of spin-active C and S isotopes is very low (Section 3.4). Coupling to more than one quadrupolar nucleus in chemically equivalent environments gives rise to $2nI + 1$ lines (n = number of nuclei with spin I). Hence coupling to two equivalent ^{14}N nuclei in 1,2,3,5-dithiadiazolyls gives a five-line (1:2:3:2:1) pattern (Figure 3.10b).

EPR spectroscopy, in conjunction with theoretical calculations, provides valuable information about the spin distribution in paramagnetic ring systems, especially with regard to electron delocalisation (see Section 4.1). Either neutral radicals or charged paramagnetic species, *i.e.* anion or cation radicals, can be studied by this technique. The latter are typically generated by chemical or electrochemical reduction or oxidation, respectively. In some inorganic heterocycles, however, the delocalisation is incomplete, *i.e.* limited to a certain segment of the ring. For example, the EPR spectrum of the six-membered P_2N_3S ring in the neutral radical $[Ph_4P_2N_3S]^{\bullet}$ consists of a 1:2:3:2:1 quintet of triplets (1:2:1) (Figure 3.11). This hyperfine pattern indicates coupling to two equivalent ^{14}N atoms (99.6%, $I = 1$) and two equivalent ^{31}P atoms (100%, $I = 1/2$), respectively, but not to the unique nitrogen atom. This interpretation of the spectrum was confirmed by the observation that the quintet collapses to a 1:2:1 triplet when the ^{14}N atoms in the ring are replaced by ^{15}N ($I = 1/2$).[32]

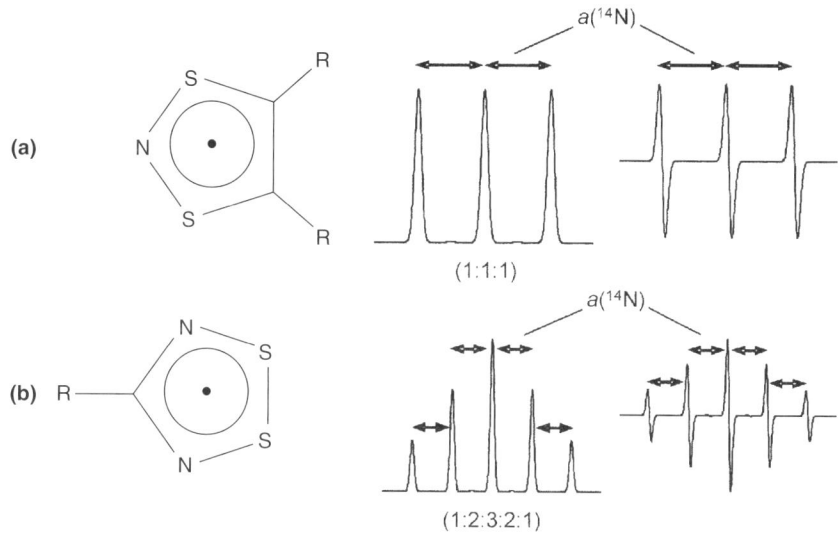

Figure 3.10 EPR spectra of (a) 1,3,2-dithiazolyl and (b) 1,2,3,5-dithiadiazolyl radicals; both absorption and first-derivative spectra are shown.

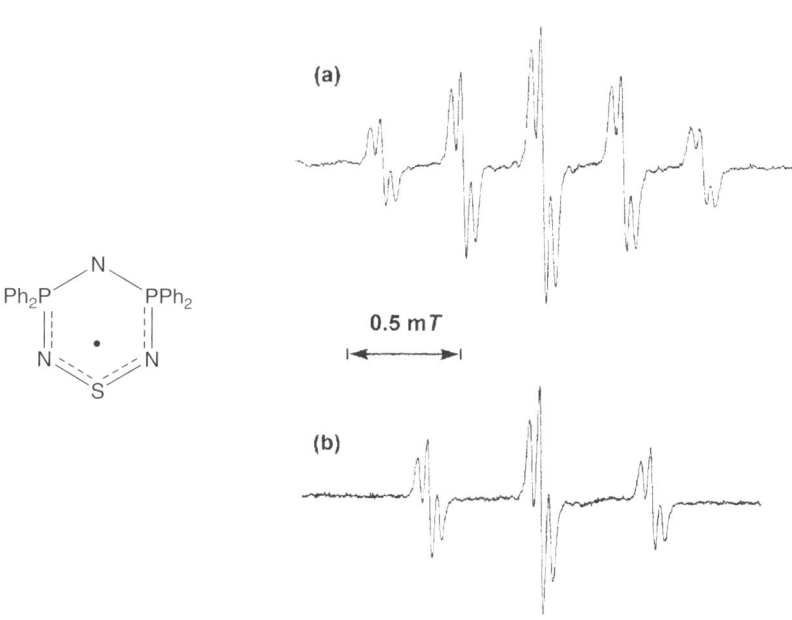

Figure 3.11 EPR spectra of (a) [Ph$_4$P$_2$N$_3$S]$^\bullet$ and (b) [Ph$_4$P$_2$^{15}N$_3$S]$^\bullet$ in CH$_2$Cl$_2$ Reproduced from R. T. Oakley, *J. Chem. Soc., Chem. Commun.*, 1986, 596, by permission of the Royal Society of Chemistry.[32]

Thus, the fine structure in the EPR spectrum reveals that delocalisation is limited to the PNSNP segment of the ring.

For a more advanced treatment of the theory and applications of EPR spectroscopy, the reader is referred to the text by Weil et al.[33]

3.7 UV–VISIBLE SPECTROSCOPY

UV–visible (or electronic absorption) spectra, in conjunction with molecular orbital calculations, can reveal information about the electronic structures of inorganic ring systems. These spectra provide the energies and molar extinction coefficients (or absorptivities) of various electronic transitions in the molecule, which can be compared with the calculated values of these parameters in order to elucidate their assignment. For practical purposes, the UV region corresponds to wavelengths of 200–400 nm, whereas the visible region spans the range 400–700 nm. Inorganic ring systems that absorb in the UV region are colourless, whereas those that exhibit visible absorption bands may range in colour from yellow to blue. A brief discussion of the UV–visible spectra of the most common saturated and unsaturated systems, both homocyclic and heterocyclic, follows.

Cyclic polysilanes $(Me_2Si)_n$ are colourless compounds that resemble aromatic hydrocarbons in that they exhibit electronic transitions in the UV region. The wavelength of the lowest energy absorption band is somewhat dependent on ring size with λ_{max} values of 272, 252 and 242 nm for $n = 5$, 6 and 7, respectively. The ionisation energies and, hence, the energies of the highest filled levels determined from photoelectron spectra are similar for the pentamer and hexamer. Consequently, the bathochromic shift of the lowest energy transition must be attributed to differences in the energies of the excited states of these inorganic ring systems.[11] Cyclic polysilanes also resemble aromatic hydrocarbons in the formation of highly coloured charge-transfer complexes. For example, a mixture of $(Me_2Si)_6$ and the π-acceptor tetracyanoethylene produces a violet solution with absorption maxima at 477 and 390 nm, which are assigned to charge-transfer transitions from the highest occupied (Si–Si) energy levels.

The members of the extensive series of sulfur homocycles S_n ($n = 6$–26) are all yellow, both in the solid state and in solution, at room temperature. UV–visible spectra are ineffectual for distinguishing between these cyclic oligomers because they all exhibit a strong absorption band at ca 254 nm; the yellow colour is attributable to a tailing of this band into the visible region.[34] However, the intense UV absorption of these sulfur rings provides an excellent method of detection in the chromatographic separation of the individual allotropes (Figure 3.3).

Both colourless and deeply coloured compounds are found among the common unsaturated heterocyclic inorganic systems. For example, borazine $B_3N_3H_6$ (see Section 4.1.2.1), like benzene, is a colourless liquid which exhibits a strong absorption band in the UV region. Similarly, cyclophosphazenes $(NPX_2)_n$ ($n = 3$–6; X = halogen, alkyl, aryl, alkoxy, dialkylamino) are also colourless compounds (see Section 4.1.2.2), regardless of the exocyclic substituents attached to the phosphorus atoms and the electronic spectra are not helpful in the identification of individual cyclic oligomers.

On the other hand, sulfur–nitrogen heterocycles exhibit a range of colours, since their π-electron richness (see Section 4.1.2.3) gives rise to low-energy electronic ($\pi^* \rightarrow \pi^*$ or n $\rightarrow \pi^*$) transitions. For example, the strong band observed for the yellow $[S_3N_3]^-$ anion at 360 nm has been assigned to the $\pi^* \rightarrow \pi^*$ (HOMO $\rightarrow$ LUMO) transition.[25,35] The attachment of a substituent X to one of the sulfur atoms gives rise to red compounds, e.g. S_3N_3X (X = O$^-$, NPPh$_3$, NAsPh$_3$), which exhibit a strong absorption band at 480–510 nm. This bathochromic shift is attributed to a loss of degeneracy of the HOMOs of $[S_3N_3]^-$ (see Figure 4.3) upon lowering the symmetry of the ring system from D_{3h} to C_{2v}, which results in a lower energy HOMO $\rightarrow$ LUMO transition.[36] A similar phenomenon is observed for the replacement of one sulfur atom in the six-membered ring of $[S_3N_3]^-$ by an R_2P group. Six-membered rings of the type $R_2PN_3S_2$ (R = Ph, Me) are purple and exhibit absorption bands in the region 560–590 nm.[37]

REFERENCES

1. W. L. Jolly, *The Synthesis and Characterisation of Inorganic Compounds*, Waveland Press, Prospect Heights, IL, 1991.
2. R. S. Drago, *Physical Methods in Chemistry*, W. B. Saunders, Philadelphia, PA, 1977.
3. E. A. V. Ebsworth, D. W. H. Rankin and S. Cradock, *Structural Methods in Inorganic Chemistry*, Blackwell Scientific, Oxford, 1987.
4. T. Chivers, *A Guide to Chalcogen–Nitrogen Chemistry*, World Scientific, Singapore, 2005, pp. 214–221.
5. (a) L. Pauling, *The Nature of the Chemical Bond*, 3rd edn., Cornell University Press, Ithaca, NY, 1960; (b) C. E. Housecroft and A. G. Sharpe, *Inorganic Chemistry*, 2nd edn., Pearson Education, Harlow, 2005.
6. N. W. Alcock, *Adv. Inorg. Chem. Radiochem.*, 1972, **15**, 1.
7. (a) W. Harshbarger, G. Lee, R. F. Porter and S. H. Bauer, *Inorg. Chem.*, 1969, **8**, 1683; (b) C. H. Chang, R. F. Porter and S. H. Bauer, *Inorg. Chem.*, 1969, **8**, 1689.
8. H. U. Höfs, J. W. Bats, R. Gleiter, G. Hartman, R. Mews, M. Eckert-Maksić, H. Oberhammer and G. M. Sheldrick, *Chem. Ber.*, 1985, **118**, 3781.
9. E. Jaudas-Prezel, R. Maggiulli, R. Mews, H. Oberhammer, T. Paust and W. -D. Stohrer, *Chem. Ber.*, 1990, **123**, 2123.
10. (a) W. Clegg, *Crystal Structure Determination*, Oxford University Press, Oxford, 1998; (b) W. Massa, *Crystal Structure Determination*, 2nd edn., Springer-Verlag, Heidelberg, 2004.
11. R. West and R. Carberry, *Science*, 1975, **189**, 179.
12. R. Steudel, *Top. Curr. Chem.*, 1982, **102**, 149.

13. R. Steudel and F. Rose, *J. Chromatogr.*, 1981, **216**, 399.
14. N. Burford, T. Chivers and J. F. Richardson, *Inorg. Chem.*, 1983, **22**, 1482.
15. J. Buchler, *Angew. Chem., Int. Ed. Engl.* 1966, **5**, 965.
16. N. C. Lloyd, H. W. Morgan, B. K. Nicholson and R. S. Ronimus, *Angew. Chem. Int. Ed.*, 2005, **44**, 941.
17. W. C. Hamilton, *Acta Crystallogr.*, 1955, **8**, 199.
18. (a) V. Chandrasekhar, *Inorganic and Organometallic Polymers*, Springer-Verlag, Berlin, 2005, pp. 106–107; (b) L. F. Brough and R. West, *J. Am. Chem. Soc.*, 1981, **103**, 3049.
19. (a) N. Logan and W. L. Jolly, *Inorg. Chem.*, 1965, **4**, 1508; (b) T. Chivers, R. T. Oakley, O. J. Scherer and G. Wolmershäuser, *Inorg. Chem.*, 1981, **20**, 914.
20. T. Maaninen, T. Chivers, R. S. Laitinen, G. Schatte and M. Nissinen, *Inorg. Chem.*, 2000, **39**, 5341.
21. J. Passmore and M. Schriver, *Inorg. Chem.*, 1988, **27**, 2751.
22. C. Lensch, W. Clegg and G. M. Sheldrick, *J. Chem. Soc. Dalton Trans.*, 1984, 723.
23. C. A. Dyker and N. Burford, *Chem. Asian J.*, 2008, **3**, 28.
24. K. Nakamoto, *Infrared and Raman Spectra of Inorganic and Coordination Compounds, Part A: Theory and Applications in Inorganic Chemistry*, 5th edn., Wiley, New York, 1997.
25. J. Bojes, T. Chivers, W. G. Laidlaw and M. Trsic, *J. Am. Chem. Soc.*, 1979, **101**, 4517.
26. H. G. von Schering, M. Somer, W. Hönle, W. Schmettow, U. Hinze, W. Bauhofer and G. Kliche, *Z. Anorg. Allg. Chem.*, 1987, **553**, 261.
27. M. Baudler, S. Akapoglu, D. Ousounis, F. Wasgestian, B. Meinigke, H. Budzikiewicz and H. Münster, *Angew. Chem. Int. Ed. Engl.*, 1988, **27**, 280.
28. R. Steudel, *J. Mol. Struct.*, 1982, **87**, 97.
29. J. S. Anderson and J. S. Ogden, *J. Chem. Phys.*, 1969, **51**, 4189.
30. R. M. Atkins and P. L. Timms, *Spectrochim. Acta, Part A*, 1977, **33**, 853.
31. J. Siivari, T. Chivers and R. S. Laitinen, *Inorg. Chem.*, 1993, **32**, 4391.
32. R. T. Oakley, *J. Chem. Soc., Chem. Commun.*, 1986, 596.
33. J. A. Weil, J. R. Bolton and J. E. Wertz, *Electron Paramagnetic Resonance: Elementary Theory and Practical Applications*, Wiley-Interscience, New York, 1994.
34. R. Steudel, H.-J. Mäusle, D. Rosenbauer, H. Möckel and T. Freyhold, *Angew. Chem. Int. Ed. Engl.*, 1981, **20**, 394.
35. J. W. Waluk and J. Michl, *Inorg. Chem.*, 1981, **20**, 963.
36. (a) T. Chivers, A. W. Cordes, R. T. Oakley and W. T. Pennington, *Inorg. Chem.*, 1983, **22**, 2429; (b) J. W. Waluk, T. Chivers, R. T. Oakley and J. Michl, *Inorg. Chem.*, 1982, **21**, 831.
37. N. Burford, T. Chivers, A. W. Cordes, W. G. Laidlaw, M. C. Noble, R. T. Oakley and P. N. Swepston, *J. Am. Chem. Soc.*, 1982, **104**, 1282.

CHAPTER 4
Electronic Structure and Bonding

4.1 DELOCALISATION (AROMATICITY) IN INORGANIC RINGS

4.1.1 Introduction

By analogy with cyclic organic molecules, inorganic ring systems can be divided into two categories: (a) saturated and (b) unsaturated. As in the case of alicyclic hydrocarbons, the former type involves single bonds between neighbouring atoms. For example, the polysilanes $(R_2Si)_n$ ($n = 3-35$) can be viewed as silicon analogues of alicyclic hydrocarbon systems. Other common examples of saturated inorganic ring systems include the cyclophosphines $(RP)_n$ ($n = 3-6$) and the sulfur allotropes cyclo-S_n ($n = 6-24$), which have one and two lone pairs of electrons, respectively, on each p-block atom.

In a similar vein, unsaturated inorganic ring systems, which have some degree of multiple bonding between adjacent atoms, are comparable to aromatic organic molecules such as benzene. Each carbon atom in planar unsaturated rings of the type C_nH_n contributes one electron to the π-system in a p-orbital perpendicular to the plane of the molecule. A useful concept for rationalising the stability and properties of these carbocyclic systems is the well-known Hückel rule, which asserts that stable molecules will have $4n + 2$ π-electrons. This rule is a direct consequence of the presence of degenerate levels in systems of high symmetry. From a glance at the molecular orbital level diagram in Figure 4.1, it can be seen that benzene, C_6H_6, with six π-electrons is a closed shell system, *i.e.* all energy levels are doubly occupied. By contrast, in square-planar cyclobutadiene, C_4H_4, or planar cyclooctatetraene, C_8H_8, which are both $4n$ π-systems, a degenerate pair of π levels is half-filled, leading to diradical character and high reactivity. As a consequence, cyclobutadiene undergoes Jahn–Teller distortion from a square to a rectangle in order to remove the orbital degeneracy, whereas the eight-membered cyclooctatetraene ring adopts a tub shape with four double bonds that are not conjugated to each other.

Inorganic Rings and Polymers of the p-Block Elements: From Fundamentals to Applications
By Tristram Chivers and Ian Manners
© Tristram Chivers and Ian Manners 2009
Published by the Royal Society of Chemistry, www.rsc.org

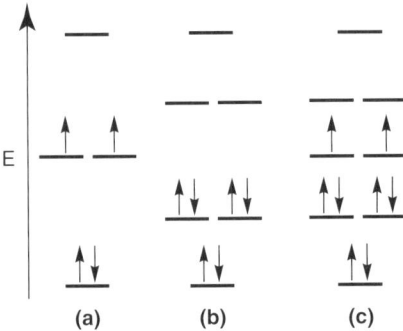

Figure 4.1 π-Molecular orbital energy levels of (a) C_4H_4, (b) C_6H_6 and (c) C_8H_8.

4.1.2 π-Electron Delocalisation

In contrast to carbocylic systems, unsaturated inorganic rings involve two (or more) elements, since homocycles such as cyclopolysilanes, cyclopolyphosphines and cyclic sulfur allotropes are all singly bonded systems (Section 4.1.1). The fundamental question of the existence of delocalised bonding ('aromaticity') in inorganic heterocycles is a subject of longstanding and ongoing debate. The concept of aromaticity in organic ring systems and, specifically, the Hückel $4n+2$ π-electron rule provides a useful starting point for this discussion. From a historical perspective, the inorganic heterocycles of particular interest in this regard are borazine, $B_3N_3H_6$ (**4.1**), the cyclophosphazenes, $(NPX_2)_3$ (**4.2**), and binary sulfur–nitrogen rings (**4.3–4.8**). These unsaturated ring systems will be discussed individually, since the bonding interactions in each of them involves different considerations.

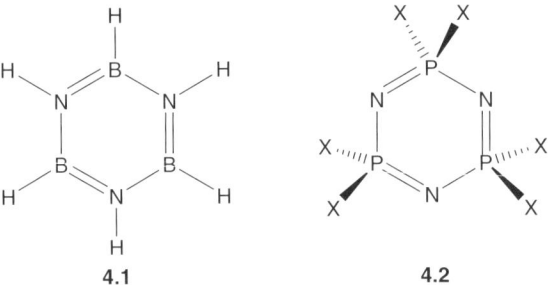

4.1.2.1 Borazine. Borazine (**4.1**) is a planar six-membered ring with endocyclic bond angles of *ca* 118° at boron and *ca* 121° at nitrogen and equal B–N bond lengths of 1.43 Å, which lie between those of typical single and double boron–nitrogen bonds. It is a volatile liquid (b.p. 55 °C). In view of the similarity in structures and physical properties, borazine is often referred to as 'inorganic benzene'. The trigonal geometry at B and N is consistent with sp^2 hybridisation of the ring atoms and, as is the case for benzene, the π-system is developed by overlap of p-orbitals perpendicular to the plane of the ring.[1] In contrast to benzene, for which each carbon atom contributes one electron to the π-system, each nitrogen in borazine has two electrons in the p_π-orbital while the boron p_π-orbitals are vacant. Thus borazine (like benzene) is a six π-electron system, which we will refer to as π-*electron precise*, *i.e.* the number of π-electrons is equal to the number of atoms in the ring. However, owing to the large electronegativity difference between boron (2.0) and nitrogen (3.0), the π-electron delocalisation is reduced in borazine compared with that in benzene and the B–N bonds are highly polar.

The extent of aromaticity of borazine is still a matter of debate because aromaticity is not an experimentally observable quantity.[2-5] The conclusions depend on the criterion that is used to estimate aromaticity. The structural data are consistent with delocalised bonding. A recent *ab initio* valence bond calculation has estimated the delocalisation energy of borazine to be about 8.3 kcal mol^{-1} higher than that in benzene.[4] On the other hand, an energy decomposition analysis estimates the value of the total π-bonding energy in borazine to be approximately 60% that of benzene.[5] These calculations suggest that borazine can be considered to have substantial aromatic character if the assumption that resonance energy can provide a criterion for aromaticity is correct. Support for some aromatic character also comes from recent experiments that have shown that borazine undergoes electrophilic aromatic substitution in the gas phase in a manner similar to benzene.[6]

On the other hand, the existence of a ring current and the resulting magnetic properties have also been used to describe aromaticity in benzenoid systems. Neither the diamagnetic susceptibility exaltation (−1.7 for borazine and 13.7 for benzene) nor the ^{1}H NMR chemical shifts of H(B) and H(N) (δ 5.0 and 5.5, respectively) indicate the presence of an appreciable ring current. A more recent magnetic criterion for probing aromaticity in planar rings is the nucleus-independent chemical shift (NICS), which is defined as the negative of the absolute magnetic shielding computed at the geometric centre of the ring. The calculated NICS values of −2.1 for borazine and −11.5 for benzene support the previous conclusion of the lack of significant aromaticity in borazine based on the ring current criterion.[7]

4.1.2.2 Cyclophosphazenes. The nature of the bonding in cyclophosphazenes has also been a source of controversy. In contrast to unsaturated B–N systems, for which ring sizes other than six are only observed when certain bulky substituents are attached to the B or N atoms (see Section 9.2.2), cyclophosphazenes form an extensive homologous series $(NPX_2)_n$ ($n = 3$–40; X = F). The six-membered rings are usually planar (or close to planar), whereas the larger ring systems adopt puckered conformations, with the exception of the tetramer $(NPF_2)_4$, which is almost planar (see Figure 11.19a). For direct comparison with borazine, the discussion here will be limited to the six-membered ring $(NPX_2)_3$ (**4.2**). The parent system (X = H) is not known, but a wide variety of derivatives where X = Cl, F, CF_3, alkyl, OR, NR_2 have been characterised (see Scheme 11.5).[8]

The phosphorus atoms in cyclophosphazenes use four valence electrons in forming σ-bonds to their four nearest neighbours, leaving one electron available for π-bonding. The nitrogen atoms utilize two electrons to bond to the two adjacent phosphorus atoms and also accommodate a lone pair of electrons in an 'sp^2' orbital in the plane of the P$_3$N$_3$ ring. Thus, each nitrogen atom has one electron available for π-bonding in a p-orbital perpendicular to the plane of the ring. Consequently, like borazine, the cyclotriphosphazenes, (NPX$_2$)$_3$ (**4.2**), are *π-electron precise* systems with six π-electrons for six ring atoms.

The P–N bond lengths in (NPX$_2$)$_3$ are approximately equal and fall between typical single and double bond values, with the shortest bonds found for X=F, indicative of multiple-bond character. The traditional bonding description for cyclophosphazenes, which is still used in many inorganic chemistry textbooks, invokes Dewar's island model.[9] In this representation, delocalisation occurs via d$_\pi$(P)–p$_\pi$(N) overlap resulting in 'islands' of electron density over P–N–P units with nodes at the phosphorus centres. It is now generally agreed, however, that valence d orbitals play only a minor role in bonding of the main group elements; they serve primarily as polarisation functions.[10]

A topological analysis of electron density distribution in cyclophosphazenes reveals that there is significant ionic character in the P–N bonds and, consequently, the zwitterionic model is the dominant contributor to the bonding in these classic inorganic ring systems (Figure 4.2a).[11] Concomitantly, according to a natural bond order analysis, a subsidiary bonding contribution is provided by negative hyperconjugation. As illustrated in Figure 4.2b, this involves the interaction of the in-plane and out-of-plane lone-pair orbitals with the σ*(P–N) and σ*(P–X) orbitals, respectively.[10] The latter contribution is enhanced as the electronegativity of the substituents X attached to phosphorus increases. In summary, the bonding in cyclophosphazenes is best described by a combination of ionic contributions and negative hyperconjugation, not by valence d-orbital participation. Consistent with the existence of an extensive homologous series of rings of different sizes, the cyclophosphazenes are not considered to be aromatic systems.

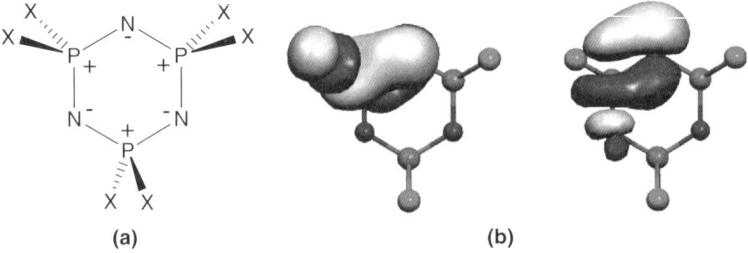

Figure 4.2 (a) The zwitterionic model and (b) negative hyperconjugation for cyclophosphazenes. Reproduced with permission from A. B. Chaplin *et al.*, *Inorg. Chem.*, 2005, **44**, 8407.[10] Copyright (2005) American Chemical Society.

4.1.2.3 Sulfur–Nitrogen Rings. Sulfur and nitrogen form a variety of cyclic systems with planar structures. These binary compounds may be neutral species, *e.g.* S_2N_2 (**4.3**), cations such as $[S_2N_3]^+$ (**4.4**), $[S_4N_3]^+$ (**4.6**), $[S_4N_4]^{2+}$ (**4.7**) and $[S_5N_5]^+$ (**4.8**) and the anion $[S_3N_3]^-$ (**4.5**). In 1972, it was proposed that these S–N heterocycles belong to a class of '*electron-rich aromatics*' which conform to the Hückel $4n+2$ π-electron rule.[12] On the reasonable assumption that each sulfur and each nitrogen atom uses two valence electrons for bonding in the σ-system and that there is a lone pair on each atom in the plane of the ring, this proposal was based on the contribution of two electrons from each sulfur and one electron from each nitrogen to the π-system. At the time, the known species **4.3**, **4.6** and **4.8** were cited as examples of 6π-, 10π- and 14π-systems, respectively. The species **4.4** (6π), **4.5** (10π) and **4.7** (10π), all of which formally conform to the Hückel rule, were discovered subsequently.

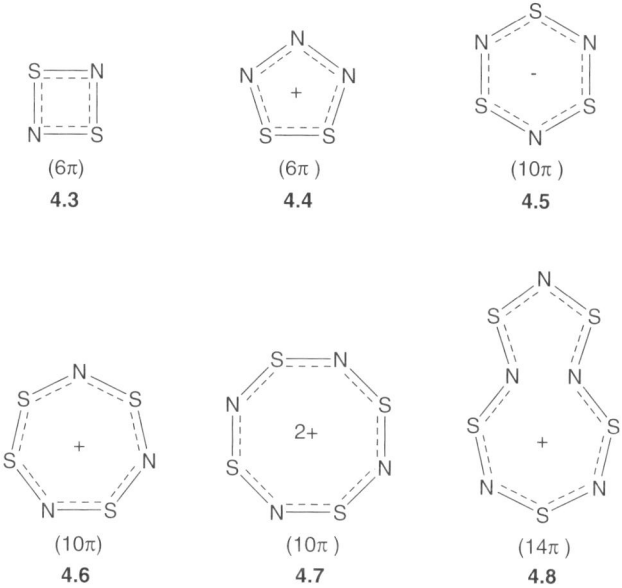

At first sight, the existence of this extensive series of sulfur–nitrogen rings would appear to provide justification for the proposal that planar, binary S–N systems can be regarded as π-*electron-rich* aromatics. In general, these ring systems exhibit approximately equal S–N bond lengths, although some distortions from equality may occur for the charged species as a result of cation–anion interactions. The S–N distances are intermediate between typical single and double bond values. Although these geometric parameters are consistent with delocalised structures, the absence of physical data such as heats of formation or diamagnetic ring currents makes it difficult to assess whether the term 'aromatic' is appropriate for these inorganic ring systems. However, the criterion of ipsocentric ring current has been used recently for this purpose.[13–15] Current density maps indicate that the 10 π-electron systems **4.5**, **4.6** and **4.7**

and the 14 π-electron system **4.8** support diatropic π currents, reinforced by σ circulations.

In order to explore the concept of aromaticity applied to sulfur–nitrogen ring systems further, it is instructive to compare the π-manifolds of cyclic S–N species with those of their aromatic hydrocarbon counterparts. In Figure 4.3, this comparison is made for the pairs S_2N_2 and cyclobutadiene, $[S_3N_3]^-$ and benzene, and $[S_4N_4]^{2+}$ and cyclooctatetraene on the basis of simple Hückel calculations. In the case of S_2N_2, the lowering of symmetry from D_{4h} in C_4H_4 to D_{2h} results in a loss of degeneracy. Consequently, the hypothetical 4 π-electron system $[S_2N_2]^{2+}$ does not violate Hund's rule. The formation of the neutral 6 π-electron ring S_2N_2 illustrates the ability of S–N heterocycles to accommodate more π-electrons than their hydrocarbon counterparts. This is a reflection of the higher electronegativities of sulfur and nitrogen compared with that of carbon, which give rise to molecular orbitals of lower energies for the heterocyclic systems. This characteristic of S–N rings is especially dramatic for $[S_3N_3]^-$, which is able to accommodate 10 π-electrons compared to 6 π-electrons for benzene. The π-bond order in $[S_3N_3]^-$ is low, however, compared with the value of 0.5 for benzene because the bonding effect of the degenerate π-levels is approximately cancelled by the occupation of the

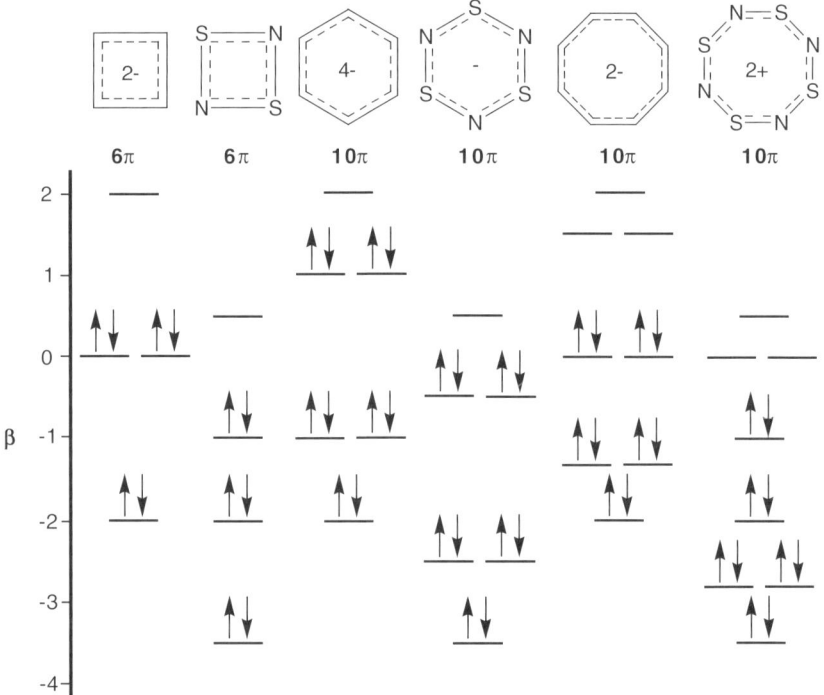

Figure 4.3 The π-manifolds for $[C_4H_4]^{2-}$, S_2N_2, $[C_6H_6]^{4-}$, $[S_3N_3]^-$, $[C_8H_8]^{2-}$ and $[S_4N_4]^{2+}$ from simple Hückel calculations.

Electronic Structure and Bonding 45

degenerate pair of antibonding π^*-orbitals. In the eight-atom systems, planar C_8H_8 is unstable because of Hund's rule (Figure 4.1), but the corresponding dianion $[C_8H_8]^{2-}$, a 10 π-electron system, which is isoelectronic with the dication $[S_4N_4]^{2+}$, is stable (Figure 4.3).

4.1.3 σ-Electron Delocalisation

The discussion in the previous sections was focused on delocalisation in the π-systems of various inorganic heterocycles. A unique characteristic of certain cyclic inorganic systems, which is not manifested in the corresponding hydrocarbon rings, is the ability to exhibit σ-electron delocalisation. The best known examples of this phenomenon involve cyclopolysilanes $(R_2Si)_n$ $(n = 5$–$35)$ (see Section 10.1.1), which are inorganic analogues of alicyclic compounds such as cyclopentane or cyclohexane. The semiconducting properties of silicon are well known from its pervasive use in transistors. However, it is difficult to design a chemical experiment to demonstrate the conducting properties of silicon, since it is insoluble in organic solvents. Cyclic polysilanes $(R_2Si)_n$ are comprised of Si–Si bonds similar to those in the element. Significantly, the two alkyl groups attached to each silicon confer solubility on these cyclic systems so that they can be used as models to elucidate the properties of elemental silicon.

For example, the cyclopentasilane $(Me_2Si)_5$ is reduced by sodium–potassium alloy in diethyl ether (or electrolytically) at low temperature to give the dark-blue radical anion $[Me_2Si]_5^{-\bullet}$ [eqn (4.1)], which exhibits a 31-line EPR spectrum consistent with coupling of the unpaired electron with 30 equivalent hydrogen atoms.[16] This observation indicates that the unpaired electron is delocalised over all five Si atoms (and on to the 10 CH_3 groups). This delocalisation over the entire ring system is reminiscent of that observed for aromatic radical anions, *e.g.* $[C_6H_6]^{-\bullet}$. Further discussion of the concept of σ-electron delocalisation in polysilanes can be found in Section 10.1.4.1.

$$(Me_2Si)_5 \xrightarrow[\text{Et}_2\text{O, -78 °C}]{\text{Na/K}} [Me_2Si]_5^{-\bullet} \text{ dark blue} \quad (4.1)$$

4.2 FRONTIER ORBITAL CONSIDERATIONS

4.2.1 Introduction

A knowledge of the composition of frontier orbitals gleaned from molecular orbital theory provides helpful insights into the structures and formation of inorganic ring systems. For example, the existence of weak intra- or intermolecular contacts (π*–π* interactions) can be understood from an inspection of the highest occupied molecular orbitals (HOMOs). This information can also aid in the rationalisation or prediction of the formation and reactivity of unsaturated inorganic ring systems. In the following sections, some examples of this aspect of the chemistry of inorganic heterocycles are examined. Frontier orbital considerations are also valuable for forecasting the possible existence of π-complexes of inorganic ring systems with metals of the types that are well known for a wide variety of unsaturated hydrocarbon ligands, as discussed in Section 7.2.

4.2.2 π*–π* Interactions

The formation of a weak σ-bond by overlap of two singly occupied π*-orbitals is a characteristic feature of certain π-electron-rich inorganic systems. This structural feature can occur via an intramolecular process resulting in a transannular interaction that converts a monocyclic ring into a bicyclic system. Alternatively, it may involve an intermolecular contact in which two cyclic monomers are weakly associated to give a dimeric species.

Sulfur–nitrogen heterocycles provide the most commonly encountered examples of π*–π* interactions involving inorganic ring systems. For example, the hybrid eight-membered rings $(R_2P)_2N_4S_2$ (R = alkyl, aryl) are found as 1,3- or 1,5-isomers (see Figure 3.4).[17] The 1,3-isomer **4.9** is an almost planar ring whereas the 1,5-isomer **4.10** adopts a folded structure with a weak transannular S···S contact of 2.50 Å (ca 0.46 Å longer than an S–S single bond).

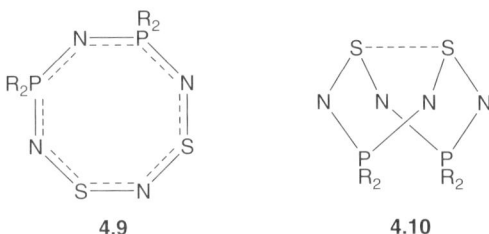

4.9   4.10

The HOMO in **4.10** is primarily antibonding with respect to the NSN units on opposite sides of the ring. Folding of the ring allows an intramolecular π*–π* interaction as illustrated in Figure 4.4. This bonding arrangement is similar to that found in the dithionite dianion $[S_2O_4]^{2-}$ $[d(S···S) = 2.39$ Å] in which a dimer is formed by a similar interaction of the π* SOMOs (singly occupied molecular orbitals) of two sulfur dioxide radical anions $[SO_2]^{-•}$. A similar transannular interaction is also found in the dication S_8^{2+} (see Figure 12.12).

In contrast to the ring folding observed in 1,5-$(R_2P)_2N_4S_2$, the π*–π* interaction in $[S_2O_4]^{2-}$ is an intermolecular process involving the association of two species. An analogy in inorganic heterocycles is provided by the thiatriazinyl

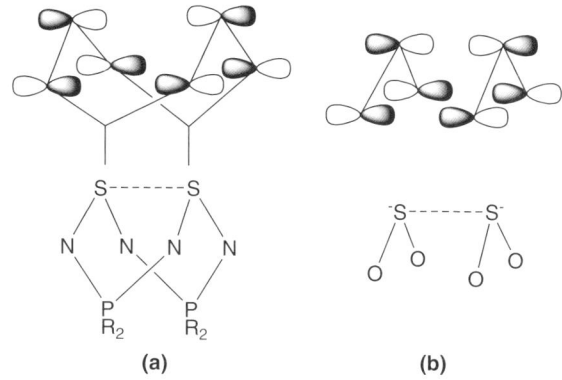

Figure 4.4 The π*–π* interactions in (a) 1,5-(R$_2$P)$_2$N$_4$S$_2$ and (b) [S$_2$O$_4$]$^{2-}$.

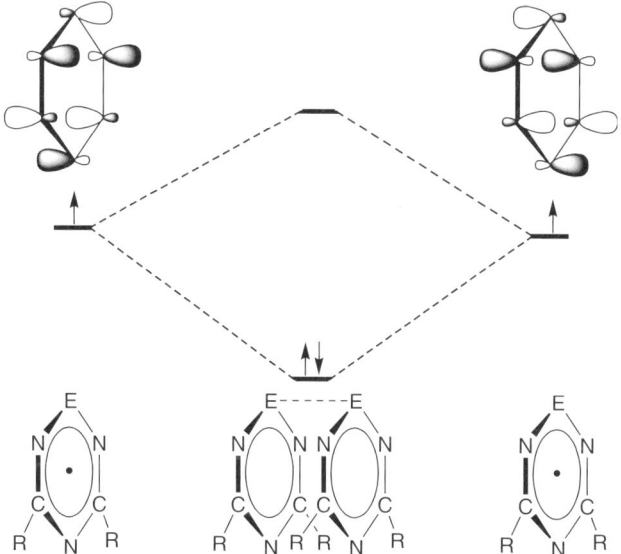

Figure 4.5 Cofacial overlap of two [(PhC)$_2$N$_3$S]$^•$ radicals.

radical [(PhC)$_2$N$_3$S]$^•$, which forms a weakly associated dimer [d(S···S) = 2.67 Å] via a π*–π* interaction (Figure 4.5).[18a] The electron density in the SOMO of this cyclic radical is primarily located on the NSN unit of the ring. The selenium analogue adopts a similar structure with a stronger intermolecular association, i.e. 4p–4p overlap is more effective than 3p–3p overlap.[18b]

A second important type of intermolecular π*–π* interactions is exhibited by the seven π-electron cation radical [S$_3$N$_2$]$^{+•}$ (**4.11**, E = S), which exists in the solid state as the transoid dimer. In this case, dimerisation occurs via two weak S···S interactions [d(S···S) = 3.00–3.10 Å].[19a] A similar structure is found for the selenium analogue [Se$_6$N$_4$]$^{2+}$ (**4.11**, E = Se) [d(Se···Se) = 3.12–3.14 Å].[19b]

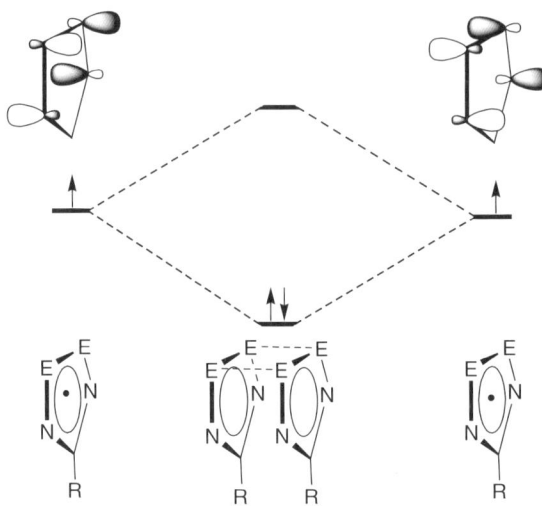

Figure 4.6 Structure and bonding in [PhCN$_2$E$_2$]$_2$ (E = S, Se).

4.11

An analogous mode of dimerisation is exhibited by dichalcogenadiazolyl radicals [PhCN$_2$E$_2$]• (E = S, Se) (isoelectronic with [E$_3$N$_2$]$^{+•}$), which are important building blocks for the construction of molecular conductors.[20] In these neutral radicals the substituent attached to carbon influences the mode of dimerisation (see Figure 3.1). With a flat substituent, e.g. a phenyl group, association occurs in a cisoid cofacial manner with two weak E···E interactions. This mode of dimerisation is readily appreciated from consideration of the composition of the SOMOs of the two interacting cyclic radicals (Figure 4.6). In some derivatives, a substituent on the phenyl ring in [PhCN$_2$E$_2$]• (E = S, Se) can give materials that are monomeric in the solid state and have interesting magnetic properties (see Section 5.1.2).

4.2.3 Formation of Inorganic Ring Systems

Although not as common as in organic chemistry, cycloaddition reactions provide a useful approach to the synthesis of inorganic heterocycles when the appropriate building blocks may be obtained readily (see Section 2.5). For

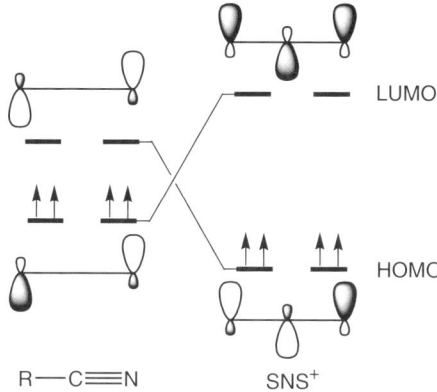

Figure 4.7 The frontier orbitals of [NS$_2$]$^+$ and RC≡N.

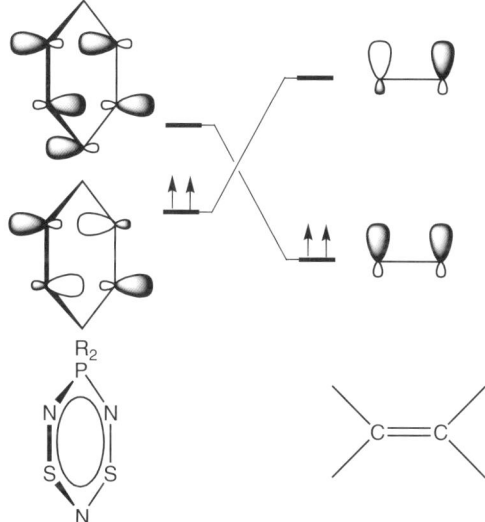

Figure 4.8 Cycloaddition of R$_2$PN$_3$S$_2$ and alkenes.

example, the linear [NS$_2$]$^+$ cation undergoes quantitative cycloaddition reactions with nitriles to give the cyclic 1,3,2,4-dithiadiazolium cations [RCN$_2$S$_2$]$^+$ (see Scheme 2.8).[21] The frontier orbitals involved in this interaction are illustrated in Figure 4.7. The HOMO (π_n) and LUMO (π^*) of the sulfur–nitrogen reagent are of the correct symmetry to interact with the LUMO (π^*) and HOMO (π) of a typical nitrile, respectively, resulting in the formation of the five-membered rings 1,3,2,4-[RCN$_2$S$_2$]$^+$. The chemical one-electron reduction of these cations generates the corresponding radicals 1,3,2,4-[RCN$_2$S$_2$]$^{\bullet}$, which rearrange into the 1,2,3,5-isomers both in solution and in the solid state (see Section 5.1.2 and Figure 5.1).

4.2.4 Reactivity of Inorganic Ring Systems

The S,S''-mode of the cycloaddition of alkenes, e.g. norbornadiene, to certain sulfur–nitrogen ring systems can be explained from a consideration of frontier orbitals. For example, both the HOMO and LUMO of the six-membered rings $R_2PN_3S_2$ are sulfur based and of the correct symmetry to overlap with the LUMO and HOMO, respectively, of an alkene thus accounting for the observed regiochemistry (Figure 4.8). The formation of norbornadiene adducts occurs readily and is of practical value for the characterisation and storage of unstable derivatives, e.g. $Me_2PN_3S_2$.[22] An interesting application of this adduct formation involves the stabilisation of the eight π-electron cyclic cation $S_3N_3^+$, which, in the absence of Jahn–Teller distortion, would be a biradical with two singly occupied orbitals (Figure 4.3).[23]

REFERENCES

1. C. E. Housecroft and A. G. Sharpe, *Inorganic Chemistry*, 2nd edn., Pearson Education, Harlow, 2005, pp. 319–320.
2. A. Soncini, C. Domene, J. J. Engelberts, P. W. Fowler, A. Rassat, J. H. van Lenthe, R. W. A. Havenith and L. W. Jenneskens, *Chem. Eur. J.*, 2005, **11**, 1257.
3. J. J. Engelberts, R. W. A. Havenith, J. H. van Lenthe, L. W. Jenneskens and P. W. Fowler, *Inorg. Chem.*, 2005, **44**, 5266.
4. D. Benker, T. M. Klapötke, G. Kuhn, J. Li and C. Miller, *Heteroat. Chem.*, 2005, **16**, 311.
5. I. Fernández and G. Frenking, *Faraday Discuss.*, 2006, 1.
6. (a) B. Chiavarino, M. E. Crestoni and S. Fornarini, *J. Am. Chem. Soc.*, 1999, **121**, 2619; (b) B. Chiavarino, M. E. Crestoni, A. Di Marzio, S. Fornarini and M. Rosi, *J. Am. Chem. Soc.*, 1999, **121**, 11204.
7. B. Kiran, A. K. Phukan and E. D. Jemmis, *Inorg. Chem.*, 2001, **40**, 3615.
8. V. Chandrasekhar, *Inorganic and Organometallic Polymers*, Springer-Verlag, Berlin, 2005, pp. 82–111.
9. R. Jaeger, M. Debowski, I. Manners and G. J. Vancso, *Inorg. Chem.*, 1999, **38**, 1153.
10. A. B. Chaplin, J. A. Harrison and P. J. Dyson, *Inorg. Chem.*, 2005, **44**, 8407.
11. V. Luaña, M. Pendás, A. Costales, G. A. Carriedo and F. J. García-Alonso, *J. Phys. Chem.*, 2001, **105**, 5280.
12. T. Chivers, *A Guide to Chalcogen–Nitrogen Chemistry*, World Scientific, Singapore, 2005, Chapter 4, pp. 54–60.
13. Y. Jung, T. Heine, P. v. R. Schleyer and M. Head-Gordon, *J. Am. Chem. Soc.*, 2004, **126**, 3132.
14. H. M. Tuononen, R. Suontamo, J. Valkonen and R. S. Laitinen, *J. Phys. Chem. A*, 2004, **108**, 5670.
15. F. De Proft, P. W. Fowler, R. W. A. Havenith, P. v. R. Schleyer, G. Van Lier and P. Geerlings, *Chem. Eur. J.*, 2004, **10**, 940.

16. E. Carberry, R. West and G. E. Glass, *J. Am. Chem. Soc.*, 1969, **91**, 5446.
17. N. Burford, T. Chivers, P. W. Codding and R. T. Oakley, *Inorg. Chem.*, 1982, **21**, 982.
18. (a) P. J. Hayes, R. T. Oakley, A. W. Cordes and W. T. Pennington, *J. Am. Chem. Soc.*, 1985, **107**, 1346; (b) R. T. Oakley, R. W. Reed, A. W. Cordes, S. L. Craig and J. B. Graham, *J. Am. Chem. Soc.*, 1987, **109**, 7745.
19. (a) R. J. Gillespie, J. P. Kent and J. F. Sawyer, *Inorg. Chem.*, 1981, **20**, 3784; (b) E. G. Awere, J. Passmore, P. S. White and T. Klapötke, *J. Chem. Soc., Chem. Commun.*, 1989, 1415.
20. J. M. Rawson, A. J. Banister and I. Lavender, *Adv. Heterocycl. Chem.*, 1995, **62**, 137.
21. S. Parsons and J. Passmore, *Acc. Chem. Res.*, 1994, **27**, 101.
22. N. Burford, T. Chivers, A. W. Cordes, W. G. Laidlaw, M. Noble, R. T. Oakley and P. N. Swepston, *J. Am. Chem. Soc.*, 1982, **104**, 1282.
23. A. Apblett, T. Chivers, A. W. Cordes and R. Vollmerhaus, *Inorg. Chem.*, 1991, **30**, 1392.

CHAPTER 5
Paramagnetic Inorganic Rings

Paramagnetism is observed for species that contain unpaired electrons. For the p-block elements, such species commonly have only one unpaired electron and they are referred to as radicals. Such radicals may either be neutral or bear a positive (cation radical) or negative (anion radical) charge. In recent years, a number of stable radicals involving cyclic ring systems of the p-block elements have been characterised. In this context, the term 'stable' describes a radical that can be *isolated* under ambient conditions and stored for long periods in the absence of air at room temperature.[1] Thus, the solid-state structures of stable radicals can be determined by X-ray crystallography (see Section 3.1). By contrast, a 'persistent' radical has a sufficiently long lifetime under the conditions it is generated to allow characterisation by spectroscopic methods (most commonly EPR spectroscopy, (see Section 3.6). The stability of cyclic main group radicals is enhanced by one or more of the following characteristics: (a) the high electronegativity of the p-block atom, (b) steric protection provided by bulky substituents and (c) delocalisation of the unpaired electron over the ring systems and, in some cases, on to substituents.

In this chapter, the formation and structures of cyclic radicals involving p-block elements are described starting with neutral species and followed by cation and anion radicals. In each classification, homocyclic radicals are discussed before heterocyclic systems. The emphasis is on those paramagnetic molecules that are sufficiently stable to be characterised in the solid state. Persistent radicals are included when their identification has led to important insights into the behaviour or properties of inorganic ring systems, *e.g.* cyclic delocalisation. In recent years, there has been increasing interest in biradicals,[2] either as stable species or as reaction intermediates, and in ring systems that exhibit biradicaloid character. The final part of this chapter deals with these interesting paramagnetic systems.

Inorganic Rings and Polymers of the p-Block Elements: From Fundamentals to Applications
By Tristram Chivers and Ian Manners
© Tristram Chivers and Ian Manners 2009
Published by the Royal Society of Chemistry, www.rsc.org

5.1 NEUTRAL RADICALS

5.1.1 Homocyclic Systems

The three-membered ring $[Al_3R_4]^{\bullet}$ (R = Si^{t}Bu$_3$) (**5.1**) provides an example of the ability of bulky substituents to stabilise cyclic radicals of the main group elements. This homocyclic radical is obtained as black–green crystals by thermolysis of the dimer R_2AlAlR_2 [eqn (5.1)].[3] One of the Al–Al bonds in **5.1** is slightly shorter than the other two (2.703 vs 2.756 Å, cf. 2.751 Å for the single Al–Al bond in R_2AlAlR_2). Interestingly, the corresponding Ga–Ga bond distances in the gallium analogue $[Ga_3R_4]^{\bullet}$ differ by 0.37 Å, implying the onset of Ga–Ga bond cleavage to give an acyclic species.[4]

$$(5.1)$$

The triangular cyclotrigermenyl $[Ge_3Ar_3]^{\bullet}$ (Ar = C$_6$H$_3$-2,6-Mes$_2$) (**5.2**) is a homocyclic group 14 radical stabilised by the steric protection of bulky aryl groups. This blue compound is obtained by reduction of the Ge(II) species Ge(Cl)Ar [eqn (5.2)].[5] The small value of the hyperfine coupling to ^{73}Ge ($I = 9/2$, 7.8%) in the EPR spectrum is consistent with a π-radical with the unpaired electron localised on one of the germanium atoms (see Section 3.6).

$$(5.2)$$

The red–purple cyclotetrasilenyl radical $[(SiR)_3(SiR'_2)]^{\bullet}$ (**5.3**, R = SiMe^{t}Bu$_2$, R' = ^{t}Bu), prepared by one-electron reduction of the corresponding cation [eqn (5.3); see also Figure 10.5 and Scheme 10.11], is an unusual example of a stable silyl radical.[6] The four-membered ring in **5.3** is almost planar. Two Si–Si bond lengths are intermediate between Si–Si single bond and Si=Si double bond values, whereas the other two are close to that of a single bond, suggesting an allyl-type radical. Consistently, the solution EPR spectrum shows hyperfine coupling (hfc) to two Si atoms (4.07 and 3.74 mT) that is larger than that to the unique Si (1.55 mT). The small values of the hfc constants are consistent with a π-type radical.

$$\text{(scheme 5.3: R-Si cluster cation with X}^-\text{, treated with }^t\text{Bu}_3\text{SiNa / Et}_2\text{O, -NaX} \rightarrow \textbf{5.3})$$

(R = SiMe^{t}Bu$_2$; R′ = ^{t}Bu)
(X = B(C$_6$F$_5$)$_4$)

(5.3)

5.1.2 Heterocyclic Systems

The most extensively studied neutral heterocyclic monoradicals are those based on the 1,2,3,5-dichalcogenadiazolyl ring systems [RCN$_2$E$_2$]$^{\bullet}$ (E = S, Se), which exhibit considerable potential for the development of non-metallic materials with unique conducting or magnetic properties. These cyclic radicals are prepared by one-electron reduction of the corresponding cations [eqn (5.4)].[7] They are formally seven π-electron systems in which the unpaired electron occupies an orbital that is localised on the chalcogen and, to a lesser extent, the nitrogen atoms with a node at carbon. Consistently, the EPR spectra of 1,2,3,5-dithiadiazolyls exhibit a characteristic five-line pattern as a result of coupling to two equivalent nitrogen nuclei (^{14}N, $I = 1$, 99.6%; the only spin-active sulfur nucleus has an abundance of <1%) (see Figure 3.10b).

$$\text{[RCN}_2\text{E}_2\text{]}^+ \text{ X}^- \xrightarrow[\text{or Zn/Cu in SO}_2]{\text{SbPh}_3} \text{[RCN}_2\text{E}_2\text{]}^{\bullet} \quad \textbf{5.4}$$

(X = AsF$_6$, SbF$_6$)

(5.4)

In the solid state, the radicals **5.4** usually form dimers with weak chalcogen–chalcogen contacts. The structures of these dimers are influenced by the nature of the substituent attached to the carbon atom of the ring of R. With a flat phenyl substituent, a cofacial arrangement of monomers with two E···E contacts is preferred (see Figure 4.6), whereas the R = Me$_2$N, CF$_3$ and Me derivatives adopt a twisted conformation with only one E···E contact (see Figure 3.1). The energy differences between these structures are small and the structures are determined primarily by packing effects. Dimeric 1,2,3,5-dichalcogenadiazolyls, including bi- and trifunctional radicals, have been studied extensively as building blocks for molecular conductors (see Section 5.4.1).

When R = ^{t}Bu or a polyfluorinated aryl group (4-XC$_6$F$_4$, X = Br, CN, NO$_2$), the 1,2,3,5-dithiadiazolyl radicals [RCN$_2$S$_2$]$^{\bullet}$ exist as monomers, whose magnetic properties are of particular interest. The *tert*-butyl derivative **5.5** is an unusual example of a liquid that is paramagnetic at room temperature;[8a] the trifluoromethyl derivative (R = CF$_3$) melts at 35 °C to give a paramagnetic liquid.[8b] The crystalline aryl derivatives **5.6** exhibit ferromagnetic behaviour at very low temperatures,[9,10] whereas the bromo derivative (R = 4-BrC$_6$F$_4$) does not undergo long-range magnetic order as a result of a different packing arrangement in the crystal lattice.[9]

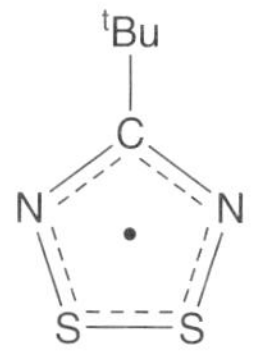

Paramagnetic liquid
5.5

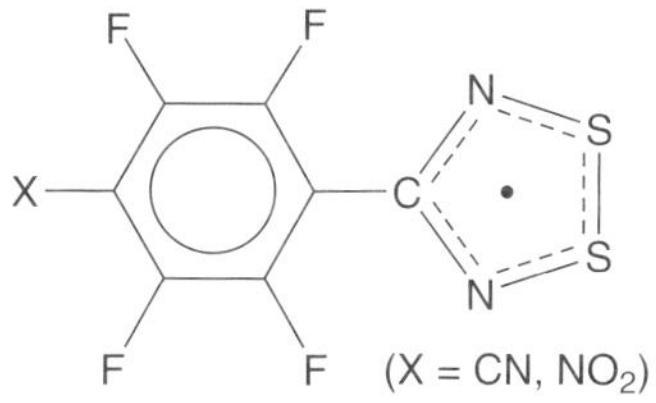

Ferromagnetic
5.6

The 1,3,2,4-dithiadiazolyl radicals (**5.7**), which are isomers of **5.4**, are also prepared by the one-electron reduction of the corresponding cation, which is obtained by the cycloaddition of the [S$_2$N]$^+$ cation to the C≡N triple bond of an organic nitrile (see Scheme 2.8). This synthetic approach can be adapted to the synthesis of biradicals (see Section 5.4.1). 1,3,2,4-Dithiadiazolyl radicals are unstable with respect to isomerisation to the 1,2,3,5-isomers both in solution and in the solid state. This isomerisation is a photochemically symmetry-allowed process, which is thermally symmetry forbidden. A bimolecular head-to-tail rearrangement has been proposed to account for this isomerisation (Figure 5.1).[11] This process is conveniently monitored by EPR spectroscopy, since the 1:1:1 triplet (coupling to a single ^{14}N centre) exhibited by the 1,3,2,4-dithiadiazoles is replaced by the 1:2:3:2:1 quintet (coupling to two equivalent ^{14}N nuclei) of the 1,2,3,5-isomers (see Figure 3.10b).

5.2 CATION RADICALS

5.2.1 Homocyclic Systems

The one-electron oxidation of homocyclic inorganic ring systems is expected to produce the corresponding cation radicals. An effective reagent for this process is

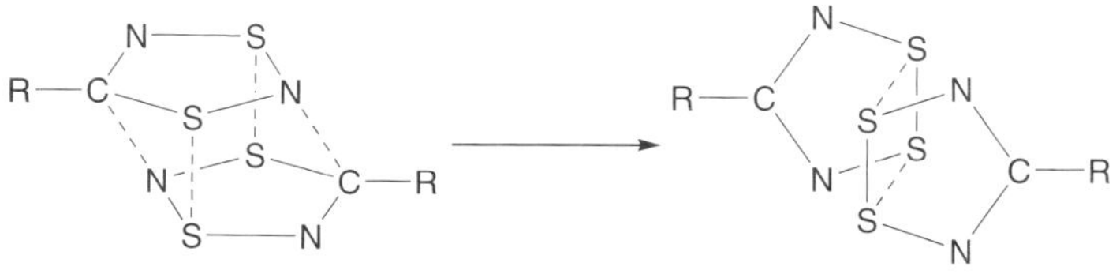

5.7

Figure 5.1 Rearrangement of 1,3,2,4-dithiadiazoles into 1,2,3,5-isomers.

a solution of $AlCl_3$ in dichloromethane, provided that the ionisation potential of the ring system is <8.0 eV. For example, the hexameric cation radical $[(Me_2Si)_6]^{+\bullet}$ is obtained in this manner [eqn (5.5)].[12] This homocyclic system is sufficiently stable to be characterised at 190 K by an EPR spectrum, which reveals coupling of the unpaired electron to two sets of 18 equivalent protons, corresponding to six axial and six equatorial methyl groups, which do not become equivalent on the EPR time scale at this temperature. Similarly to the formation of the anion radical $[(Me_2Si)_5]^{-\bullet}$ [see eqn (4.1)], this experiment provides additional evidence for σ-electron delocalisation in the cyclic polysilane framework.

$$\text{(Me}_2\text{Si)}_6 \xrightarrow{AlCl_3/CH_2Cl_2} [(Me_2Si)_6]^{+\bullet} \; AlCl_4^- \quad (5.5)$$

The oxidation of cyclo-S_8 with strong oxidising agents, *e.g.* AsF_5, produces polysulfur cations, which are isolated as lattice-stabilised salts of dications, *e.g.* S_8^{2+}, in the solid state (Section 12.3.1). In dilute solutions (HSO_3F or SO_2), the S_8^{2+} dication is in equilibrium with other polysulfur cations, including the cyclic $S_5^{+\bullet}$ radical cation [eqn (5.6)], which has been identified by EPR studies of the 92% ^{33}S-enriched material (see Figure 12.13).[13]

$$S_8^{2+} \leftrightarrow \tfrac{1}{2} S_6^{2+} + S_5^{+\bullet} \quad (5.6)$$

5.2.2 Heterocyclic Systems

The most extensively studied stable cyclic cation radicals of the p-block elements are the seven π-electron systems $[E_3N_2]^{+\bullet}$. The all-sulfur (E = S) and the all-selenium (E = Se) cations have been structurally characterised as dimers in the solid state (see **4.11** in Section 4.2.2).[14] In solution, the monomeric radicals, including all the possible mixed S–Se systems, have been identified by their characteristic five-line (1:2:3:2:1) EPR spectra with additional hyperfine coupling to ^{77}Se for the selenium-containing radicals.[15]

5.3 ANION RADICALS

The formation of anion radicals by chemical or electrochemical reduction may provide information about electron delocalisation in an inorganic ring system through the determination of the EPR spectrum in conjunction with molecular orbital calculations. In practice, however, the LUMO of the homo- or heterocycle is often an antibonding orbital. Consequently, occupation of the LUMO by an electron will result in the elongation or, in some cases, complete cleavage of a bond. Some examples are discussed in this section.

As discussed in Section 4.1.3, the cyclopentasilane $(Me_2Si)_5$ is reduced either chemically or electrochemically to give the dark-blue radical anion $[(Me_2Si)_5]^{-\bullet}$ [eqn (4.1)]. Interestingly, the same radical anion is produced upon reduction of the six-membered ring $(Me_2Si)_6$ or the seven-membered ring $(Me_2Si)_7$ and identified by its characteristic 31-line EPR spectrum, demonstrating that a ring contraction process occurs in solution for the radical anions of the larger cyclosilanes.[16]

The electrochemical reduction of cyclo-S_8 involves a two-electron process that, at low temperatures, produces cyclo-S_8^{2-}. Subsequent ring opening generates the linear species *catena*-S_8^{2-} (see Figure 12.22b), which dissociates and/or disproportionates into the radical anions $S_3^{-\bullet}$ and $S_4^{-\bullet}$ (see Section 12.4.1).[17] Direct evidence for this ring opening upon reduction of sulfur homocycles comes from the structural characterisation of the cyclic radical anion $S_6^{-\bullet}$ as the $[PPh_4]^+$ salt in the solid state.[18] The structure of cyclo-$S_6^{-\bullet}$ is characterised by two long central S–S bonds (2.633 Å, *cf.* 2.057 Å in cyclo-S_6), which connect the two S_3 units in a chair conformation (Figure 5.2). The average S–S distance in these units is 2.060 Å. A molecular orbital analysis reveals that the bonding interaction between the two S_3 units in cyclo-$S_6^{-\bullet}$ involves two components (Figure 5.3): (a) an electron-pair bond between the $2b_1$ SOMOs of both

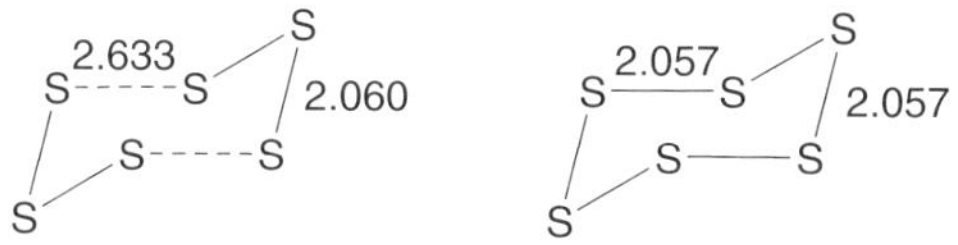

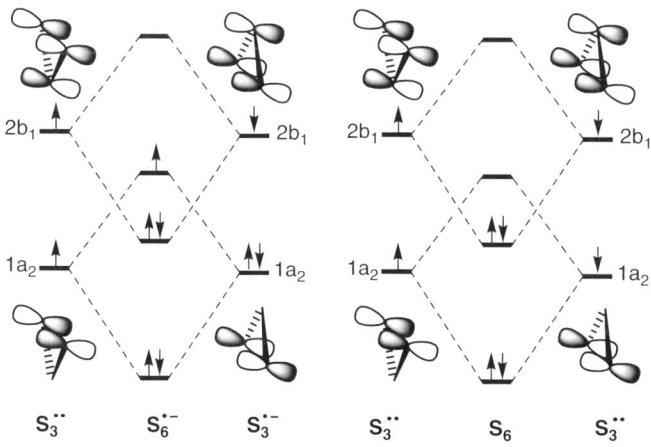

Figure 5.2 Structures of cyclo-$S_6^{-\bullet}$ and cyclo-S_6.

Figure 5.3 Orbital interaction diagram for formation of cyclo-$S_6^{-\bullet}$ and cyclo-S_6 from two S_3 fragments. Reproduced with permission from B. Neumüller *et al.*, *Angew. Chem. Int. Ed.*, 2000, **39**, 4580.[18]

fragments and (b) a three-electron bond between the $1a_2$ SOMO of the biradical $S_3^{\bullet\bullet}$ and the $1a_2$ HOMO of $S_3^{-\bullet}$. Neutral S_3 has a closed-shell ground state, but becomes a biradical after valence excitation of an electron from $1a_2$ to $2b_1$. The corresponding interaction for neutral cyclo-S_6 is shown in Figure 5.3 for comparison. The loss of the antibonding electron converts the three-electron bond of cyclo-$S_6^{-\bullet}$ into a more stabilising electron-pair bond in cyclo-S_6 and accounts for the disparity in the central S–S bond lengths for these two cyclic species.

5.4 BIRADICALS

Biradicals are molecules that contain two unpaired electrons in two (approximately) degenerate non-bonding molecular orbitals.[2] If the two electrons behave independently, *i.e.* there is spatial separation and negligible overlap of orbitals, they are known as *disjoint* biradicals. Their ground state can either be a low-spin singlet ($S = 1$, antiparallel spins, antiferromagnetic coupling of the spins) or high-spin triplet ($S = 3$, parallel spins, ferromagnetic coupling).

In organic systems, biradicals are usually very short-lived species, which play a central role in bond-breaking and -formation processes. The replacement of some or all of carbon atoms in organic biradicals by other p-block elements can, in selected cases, produce inorganic analogues that can be isolated and structurally characterised. The penalty for the increased stability of these inorganic ring systems is a reduction of the biradical character. Hence, such species are referred to as biradicaloids. This section begins with a brief discussion of stable biradicals that incorporate inorganic ring systems. This is followed by some examples of the fascinating inorganic heterocycles, both homocyclic and heterocyclic, that exhibit biradicaloid character. The chapter concludes with a consideration of the role of biradicals in ring-opening polymerisation processes.

5.4.1 Stable Biradicals

The most extensively studied biradicals involving inorganic ring systems are thiazyl-based materials related to the dithiadiazolyl monoradicals described in Section 5.1.2, *e.g.* bis(dithiadiazolyl) biradicals containing a phenylene (**5.8**)[19] or heterocyclic (**5.9**)[20] spacer group. In the C_6H_4-bridged systems, the radical centres are pushed so far away from each other that no exchange contribution is likely. Consistently, the EPR spectrum (a 1:1:1 triplet) and calculations predict a disjoint biradical with a triplet ground state (*i.e.* the two unpaired electrons are confined to separate spatial domains).

5.8 **5.9**

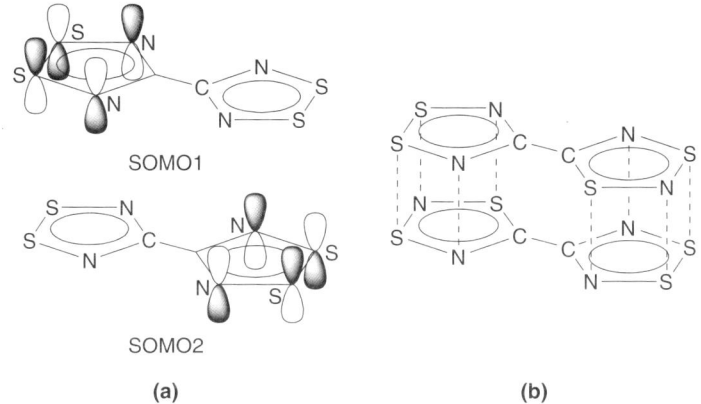

Figure 5.4 (a) SOMOs of **5.11** and (b) the solid-state structure of **5.12**.

Biradicals in which two dithiadiazolyl rings are directly linked are highly coloured. The known examples include those containing two 1,2,3,5- or two 1,3,2,4-dithiadiazolyl radicals connected by a C–C bond, **5.10** and **5.11**, respectively, and also the hybrid system **5.12**, in which the two isomeric dithiadiazolyl rings are joined.[21,22] Biradical **5.11** is a rare example of a disjoint biradical in which the molecular orbitals for the two unpaired electrons are localised on separate groups of atoms (Figure 5.4). For disjoint biradicals, exchange interactions between the two centres are very small (<1 kcal mol^{-1}), hence the two states are essentially degenerate. The five-line EPR pattern observed for **5.11** in SO$_2$ at 273 K is consistent with essentially no interaction, although some exchange occurs at 303 K. In the solid state, the hybrid system **5.12** provides a rare example of the formation of dimers via a π^*–π^* interaction between different (isomeric) ring systems. This dimer is essentially diamagnetic at room temperature, but undergoes a pressure-induced change to a paramagnetic phase upon grinding.

5.10   **5.11**   **5.12**

5.4.2 Biradicaloids

5.4.2.1 Homocyclic Systems. Cyclobutanediyls exhibit spin states that are very close in energy. The triplet state is preferred by only 1.7 kcal mol^{-1}; it can be observed by EPR spectroscopy because the ring closure to a bicyclo[1.1.0]butane is spin-forbidden. Singlet cyclobutanediyls are predicted as very short-lived transition states for the ring inversion of bicyclo[1.1.0]butanes ($\Delta E \approx 50$ kcal mol^{-1}).[2b] Quantum chemical calculations predict that the heavier group 14 analogues of bicyclo[1.1.0]butanes will exhibit bond stretch isomerism, *i.e.* there are two distinct minima on the potential energy surface which differ mainly in the length of a transannular E–E bond as a result of the high ring strain and low σ-bond energies.[23] As illustrated in Figure 5.5, the bridge bonds (E_b–E_b) are formed by almost pure p-orbitals in the short-bond isomer, whereas in the long-bond isomer the s contribution to these orbitals leads to a less effective overlap and, hence, a weaker E_b–E_b bond. Recent calculations reveal that the relative energies of these two isomers depend strongly on the substituents R due to a competition between ring strain and steric effects.[24] Small R groups favour the formation of the long-bond isomer (larger angle Φ) whereas steric repulsion (larger angle Θ) destabilizes this isomer (Figure 5.5).

In practice, tetrasilabicyclo[1.1.0]silanes are rare. The derivative $R_4{}^tBu_2Si_4$ (**5.13**, R = 2,6-$Et_2C_6H_3$) has a normal transannular Si–Si single bond length and an interflap angle Φ of 121°, consistent with a short-bond isomer (Figure 5.6).[25] However, this bicyclic compound is thermochromic and the ring-inversion barrier is low ($E_a = 15$ kcal mol^{-1}), suggesting a close relationship to the biradicaloid

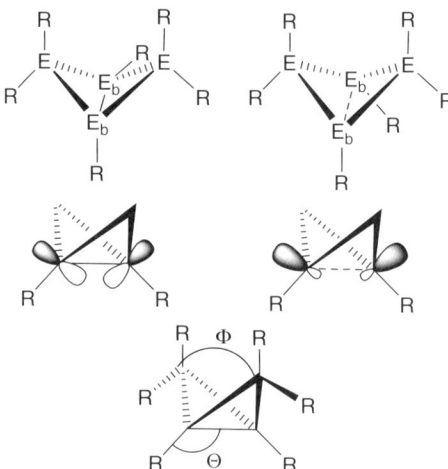

Figure 5.5 Bond-stretch isomerism for tetrametallabicyclo[1.1.0]butanes (E = Si, Ge).

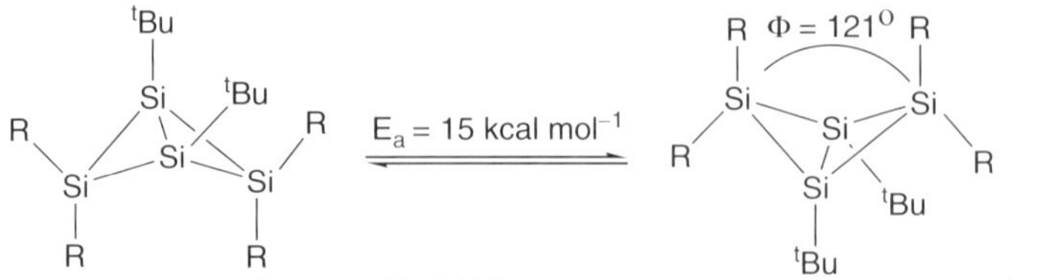

Figure 5.6 Ring inversion of a tetrasilabicyclo[1.1.0]silane.

form. Furthermore, the central Si–Si bond in **5.13** is unusually reactive towards water.

For a discussion of two intriguing biradical systems based on a four-membered M_2N_2 ring (M = Ge, Sn), see Section 10.3.2 and Figure 10.9.

5.4.2.2 Heterocyclic Systems. The 1,3-diphosphacyclobutane-2,4-diyls $(RPCR)_2$, **5.14** and **5.15**, are related to the carbon-based cyclobutanediyls by replacement of the two CH_2 units by PR groups. These biradicaloid species are obtained from the reaction of *C*-dichlorophosphaalkenes with *n*-butyllithium (Scheme 5.1).[26] The *P*-amino derivative **5.14** (R = TMP = tetramethylpiperidyl) is unstable at room temperature and isomerises, even in the solid state, to **5.16**. However, the Mes* derivative **5.15** is stable up to 150 °C. The P_2C_2 ring in these four-membered systems is planar, but both the carbon and the phosphorus centres are pyramidal. The comparatively high inversion barriers at phosphorus impede the formation of a planar six π-electron ring system and, hence, **5.14** and **5.15** exhibit high biradicaloid character.[27]

The electronic ground state of 1,3-diphosphacyclobutane-2,4-diyls may be represented approximately by the resonance structures **A** and **B** (Figure 5.7); consequently, they are described as weakly π-conjugated biradicaloids.[27] By varying the substituents on the carbon centres, the spin state of these biradicaloids can be controlled. For example, *C*-bonded silyl groups (a σ-donor and π-acceptor) stabilise the singlet state, whereas the combination of an NR_2 group at phosphorus and an alkyl group at carbon lowers the energy of the triplet state.[27]

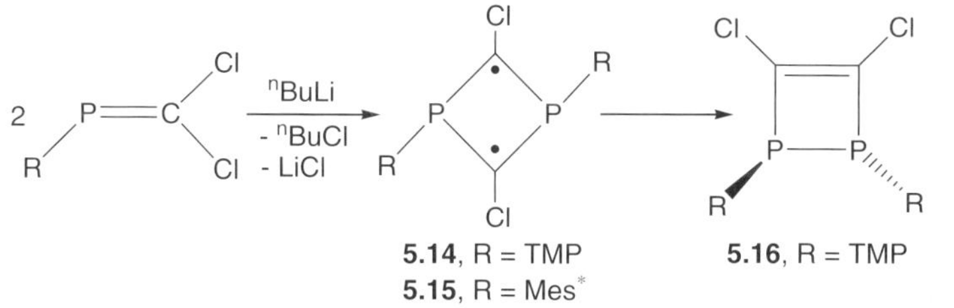

Scheme 5.1 Synthesis and isomerisation of 1,3-diphosphacyclobutane-2,4-diyls.

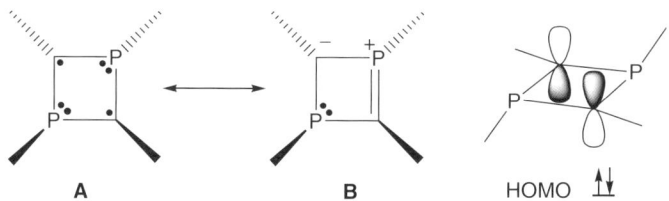

Figure 5.7 Resonance structures and HOMO for 1,3-diphosphacyclobutane-2,4-diyls.

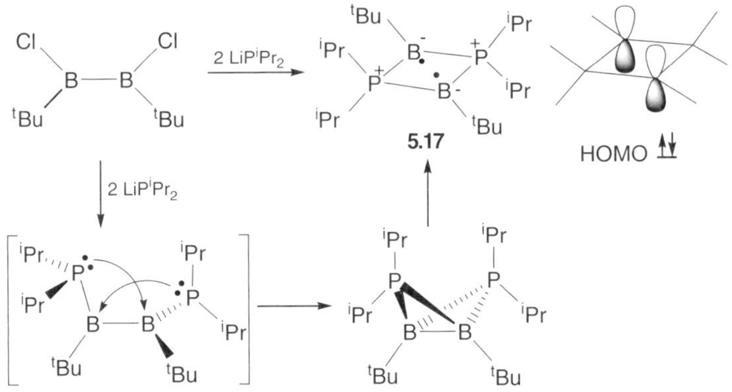

Scheme 5.2 Synthesis of the biradicaloid (^{i}Pr$_2$PB^{t}Bu)$_2$.

The boron-centred biradicaloid (R$_2$PBR′)$_2$ (**5.17**, R = ^{i}Pr, R′ = ^{t}Bu) is an isoelectronic analogue of the (RPCR)$_2$ biradicaloids, **5.14** and **5.15**. Both of these four-membered rings contain 22 valence electrons for R=H. The P$_2$B$_2$ system **5.17** is synthesised by the reaction of a 1,2-dichlorodiborane with two equivalents of LiP^{i}Pr$_2$.[28] The P$_2$B$_2$ ring is presumably formed by rearrangement of the initially formed acyclic P–B–B–P skeleton (Scheme 5.2).

The biradicaloid **5.17** is obtained as yellow, air-sensitive, but thermally stable (>200 °C) crystals. The P$_2$B$_2$ ring is perfectly planar with a transannular B–B distance of 2.57 Å, *cf.* 1.76 Å for a single B–B bond. Compared with the isoelectronic P$_2$C$_2$ systems, the contribution of resonance structure **B** (Figure 5.7) to the electronic ground state in **5.17** is significantly reduced by the transformation of the phosphorus lone pair into a P–C σ-bond. In addition, the heterocyclic ring in **5.17** is enlarged because the P–B bonds are *ca* 0.15 Å longer than the P–C bonds in **5.14** and **5.15**. Both factors favour an open singlet-biradical form for **5.17**.

Interestingly, for the parent molecule (H$_2$PBH)$_2$ calculations show that the planar form would be the transition state (at 16.4 kcal mol^{-1}) for the inversion of the 1,3-diphospha-2,4-diborabicyclo[1.1.0]butane **5.17**. Hence the choice of the sterically demanding substituents on P and B (^{i}Pr and ^{t}Bu, respectively) appears to be the determining factor in the isolation of the biradicaloid **5.17**. A more detailed discussion of the influence of the exocyclic substituents on the relative stabilities of cyclic B$_2$P$_2$ biradicaloid systems can be found in Section 9.3.2.

5.4.3 Biradicals as Reaction Intermediates

In some well-known transformations involving inorganic ring systems, biradicals act as reaction intermediates that are sufficiently persistent to be detected by EPR spectroscopy (see Section 3.6). For example, the polymerisation of cyclo-S$_8$ to form the helical sulfur polymer S$_\infty$ (plastic sulfur) occurs via the

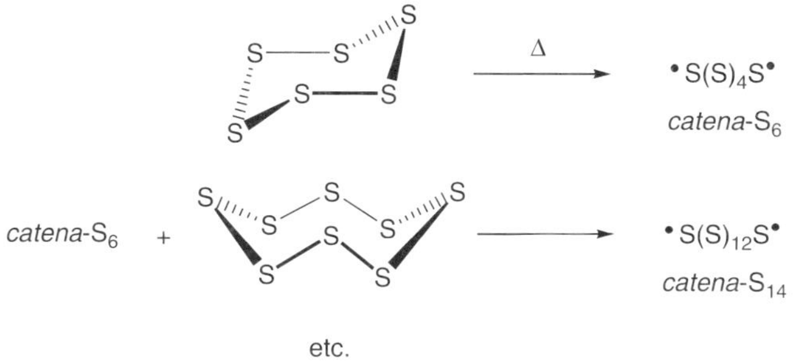

Paramagnetic Inorganic Rings

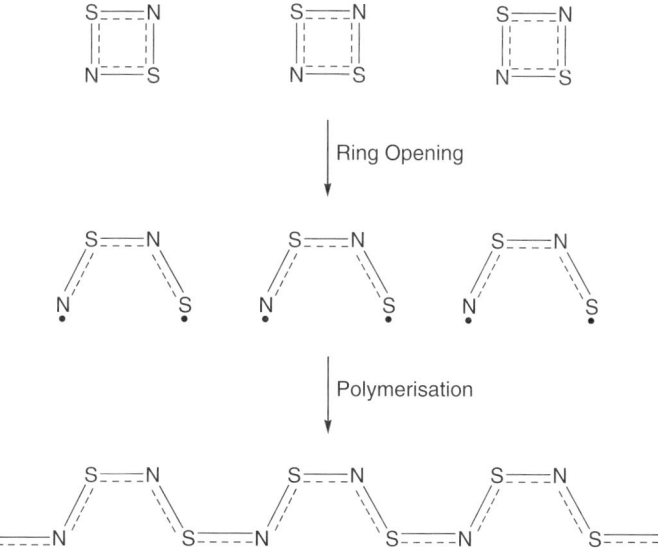

Figure 5.8 Polymerisation of cyclo-S_8 (cyclo-S_7 may also serve as the initiator).

Figure 5.9 Topotactic polymerisation of S_2N_2.

formation of biradicals generated by the thermally initiated homolytic cleavage of an S–S bond (see Section 12.5). The mean S–S bond strength in cyclo-S_6 and cyclo-S_7 (see Section 12.1.1) is significantly lower than that in cyclo-S_8. Consequently, the thermal or photochemical cleavage of an S–S bond in these smaller cyclic oligomers occurs more easily and the presence of cyclo-S_6 or cyclo-S_7 as impurities in cyclo-S_8 will initiate the polymerisation process (Figure 5.8).[29] Chain propagation occurs via reaction of the catenated biradicals formed in the initiation step with cyclo-S_8.

The oxidation of phosphorus by elemental sulfur to give binary phosphorus sulfides, which is initiated by mild heating (<100 °C) or upon photolysis at 0 °C, may involve the formation of the *catena*-S_8 biradical in the rate-determining first step.[30]

EPR spectroscopic evidence indicates that the topotactic polymerisation of the four-membered ring S_2N_2 to produce the conducting polymer poly(sulfur nitride) $(SN)_x$ (Section 12.7.1) also occurs via the intermediacy of biradicals formed by homolytic cleavage of an S–N bond (Figure 5.9). It has been suggested that the polymerisation process involves excitation of the square-planar singlet S_2N_2 molecule to the triplet surface followed by puckering of the triplet species and polymerisation in a direction approximately perpendicular to the S_2N_2 plane.[31] Recent *ab initio* calculations reveal that S_2N_2 has significant biradical character.[32]

REFERENCES

1. P. P. Power, *Chem. Rev.*, 2003, **103**, 789.
2. (a) H. Grützmacher and F. Breher, *Angew. Chem. Int. Ed.*, 2002, **41**, 4006; (b) F. Breher, *Coord. Chem. Rev.*, 2007, **251**, 1007.
3. N. Wiberg, T. Blank, W. Kaim, B. Schwederski and G. Lintl, *Eur. J. Inorg. Chem.*, 2000, 1475.
4. N. Wiberg, T. Blank, K. Amelunxen, H. Nöth, J. Knizek, T. Habereden, W. Kaim and M. Wanner, *Eur. J. Inorg. Chem.*, 2001, 1719.
5. M. M. Olmstead, L. Pu, R. S. Simons and P. P. Power, *Chem. Commun.*, 1997, 1595.
6. A. Sekiguchi, T. Matsuno and M. Ichinohe, *J. Am. Chem. Soc.*, 2001, **123**, 12435.
7. (a) J. M. Rawson, A. J. Banister and I. Lavender, *Adv. Heterocycl. Chem.*, 1995, **62**, 137; (b) J. M. Rawson and A. Alberola, in: *Handbook of Chalcogen Chemistry: New Perspectives in Sulfur, Selenium and Tellurium*, ed. F. A. Devillanova, Royal Society of Chemistry, Cambridge, 2007, pp. 737–742.
8. (a) W. V. F. Brooks, N. Burford, J. Passmore, M. J. Schriver and L. H. Sutcliffe, *J. Chem. Soc. Chem. Commun.*, 1987, 69; (b) H. Du, R. C. Haddon, I. Krossing, J. Passmore, J. M. Rawson and M. J. Schriver, *Chem. Commun.*, 2002, 1836.
9. J. M. Rawson and F. Palacio, *Struct. Bonding*, 2001, **100**, 93.
10. A. Alberola, R. J. Lees, C. M. Pask, J. M. Rawson, F. Palacio, P. Oliete, C. Paulsen, A. Yamaguchi, R. D. Farley and D. M. Murphy, *Angew. Chem. Int. Ed.*, 2003, **42**, 4782.
11. J. Passmore and X. Sun, *Inorg. Chem.*, 1996, **35**, 1313.
12. H. Bock, W. Kaim, M. Kira and R. West, *J. Am. Chem. Soc.*, 1979, **101**, 7667.
13. (a) I. Krossing and J. Passmore, *Inorg. Chem.*, 2004, **43**, 1000; (b) H. S. Low and R. A. Beaudet, *J. Am. Chem. Soc.*, 1976, **98**, 3849.
14. E. G. Awere, J. Passmore and P. S. White, *J. Chem. Soc. Dalton Trans.*, 1993, 299.
15. E. G. Awere, J. Passmore, K. F. Preston and L. H. Sutcliffe, *Can. J. Chem.*, 1988, **66**, 1776.

16. E. Carberry, R. West and G. E. Glass, *J. Am. Chem. Soc.*, 1969, **91**, 5446.
17. (a) R. Steudel, *Top. Curr. Chem.*, 2003, **231**, 127; (b) A. Evans, M. I. Montenegro and D. Pletcher, *Electrochem. Commun.*, 2001, **3**, 514.
18. B. Neumüller, F. Schmock, R. Kirmse, A. Voigt, A. Diefenbach, F. M. Bickelhaupt and K. Dehnicke, *Angew. Chem. Int. Ed.*, 2000, **39**, 4580.
19. A. W. Cordes, R. C. Haddon, R. G. Hicks, D. K. Kennepohl, R. T. Oakley, T. T. M. Palstra, L. F. Schneemeyer, S. R. Scott and J. V. Wasczak, *Chem. Mater.*, 1993, **5**, 820.
20. A. W. Cordes, R. C. Haddon, C. D. MacKinnon, R. T. Oakley, G. W. Patenaude, R. W. Reed, T. Rietveld and K. E. Vajda, *Inorg. Chem.*, 1996, **35**, 7625.
21. C. D. Bryan, A. W. Cordes, J. D. Goddard, R. C. Haddon, R. G. Hicks, C. D. MacKinnon, R. C. Mawhinney, R. T. Oakley, T. T. M. Palstra and A. S. Perel, *J. Am. Chem. Soc.*, 1996, **118**, 330.
22. (a) T. S. Cameron, M. T. Lemaire, J. Passmore, J. M. Rawson, K. V. Shuvaev and L. K. Thompson, *Inorg. Chem.*, 2005, **44**, 2576; (b) G. Antorrena, S. Brownridge, T. S. Cameron, F. Palacio, S. Parsons, J. Passmore, L. K. Thompson and F. Zarlaida, *Can. J. Chem.*, 2002, **80**, 1568.
23. M.-M. Rohmer and M. Bénard, *Chem. Soc. Rev.*, 2001, **30**, 340.
24. R. Koch, T. Bruhn and M. Weidenbruch, *J. Mol. Struct. (Theochem.)*, 2004, **680**, 91.
25. R. Jones, D. J. Williams, Y. Kabe and S. Masamune, *Angew. Chem. Int. Ed. Engl.*, 1986, **98**, 176.
26. E. Niecke, A. Fuchs, F. Baumeister, M. Nieger and W. W. Schoeller, *Angew. Chem. Int. Ed. Engl.*, 1995, **34**, 555.
27. W. W. Schoeller, C. Begeman, E. Niecke and D. Gudat, *J. Phys. Chem. A*, 2001, **105**, 10731.
28. D. Scheschkewitz, H. Amii, H. Gornitzka, W. W. Schoeller, D. Bourisssou and G. Bertrand, *Science*, 2002, **295**, 1880.
29. R. Steudel, S. Passlack-Stephan and G. Holdt, *Z. Anorg. Allg. Chem.*, 1984, **517**, 7.
30. M. E. Jason, T. Ngo and S. Rahman, *Inorg. Chem.*, 1997, **36**, 2633.
31. R. Mawhinney and J. D. Goddard, *Inorg. Chem.*, 2003, **42**, 6223.
32. H. M. Tuononen, R. Suontamo, J. Valkonen and R. S. Laitinen, *J. Phys. Chem. A*, 2004, **108**, 5670.

CHAPTER 6
Inorganic Macrocycles

Alicyclic hydrocarbons of the type $(-CH_2-)_n$ form rings that range in size from three to over 30 carbon atoms and in principle there appears to be no limit to the size of these rings. The introduction of heteroatoms, *e.g.* O, S, NR, into these saturated cyclic systems produces a more limited series of cyclic oligomers. Cyclic ethers of the type $(OCH_2CH_2)_n$ ($n = 4-7$) are perhaps the most widely known members of this class of ring system. Common examples include 12-crown-4 (**6.1**, $n = 4$) and 18-crown-6 (**6.2**, $n = 6$); this nomenclature gives the total number of atoms (C + O) and the number of O atoms in the ring as the numerical prefix and suffix, respectively. Crown ethers are widely used for the encapsulation of alkali metal ions. Their different cavity sizes allow for selective coordination of metal ions that can be easily accommodated within the hole. Thus, 12-crown-4 comfortably houses Li^+ ions, and 18-crown-6 provides a fit for the larger K^+ ion.[1]

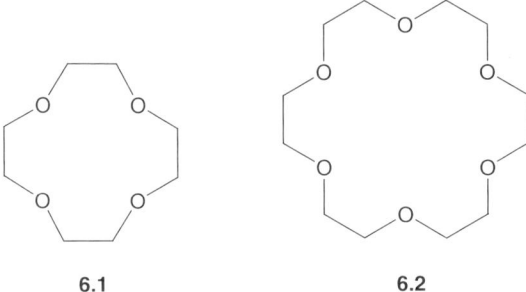

      6.1   **6.2**

An extensive homologous series is also known for unsaturated organic ring systems of the type $(-HC=)_n$, which are known as annulenes.[2] The most important example is benzene or [6]annulene. The stability of annulenes is governed by Hückel's rule, which states that planar ring systems of this type

---

Inorganic Rings and Polymers of the p-Block Elements: From Fundamentals to Applications
By Tristram Chivers and Ian Manners
© Tristram Chivers and Ian Manners 2009
Published by the Royal Society of Chemistry, www.rsc.org

Inorganic Macrocycles

with $(4n + 2)$ π-electrons will be aromatic (see Section 4.1.1). In practice, however, medium-sized rings such as [10]annulene and [14]annulene (**6.3**) are unstable because of interactions between inner hydrogen atoms in a strain-free planar arrangement. These interactions are less pronounced in larger rings; [18]annulene (**6.4**) has a nearly planar structure and can be distilled without decomposition under reduced pressure. The introduction of a heteroatom into annulenes, *e.g.* $(-HC=N-)_n$, results in a more limited homologous series involving primarily trimeric (*i.e.* six-membered) ring systems.

6.3   **6.4**

In the following sections, inorganic macrocycles of the p-block elements are discussed with emphasis on larger rings. The term macrocycle is (arbitrarily) defined here as a ring system containing more than eight atoms. Smaller rings, *i.e.* those containing 3–8 atoms, are discussed in Chapters 9–12. However, molecules that are created by linking together small inorganic rings with appropriate spacer groups are also described in this chapter. By analogy with crown ethers, some of these inorganic macrocycles have a potentially useful host–guest chemistry.

6.1 HOMOCYCLIC SYSTEMS

This section begins with an account of large homocyclic systems, *i.e.* those in which only one p-block element is present. These rings are exclusively saturated systems, *i.e.* they involve single bonds between the p-block elements. By contrast, the subsequent discussion of heterocyclic systems includes both saturated and unsaturated rings. The latter involve some form of multiple bonding between different *p*-block elements.

Two well-known classes of homocyclic inorganic ring systems that are formally isovalent with alicyclic hydrocarbons are cyclosilanes of the type $(R_2Si)_n$ and the cyclic sulfur allotropes, cyclo-S_n. Although the parent cyclosilanes (R = H) are unstable, an extensive homologous series is known for alkyl derivatives (*e.g.* R = Me). These cyclic oligomers can be separated by high-performance liquid chromatography (HPLC) and the entire series $n = 5$–19 has been characterised by spectroscopic methods (see Section 3.4.2). The 13- and 16-membered rings, $(Me_2Si)_n$ (**6.5**, $n = 13$; **6.6**, $n = 16$), have been structurally characterised in the solid state by X-ray crystallography (Figure 6.1).[3] Their structures are completely different from those of other 13- or 16-membered

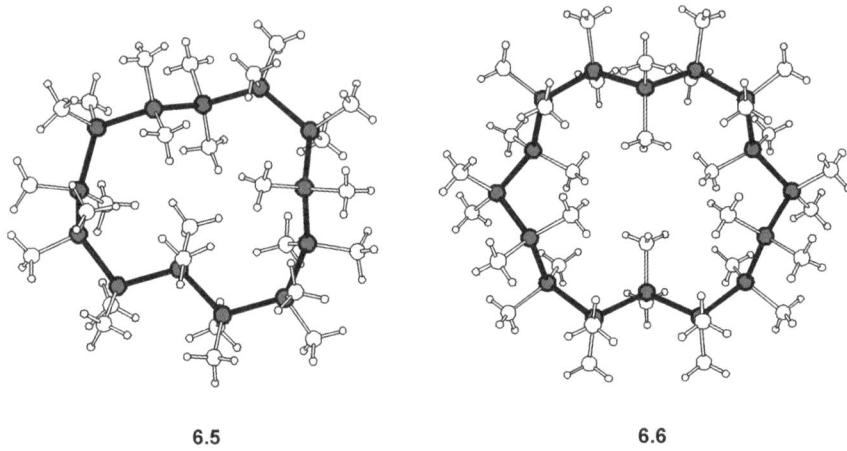

Figure 6.1 Molecular structures of cyclo-(Me$_2$Si)$_n$ (**6.5**, $n = 13$; **6.6**, $n = 16$).

rings as a result of distortions that occur to relieve unfavourable endocyclic methyl–methyl interactions.

An extensive series of cyclic sulfur allotropes, cyclo-S$_n$, have been identified. For example, all the individual allotropes ($n = 6$–22) have been detected in the mixture obtained by reaction of SCl$_2$ with potassium iodide in CS$_2$ by HPLC by using methanol as eluent and a dodecylsiloxane stationary phase (see Figure 3.3). The larger sulfur rings, cyclo-S$_{10}$, -S$_{15}$, -S$_{20}$ and -S$_{25}$, are obtained from the reactions of titanocene pentasulfide with sulfuryl chloride in CS$_2$ [eqn (6.1)].[4] The yields of the individual allotropes decrease with increasing ring size, but this method can be used for the synthesis of gram quantities of cyclo-S$_{20}$ by fractional crystallisation. The metallocyclic reagent Cp$_2$TiS$_5$ also serves as a source of the odd-numbered rings, cyclo-S$_9$, -S$_{11}$ and -S$_{13}$, by reaction with dichlorosulfanes S$_n$Cl$_2$ ($n = 4$, 6 or 8, respectively), *cf.* eqn (2.1).[5a]

$$\text{Cp}_2\text{TiS}_5 + \text{SO}_2\text{Cl}_2 \rightarrow (1/n)\text{cyclo-S}_{5n} + \text{Cp}_2\text{TiCl}_2 \qquad (6.1)$$

The larger sulfur rings are all thermodynamically unstable with respect to the formation of cyclo-S$_8$. The well-known crown structure of this eight-membered ring minimises the repulsions between the lone pairs on neighbouring sulfur atoms. For the same reason, all the larger cyclic sulfur allotropes exhibit puckered ring conformations. Some representative examples are shown in Figure 6.2.

6.2 HETEROCYCLIC SYSTEMS

6.2.1 Saturated Rings

Cyclic sulfur imides are molecules in which one (or more) of the sulfur atoms in a cyclic sulfur allotrope is replaced by an NR (R = H, alkyl) group (see **2.1**–**2.4**

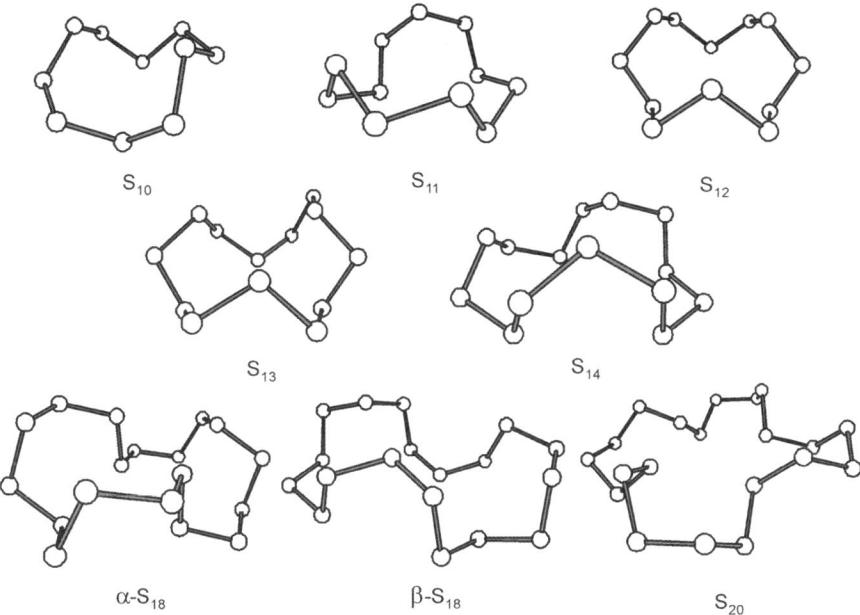

Figure 6.2 Conformations of the cyclic allotropes S_{10}, S_{11}, S_{12}, S_{13}, S_{14}, two isomers of S_{18} and S_{20}.

in Section 2.1). The most common examples are eight-membered rings, *e.g.* cyclo-S_7NH and $S_4N_4H_4$ (see Section 12.6.1), which have crown-shaped structures similar to that of cyclo-S_8. Larger sulfur imide rings are prepared by utilising the methodology employed for the synthesis of the larger sulfur allotropes. For example, the metathesis reaction between the metallacyclic complex Cp$_2$Ti(S$_7$NH) and the appropriate dichlorosulfane yields cyclo-S_8NH or cyclo-S_9NH [eqn (6.2)].[5b] The structure of the nine-membered ring, cyclo-S_8NH (**6.7**), shows a sequence of torsion angles similar to that of cyclo-S_9, whereas the 10-membered ring, cyclo-S_9NH (**6.8**), resembles the crown structure of cyclo-S_8, with the insertion of an SN(H) unit, rather than that of cyclo-S_{10}. The 12-membered ring cyclo-S_{11}NH has only been characterised spectroscopically (IR, Raman and mass spectra).

$$Cp_2Ti(S_7NH) + S_nCl_2 (n = 1, 2) \rightarrow Cp_2TiCl_2 + cyclo\text{-}S_{7+n}NH \qquad (6.2)$$

The 15-membered ring Se$_9$(N^{t}Bu)$_6$ (**6.9**) is a unique member of the series of cyclic chalcogen imides (see Section 12.8.1). It was first obtained as the major product of the reaction of LiN(^{t}Bu)SiMe$_3$ with Se$_2$Cl$_2$ or SeOCl$_2$.[6] It is also formed, in addition to smaller cyclic selenium imides, in the cyclocondensation of SeCl$_2$ with *tert*-butylamine in THF or from the decomposition of the selenium(IV) diimide ^{t}BuN=Se=N^{t}Bu.[7]

6.7

6.8

6.9

Cyclodimethylsiloxanes (Me$_2$SiO)$_n$ represent an extensive series of cyclic molecules in which Si and O alternate in the ring. In principle, there is no limit to the value of n. In practice, ring sizes up to at least $n = 25$ have been detected by chromatographic methods.[8] They are prepared by the hydrolysis of Me$_2$SiCl$_2$. Cyclodimethylsiloxanes exhibit puckered ring systems, as exemplified by the hexamer (Me$_2$SiO)$_6$, for which the electron diffraction structure reveals that some of the methyl groups point inwards.[9] The behaviour of these silicon-containing polyethers (Me$_2$SiO)$_n$ ($n = 5, 6, 7$) as pseudo-crown ether ligands in metal complexes is discussed in Section 7.1.3.

Cyclopolyphosphate anions of the general formula [PO$_3^-$]$_n$ form a homologous series of saturated heterocycles. Although the most important members of this series are the cyclotriphosphates ($n = 3$) and cyclotetraphosphates ($n = 4$), several examples of cyclooctaphosphates, involving a 16-membered P$_8$O$_8$ ring (**6.10**) and one cyclodecaphosphate (**6.11**) containing a 20-membered P$_{10}$O$_{10}$ ring, are known (Figure 6.3).[10]

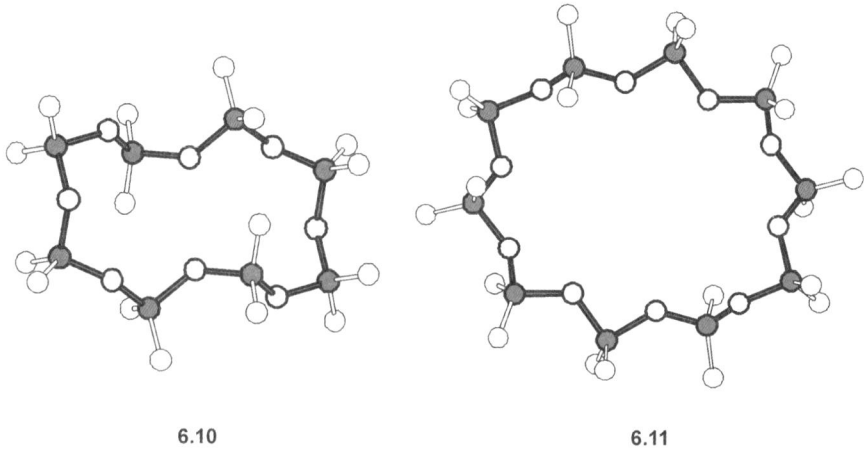

6.10 **6.11**

Figure 6.3 Structures of P$_8$O$_{24}^{8-}$ and P$_{10}$O$_{30}^{10-}$ anions; the cations are (Cu^{2+})$_3$ and (NH$_4^+$)$_2$ (**6.10**) and [C(NH$_2$)$_3)^+$]$_{10}$ (guanidinium) (**6.11**).

A unique example of an inorganic macrocycle that involves the linkage of saturated heterocycles by bridging groups is B_8S_{16} (boron disulfide) (**6.12**), which is obtained in low yield by the route shown in eqn (6.3).[11] The moisture-sensitive macrocycle **6.12** is comprised of four five-membered B_2S_3 rings connected by sulfur atoms to give a completely planar porphin-like molecule with D_{4h} symmetry. The compositionally similar polymer $(BS_2)_x$ also incorporates B_2S_3 rings that are linked by sulfur atoms to form infinite chains.

$$4 \; \text{[B}_2\text{S}_3\text{Br}_2\text{]} \xrightarrow[\substack{-4 \text{ CS}_2 \\ -8 \text{ HBr}}]{4 \text{ SC(SH)}_2} \textbf{6.12} \tag{6.3}$$

6.2.2 Unsaturated Rings

The inorganic skeleton in cyclophosphazenes $(NPX_2)_n$ is formally isoelectronic with that in cyclosiloxanes, although the former are unsaturated heterocycles (*i.e.* the P–N bonds have some degree of multiple bonding), whereas the $(Me_2SiO)_n$ systems are saturated with formally single Si–O bonds. The cyclophosphazenes form a very extensive homologous series; oligomers from $n = 3$ to 40 have been claimed for the difluoro derivatives $(NPF_2)_n$. The dichloro derivatives $(NPCl_2)_n$ are prepared by the cyclocondensation reaction of PCl_5 with ammonium chloride in a high-boiling solvent (see Scheme 2.1). Other derivatives are usually obtained by nucleophilic substitution of the exocyclic Cl substituents (see Scheme 11.5). The largest structurally characterised ring is the 24-membered ring $(NPMe_2)_{12}$ (**6.13**) (Figure 6.4).[12]

Hybrid (P–N)–(S–N) systems also form a variety of larger heterocycles. It is convenient to categorise this class of mixed ring system by the coordination number of the sulfur centres. The 12-membered ring $[(Ph_2PN)_2(SN)]_2$ (**6.14**) formally involves two-coordinate sulfur. However, the two sulfur atoms are involved in a weak transannular S···S interaction (2.39 Å) (*cf.* Figure 4.4).[13a] In solution, this ring dissociates into the six-membered cyclic radical $[(Ph_2PN)_2(SN)]$, which has been identified by EPR spectroscopy (see Figure 3.11). Interestingly, related 12-membered rings with three-coordinate sulfur $[(Ph_2PN)_2(SNR_2)]_2$ (R = alkyl) (**6.15**) are obtained from the corresponding six-membered rings $[(Ph_2PN)_2(SNR_2)]$ either upon mild heating or after prolonged standing in acetonitrile.[13b] The electron-donating dialkylamino substituent on sulfur promotes ring expansion.

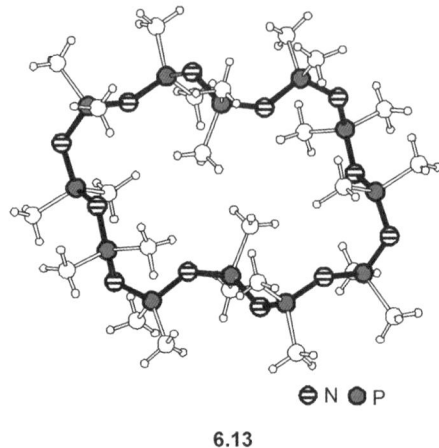

6.13

Figure 6.4 Molecular structure of $(NPMe_2)_{12}$.

6.14 **6.15**

Mixed phosphazene–sulfanuric ring systems of the type $[NS(O)Cl]_x[NPCl_2]_y$ are well known and the six-membered ring ($x = 1$, $y = 2$) has been used to generate the corresponding polymer by ring-opening polymerisation [see eqn (8.2)]. The production of the polymer is accompanied by the formation of a 12-membered ring $[NS(O)Cl(NPCl_2)_2]_2$, which exists as a *cis* and a *trans* isomer, **6.16a** and **6.16b**, respectively, together with the 24-membered ring $[NS(O)Cl(NPCl_2)_2]_4$ (**6.17**).[14]

6.16a **6.16b**

Inorganic Macrocycles

6.17

The incorporation of carbon into P–N or S–N ring systems generates macrocycles with unusual ring conformations. The largest structurally characterised carbophosphazene ring is the 16-membered ring [NCCl(NPCl$_2$)$_3$]$_2$ (**6.18**) (Figure 6.5), which adopts a double crown conformation with trigonal planar geometry at the two carbon atoms. The transannular distances within the large cavity of **6.18** range from 5.1 to 7.6 Å.[15] Sixteen-membered rings may also be obtained for carbon-containing S–N ring systems.[16a] The macrocycle [(RCN)(NSR′)]$_4$ (**6.19**, R = 4-XC$_6$H$_4$, X = Br, CH$_3$; R′ = C$_6$H$_5$) (Figure 6.5) adopts a cradle-like structure with approximate S_4 symmetry. The sulfur centres in the thiazyl

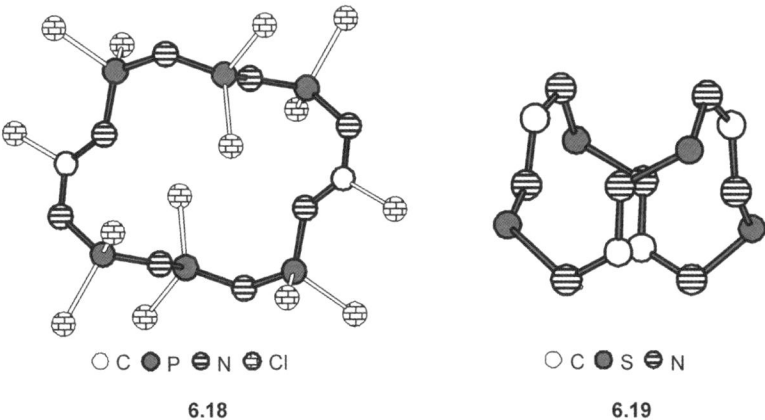

Figure 6.5 Molecular structures of the 16-membered rings [NCCl(NPCl$_2$)$_3$]$_2$ and [(RCN)(NSR′)]$_4$ (R and R′ groups on C and S, respectively, have been omitted in **6.19**).

system **6.19** are oxidised by *m*-chloroperbenzoic acid to the corresponding sulfanuric macrocycle [(RCN)(NS(O)R')]$_4$, which retains the cradle conformation of the 16-membered $C_4S_4N_8$ ring, but with a more open cavity.[16b]

6.3 HOST–GUEST CHEMISTRY

The term host–guest chemistry is used to describe complexes in which an electrophile (the guest), *e.g.* a metal ion, is encapsulated by a macrocyclic host with electron-donating sites. The best known examples of such complexes are the alkali metal complexes of crown ethers, e.g. **6.1** and **6.2**. More recently, a new class of host–guest complexes in which the host molecule is a macrocycle with multiple electron-accepting sites and the guests are anions such as halides or oxide has been investigated extensively.[17,18] These macrocycles are described as 'inverse crowns', *i.e.* the term 'ether' is omitted since they do not contain oxygen sites.[18] Typically, these inverse crowns involve s-block metals, as the electrophilic sites, bridged by amido (NR$_2$) ligands. For example, the heterometallic complex [Li$_2$Mg$_2${N(SiMe$_3$)$_2$}$_4$(μ-O)] (**6.20**) is comprised of an eight-membered ring made up of alternating nitrogen and metal atoms which entraps an oxide anion, thus rendering the molecule neutral.[19] The potential for an extensive series of inverse crowns is readily evident, since (a) the monopositive cation in **6.20** can be replaced by a larger alkali metal ion, (b) another 2^+ cation could be introduced for Mg^{2+} and (c) the secondary amide ligand could also be varied.

The introduction of larger alkali metal ions not only produces larger electron-accepting macrocycles, but also leads to complexes that exhibit unprecedented chemical reactivity in deprotonation reactions. The complex [Na$_4$Mg$_2$(TMP)$_6$(C$_6$H$_4$)] (**6.21**) (TMP = tetramethylpiperidyl) provides a dramatic example of this enhanced reactivity.[20] The 12-membered Na$_4$Mg$_2$N$_6$ ring in **6.21** encapsulates a doubly deprotonated benzene ring. Since neither NaTMP nor Mg(TMP)$_2$ deprotonates benzene on its own, it is clear that these heterometallic ring systems exhibit a powerful synergic effect. When K^+ is used as the alkali metal, ring expansion to give the 24-membered [K$_6$Mg$_6$(TMP)$_{12}$(C$_6$H$_5$)$_6$] is observed. In this complex, the polymetallic ring acts as a host to six singly deprotonated arene anions.

An equally impressive demonstration of this synergic reactivity of s-block metal amides is observed in the reaction of ferrocene with a mixture of ^{n}BuNa, Bu$_2$Mg and diisopropylamine. In this case, tetradeprotonation occurs regioselectively to give the 1,1,3,3-tetraanion of ferrocene, which is trapped in the 16-membered heterometallic amide ring (**6.22** in Figure 6.6).[21]

Saturated inorganic heterocycles known as cyclophosph(III)azanes (XENR)$_n$ (E = P)[22] and, to a lesser extent, their heavier analogues (E = As, Sb, Bi)[23] have been studied extensively. The chemistry of these pnictogen–nitrogen ring systems is limited primarily to four- and six-membered rings (see Section 11.4.1). An interesting recent development in this area has been the use of E$_2$N$_2$ building blocks for the synthesis of macrocyclic molecules.[24] As an illustrative example, the cyclocondensation of [ClP(μ-N^{t}Bu)]$_2$ and [H$_2$NP(μ-N^{t}Bu)]$_2$ produces the tetrameric macrocycle [{P(μ-N^{t}Bu)}$_2$(μ-NH)]$_4$ (**6.23**) with imido (NH) linkages between the four-membered P$_2$N$_2$ rings [eqn (6.4)].[25] The N–H protons are directed *endo* to the cavity in **6.23** because of the *cis* conformation of the P$_2$N$_2$ rings.

$$ (6.4) $$

The cavity size in **6.23** is ca. 5.22 Å, considerably larger than those in comparable organic ligands, such as those based on [CH$_2$CH$_2$NH]$_4$, which range from 4.07 to 4.29 Å, suggesting that this class of inorganic macrocycle is a good candidate for host–guest chemistry. A complex in which a Cl$^-$ ion is encapsulated by the pentamer [{P(μ-N^{t}Bu)}$_2$(μ-NH)]$_5$, a 20-membered ring (**6.24**) (Figure 6.7), is also formed in very low yields in the reaction depicted in eqn (6.4).[26] The tetramer **6.23** and the pentamer **6.24** do not interconvert, suggesting that the production of these macrocycles is kinetically driven. The formation of the tetramer is favoured if the cyclocondensation reaction is performed in the presence of lithium chloride or bromide, whereas the pentamer **6.24** is obtained exclusively in the presence of lithium iodide.

Even larger rings are generated by changing either the bridging group or the pnictogen in the E$_2$N$_2$ rings of these macrocyclic systems. For example, the reaction of the dimer [Cl(Se)P(μ-N^{t}Bu)$_2$P(Se)Cl] with an excess of sodium metal in boiling toluene produces the selenium-bridged hexamer

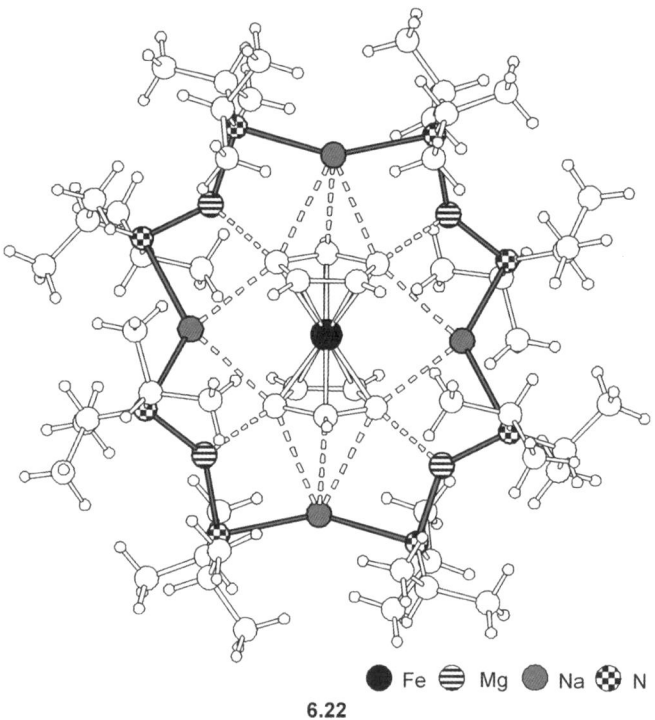

6.22

Figure 6.6 Molecular structure of $[Na_4Mg_4(^iPr_2N)_8\{Fe(C_5H_3)_2\}]$.

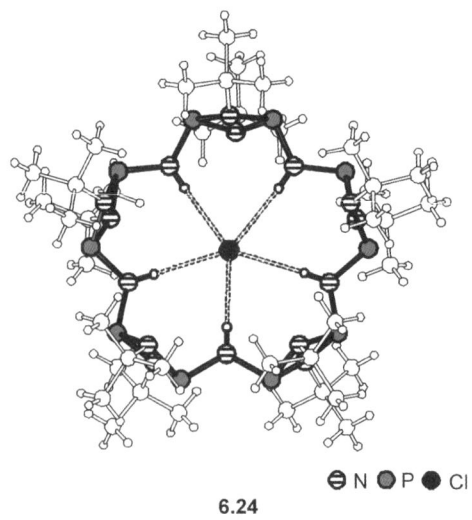

6.24

Figure 6.7 Encapsulation of Cl^- ion by $[\{P(\mu-N^tBu)\}_2(\mu-NH)]_5$.

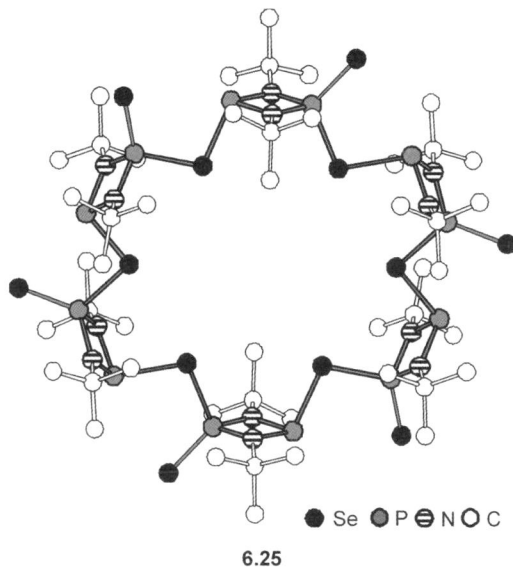

Figure 6.8 Molecular structure of the 24-membered macrocycle [(Se)P(μ-N^{t}Bu)$_2$ P(μ-Se)]$_6$.

[(Se)P(μ-N^{t}Bu)$_2$P(μ-Se)]$_6$ (**6.25**) (Figure 6.8).[27] The bridging selenium atoms in this intriguing macrocycle are favourably disposed for coordination to softer metal centres within the large cavity of this 24-membered macrocycle.

Twenty-four-membered rings have also been characterised for macrocycles involving the Sb$_2$N$_2$ building block bridged by imido (NR) groups.[28] The phenylimido derivative [{Sb(μ-NPh)}$_2$(μ-NPh)]$_6$ (**6.26a**) is obtained in high yield from the cyclocondensation reaction of SbCl$_3$ and three equivalents of Li[N(H)Ph];[28a] the related *ortho*-substituted aryl derivative [{Sb(μ-N-2-MeOC$_6$H$_4$)}$_2$(μ-N-2-MeOC$_6$H$_4$)]$_6$ (**6.26b**) has a similar structure (Figure 6.9).[28b] However, in contrast to the selenium-bridged system **6.25**, the orientation of the bridging μ-NAr groups in **6.26a** and **6.26b** is such that there are no donor atoms available for metal coordination within the framework of these imido-bridged macrocycles.

6.4 SUPRAMOLECULAR ASSEMBLIES

The synthesis and structures of transition metal complexes of anionic cyclopolypnictogen ligands, *e.g.* cyclo-P$_5^-$, are discussed in Section 7.2.1 and 11.2. An exciting recent development involves the use of these complexes as building blocks for the construction of novel supramolecular assemblies, including one-dimensional (1D) and two-dimensional (2D) polymers and even soluble spherical fullerene-like aggregates.[29] Complexes that have been used for this

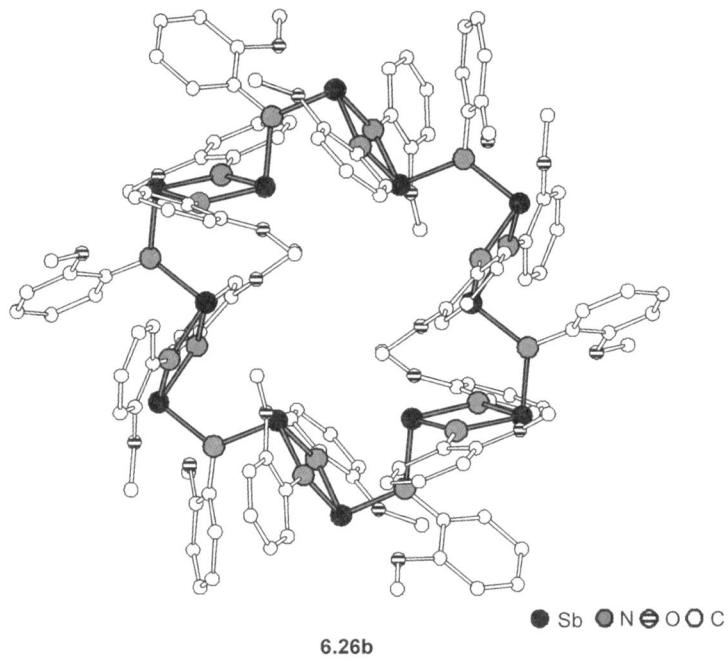

6.26b

● Sb ● N ⊖ O ○ C

Figure 6.9 Molecular structure of the 24-membered macrocycle [{Sb(μ-N-2-MeOC$_6$H$_4$)}$_2$(μ-N-2-MeOC$_6$H$_4$)]$_6$.

purpose include Cp*Fe(η^5-P$_5$) (**6.27**, Cp* = C$_5$Me$_5$),[30] Cp′Ta(CO)$_2$(η^4-P$_4$) (**6.28**, Cp′ = 2,5-^{t}Bu$_2$C$_5$H$_3$)[31] and Cp*Mo(CO)$_2$(η^3-As$_3$) (**6.29**).[32]

The most spectacular of these assemblies is the spherical complex [{Cp*Fe(η^5-P$_5$)}$_{12}${CuCl}$_{10}${Cu$_2$Cl$_3$}$_5${Cu(CH$_3$CN)$_2$}$_5$] (**6.30**), which is formed by reaction of **6.27** with CuCl in CH$_2$Cl$_2$–CH$_3$CN.[30b] The 'inorganic fullerene' that is isolated from this reaction has 90 core atoms comprised of cyclo-P$_5$ rings surrounded by six-membered P$_4$Cu$_2$ rings in an arrangement that mimics that observed in fullerene molecules (Figure 6.10). However, the outside diameter of this inorganic sphere is about three times larger than that of C$_{60}$.

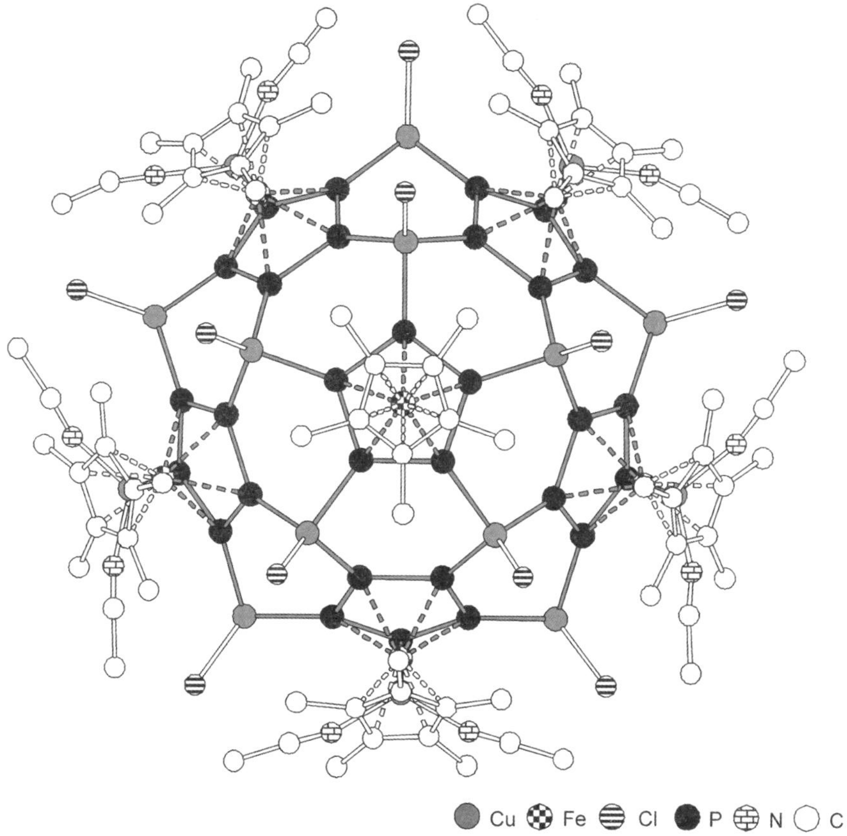

○ Cu ⊕ Fe ⊜ Cl ● P ⊞ N ○ C

6.30

Figure 6.10 View of the half-shell of the spherical complex **6.30** showing the five-membered P$_5$ and six-membered P$_4$Cu$_2$ rings.

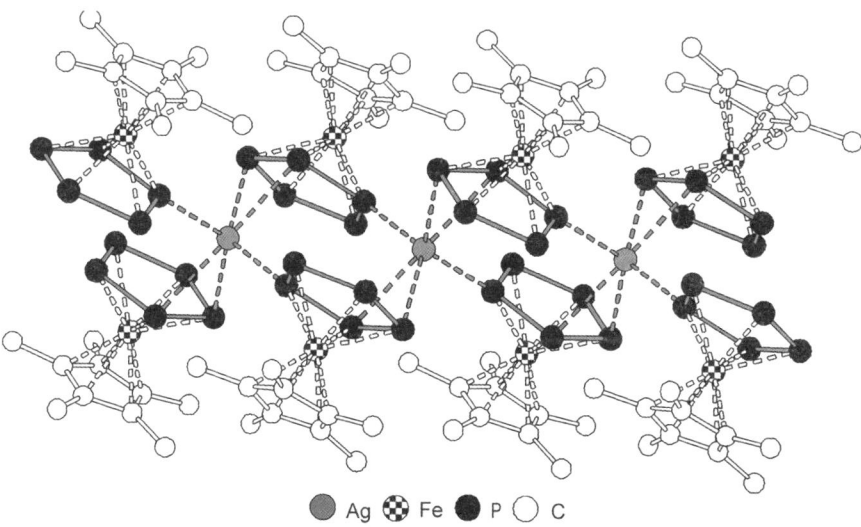

○ Ag ⊞ Fe ● P ○ C

Figure 6.11 Part of the cationic chain in the one-dimensional polymer **6.32** showing the 1,2,3-coordination mode of the cyclo-P$_5$ ligands.

In a similar manner, the reaction of **6.28** with CuCl in CH_2Cl_2–CH_3CN produces the spherical complex [{Cp′Ta(CO)$_2$(η^4-P$_4$)}$_6${CuCl}$_8$] (**6.31**). The tantalum complex **6.31** is comprised of 32 core atoms forming four-membered P$_4$ rings and six-membered P$_4$Cu$_2$ rings in an extended cubic arrangement with overall O_h symmetry.[32]

The reaction of **6.27** with silver salts of the weakly coordinating anion [Al{OC(CF$_3$)$_3$}$_4$]$^-$ produces the soluble complex [Ag{Cp*Fe(η^5-P$_5$)}$_2$]$_n$[Al{OC(CF$_3$)$_3$}$_4$]$_n$ (**6.32**).[30d] The cyclo-P$_5$ ligand in **6.32** adopts a 1,2,3-coordination mode, leading to a one-dimensional polymer (Figure 6.11).

REFERENCES

1. C. E. Housecroft and A. G. Sharpe, *Inorganic Chemistry*, 2nd edn., Pearson Education, Harlow, 2005, pp. 268–269.
2. J. March, *Advanced Organic Chemistry*, 5th edn., Wiley, New York, 2001, pp. 62–69.
3. F. Shakiee, K. J. Haller and R. West, *J. Am. Chem. Soc.*, 1986, **108**, 5478.
4. R. Steudel, in *The Chemistry of Inorganic Ring Systems*, ed. R. Steudel, Elsevier, Amsterdam, 1992, pp. 233–254.
5. (a) R. Steudel, J. Steidel and T. Sandow, *Z. Naturforsch., Teil B*, 1986, **41**, 958; (b) R. Steudel, K. Bergemann, J. Buschman and P. Luger, *Angew. Chem. Int. Ed. Engl.*, 1996, **35**, 2537.
6. H. W. Roesky, K. -L. Weber and J. W. Bats, *Chem. Ber.*, 1984, **117**, 2686.
7. T. Maaninen, T. Chivers, R. Laitinen, G. Schatte and M. Nissinen, *Inorg. Chem.*, 2000, **39**, 5341.
8. (a) J. F. Brown Jr and G. M. J. Slusarczhuk, *J. Am. Chem. Soc.*, 1965, **87**, 931; (b) D. Seyferth, C. Prud'homme and G. H. Wiseman, *Inorg. Chem.*, 1983, **22**, 2163.
9. H. Oberhammer and G. Fogarasi, *J. Mol. Struct.*, 1973, **18**, 309.
10. U. Schülke, M. T. Averbuch-Pouchot and A. Durif, *Z. Anorg. Allg. Chem.*, 1992, **612**, 107.
11. W. Siebert, in *The Chemistry of Inorganic Homo- and Heterocycles*, ed. I. Haiduc and D.B. Sowerby, Academic Press, London, 1987, p. 143.
12. R. T. Oakley, S. J. Rettig, N. L. Paddock and J. Trotter, *J. Am. Chem. Soc.*, 1985, **107**, 6923.
13. (a) T. Chivers, M. N. S. Rao and J. F. Richardson, *J. Chem. Soc., Chem. Commun.*, 1983, 186; (b) T. Chivers, M. N. S. Rao and J. F. Richardson, *J. Chem. Soc., Chem. Commun.*, 1983, 702.
14. Y. Ni, A. J. Lough, A. L. Rheingold and I. Manners, *Angew. Chem. Int. Ed. Engl.*, 1995, **34**, 998.
15. E. Rivard, A. J. Lough and I. Manners, *Inorg. Chem.*, 2004, **43**, 2765.
16. (a) T. Chivers, M. Parvez, I. Vargas-Baca, T. Ziegler and P. Zoricak, *Inorg. Chem.*, 1997, **36**, 1669; (b) T. Chivers, M. P. Gibson, M. Parvez and I. Vargas-Baca, *Inorg. Chem.*, 2000, **39**, 1697.
17. R. E. Mulvey, *Chem. Commun.*, 2001, 1049.

18. M. Driess, R. E. Mulvey and M. Westerhausen, in *Molecular Cluster of the Main Group Elements*, ed. M. Driess and H. Nöth, Wiley-VCH, Weinheim, 2004, pp., 398–403.
19. A. R. Kennedy, R. E. Mulvey and R. B. Rowlings, *J. Am. Chem. Soc.*, 1998, **120**, 7816.
20. P. C. Andrews, A. R. Kennedy, R. E. Mulvey, C. L. Raston, B. A. Roberts and R. B. Rowlings, *Angew. Chem. Int. Ed.*, 2000, **39**, 1960.
21. W. Clegg, K. W. Henderson, R. E. Mulvey, C. T. O'Hara, R. B. Rowlings and D. M. Tooke, *Angew. Chem. Int. Ed.*, 2001, **40**, 3902.
22. L. Stahl, *Coord. Chem. Rev.*, 2000, **210**, 203.
23. M. S. Balakrishna, D. J. Eisler and T. Chivers, *Chem. Soc. Rev.*, 2007, **36**, 650.
24. E. L. Doyle, L. Riera and D. S. Wright, *Eur. J. Inorg. Chem.*, 2003, 3279.
25. A. Bashall, A. D. Bond, E. L. Doyle, F. Garcia, S. Kidd, G. T. Lawson, M. C. Parry, M. McPartlin, A. D. Woods and D. S. Wright, *Chem. Eur. J.*, 2002, **8**, 3377.
26. F. Garcia, J. M. Goodman, R. A. Kowenicki, I. Kuzu, M. McPartlin, M. A. Silva, L. Riera, A. D. Woods and D. S. Wright, *Chem. Eur. J.*, 2004, **10**, 6066.
27. S. González-Calera, D. J. Eisler, J. V. Morey, M. McPartlin, S. Singh and D. S. Wright, *Angew. Chem. Int. Ed.*, 2008, **47**, 1111.
28. (a) R. Bryant, S. C. James, J. C. Jeffery, N. C. Norman, A. G. Orpen and U. Weckenmann, *J. Chem. Soc., Dalton Trans.*, 2000, 4007; (b) M. A. Beswick, M. K. Davies, M. A. Paver, P. R. Raithby, A. Steiner and D. S. Wright, *Angew. Chem. Int. Ed. Engl.*, 1996, **35**, 1508.
29. M. Scheer, L. J. Gregoriades, R. Merkle, B. P. Johnson and F. Dielmann, *Phosphorus Sulfur Silicon*, 2008, **183**, 504.
30. (a) J. Bai, A. V. Virovets and M. Scheer, *Angew. Chem. Int. Ed.*, 2002, **41**, 1737; (b) J. Bai, A. V. Virovets and M. Scheer, *Science*, 2003, **300**, 781; (c) M. Scheer, J. Bai, B. P. Johnson, R. Merkle, A. V. Virovets and C. E. Anson, *Eur. J. Inorg. Chem.*, 2005, 4023; (d) M. Scheer, L. J. Gregoriades, A. V. Virovets, W. Kunz, R. Neueder and I. Krossing, *Angew. Chem. Int. Ed.*, 2006, **45**, 5689.
31. B. P. Johnson, F. Dielmann, G. Balázs, M. Sierka and M. Scheer, *Angew. Chem. Int. Ed.*, 2006, **45**, 2473.
32. L. J. Gregoriades, H. Krauss, J. Wachter, A. V. Virovets, M. Sierka and M. Scheer, *Angew. Chem. Int. Ed.*, 2006, **45**, 4189.

CHAPTER 7
Ligand Chemistry

This chapter is intended provide an overview of the behaviour of inorganic heterocycles as ligands towards metal centres. Such coordination complexes can be divided into two major types. The first involves Lewis base–Lewis acid adducts in which the lone pairs of the heteroatoms in the ring system act as two-electron donors towards metal ions such as the oxygen atoms of crown ethers, *i.e.* σ-complexes. In the second type of coordination complex, the cyclic systems are unsaturated and use their π-electrons in the interaction with metal centres. The latter types of complex are comparable to organometallic complexes involving carbocyclic π-ligands such as cyclobutadiene or benzene. In contrast to carbocyclic systems, however, some unsaturated inorganic rings may also form σ-complexes because of the presence of lone pairs on the heteroatoms.

7.1 σ-COMPLEXES

7.1.1 Homocyclic Ligands

The formation of metal complexes of homocyclic inorganic ring systems that contain a lone pair of electrons on the p-block element, *e.g.* $(PhP)_n$, can be related to the behaviour of two-electron σ-donor ligands such as R_3P. However, the presence of multiple donor sites offers the possibility of coordination to more than one metal site and, consequently, these rings may serve as monodentate, bridging or chelating ligands.[1] Thus, the three-membered ring $(PhP)_3$ forms mononuclear, dinuclear or trinuclear complexes in which each phosphorus centre acts as a two-electron donor to a metal carbonyl fragment, *e.g.* the mononuclear complex $(P^tBu)_3[Cr(CO)_5]$ (**7.1**). Dinuclear complexes in which two cyclophosphine rings bridge two metal centres, *e.g.* $[(\mu-P^iPr)_3Ni(CO)_2]_2$ (**7.2**), are also known. Another mode of coordination mode is exhibited by the six-membered ring $(MeP)_6$, which adopts a boat conformation in acting as a

---

Inorganic Rings and Polymers of the p-Block Elements: From Fundamentals to Applications
By Tristram Chivers and Ian Manners
© Tristram Chivers and Ian Manners 2009
Published by the Royal Society of Chemistry, www.rsc.org

chelating ligand towards a single metal site, *e.g.* (MeP)$_6$W(CO)$_4$ (**7.3**).

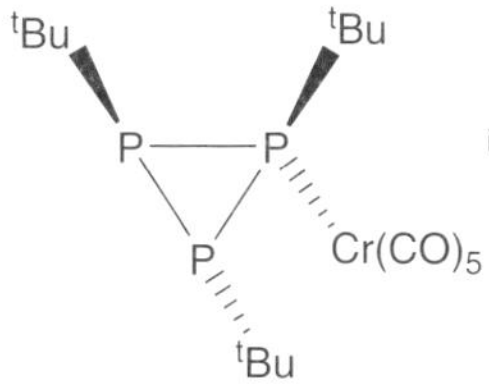

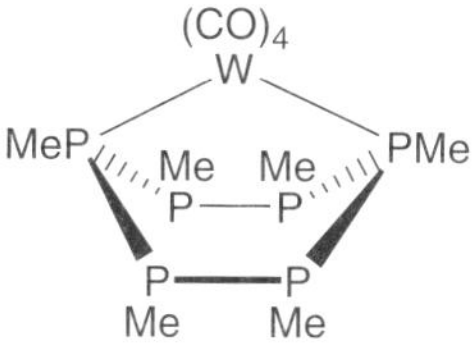

The treatment of cyclophosphines with transition metal complexes may also result in activation of a P–P bond. An interesting example of this behaviour is provided by the reaction of the cyclic pentamers (RP)$_5$ (R = Ph, Et) with the β-deketiminate rhodium(I) complex (NacNac)Rh(C$_8$H$_{14}$)(N$_2$).[2] As illustrated in Scheme 7.1, the outcome of this procedure depends on the electronic properties of the R groups attached to phosphorus. When R = Ph, the integrity of the five-membered P$_5$ ring is retained in the formation of the η^2-complex **7.4**. With electron-donating Et substituents on phosphorus, however, oxidative addition occurs via insertion of the metal into a P–P bond to give the rhodium(III) complex **7.5**.

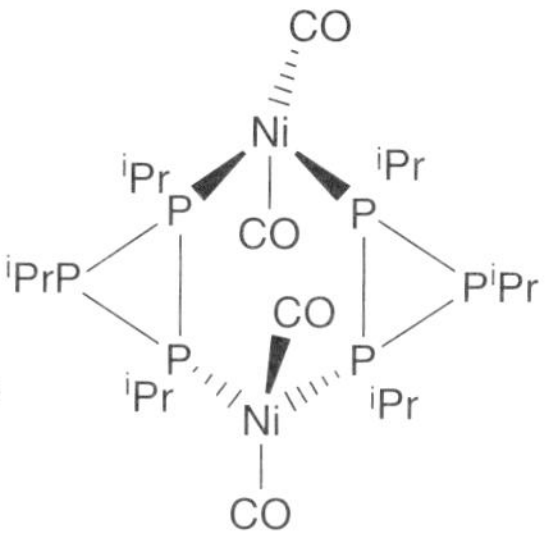

The occurrence of E–E and, in some cases, E–C bond activation in the reaction of the cyclic ligands (RE)$_n$ with transition metal complexes becomes more prevalent for the heavier pnictogens (E = As, Sb, Bi). The reaction of

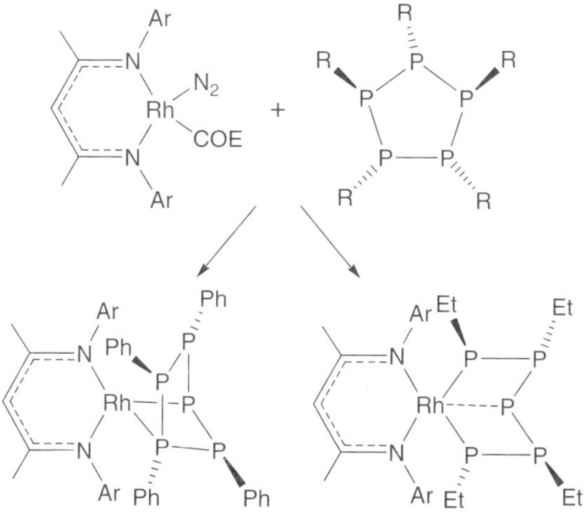

Scheme 7.1 Activation of a P–P bond in (EtP)$_5$ by a rhodium(I) complex.

the five-membered ring (MeAs)$_5$ with metal carbonyls results in ring expansion to give nine- or 10-membered cycloarsine rings in dinuclear complexes, *e.g.* [cyclo-(MeAs)$_9$Cr$_2$(CO)$_6$] and [cyclo-(MeAs)$_{10}$Mo$_2$(CO)$_6$] (see Figure 11.12).[3] In the case of antimony and bismuth, the increasing lability of the E–C bonds becomes apparent in the formation of metal complexes of triatomic Sb$_3$ and diatomic E$_2$ (E = Sb, Bi) from (^{t}BuSb)$_4$ or (RBi)$_n$ (n = 3, 5; R = Me$_3$SiCH$_2$) [see Section 11.3.2 and eqns (11.18) and (11.19)].

Homocyclic chalcogen molecules are also capable of acting as multidentate Lewis bases and, in this case, the donor centres (chalcogen atoms) have two available electron pairs. In practice, the interaction of cyclic sulfur allotropes, most commonly cyclo-S$_8$, with metal centres often gives rise to S–S bond scission as a result of an oxidative-addition process. There are, however, a few examples of structurally characterised complexes in which the homocyclic chalcogen molecule is retained as an intact ligand. Cyclo-S$_8$ may function as either a bidentate, tridentate or tetradentate ligand towards a single metal centre or as a bridging ligand between two metal centres while retaining the crown-shaped ring conformation of cyclo-octasulfur. In the 1:1 complex [Ag(cyclo-S$_8$)][Al{OCH(CF$_3$)$_2$}$_4$] (**7.6**), the silver ion binds to cyclo-S$_8$ in a 1,3,5,7-tetradentate (η^4) fashion and the cation adopts close to C_{4v} symmetry (Figure 7.1a).[4] By contrast, the two S$_8$ rings in the 1:2 complex [Ag(cyclo-S$_8$)$_2$][AsF$_6$] (**7.7**) behave as chelating 1,3-bidentate ligands (Figure 7.1b).[5] The structure of **7.7** is, however, strongly influenced by two Ag–F contacts with the relatively basic [AsF$_6$]$^-$ anion. The distortion caused by these interactions is evident when the [AsF$_6$]$^-$ anion is replaced by the weakly coordinating [Al{OC(CF$_3$)$_3$}$_4$]$^-$ anion.[6] The structure of the sandwich complex [Ag(cyclo-S$_8$)$_2$][Al{OC(CF$_3$)$_3$}$_4$] (**7.8**) contains

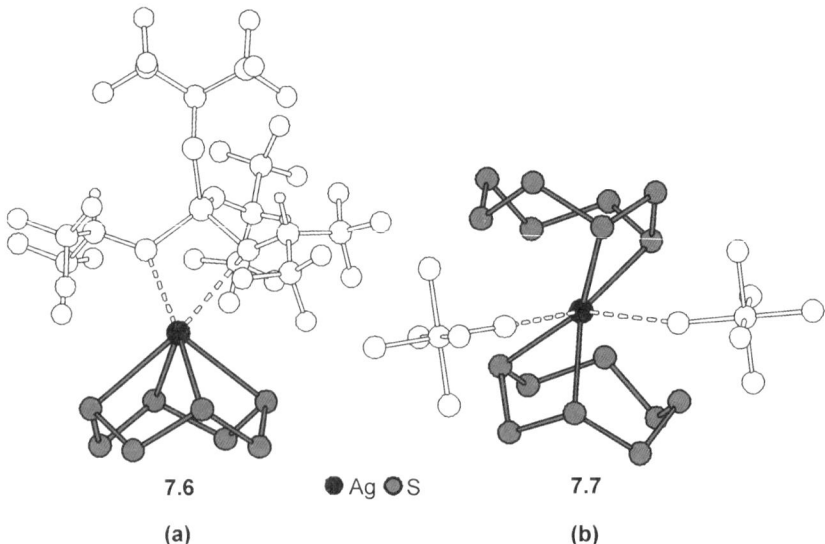

Figure 7.1 (a) Tetradentate and (b) bidentate coordination of cyclo-S$_8$ with Ag$^+$.

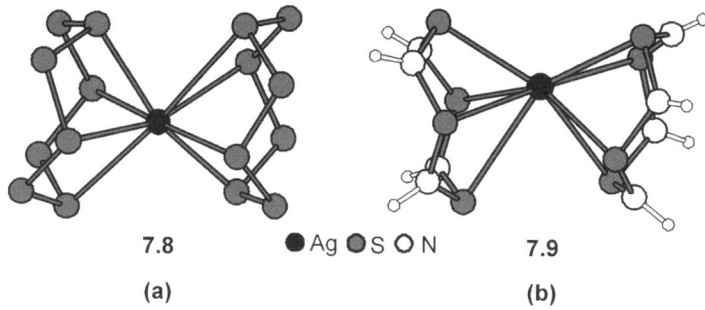

Figure 7.2 Sandwich complexes of Ag^+ with (a) cyclo-S_8 and (b) $S_4(NH)_4$.

an approximately centrosymmetric $[Ag(\eta^4\text{-cyclo-}S_8)_2]^+$ cation (Figure 7.2a) with Ag–S distances that range from 2.68 to 3.31 Å.[5] Interestingly, the related heterocyclic eight-membered ring $S_4(NH)_4$ (see Section 12.6.1) also behaves as a tetradentate ligand (through the four S atoms) towards a silver ion in the complex $[Ag\{S_4(NH)_4\}_2][ClO_4]$ (**7.9**, Figure 7.2b).[7a] Tetrasulfur tetraimide acts as an S-monodentate ligand in complexes of the type $[S_4(NH)_4]M(CO)_5$ (M = Cr, W).[7b]

Cyclo-S_8 also serves as a bridging ligand in complexes with dirhodium tetra(trifluoroacetate).[8] The 1:1 complex $\{[Rh_2(O_2CCF_3)_4](\text{cyclo-}S_8)\}_\infty$ forms a zig-zag one-dimensional chain, in which the S_8 rings exhibit 1,3-coordination to the metal centres of two different dirhodium units. By contrast, the S_8 rings act as tridentate ligands in the 3:2 complex $\{[Rh_2(O_2CCF_3)_4]_3(\text{cyclo-}S_8)_2\}_\infty$ (**7.10**). The structure of **7.10** has a two-dimensional architecture in which polymeric chains of the 1:1 complex are linked together by an additional $[Rh_2(O_2CCF_3)_4]$ molecule via the third sulfur donor site (Figure 7.3).

Metal complexes of cyclo-Se_6 and cyclo-Se_8 are discussed in Section 12.1.1 and the structures of $PdBr_2(\text{cyclo-}Se_6)$ and the $[Rb(\text{cyclo-}Se_8)]^+$ chain are depicted in Figure 12.1. The structure of the dinuclear rhenium complex $[Re_2I_2(CO)_6(\text{cyclo-}Se_7)]$ (**7.11**) incorporates cyclo-Se_7 as a ligand that bridges two metal sites through adjacent Se atoms (Figure 7.4).[9] The 1,2-dihapto (η^2) coordination of the Se_7 ring results in an elongation of the Se–Se bond by ca 0.15 Å compared with the corresponding value in lattice-trapped cyclo-Se_7 (2.56 vs 2.41 Å) (see Figure 12.2).

The interaction of various isomers of the homocyclic sulfur ligands, S_6, S_7 and S_8 with Li^+ has been probed by *ab initio* molecular orbital methods.[10] The global minimum structure of $[LiS_6]^+$ was found to involve a 1,3,5-tridentate six-membered S_6 ring in a chair conformation,[10a] whereas that of $[LiS_7]^+$ is comprised of an Li^+ cation linked to the four sulfur atoms that are negatively charged in the free S_7 ligand.[10b] The global minimum structure of $[LiS_8]^+$ depicts a complex of C_{4v} symmetry with the S_8 ring in a crown conformation and four equal Li–S bonds, *cf.* **7.6**.[10c] The coordination of Li^+ to cyclic sulfur homocycles activates the S–S bonds by polarisation of the valence electrons.

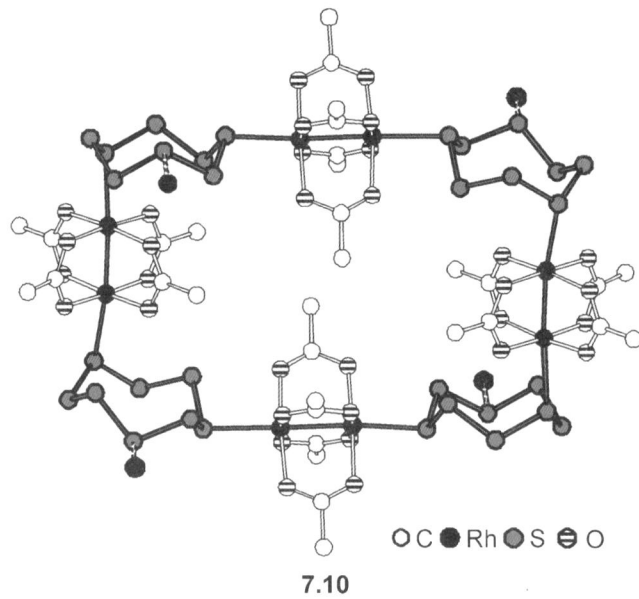

7.10

Figure 7.3 The octagonal unit in the 3:2 complex $\{[Rh_2(O_2CCF_3)_4]_3(S_8)_2\}_\infty$.

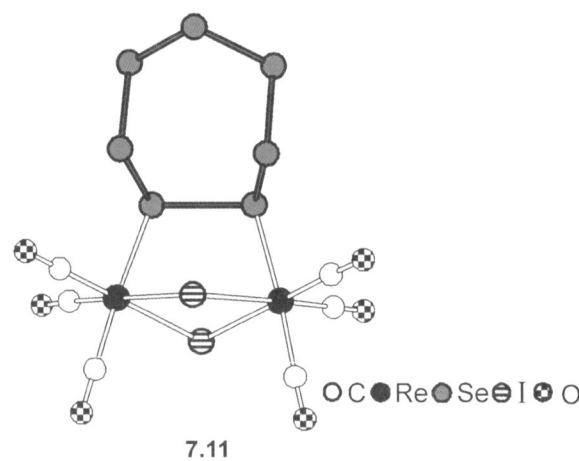

7.11

Figure 7.4 Molecular structure of $[Re_2I_2(CO)_6(cyclo\text{-}Se_7)]$.

7.1.2 Heterocyclic Ligands

There are numerous examples of the use of cyclophosphazenes with exocyclic dialkylamino or alkyl/aryloxy substituents on phosphorus as scaffolds for the formation of metal complexes.[11] However, the number of complexes in which the endocyclic nitrogen atoms are involved in adduct formation with electrophilic acceptors is more limited. Electron-donating substituents on phosphorus favour the formation of such complexes. For example, $P_3N_3Me_6$ readily forms 1:1 adducts with MCl_4 (M = Ti, Sn). Transition metal complexes of larger cyclophosphazenes $(NPMe_2)_n$ (n = 6 or 8) are discussed in Section 7.1.3.

The formation of the cation $[N_3P_3Cl_5]^+$ is of particular interest in the context of the mechanism of ring-opening polymerisation of $N_3P_3Cl_6$ (see Scheme 8.2). *N*-Bonded 1:1 complexes are formed between $N_3P_3Cl_6$ and the Lewis acids MCl_3 (M = Al, Ga) (see Scheme 11.6).[12a] This coordination mode has also been established in the σ-bonded complex $[Ag(N_3P_3Cl_6)_2]^+$ (**7.12**), which was isolated as the $[Al\{OC(CF_3)_3\}_4]^-$ salt (Figure 7.5).[12b] The two $N_3P_3Cl_6$ rings in this cation are coplanar and the mean P–N bond distance involving the coordinated N atom is *ca* 0.04 Å shorter than the average length of the other P–N bonds.

Although the formation of π-complexes in which the four-membered S_2N_2 ring acts as a six π-electron ligand (see Section 4.1.2.3), *e.g.* $(\eta^4\text{-}S_2N_2)M(CO)_3$ (M = Cr, Mo, W) can be envisaged from frontier orbital considerations,[13] only *N*-bonded adducts are observed. Both mono- and di-adducts of the type $S_2N_2 \cdot L$ and $S_2N_2 \cdot 2L$ are formed with Lewis acids such as AlX_3 (X = Cl, Br), $SbCl_5$ and with various transition-metal halides.[14] Coordination to a metal centre has little effect on the square-planar geometry of the S_2N_2 ring. N-bonded di-adducts are also formed between Se_2N_2 and $AlBr_3$ and $[PdX_3]^-$ (X = Cl, Br) (see Section 12.8.2);[15,16a] the palladium complex $[Br_3Pd(\mu\text{-}Se_2N_2)PdBr_3]^-$ (**7.13**) has been used as an *in situ* source of the highly reactive free ligand Se_2N_2, which is quickly transformed into Se_4N_4, but can be trapped as the platinum complex **7.14** by using the strategy illustrated in Scheme 7.2.[16b]

The binary sulfur nitride S_4N_4 also forms *N*-bonded monoadducts $S_4N_4 \cdot L$ with a variety of Lewis and Brønsted acids and with some transition metal halides, *e.g.* $FeCl_3$ and VCl_5. In these complexes, the eight-membered ring is flattened into a puckered boat conformation with no cross-ring interactions.[17] The reactions of S_4N_4 with metal halides frequently give rise to ring systems in which the metal is incorporated into a sulfur–nitrogen ring (cyclometallathiazenes). However, the replacement of the sulfur atoms in the 1,5-positions of S_4N_4 by $NPPh_2$ units serves to act as a structural brace. As a consequence, the

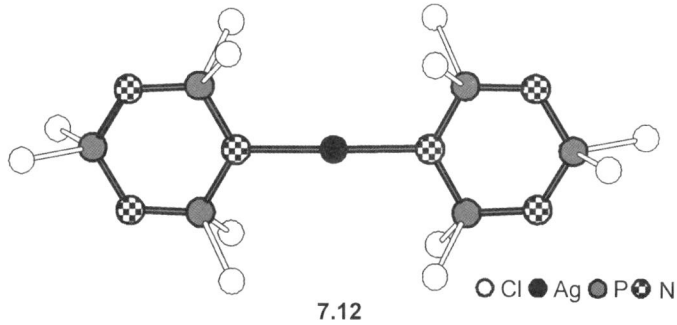

Figure 7.5 The planar centrosymmetric cation $[Ag(N_3P_3Cl_6)_2]^+$.

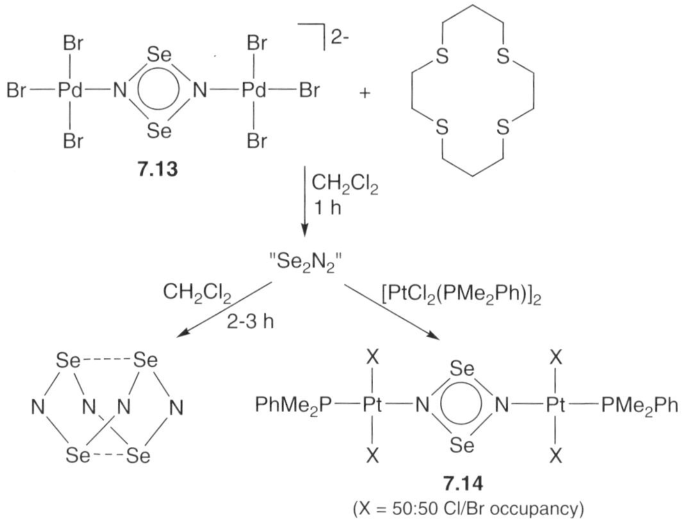

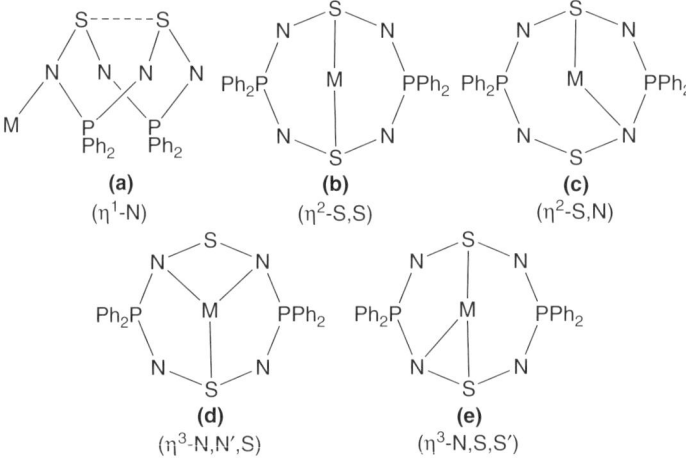

Scheme 7.2 Formation of Se$_2$N$_2$ from an *N*-bonded Pd complex.

Figure 7.6 Coordination modes in metal complexes of the 1,5-P$_2$N$_4$S$_2$ ring.

integrity of the heterocyclic ring in 1,5-(Ph$_2$P)$_2$N$_4$S$_2$ is retained in reactions with metal-containing reagents. The combination of hard (*N*) and soft (*S*) centres in the P$_2$N$_4$S$_2$ ring gives rise to a variety of coordination modes for this multi-dentate, cyclic ligand, as illustrated in Figure 7.6.[18]

7.1.3 Macrocyclic Ligands

Cyclic siloxanes (R$_2$SiO)$_n$ (see Section 6.2.1) may be considered as silicon analogues of crown ethers (see the introduction to Chapter 6). Note, however,

Ligand Chemistry 89

that the neighbouring oxygen atoms are only separated by one silicon atom in siloxanes whereas the donor sites are linked by two-carbon units ($-H_2C-CH_2-$) in crown ethers.[19] The first examples of structurally characterised metal complexes of cyclic siloxanes involved their serendipitous formation from the interaction of highly reactive reagents or products of various chemical reactions with the silicon grease used to lubricate ground-glass joints [silicone grease is a linear polymer of repeating ($-OSiMe_2-$) units terminated by Si–OH groups (see Section 10.6)]. Potassium complexes of the cycloheptasiloxane $(Me_2SiO)_7$ (**7.15**) are obtained in this way.[20] The potassium cation in **7.15** resides in the middle of the cavity provided by the siloxane ligand forming an approximately planar arrangement [$d(K \cdots O) = 2.86(3)–2.99(4)$ Å]. Interestingly, the silver complex of the heptamer $(Me_2SiO)_7$ (**7.16**) is obtained as a result of ring expansion upon treatment of the cyclic pentamer $(Me_2SiO)_5$ with $Ag[AsF_6]$ in liquid SO_2.[21] Unlike the encapsulated K^+ ion in **7.15**, in which all seven oxygen atoms of the 14-membered ring are strongly coordinated to the metal ion, only three strong and two weak metal–oxygen interactions are observed in the Ag^+ complex **7.16**; two of the oxygen atoms are non-coordinating.

7.15   **7.16**

By contrast, host–guest complexes between lithium and cyclosiloxanes **7.17** and **7.18** are obtained by direct reactions of lithium salts with $(Me_2SiO)_n$ ($n = 5$, 6, respectively) without change in the size of the polysiloxane ring. Only four out of the six oxygen atoms in **7.18** are coordinated to the Li^+ ion, which is located 0.13(1) Å out of the plane of the ligand.[22] The use of a weakly coordinating anion, *e.g.* $[Al\{OC(CF_3)_3\}_4]^-$, is necessary for the formation of stable complexes of this type. By comparison, similar complexes in which iodide is the counterion are thermodynamically unstable because of unfavourable lattice energy effects.[22]

7.17   **7.18**

The counterion [Al{OC(CF$_3$)$_3$}$_4$]$^-$ has also facilitated the isolation of a complex in which a lithium ion is encapsulated by an inorganic cryptand that incorporates of a siloxane framework.[23] The Li$^+$ cation in complex **7.19** is coordinated to only three of the O atoms (two short and one long Li–O contacts) and also exhibits two weak Li···P interactions. The bonding interactions in these lithium complexes of polyethers and cryptands based on siloxane ring systems primarily involve electrostatic interactions with the lone electron pairs on donor centres (O or P). Calculations indicate that, consistent with the lower basicity of siloxanes, the cyclosiloxane (Me$_2$SiO)$_6$ has a binding energy towards Li$^+$ that is *ca* 24 kcal mol^{-1} smaller than that of 18-crown-6, whereas the binding energy of the inorganic cryptand in **7.19** is comparable to that of (Me$_2$SiO)$_6$.[22,23]

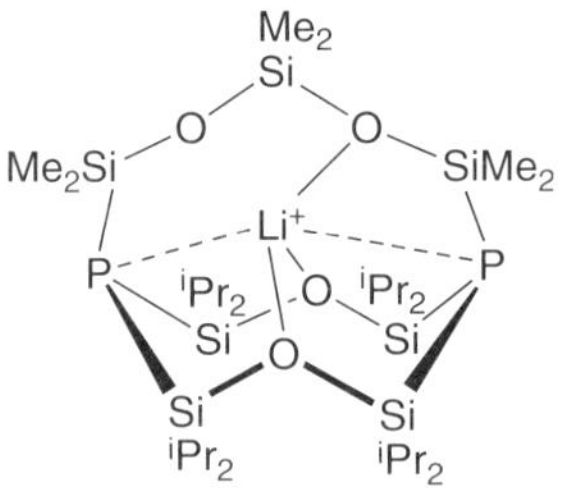

7.19

The larger cyclophosphazenes (NPR$_2$)$_n$ ($n > 4$), which are formally isoelectronic with cyclosiloxanes, may also act as macrocyclic ligands. This behaviour is facilitated by the presence of electron-donating groups, *e.g.* Me or NMe$_2$, on the phosphorus centres. Some examples involving 12- or 16-membered ring systems in which the endocyclic nitrogen atoms are coordinated to metal centres include the adducts N$_6$P$_6$Me$_{12}$·MCl$_2$ (**7.20**, M = Pt or Pd)[24a] and the ionic complex [N$_8$P$_8$Me$_{16}$Co(NO$_3$)][NO$_3$] (**7.21**).[24b]

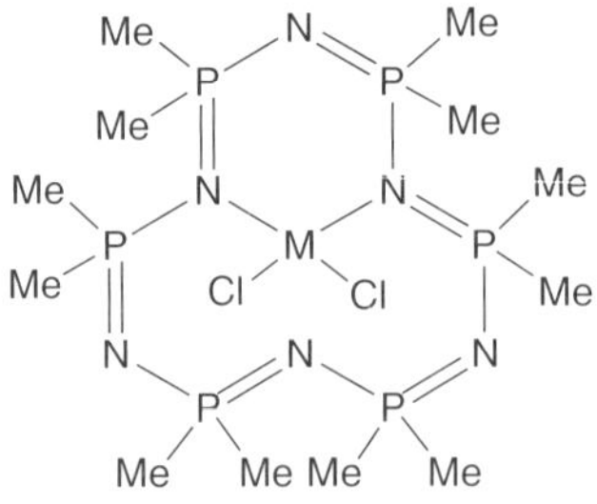

7.20 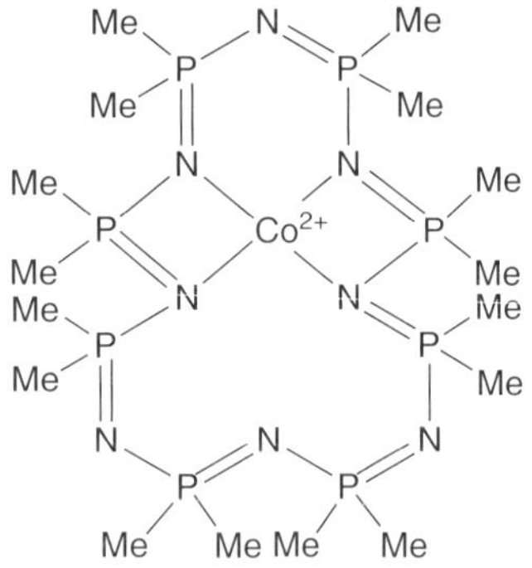   **7.21**

7.2 π-COMPLEXES

The formation of π-complexes between neutral or charged carbocyclic ligands is a well-developed aspect of organometallic chemistry. Although these complexes commonly involve stable carbon-containing ligands such as alkenes or arenes, complexation to a metal centre may also be used to anchor unstable carbocyclic ligands, *e.g.* cyclobutadiene or trimethylenemethane. By analogy with these organometallic systems, the formation of π-complexes between 'aromatic' inorganic ring systems (see Section 4.1) and metal centres might be expected. However, the replacement of the carbon atoms in a carbocyclic ligand by another p-block atom introduces an alternative type of interaction with metals as a result of the presence of available lone pairs on the heteroatom. π-Complexes of inorganic ring systems are much less common than those of their organic counterparts. The difference in electronegativities of neighbouring heteroatoms gives rise to more polar and labile bonds in inorganic ring systems. Consequently, insertion reactions to give metallacyclic systems or ring transformations catalysed by metal centres may occur in preference to the formation of π-complexes.

This section begins with a discussion of metal π-complexes of homocyclic inorganic ring systems, followed by consideration of the types of interactions that are observed between metals and unsaturated inorganic heterocycles. The discussion is limited to inorganic ligands in which all carbon atoms have been replaced by another p-block element. π-Complexes of organic heterocycles containing one or two heteroatoms are well covered in the book by Elschenbroich.[25]

7.2.1 Homocyclic Systems

A number of complexes in which a diphosphene $RP=PR'$ acts as a two-electron ligand towards a transition metal are known. The interaction between the ligand and the metal in such complexes is comparable to the well-known synergic bonding established for alkene–metal complexes. The complex $(^tBuPP^tBu)Ni(^tBuP)_4$ is an interesting example in which, in addition to forming an η^2-complex with the diphosphene ligand, the metal centre is part of a five-membered ring.[26] The anionic $[P_5{}^tBu_4]^-$ reagent, as the sodium salt [see eqn (2.8) for synthesis and Figure 11.10 for structure], is a particularly fruitful source of metal complexes. For example, reaction with $NiCl_2(PEt_3)_2$ produces the cyclopentaphosphene complex $Ni(PEt_3)_2(P_5{}^tBu_3)$ (**7.22**) via elimination of ^tBuCl. The P–P bond length of 2.11 Å in **7.22** is indicative of multiple bond character. Complexes similar to **22** are also formed by palladium.[27] By contrast, reaction of $Na[P_5{}^tBu_4]$ with $NiCl_2(PPh_3)_2$, which contains the more labile triphenylphosphine ligand, gives $[NiCl(cyclo-P_5{}^tBu_4)(PPh_3)]$ (**7.23**) in which the *tert*-butyl substituent is retained (Scheme 7.3).[27]

The closest analogy between carbocyclic and inorganic ligands involves the replacement of all carbon atoms in the ring by silicon. One of the classic developments in organometallic chemistry was the discovery of cyclobutadienyliron tricarbonyl $(\eta^4\text{-}C_4H_4)Fe(CO)_3$ in which the short-lived cyclobutadiene molecule is stabilised by coordination to a metal. More than 40 years later, a silicon analogue of this class of complex, in which the substituent on the Si atoms is the bulky group $^tBu_2MeSi$, was prepared by the route shown in

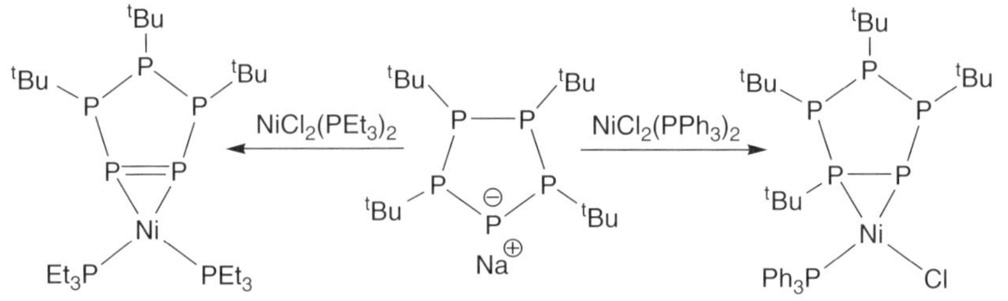

Scheme 7.3 Formation of cyclopentaphosphene and -phosphane complexes from reactions of Na[P$_5^t$Bu$_4$].

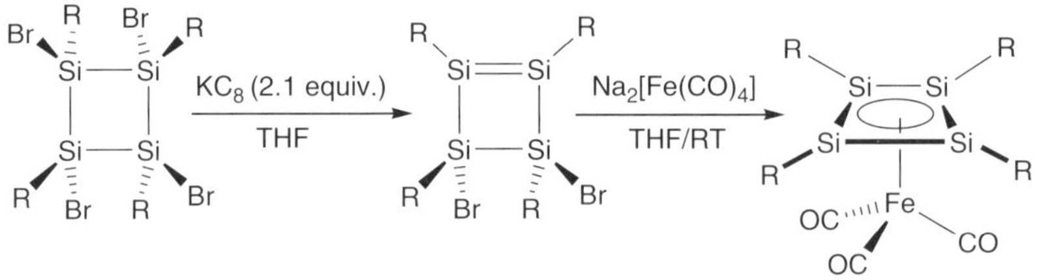

Scheme 7.4 A tetrasilacyclobutadiene complex (R=^{t}Bu$_2$MeSi).

Scheme 7.4.[28] The Si$_4$ ring in the tetrasilacyclobutadiene complex **7.24** is almost planar and coordinated to the Fe atom in a tetrahapto (η^4) fashion. The endocyclic Si–Si bond lengths are approximately equal (2.26–2.28 Å) and the mean value is intermediate between those of single and double bonds. The primary ligand–metal interaction in **7.24** involves the singly occupied π-orbitals of the ligand and the d$_{xz}$ and d$_{yz}$ orbitals of the metal, cf. [(η^4-C$_4$H$_4$)Fe(CO)$_3$].[25] The lower IR stretching frequencies observed for the CO vibrations in **7.24**, compared with those in the cyclobutadiene analogue, indicate that the tetrasilacyclobutadiene ligand is a stronger π-donor than the parent ligand in these Fe(CO)$_3$ complexes.

Homocyclic phosphorus species (and their heavier pnictogen analogues) constitute another class of inorganic ring that can be compared directly with carbocyclic systems. Since a phosphorus atom is formally isovalent with a CH fragment, the cyclic ligands P$_5^-$ and P$_6$ can be viewed as inorganic analogues of the [C$_5$H$_5$]$^-$ anion and C$_6$H$_6$, respectively. In support of this analogy, a variety of sandwich complexes involving these polyphosphorus ligands are known. Some leading examples include the mixed ligand complex Cp*FeP$_5$ (**7.25**), the surprisingly inert homoleptic dianion [(P$_5$)$_2$Ti]$^{2-}$ (**7.26**) and the dinuclear triple-decker complex Cp*Mo(P$_6$)MoCp* (**7.27**).[29,30] The P–P bond lengths in these complexes are intermediate between single and double bond values and computational studies indicate that the inorganic ligand P_5^- is almost as aromatic as $[C_5H_5]^-$. The cyclic arsenic ligands As_5^- and As_6 form similar coordination complexes and also a mononuclear complex $Co(CO)_3(\eta^3\text{-}As_3)$ in which the three As atoms form an equilateral triangle.[31] A more detailed discussion of metal π-complexes of homocyclic anions of P, As, Sb and Bi is given in Section 11.2. The construction of supramolecular assemblies by using complexes of cyclo-P_n ($n = 4$, 5) ligands as building blocks is presented in Section 6.4 (see Figures 6.10 and 6.11).

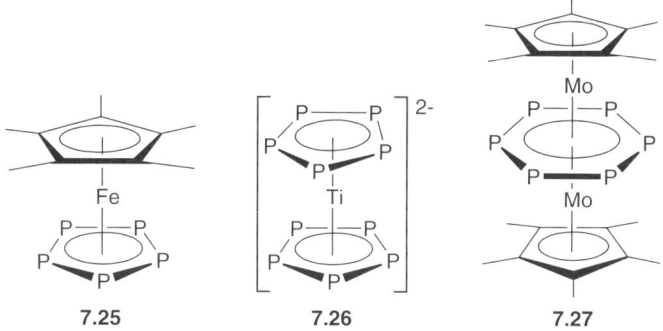

7.25   **7.26**   **7.27**

7.2.2 Heterocyclic Systems

Borazine, $B_3N_3H_6$, is sometimes referred to as 'inorganic benzene', although the extent of aromaticity in this inorganic ring system is still matter of debate (see Section 4.1.2.1). Nevertheless, it is a six π-electron system that might be expected to form stable complexes with 12-electron fragments such as $M(CO)_3$ (M = Cr, Mo, W). In practice, however, stabilisation of borazine–metal complexes requires the presence of electron-donating alkyl substituents on the heteroatoms, e.g. in $(EtBNEt)_3Cr(CO)_3$ (**7.28**) (Figure 7.7).[32] The nitrogen

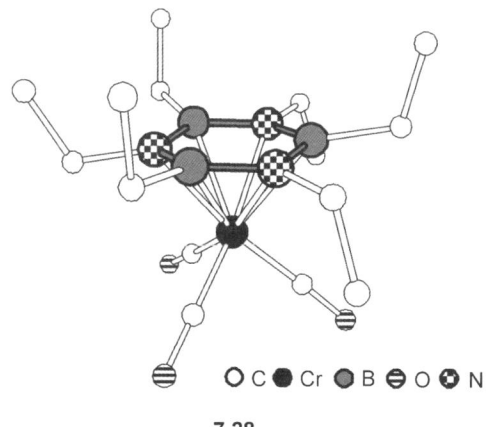

7.28

Figure 7.7   An η^6-borazine–metal complex.

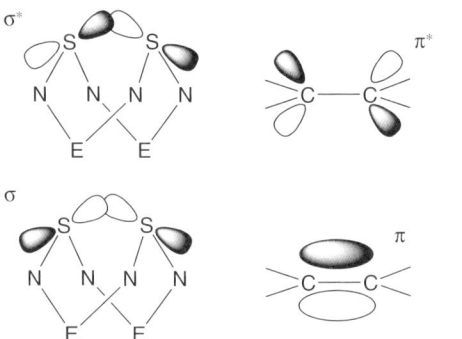

Figure 7.8 Comparison of the σ(S···S) and σ*(S···S) orbitals of eight-membered rings $E_2N_4S_2$ (E = Me$_2$NC, R$_2$P) and the π- and π*-orbitals of an alkene.

atoms are *ca* 0.1 Å closer to the metal than the boron atoms in **7.28**, leading to a slight deviation from planarity towards a chair confirmation.

The majority of the metal complexes formed by the eight-membered $P_2N_4S_2$ ligand (Figure 7.6) are σ-complexes in which a lone pair on the coordinated N or S atoms is donated to the metal centre. The single exception is the (*S,S'*) adduct (Figure 7.6b). This type of η^2-complex is formed by oxidative-addition of the neutral heterocyclic ligands 1,5-(Ph$_2$P)$_2$N$_4$E$_2$ (E = S, Se)[33] or 1,5-(Me$_2$NC)$_2$N$_4$S$_2$[34] to zerovalent Pt and Pd complexes. The bonding in these complexes is comparable to that in the well-known η^2-alkene complexes of d^{10} metals. As indicated in Figure 7.8, the σ- and σ*-orbitals of the transannular S···S bond in 1,5-(Me$_2$NC)$_2$N$_4$S$_2$ are isolobal with the π- and π*-orbitals of an electron-deficient alkene.[35] Thus, the formation of metal complexes involves the synergic interaction of the σ(S···S) and σ*(S···S) orbitals of the inorganic ligand with metal d-orbitals of suitable symmetry; the back-donation from the electron-rich metal to σ*(S···S) is the strongest contributor to the overall bond energy.[34]

REFERENCES

1. M. Baudler and K. Glinka, *Chem. Rev.*, 1993, **93**, 1623.
2. S. J. Geier and D. W. Stephan, *Chem. Commun.*, 2008, 2779.
3. A-J. Dimaio and A. L. Rheingold, *Chem. Rev.*, 1990, **90**, 169.
4. T. S. Cameron, A. Decken, I. Dionne, M. Fang, I. Krossing and J. Passmore, *Chem. Eur. J.*, 2002, **8**, 3386.
5. H. W. Roesky, M. Thomas, J. Schimkowiak, P. G. Jones, W. Pinkert and G. M. Sheldrick, *J. Chem. Soc., Chem. Commun.*, 1982, 895.
6. For a review on weakly coordinating anions, see I. Krossing and I. Raabe, *Angew. Chem. Int. Ed.*, 2004, **43**, 2066.

7. (a) M. B. Hursthouse, K. M. A. Malik and S. N. Nabi, *J. Chem. Soc., Dalton Trans.*, 1980, 355; (b) G. Schmid, R. Greese and R. Boese, *Z. Naturforsch., Teil B*, 1982, **37**, 620.
8. F. A. Cotton, E. V. Dikarev and M. A. Petrukhina, *Angew. Chem. Int. Ed.*, 2001, **40**, 1521.
9. A. Bacchi, W. Baratta, F. Calderazzo, F. Marchetti and G. Pelizzi, *Angew. Chem. Int. Ed. Engl.*, 1994, **33**, 193.
10. (a) Y. Steudel, M. W. Wong and R. Steudel, *Chem. Eur. J.*, 2005, **11**, 1281; (b) M. W. Wong, Y. Steudel and R. Steudel, *Inorg. Chem.*, 2005, **44**, 8908; (c) Y. Steudel, M. W. Wong and R. Steudel, *Eur. J. Inorg. Chem.*, 2005, 2514.
11. V. Chandrasekhar and S. Nagendran, *Chem. Soc. Rev.*, 2001, **30**, 193.
12. (a) A. J. Heston, M. J. Panzer, W. J. Youngs and C. A. Tessier, *Inorg. Chem.*, 2005, **44**, 6518; (b) M. Gonsior, S. Antonijevic and I. Krossing, *Chem. Eur. J.*, 2006, **12**, 1997.
13. M. Bénard, *New. J. Chem.*, 1986, **10**, 539.
14. K. Dehnicke and U. Müller, *Transition Met. Chem.*, 1985, **10**, 361.
15. P. F. Kelly and A. M. Z. Slawin, *J. Chem. Soc., Dalton Trans.*, 1996, 4027.
16. (a) P. F. Kelly and A. M. Z. Slawin, *Angew. Chem. Int. Ed. Engl.*, 1995, **34**, 1758; (b) S. M. Aucott, D. Drennan, S. L. James, P. F. Kelly and A. M. Z. Slawin, *Chem. Commun.*, 2007, 3054.
17. (a) T. Chivers and F. Edelmann, *Polyhedron*, 1986, **5**, 1661; (b) P. F. Kelly and J. D. Woollins, *Polyhedron*, 1986, **5**, 607.
18. T. Chivers and R. W. Hilts, *Coord. Chem. Rev.*, 1994, **137**, 201.
19. J. S. Ritch and T. Chivers, *Angew. Chem. Int. Ed.*, 2007, **46**, 4610.
20. (a) M. R. Churchill, C. H. Lake, S-H. L. Chao and O. T. Beachley, *J. Chem. Soc., Chem. Commun.*, 1993, 1577; (b) C. Eaborn, P. B. Hitchcock, K. Izod and J. D. Smith, *Angew. Chem. Int. Ed. Engl.*, 1995, **34**, 2679.
21. A. Decken, F. A. Leblanc, J. Passmore and X. Wang, *Eur. J. Inorg. Chem.*, 2006, 4033.
22. A. Decken, J. Passmore and X. Wang, *Angew. Chem. Int. Ed.*, 2006, **45**, 2773.
23. C. von Hänisch, O. Hampe, F. Wigend and S. Stahl, *Angew. Chem. Int. Ed.*, 2007, **46**, 4775.
24. (a) N. L. Paddock, T. N. Ranganathan, S. J. Rettig, R. D. Sharma and J. Trotter, *Can. J. Chem.*, 1981, **59**, 2429; (b) K. D. Gallicano, N. L. Paddock, S. J. Rettig and J. Trotter, *Can. J. Chem.*, 1981, **59**, 2435.
25. C. Elschenbroich, *Organometallics*, 3rd edn., Wiley-VCH, Weinheim, 2006.
26. R. A. Jones, M. H. Seeberger and B. R. Whittlesey, *J. Am. Chem. Soc.*, 1985, **107**, 6424.
27. S. Gómez-Ruiz, A. Schisler, P. Lönnecke and E. Hey-Hawkins, *Chem. Eur. J.*, 2007, **13**, 7974.
28. K. Takanashi, V. Y. Lee, M. Ichinohe and A. Sekiguchi, *Angew. Chem. Int. Ed.*, 2006, **45**, 3268.
29. O. J. Scherer, *Angew. Chem. Int. Ed.*, 2000, **39**, 1029.

30. E. Urnezius, W. W. Brennessel, C. J. Cramer, J. E. Ellis and P. v. R. Schleyer, *Science*, 2002, **295**, 832.
31. A-J. Dimaio and A. l. Rheingold, *Chem. Rev.*, 1990, **90**, 169.
32. G. Huttner and B. Krieg, *Chem. Ber.*, 1972, **105**, 3437.
33. T. Chivers, D. D. Doxsee and R. W. Hilts, *Inorg. Chem.*, 1993, **32**, 3244.
34. T. Chivers, K. S. Dhathathreyan and T. Ziegler, *J. Chem. Soc. Chem. Commun.*, 1989, 86.
35. R. T. Boeré, A. W. Cordes, S. L. Craig, R. T. Oakley and R. W. Reed, *J. Am. Chem. Soc.*, 1987, **109**, 868.

CHAPTER 8
Synthesis and Characterisation of Inorganic Polymers

Organic polymers are manufactured and used on a massive scale as plastics and elastomers, films and fibres in areas as diverse as clothing, car tyres, compact discs, packaging materials, prostheses and most recently electroluminescent and electronic devices and sensors.[1-4] The enormous growth in the use of organic polymeric materials since the 1930s can be mainly attributed to their ease of preparation, lightweight nature and unique ease of fabrication.

The incorporation of inorganic elements in the main chain of a polymer is an attractive and complementary method for accessing easily processed materials with novel properties. The most well-developed classes of inorganic polymers with main group elements in the main chain are the polysiloxanes or 'silicones' (**8.1**), polyphosphazenes (**8.2**) and polysilanes (**8.3**).[5,6] Since the latter part of the 20th century, a substantial expansion of the inorganic polymer area has taken place and a variety of other materials has become available. Significantly, these polymers all possess remarkable properties, which differ substantially from those of conventional organic macromolecules.

It should be noted that inorganic rings play a crucial role as monomers for the preparation of both polysiloxanes and polyphosphazenes and they are also utilised for making polysilanes. Inorganic rings have also been used as key precursors to several other inorganic polymer systems [*e.g.* poly(sulfur nitride) and polythionylphosphazenes].

$$\left[\begin{array}{c}R\\|\\-Si-O-\\|\\R\end{array}\right]_n \quad \left[\begin{array}{c}R\\|\\-P=N-\\|\\R\end{array}\right]_n \quad \left[\begin{array}{c}R\\|\\-Si-\\|\\R\end{array}\right]_n$$

8.1   8.2   8.3

Inorganic Rings and Polymers of the p-Block Elements: From Fundamentals to Applications
By Tristram Chivers and Ian Manners
© Tristram Chivers and Ian Manners 2009
Published by the Royal Society of Chemistry, www.rsc.org

This chapter initially provides an overview of the considerations associated with the synthesis of inorganic polymers and the reasons why inorganic rings are so important as polymer precursors. The methods commonly used to characterise polymers are then discussed. As in Chapter 3, which describes the techniques used for the characterisation of inorganic rings, this section focuses on utility rather than on theoretical and practical details of the different methods. The reader is referred to a variety of texts for further details about these polymer structural characterisation techniques.[7–10]

8.1 SYNTHESIS OF INORGANIC POLYMERS

8.1.1 Challenges

It might be expected that transposing the synthetic methods that have worked so well for the preparation of organic polymers would allow similarly facile access to inorganic macromolecules. However, consideration of the main synthetic routes to organic polymers (Scheme 8.1) illustrates the problems associated with this approach.

The most important method for preparing organic polymers is addition polymerisation. This involves a 'chain growth' mechanism with initiation, propagation and, in most cases, also chain transfer and chain termination steps. Chain growth polymerisations involve highly reactive intermediates as initiating or propagating species such as a radical, a cation, an anion or an organometallic

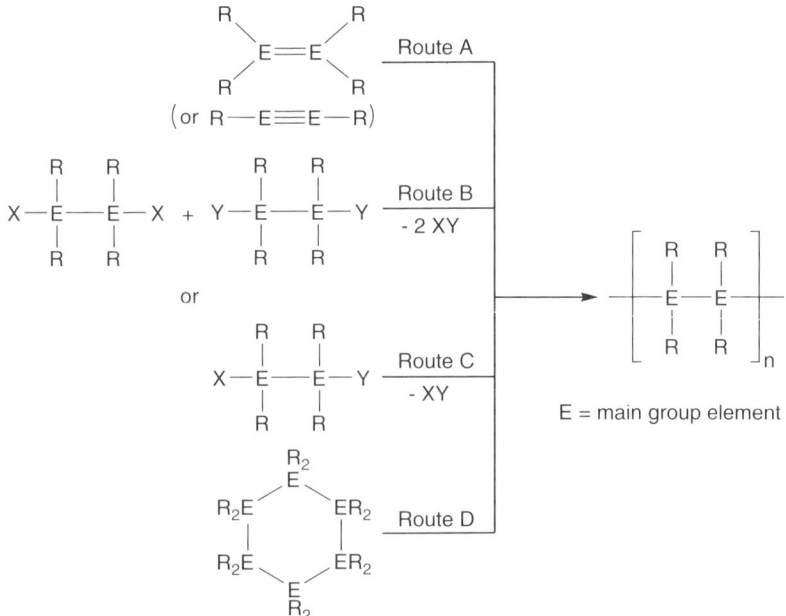

Scheme 8.1 Different potential routes to polymers based on main group elements.

species. The propagation step is usually very rapid and efficient and high molecular weight polymers are easily prepared. Unfortunately, addition polymerisation (Scheme 8.1, Route A) is very difficult to use for inorganic polymer synthesis as suitably reactive but stable multiple bonds involving inorganic elements are generally difficult to prepare. In contrast to the situation for unsaturated organic molecules (α-olefins, acetylenes, isocyanides, *etc.*), the isolation of stable species with multiple bonds between inorganic elements usually requires the presence of sterically demanding, oligomerisation-inhibiting ancillary groups. Indeed, the first well-characterised example of the addition polymerisation of an unsaturated inorganic monomer, the phosphaalkene MesP=CPh$_2$, which affords the polyphosphine (MesP–CPh$_2$)$_n$, was not reported until 2003.[11,12]

Most of the early attempts to prepare inorganic polymers have focused on the use of polycondensation processes (Scheme 8.1, Route B). In these reactions, the build-up of high molecular weight material occurs gradually by a 'step-growth' process with an efficiency that depends critically on reactant stoichiometry and the degree of conversion. These step-growth routes work well for the synthesis of carbon-based polymers when difunctional organic monomers are used because such species are generally easily accessible in a high degree of purity. This allows the stringent stoichiometric balance and conversion requirements that need to be fulfilled for the formation of high molecular weight condensation polymers to be satisfied.[5,13] However, the importance of these criteria, which were first elucidated by Carothers during the 1930s, cannot be overestimated. For example, to prepare a polymer of molecular weight 10 000 by a step-growth polycondensation, the stoichiometric balance between the difunctional monomers with reactive functional groups needs to be 0.98:1.00 or better. Moreover, this assumes that the conversion is 100%. If the reaction does not proceed to completion, then an even more exact stoichiometric balance is required.

Unfortunately, compared with the case for organic chemistry, inorganic functional group chemistry is poorly developed and difunctional inorganic monomers (dilithiated species, for example) are often so reactive that they are difficult to prepare and purify. Therefore, exact reactant stoichiometries necessary for the preparation of high molecular weight polymers using polycondensation reactions cannot be achieved. This generally results in the formation of low molecular weight oligomeric products, which are well below the critical molar mass (typically *ca* 10 000) needed to obtain the necessary interchain entanglement and materials mechanical strength for fabrication into freestanding films, fibres, *etc.* The advantageous processability characteristics associated with macromolecules, which generally constitute the major motivation for making polymers in the first place, cannot therefore be realised.[14]

A very promising variant on this type of condensation polymerisation is to use monomers that possess groups X and Y which can be eliminated from the same molecule (Scheme 8.1, Route C). This circumvents the need for careful control of reaction stoichiometry as an equal number of the different functional groups are 'built in' to the monomer. Furthermore, in certain cases, polymerisation of monomers of this type can follow a 'chain-growth polycondensation' type of mechanism, which, as noted above, leads to high molecular weights much more easily as chain propagation is generally efficient. However, the most common approach to inorganic polymers to date is ring-opening polymerisation (ROP) (Scheme 8.1, Route D). ROP processes usually occur via chain-growth mechanisms that yield high molecular weights and, as inorganic ring chemistry is very well developed, many potential ROP monomers are available.

8.1.2 Ring-opening Polymerisation (ROP)

8.1.2.1 Thermodynamic Considerations. Most polymerisations are characterised by a reduction in entropy as a large number of monomer molecules with the freedom to move in three dimensions are joined together and ultimately constrained to a linear 1D polymer chain, where motion is much more restricted. Hence, for a ROP to be thermodynamically favourable under such circumstances, ring strain is usually needed to provide an enthalpic driving force, ΔH_{ROP}, that overcomes the unfavourable $T\Delta S_{ROP}$ term that contributes to ΔG_{ROP} [eqn (8.1)].

$$\Delta G_{ROP} = \Delta H_{ROP} - T\Delta S_{ROP} \qquad (8.1)$$

In the case of some cyclics, such as the thionylphosphazene six-membered ring [NS(O)Cl(NPCl$_2$)$_2$], the small degree of ring-strain makes the favourable ΔH_{ROP} and unfavourable $T\Delta S_{ROP}$ terms finely balanced.[15] In the presence of a catalyst such as GaCl$_3$, ROP can be favoured at high concentrations and depolymerisation to the cyclic monomer at low concentrations [eqn (8.2)].

(8.2)

A rare exception to the dominance of enthalpic driving force is provided by the unstrained eight-membered ring (Me$_2$SiO)$_4$, which undergoes ROP to give poly(dimethylsiloxane). For this process, ΔH_{ROP} is approximately zero but the remarkable skeletal flexibility of the polysiloxane backbone makes the $T\Delta S_{ROP}$ term favourable and *entropy* provides the driving force for ROP. Exceptional behaviour occurs in the extremely rare cases of monomers where the polymerisation is driven by entropy and ΔH_{ROP} is unfavourable. In these cases, the formation of polymer is favoured at elevated temperatures. For example, cyclo-S$_8$ will polymerise above 150 °C (where the favourable $T\Delta S_{ROP}$ term dominates), but slowly depolymerises back to the cyclic S$_8$ monomer at room temperature.[6]

8.1.2.2 Mechanisms. A range of mechanisms has been either proposed or established for the ROP of inorganic rings depending on the nature of the highly reactive propagating site. A cationic mechanism has been postulated for the thermal ROP of cyclic phosphazenes such as $(NPCl_2)_3$ in the melt at 250 °C (Scheme 8.2).[6] Evidence in favour of this mechanism includes the observation that the addition of halide acceptors (such as $AlCl_3$) slightly lowers the temperature required for polymerisation. In the case of the addition of silylium ions $[R_3Si]^+$, ROP can be achieved at room temperature.[16] In both cases high molecular weights (10^4–10^6 g mol^{-1}) and broad molecular weight distributions result.

Anionic ROP mechanisms have been established for cyclic siloxanes such as $(Me_2SiO)_3$ and $(Me_2SiO)_4$ (Scheme 8.3) and the four-membered cyclic silane

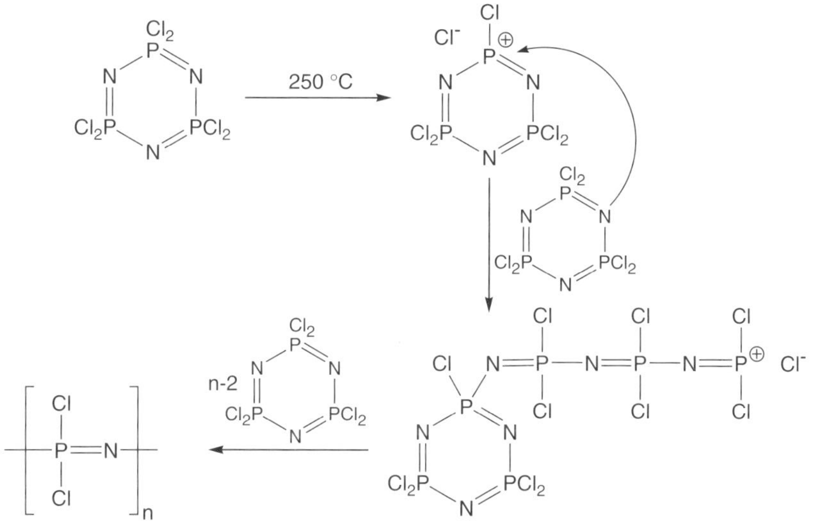

Scheme 8.2 Proposed mechanism for the ROP of the cyclic phosphazene $(NPCl_2)_3$.

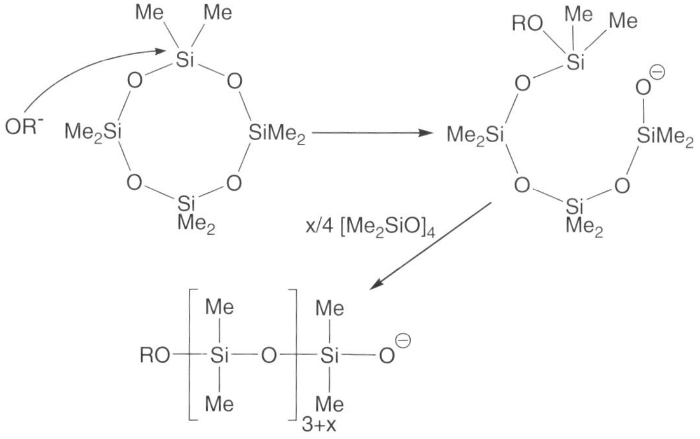

Scheme 8.3 Proposed mechanism for the ROP of the cyclic siloxane $(Me_2SiO)_4$.

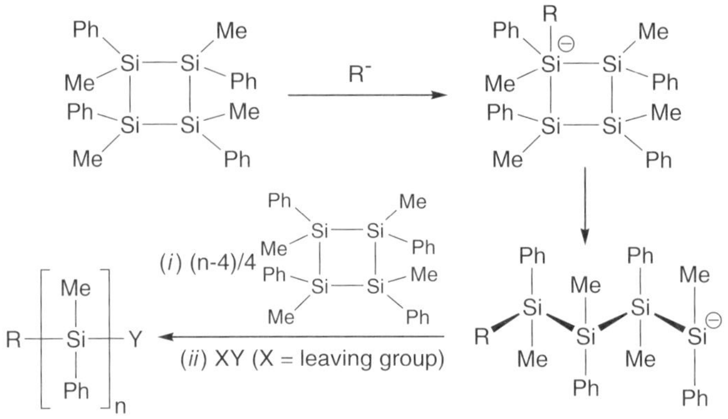

Scheme 8.4 Proposed mechanism for the ROP of the cyclic silane (MePhSi)$_4$.

(MePhSi)$_4$ (Scheme 8.4), and these yield high molecular weight materials (molecular weights up to 10^6 g mol^{-1}).[6,17] On the other hand, a radical mechanism is believed to operate for the solid-state ROP of the four-membered ring S$_2$N$_2$ (see Section 5.4.3 and Figure 5.9).

In the case of the anionic polymerisation of six-membered cyclic siloxanes (R$_2$SiO)$_3$ using initiators such as BuLi or K[OSiMe$_3$], the ROP proceeds without chain transfer and chain termination reactions that broaden molecular weight distributions. This type of polymerisation is termed a *living polymerisation*[18] and provides samples of polysiloxanes with very narrow molecular weight distributions as initiation is rapid (Section 8.2.2). Moreover, in a living polymerisation each initiator molecule generates a single polymer chain and the molecular weight can be controlled by the initiator:monomer ratio. The absence of chain transfer and chain termination reactions allows the resulting living anionic polymer to be used in further chemistry to prepare well-defined structures; for example, in the reaction with a different monomer to generate block copolymers.[19]

8.2 CHARACTERISATION OF INORGANIC POLYMERIC MATERIALS

8.2.1 Structural Characterisation

The chemical structure of a polymer can be analysed by many of the techniques used to characterise molecular species (see Chapter 3). Multinuclear NMR, IR and UV–visible spectroscopy, for example, are widely used key characterisation tools. Most polymers will dissolve in at least some readily available solvents (although the rate of dissolution may be slow due to chain entanglement effects). In cases where polymers are insoluble, solid-state NMR techniques can be used to provide excellent structural characterisation. Due to structural imperfections, unknown end groups and incomplete combustion problems as a result of ceramic formation (Section 8.2.5), elemental analysis data obtained by combustion techniques are often significantly less accurate than those for molecular species. Many of the usual forms of mass spectrometry (such as electron impact and electrospray) are not useful as polymers are involatile under most circumstances. This problem can be circumvented for high molecular weight species using matrix-assisted laser desorption/ionisation-time-of-flight (MALDI-TOF) techniques (Section 8.2.2). This method has been successfully used to characterise polymers with molecular weights up to *ca* 100 000 and often repeat units can be identified from the resulting spectra.

8.2.2 Molecular Weights and Molecular Weight Distributions

Samples of synthetic polymers are generally formed by polymerisation reactions where both the start and end of the growth of the macromolecular chain are uncontrolled and are relatively random events. Even chain transfer reactions, where, for example, one polymer chain stops growing and in the process induces another to begin, are prevalent in many systems. Synthetic polymer samples therefore contain macromolecules with a variety of different chain lengths and are termed *polydisperse*. For this reason, the resulting molecular weight distribution is characterised by *average molecular weights*. The two most common are the *weight-average molecular weight*, M_w, and the *number-average molecular weight*, M_n.

The quantity M_w/M_n is termed the *polydispersity index* (PDI) and measures the breadth of the molecular weight distribution and takes a value ≥ 1. In the case where long polymer chains are of the same length, $M_w = M_n$ (*i.e.* PDI = 1), the sample is termed *monodisperse*. Such perfect situations are rare except for the case of biological macromolecules, but essentially monodisperse systems also occur with synthetic polymers where the polymerisation used for the preparation is termed *living*. In such cases, initiation is rapid and no termination or chain transfer reactions occur; under such conditions, the polymer chains initiate together at the same instant and grow until the monomer is completely consumed, resulting in macromolecular chains of the same length.[18] In practice, living systems are not perfect; for example, very slow termination reactions generally occur. This leads to polymer samples that are of narrow polydispersity (1.0 < PDI < 1.2), rather than perfectly monodisperse (PDI = 1.0). Living polymerisation systems are of widespread interest because they allow the formation of controlled polymer architectures. For example, unterminated chains can be subsequently reacted with a different monomer to form block copolymers. These materials exhibit many fascinating properties as a result of their ability to self-assemble into phase-separated nanoscopic domains derived from the constituent blocks.[19]

A variety of different experimental techniques exist for the measurement of M_w and M_n.[7-10] Some afford absolute values and others give estimates that are relative to calibration standards. One of the simplest techniques for obtaining a measurement of the complete molecular weight distribution of a polymer is gel permeation chromatography (GPC) [also known as size-exclusion chromatography (SEC)]. This method affords information on the complete molecular weight distribution and also values of both M_w and M_n (and hence the PDI). Unfortunately, the molecular weights obtained are relative to the polymer standard used to calibrate the instrument (such as narrow polydispersity samples of polystyrene) unless special adaptations of the experiment are made or if standard monodisperse samples of the polymer being studied are also available as references.

Light-scattering measurements are generally time consuming but permit absolute values of M_w to be obtained in addition to a wealth of other information concerning the effective radii of polymer coils in the solvent used, polymer–solvent interactions and polymer diffusion coefficients. The introduction of light-scattering detectors for GPC instruments has now made it possible for both absolute molecular weights and molecular weight distributions to be determined routinely.

As noted above, mass spectrometric techniques such as MALDI-TOF have now been developed to the stage where they are extremely useful for analysis of the molecular weights of polymers. This technique can give molecular ions for macromolecules with molecular weights substantially greater than 10 000 and values up to 100 000 have been successfully determined.[20]

Although most polymer samples possess a single molecular weight distribution by GPC and are termed *monomodal*, for some, the molecular weight distribution actually consists of several individual, resolvable distributions. In such cases, the molecular weight distribution is referred to as *multimodal*. For example, if a higher and a lower molecular weight fraction can be distinguished, then the distribution is termed *bimodal*. Examples of broad and narrow monomodal molecular weight distributions are shown in Figure 8.1a and b, respectively.

8.2.3 Thermal Transitions: Amorphous and Crystalline Polymers

As polymer chains are usually long and flexible, they would be expected to pack randomly in the solid state to give an amorphous material. This is true for

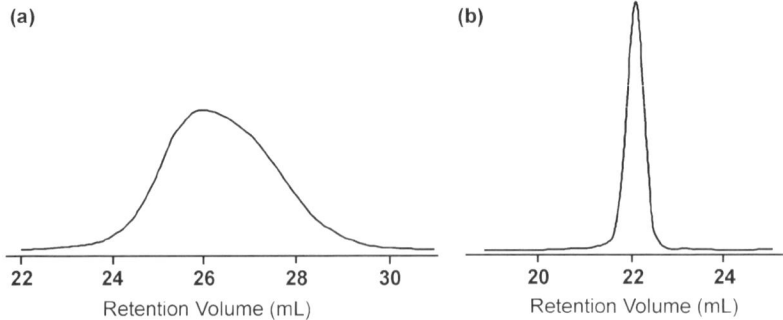

Figure 8.1 Examples of GPC traces showing (a) a broad monomodal molecular weight distribution (PDI = 2.3) and (b) a narrow monomodal molecular weight distribution (PDI = 1.05). The *x*-axis shows the elution volume for the GPC instrument with molecular weight increasing from right to left.

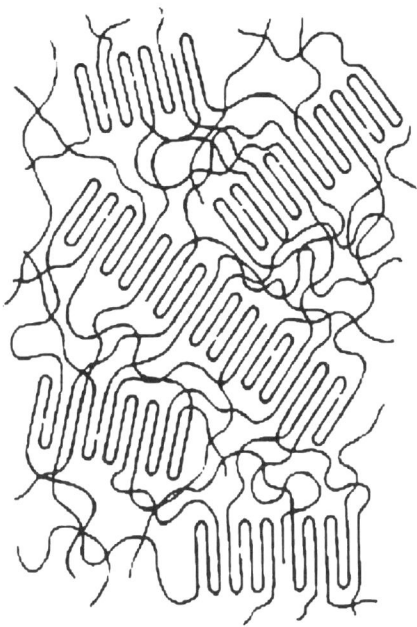

Figure 8.2 Model of a semicrystalline polymer showing ordered crystalline regions and disordered amorphous regions. Reproduced with permission from S.L. Rosen, *Fundamental Principles of Polymeric Materials*, 2nd edn., Wiley-Interscience, New York, 1993.[13]

many polymers, particularly those with an irregular chemical structure. However, polymer chains that have highly regular structures, *e.g.* linear polyethylene and inorganic polymers such as poly(dimethylsiloxane) or poly[bis(trifluoroethoxyphosphazene)], or intermolecular interactions such as hydrogen bonding, *e.g.* polyamides as illustrated by nylons, can pack together in an ordered manner to give crystallites. In general, perfect single crystals are not formed by long polymer chains for entropic reasons and polymers that crystallize are therefore often more correctly referred to as *semicrystalline*, as amorphous regions are also present (see Figure 8.2).[7]

Information on the morphology of polymers is revealed by techniques such as powder X-ray diffraction (PXRD), which is often called wide-angle X-ray scattering (WAXS) by polymer scientists, and small-angle X-ray scattering (SAXS).[8,10,21] The crystallites exist in a polymer sample below the *melting temperature* (T_m), an order–disorder transition, above which a viscous melt is formed.

The presence of crystallites is important as it can lead to profound changes in the properties and applications of a polymeric material. For example, crystallites are often of the appropriate size to lead to the scattering of visible light and cause the material to appear opaque. In addition, crystallinity usually leads to a significant increase in mechanical strength, but also brittleness. Gas permeability generally decreases, as does solubility in organic solvents, because an additional lattice energy term must be overcome for dissolution to occur. Polymers that crystallise tend to form high-quality fibres, which is the basis for many applications such a nylon cables and clothing. On the other hand, for applications as clear plastics or elastomers, a high degree of crystallinity is not desirable as it leads to turbidity and rigidity.

In addition to the melting temperature (T_m), which arises from the order–disorder melting transition for crystallites in a polymer sample, amorphous regions of a polymer show a *glass transition* (T_g). This second-order thermodynamic transition is not characterised by an exotherm or endotherm, but rather a change in heat capacity, and is related to the onset of large-scale conformational motions of the polymer main chain. Generally, stiff polymer chains and large, rigid side groups generate high T_g values. Below the T_g an amorphous polymer is a glassy material, whereas above the T_g it behaves like a viscous gum, as the polymer chains are capable of reptating motion past one another. By linking the polymer chains together through crosslinking reactions, rubbery elastomers can be generated from low-T_g polymers. Purely amorphous polymers show only a glass transition whereas semicrystalline polymers show both a T_m and a T_g. Semicrystalline polymeric materials are rigid plastics below the T_g and become more flexible above the T_g. Above the T_m, a viscous melt is formed. Some illustrative examples of inorganic polymers and their corresponding thermal transition data are shown in Table 8.1.

Significantly, for macromolecular materials the rate of polymer crystallisation can be extremely slow and polymers that can potentially crystallise are often isolated in a kinetically stable, amorphous state. A potentially crystallisable polymer that is in an amorphous state can show an exothermic *crystallisation transition* (T_c) at elevated temperatures. The thermal transitions of a polymer are commonly investigated by the technique of differential scanning

Table 8.1 Thermal transitions for selected inorganic polymers.

Polymer	T_m (°C)	T_g (°C)	Type of material (at 25 °C)
$(Me_2SiO)_n$	−40	−123	Gum
$(MeHSiO)_n$	—	−137	Gum
$[Me(CH_2CH_2CF_3)SiO]_n$	—	−70	Gum
$[NP(OCH_2CF_3)_2]_n$	242	−66	Semicrystalline, film- and fibre-forming material
$[NP(OCH_2CH_2CH_3)_2]_n$	—	−100	Amorphous gum
$[NP(NHPh)_2]_n$	—	91	Amorphous glass
$[NP(OPh)_2]_n$	390	−8	Semicrystalline, film- and fibre-forming material
$(MePhSi)_n$	—	117	Glassy material
$[NSO(NHC_4H_9)\{NP(NHC_4H_9)_2\}_2]_n$	—	−15	Gum

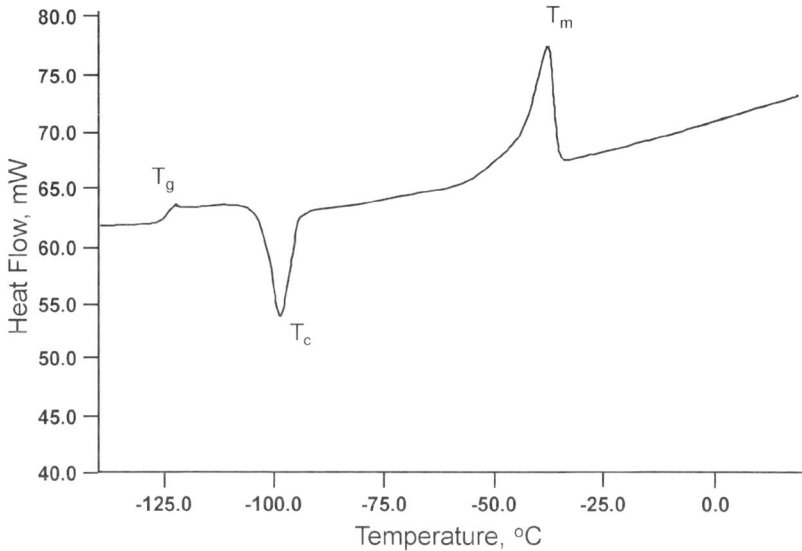

Figure 8.3 A DSC trace showing the thermal transitions T_g (−123 °C), T_m (−40 °C) and T_c (−100 °C) for poly(dimethylsiloxane) (PDMS) $(Me_2SiO)_n$ on heating a sample from −140 to 25 °C. Heating rate 10 °C min^{-1}.

calorimetry (DSC).[8] A typical DSC trace showing a T_g, a T_c and a T_m is shown for poly(dimethylsiloxane) in Figure 8.3. This material has a very symmetrical structure with identical substituents at silicon and so crystallisation is possible. However, as the backbone is so flexible the T_g is remarkably low (−123 °C) and the T_m is below ambient temperature (−40 °C).

8.2.4 Chain Conformations

X-ray diffraction techniques can be used to establish the structure of crystalline polymers. Measurements are typically made on crystalline lamellar platelets grown from dilute solution, fibres or stretched films. Such methods have been applied to several different inorganic polymers. For example, based on measurements on stretched samples of silicone rubber, poly(dimethylsiloxane) $(Me_2SiO)_n$ has been shown to possess a helical conformation (Figure 8.4).[22]

Other crystalline inorganic polymers such as poly(dichlorophosphazene), poly(aryloxyphosphazenes), liquid crystalline polysiloxanes and poly(dichlorosilane) have also been studied by X-ray diffraction methods, enabling the conformations in the crystallites in the solid state to be established.[23-26]

8.2.5 Thermal Stability: Thermogravimetric Analysis

The ability of polymers to withstand elevated temperatures is critical to many applications. For example, the excellent thermal stability of polysiloxanes is the basis for many of their uses, including their application in low molecular weight

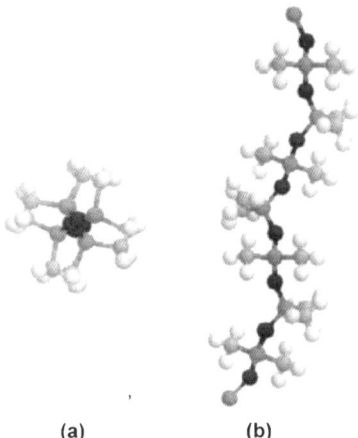

(a)   (b)

Figure 8.4 Model of a poly(dimethylsiloxane) [PDMS, $(Me_2SiO)_n$] chain with a four-fold helical conformation. (a) Top view; (b) side view (oxygen atoms are in black). Reproduced with permission from P.-A. Albouy, *Polymer*, 2000, **41**, 3083.[22]

form as fluids or 'silicone oils' in the laboratory. Significantly, an interesting feature of many inorganic polymers compared with organic counterparts involves their thermal conversion to involatile ceramic solids rather than CO_2 and water and other molecular products at elevated temperatures in air. Polymers such as polysilanes, polycarbosilanes and polyborazylenes have been developed commercially as preceramic polymers where the ceramic yield is so high ($>90\%$) that useful-shaped ceramic objects can be obtained by the pyrolysis of shaped polymer precursors. The thermal stability of a polymer can be analysed by thermogravimetric analysis (TGA), which allows weight loss to be measured as a function of temperature.[6] Care has to be taken in the interpretation of TGA data as the quantity measured is weight loss and if a polymer decomposes to afford involatile oligomeric products before losing mass this will not be made apparent. The most informative analyses therefore also require the residues from thermal treatment to be carefully characterised. In addition, information on the volatile products produced as a function of temperature can be usefully obtained by coupling a TGA to a mass spectrometer.

REFERENCES

1. J. Alper and G. L. Nelson, *Polymeric Materials: Chemistry for the Future*, American Chemical Society, Washington, DC, 1989.
2. H. Shirakawa, *Angew. Chem. Int. Ed.*, 2001, **40**, 2574.
3. A. G. MacDiarmid, *Angew. Chem. Int. Ed.*, 2001, **40**, 2581.
4. A. J. Heeger, *Angew. Chem. Int. Ed.*, 2001, **40**, 2591.
5. I. Manners, *Angew. Chem. Int. Ed. Engl.*, 1996, **35**, 1602.

6. J. E. Mark, H. R. Allcock and R. West, *Inorganic Polymers*, 2nd edn., Oxford University Press, New York, 2005.
7. L. H. Sperling, *Introduction to Physical Polymer Science*, Wiley, New York, 2001.
8. H. R. Allcock, F. W. Lampe and J. E. Mark, *Contemporary Polymer Chemistry*, 3rd edn., Pearson/Prentice Hall, Upper Saddle River, NJ, 2003.
9. I. M. Campbell, *Introduction to Synthetic Polymers*, 2nd edn., Oxford University Press, New York, 2000.
10. P. C. Hiemenz and T. P. Lodge, *Polymer Chemistry*, 2nd edn., CRC Press, Boca Raton, FL, 2007.
11. C-W. Tsang, M. Yam and D. P. Gates, *J. Am. Chem. Soc.*, 2003, **125**, 1480.
12. For a further example involving the addition polymerisation of species with Ge=C bonds, see L. C. Pavelka, S. J. Holder and K. M. Baines, *Chem. Commun.*, 2008, 2346.
13. See, for example, S. L. Rosen, *Fundamental Principles of Polymeric Materials*, 2nd edn., Wiley-Interscience, New York, 1993.
14. See, for example, I. M. Campbell, *Introduction to Synthetic Polymers*, Oxford University Press, New York, 1994, pp. 36–39.
15. A. R. McWilliams, D. P. Gates, M. Edwards, L. M. Liable-Sands, I. Guzei, A. L. Rheingold and I. Manners, *J. Am. Chem. Soc.*, 2000, **122**, 8848.
16. Y. Zhang, K. Huynh, I. Manners and C. A. Reed, *Chem. Commun.*, 2008, 494.
17. M. Cypryk, Y. Gupta and K. Matyjaszewski, *J. Am. Chem. Soc.*, 1991, **113**, 1046.
18. G. G. Odian, *Principles of Polymerisation*, Wiley, New York, 2004.
19. I. Manners, *Angew. Chem. Int. Ed.*, 2007, **46**, 1565.
20. H. Pasch and W. Schrepp, *MALDI-TOF Mass Spectrometry of Synthetic Polymers*, Springer-Verlag, Berlin, 2003.
21. H. Schnablegger and Y. Singh, *A Practical Guide to SAXS*, Anton Paar, Graz, 2006.
22. P-A. Albouy, *Polymer*, 2000, **41**, 3083.
23. Y. Chatani and K. Yatsuyanagi, *Macromolecules*, 1987, **20**, 1042.
24. T. Miyata, K. Yonetake and T. Masuko, *J. Mater. Sci.*, 1994, **29**, 2467.
25. K-J. Lee, G-H. Hsiue, J-L. Wu and Y-A. Sha, *Polymer*, 2007, **48**, 5161.
26. J. R. Koe, D. R. Powell, J. J. Buffy, S. Hayase and R. West, *Angew. Chem. Int. Ed.*, 1998, **37**, 1441.

CHAPTER 9
Group 13: Rings and Polymers

Ring systems involving group 13 elements are important both from the fundamental perspective and as precursors to useful inorganic materials. The historical significance of borazine, $B_3N_3H_6$, was mentioned in the introductory chapter and the issue of π-electron delocalisation in this seminal inorganic ring system is discussed in Section 4.1.2. In this (and the succeeding) chapter(s), the ring systems of the p-block elements (groups 13–16) will be considered in more detail. Each chapter starts with a description of the chemistry of homoatomic systems, followed by a discussion of heterocycles involving the combination of an element of the group under consideration and another p-block element. In keeping with the title and scope of the book, the coverage is limited to monocyclic systems, except in cases where the formation of bicyclic or cluster molecules is closely connected with the chemistry of the monocycles, *e.g.* the existence of structural isomers of the monocyclic ring.

9.1 HOMOATOMIC SYSTEMS

9.1.1 Neutral Rings

Boron(I) compounds BX (X = Cl, Br) are thermodynamically unstable; they readily oligomerise to form molecules such as B_4Cl_4, B_8Cl_8 and B_9Br_9.[1] In the solid state, these oligomers form polyhedral frameworks rather than ring systems involving three-coordinate boron atoms and two-electron, two-centre bonds, *e.g.* B_4Cl_4 is tetrahedral and B_8Cl_8 forms a square antiprism. These *closo* systems violate Wade's rules for the prediction of cluster structures since, in each case, there are n cluster electron pairs for n cluster atoms. Wade's rules require $n+1$ electron pairs for the formation of a *closo* system. The exceptions for these electron-deficient systems arise from a combination of (a) the ability of the halogen substituents to donate p(π) electron density to the cluster and

---

Inorganic Rings and Polymers of the p-Block Elements: From Fundamentals to Applications
By Tristram Chivers and Ian Manners
© Tristram Chivers and Ian Manners 2009
Published by the Royal Society of Chemistry, www.rsc.org

Group 13: Rings and Polymers 111

(b) the occurrence of closed three-centre bonding on the face of the polyhedron. Interestingly, however, in solution the tetramers B_4X_4 (X = Cl, Br) exhibit ^{11}B NMR chemical shifts that are consistent with three-coordinate boron, *i.e.* four-membered rings, rather than the *closo* polyhedral structures that are observed in the solid state.[2]

The alkyl derivative $(B^tBu)_4$ is prepared either by the reaction of B_4Cl_4 with Li^tBu[3a] or, in high yield as a yellow solid, by reductive coupling of $^tBuBF_2$ with Na–K alloy in pentane.[3b] In the solid state $(B^tBu)_4$ adopts a slightly distorted tetrahedral structure with B–B bond lengths in the range 1.70–1.71 Å. In solution it exhibits a single ^{11}B NMR resonance at *ca* 135 ppm. Another route to $(B^tBu)_4$ involves the reductive coupling of the diboron reagent $Cl(^tBu)B-B(^tBu)Cl$ with Na–K alloy in THF. Under these conditions, the anion radical $[(B^tBu)_4]^{-\bullet}$, identified by EPR spectroscopy, is also generated.[4] The EPR spectrum for the major isotopomer $[(^{11}B^tBu)_4]^{-\bullet}$ gives rise to a 13-line multiplet as a result of coupling of the unpaired electron with four equivalent ^{11}B nuclei ($I = 3/2$) (see Section 3.6).

The replacement of the halogen substituents in $(BX)_n$ by the strongly π-donating dialkylamino groups stabilises the ring systems cyclo-$(BNR_2)_n$. However, both theoretical (DFT) calculations[5] and experimental results[6] indicate that the energy difference between the classical non-planar rings and the isomeric cage structures is not large. The first example to be reported, the orange–red hexamer $(BNMe_2)_6$ (**9.1**, R = Me), was obtained unexpectedly in very low yields and shown to adopt a chair conformation in the solid state with mean B–B bond lengths of 1.70 Å and mean B–N bond lengths of 1.40 Å.[7] Subsequently, a designed synthesis of this class of homoatomic ring system by the reductive coupling of Et_2NBCl_2 with potassium in boiling cyclohexane produced a mixture of oligomers $(BNEt_2)_3$, $(BNEt_2)_4$ and cyclo-$(BNEt_2)_6$, and also *closo*-$(BNEt_2)_6$ (**9.2**, R = Et), which has an octahedral cage structure.[8] The structures of the trimer and tetramer have not been established. Both cyclic hexamers $(BNR_2)_6$ (**9.1**, R = Me, Et) display resonances in the ^{11}B NMR spectrum at 65 ppm. In solution at room temperature, the cyclic hexamer **9.1** (R = Et) slowly rearranges to the *closo* isomer **9.2** (R = Et).

9.1   9.2

Interestingly, an attempt to produce the trimer $(BNMe_2)_3$ by reductive coupling of monoboron and diboron fragments generated (in very low yield) a colourless compound **9.3** that had the molecular mass of the hexamer

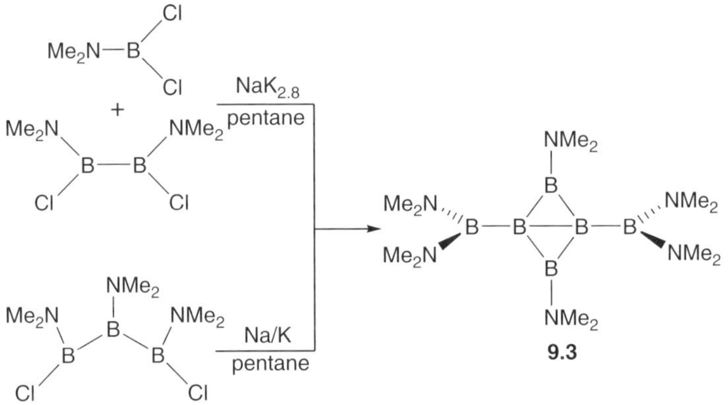

Scheme 9.1 Synthetic routes to the bicyclic isomer of $B_6(NMe_2)_6$.

$(BNMe_2)_6$ (by EI-MS), but exhibited three ^{11}B NMR signals at 6, 41 and 63 ppm.[9] The hexamer **9.3** is obtained in 40–50% yields by the reaction of the acyclic triboron building block $Cl(Me_2N)B(Me_2N)BB(NMe_2)Cl$ with Na–K alloy in pentane (Scheme 9.1).[10] A crystal structure analysis of **9.3** revealed a planar diamond-shaped B_4 ring with a short transannular B–B distance of 1.63 Å, indicating a B–B bond.[9] The pairs of B–B bonds on opposite sides of the diamond have lengths of 1.60 and 1.63 Å. The planes of the $(Me_2N)_2B$ substituents are nearly perpendicular to the B_4 ring. The B–N bonds in these substituents are ca 0.05 Å longer than the other B–NMe$_2$ bond lengths. In order to account for the formation of five B–B bonds in the B_4 diamond with only eight electrons available, a non-classical bonding description is required.[9]

Reductive coupling of the diboron reagents $Cl(R_2N)B–B(NR_2)Cl$ with Na–K alloy in hexane also generates the tetramers $(BNR_2)_4$ [**9.4**, R = ^{i}Pr; **9.5**, NR$_2$ = 2,2′,6,6′-tetramethylpiperidyl (TMP)].[6] The isopropyl derivative $(BN^iPr_2)_4$ is obtained as blue crystals ($\lambda_{max} = 620$ nm). The crystal structure analysis reveals a bent B_4 ring with a fold angle of ca 59° and equal B–B bond lengths of 1.71 Å. The exocyclic B–N bond lengths of ca 1.40 Å are indicative of a strong π-interaction. In solution, an ^{11}B NMR resonance is observed at 65 ppm, in the same region as for cyclo-$(BNR_2)_6$ (R = Me, Et) (see above). The steric influence of the dialkylamino substituents is evident upon replacement of the N^{i}Pr$_2$ groups by TMP. Like $(B^tBu)_4$, the tetramer **9.5** is yellow. In the solid state it adopts a tetrahedral structure with two short B–B bonds (1.70 Å) and four longer B–B bonds (1.75–1.76 Å).[6] The B–N bond lengths of ca 1.45 Å indicate significantly weaker π-bonding than that in **9.3**. Apparently, the bulky TMP ligands impose complete folding of the B_4 tetrahedron. These experimental results are consistent with the theoretical prediction that the cage and cyclic isomers of the model tetramer $[B(NH_2)]_4$ are close in energy.[5]

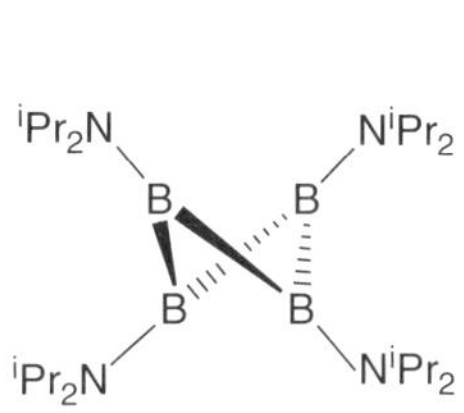

9.4 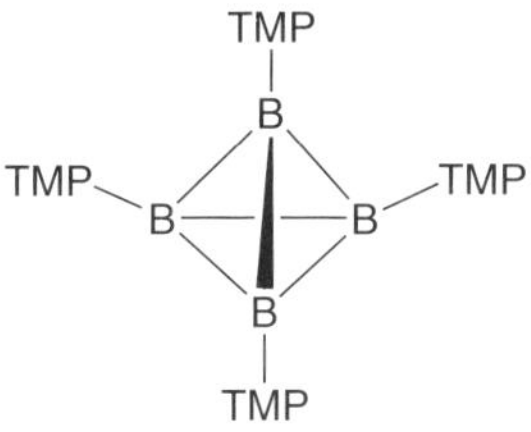 **9.5**

For the heavier group 13 elements, there is an extensive cluster chemistry of neutral oligomers of the type $(MX)_n$ (X = halide), $(MR)_n$ (R = alkyl) or $(MNR_2)_n$.[11] For example, the series of tetramers $[MC(SiMe_3)_3]_4$ (M = Al, Ga, In, Tl) all form tetrahedral structures,[12] *cf*. $(B^tBu)_4$. In the formal +1 oxidation state, these elements do not form monocyclic ring systems even with the stabilising influence of dialkylamino substituents.

9.1.2 Anionic Rings

Cyclic structures have been predicted for anionic polyboron clusters of the type B_n^- (n = 3–15) and B_n^{2-} (n = 6, 8). However, these species have only been characterised in the gas phase by photoelectron spectroscopy; they have not yet been identified in the condensed phase.[13] Highly charged boranes, *e.g.* $[B_6H_6]^{6-}$, stabilised by six Li^+ cations, have also been predicted to represent a new family of planar aromatic species on the basis of theoretical calculations.[14] Significantly, the planar cyclic B_6 fragment exists in magnesium boride MgB_2, which is a high-temperature superconductor.[15]

Sandwich complexes in which highly charged, cyclic polyboron ligands are stabilised by coordination to a transition metal have been structurally characterised in the solid state.[16] The planar B_5Cl_5 and $B_6H_4Cl_2$ ligands in the triple-decker rhenium complexes **9.6** and **9.7** are thought to acquire six electrons from the transition metal centres and may therefore be regarded as analogues of the boron hydride hexaanions $[B_nH_n]^{6-}$ (n = 5, 6).

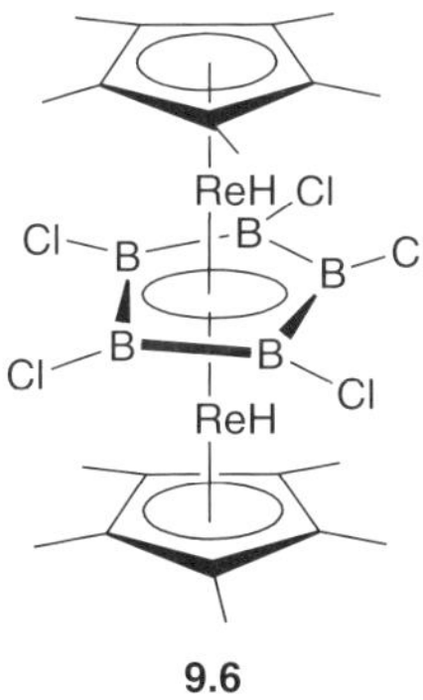

9.6 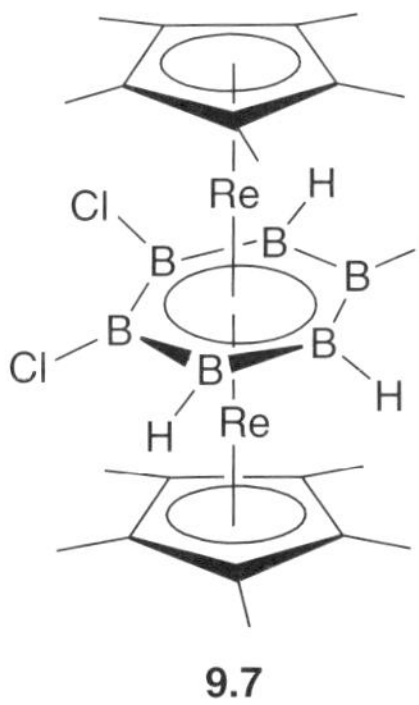 **9.7**

The square-planar dianions M_4^{2-} (M = Al, Ga, In) are predicted by *ab initio* calculations to be two π-electron aromatic species. These species are stabilised in the gas phase as the alkali metal derivatives $[NaM_4]^-$, which have been characterised by photoelectron spectroscopy.[17] Computational studies indicate that early transition metals of the first, second and third rows can form stable sandwich complexes of the type $[Al_4MAl_4]^{x-}$ ($x = 0-2$ and M = Ti, V, Cr, Zr, Nb, Mo, Hf, Ta and W) in which the properties of the square-planar Al_4^{2-} ligand remain essentially unchanged.[18] The monoanion $[Li_3Al_4]^-$, which contains the rectangular four π-electron Al_4^{4-} tetraanion, has been produced by using laser vaporisation and characterised in the gas phase by photoelectron spectroscopy.[19] In an interesting challenge to synthetic chemists, the stabilisation of the all-metal antiaromatic molecule Al_4Li_4 by complexation to a transition metal fragment, *e.g.* $Fe(CO)_3$, in a manner similar to the well-known stabilisation of cyclobutadiene, has been proposed.[20]

Cyclic anions of the type $[M_3Ar_3]^{2-}$ (M = Al, Ga) have been isolated as alkali-metal salts and structurally characterised. The first example was the trimer $K_2[(2,6\text{-}Mes_2C_6H_3)Ga]_3$, obtained as dark-red crystals by the reduction of $(2,6\text{-}Mes_2C_6H_3)GaCl_2$ with potassium in diethyl ether.[21] The triangular Ga_3 ring exhibits equal Ga–Ga bond lengths of 2.42–2.43 Å. In a similar approach, the aluminium analogue $Na_2[(2,6\text{-}Mes_2C_6H_3)Al]_3$ (Figure 9.1a) was obtained as red–orange crystals by reduction of $(2,6\text{-}Mes_2C_6H_3)AlI_2$ with an excess of sodium in diethyl ether; the Al–Al distances are 2.52 Å.[22] Theoretical studies confirm the aromatic character of the $[Ga_3Ar_3]^{2-}$ dianion.[21] The HOMO-2 in

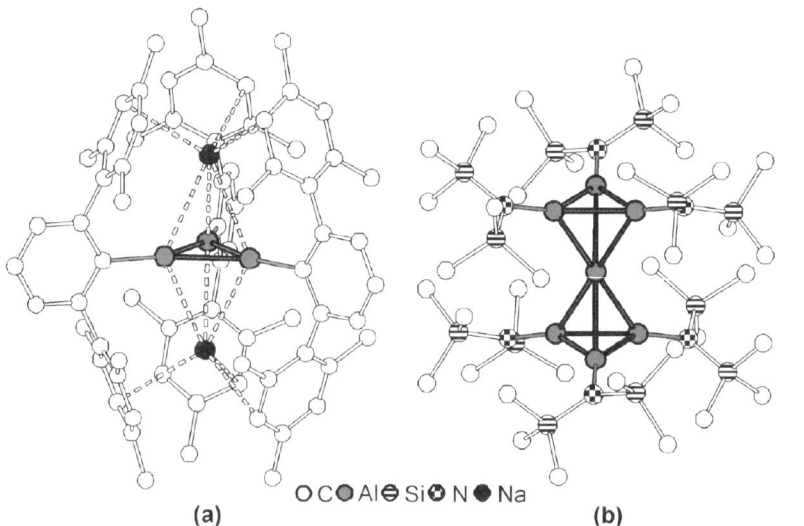

○ C ⦿ Al ⊖ Si ⊗ N ● Na
(a)　　　　　　　　　　　　(b)

Figure 9.1 Molecular structures of (a) $Na_2[(2,6\text{-}Mes_2C_6H_3)Al]_3$ and (b) the $[Al_7\{N(SiMe_3)_2\}_6]^-$ anion (the cation is $[Li(OEt_2)_3]^+$).

Group 13: Rings and Polymers

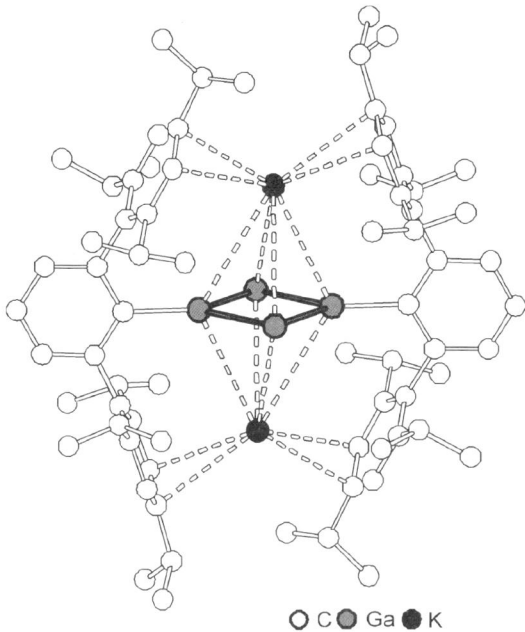

○ C ◐ Ga ● K

Figure 9.2 Molecular structure of $K_2[Ga_4(2,6\text{-Trip}_2C_6H_3)_2]$.

the aluminium analogue is a π-orbital delocalised over the three metal centres; thus, the formal bond order in these homocyclic dianions is 1.33, as expected for two π-electron systems.[22] However, as illustrated in Figure 9.1, the alkali metal–metal counterions play an important role in the structures of these complexes, which should be viewed as M/M' clusters (M = alkali metal, M' = group 13 metal).

Black crystals of the ion-separated salt $[Al_7\{N(SiMe_3)_2\}_6]^-$ $[Li(OEt_2)_3]^+$ are obtained in very low yield from the reaction of a solution of metastable AlCl in *m*-xylene–diethyl ether with $LiN(SiMe_3)_2$ at –78 °C.[23a] Although the structure of this complex can be viewed formally as the coordination of an Al^{3+} cation by two cyclic $[Al_3R_3]^{2-}$ ligands (Figure 9.1b), the ^{27}Al NMR chemical shift indicates that the central aluminium atom is similar in character to the Al atoms in aluminium metal. The dark-crimson neutral radical $[Al_7\{N(SiMe_2Ph)_2\}_6]^{\bullet}$ has also been structurally characterised.[23b]

Cyclic dianions of the type $[Ga_4Ar_2]^{2-}$ have also been structurally characterised as alkali metal salts. Red–brown crystals of $K_2[Ga_4(2,6\text{-Trip}_2C_6H_3)_2]$ are obtained by reduction of $(2,6\text{-Trip}_2C_6H_3)GaCl_2$ (Trip = 2,4,6-triisopropylphenyl) with an excess of potassium.[24] The square-planar four-membered ring in this complex has Ga–Ga bond lengths of 2.46–2.47 Å, consistent with single bonding. As indicated in Figure 9.2, the $[Aryl_2Ga_4]^{2-}$ dianion is stabilised by alkali metal–aryl interactions.

9.2 BORON–NITROGEN RINGS

Iminoboranes, $RB\equiv NR'$, are isoelectronic with alkynes. In contrast to alkynes, however, iminoboranes cyclooligomerise in a manner that is strongly influenced by the nature of the substituents on boron or nitrogen (Section 9.2.2). Boron–nitrogen rings $(RB\equiv NR')_n$ that are analogues of cyclobutadienes ($n = 2$), benzenes ($n = 3$) or cyclooctatetraenes ($n = 4$) may be formed. Since the six-membered rings (borazines) are, by far, the most commonly encountered cyclic oligomers, this section commences with a discussion of these benzene analogues.[25] The historical significance of the parent system, $B_3N_3H_6$, is mentioned in Chapter 1 and the continuing debate over the electronic structure (aromaticity) of borazine is summarised in Section 4.1.2.1. The structures of $\pi(\eta^6)$-complexes of borazine ligands with transition metals are presented in Chapter 7 (see Figure 7.7).

9.2.1 Borazines

Borazine, $B_3N_3H_6$, was first prepared by thermolysis of the diborane–ammonia adduct $[(BH_2)(NH_3)_2]^+[BH_4]^-$. More convenient procedures for the laboratory preparation of this important ring system in multigram quantities involve either (a) the decomposition of ammonia–borane [eqn (9.1)] or (b) the reaction between ammonium sulfate and sodium borohydride [eqn (9.2)].[26] The latter method provides a convenient and economical synthesis of borazine.

$$3H_3N\cdot BH_3 \rightarrow B_3N_3H_6 + 6H_2 \quad (9.1)$$

$$3(NH_4)_2SO_4 + 6NaBH_4 \rightarrow 2B_3N_3H_6 + 3Na_2SO_4 + 18H_2 \quad (9.2)$$

A versatile route to borazine derivatives with alkyl or aryl substituents attached to the boron or nitrogen atoms involves the condensation of (alkyl/aryl)boron halides with (alkyl/aryl)ammonium halides in a high-boiling solvent, e.g. chlorobenzene [eqn (9.3)].[27a] B,B',B''-Trichloroborazine, $Cl_3B_3N_3H_3$, is also prepared in this way (from BCl_3 and NH_4Cl) and may be subsequently reduced to $B_3N_3H_6$ by $LiAlH_4$. Alternatively, the reaction of BCl_3 with a trimethylsilylamine is a convenient route to B,B',B''-trichloroborazines, e.g. $Cl_3B_3N_3(SiMe_3)_3$.[27b]

$$3\ RBCl_2 + 6\ H_2NR' \xrightarrow{-3\ [R'NH_3]Cl} \text{borazine ring}$$

(9.3)

An alternative route to B-substituted mono-, di- and trialkylborazines involves the $RhH(CO)(PPh_3)_3$-catalysed reactions of borazines with alkenes, as illustrated in eqn (9.4) for the trialkyl-substituted derivatives.[26b]

Group 13: Rings and Polymers 117

$$\text{(borazine)} + RHC=CH_2 \xrightarrow{RhH(CO)(PPh_3)_3} \text{(substituted borazine)}$$

$$R' = CH_2CH_2R \qquad (9.4)$$

B,B',B''-Trialkyl, -trialkenyl- or -trialkynylborazines are also obtained by using organometallic methodology, *i.e.* the reaction of B,B',B''-trichloroborazine $Cl_3B_3N_3H_3$ (or N,N',N''-substituted derivatives) with the appropriate organolithium or Grignard reagent. An interesting recent application of this protocol is the synthesis of B,B',B''-trianthryl-N,N',N''-triarylborazines (**9.8**) in a one-pot procedure.[28a] The borazine core has a distinct advantage over benzene for the regioselective construction of such sterically crowded molecules. The three anthryl and three phenyl groups in **9.8** are aligned alternately in a C_3 gear-shaped fashion, an arrangement that allows for π-stacking of anthryl groups. The interest in this class of compounds stems from the novel physical properties, *e.g.* high carrier-transporting ability, that may accrue from these interactions. Applications of related borazine-centred derivatives in electroluminescent devices have also been suggested.[28b]

9.8

The electronegativity difference between boron and nitrogen (2.0 and 3.0, respectively, on the Pauling scale) results in a polar B–N bond. The regioselectivity of the reactions of borazines with electrophiles and nucleophiles can be understood in this context. Thus the reaction of borazine with three equivalents of HCl produces $(ClHBNH_2)_3$ in which the proton seeks out the nitrogen centres and the chloride ions become attached to boron.[1]

B,B',B''-Trialkylated borazines $R_3B_3N_3H_3$ are deprotonated by organolithium reagents, but other reaction pathways also occur.[29a] For example, the reaction of $Me_3B_3N_3H_3$ with one equivalent of methyllithium produces the solvated monolithium derivative $[(Me_3B_3N_3H_2)Li(OEt_2)]_2$ (**9.9**), which is dimeric in the solid state. The formation of di- or trilithiated derivatives, however, is accompanied by borate formation and also destruction of the borazine ring.[29a] Borate formation occurs cleanly in the reaction of $^tBu_3B_3N_3H_3$ with one equivalent of ^tBuLi to give the $[^tBu_4B_3N_3H_3]^-$ anion (**9.10**), which is isolated as the $[(TIPTA)_2Li]^+$ (TIPTA = triisopropylhexahydrotriazine) salt. The lithium reagent **9.9** can be used to make *N*-borazinyl derivatives of p-block elements by metathesis. For example, reaction of **9.9** with phosphorus trihalides in the appropriate molar ratio produces derivatives of the type $[(Me_3B_3N_3H_2)_{3-x}PX_x]$ ($x = 0, 1, 2$; X = Cl, Br).[29b]

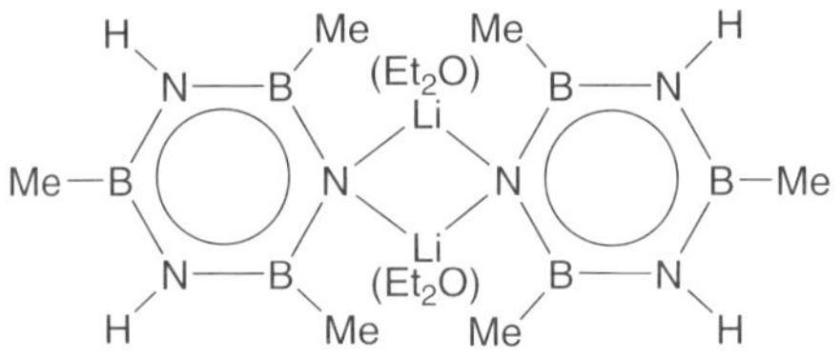

9.9 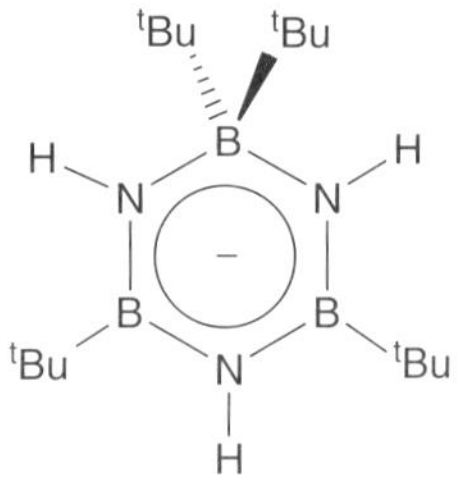   **9.10**

Boron-bonded η^1-borazine complexes of transition metals have been prepared by two different approaches: (a) nucleophilic substitution of B,B',B''-trichloroborazine with an anionic metal carbonyl reagent[30a] and (b) oxidative addition of a B–Br bond of B,B',B''-tribromoborazine to a zerovalent group 10 complex[30b] (see examples in Scheme 9.2).

Synthetic methods are available for the incorporation of other p-block elements into a borazine ring. An illustration of the potentially versatile methodology using an NBNBN building block is given for the tellurium-containing borazine **9.11** in eqn (9.5).[31]

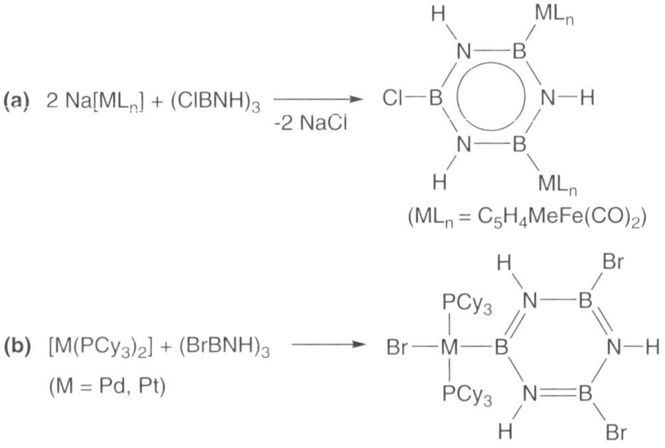

(a) $2 Na[ML_n] + (ClBNH)_3 \xrightarrow{-2 NaCl}$

($ML_n = C_5H_4MeFe(CO)_2$)

(b) $[M(PCy_3)_2] + (BrBNH)_3 \longrightarrow$
    (M = Pd, Pt)

Scheme 9.2 Syntheses of η^1-borazine–metal complexes.

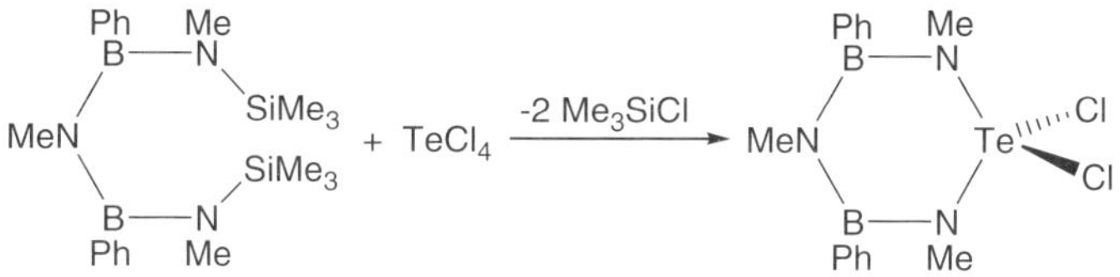

Group 13: Rings and Polymers

[Scheme showing reaction of borazine derivative with TeCl$_4$ yielding compound **9.11** with loss of 2 Me$_3$SiCl]

(9.5)

Two intriguing examples of this class of compound involve ring systems in which one boron atom of a borazine ring is replaced by the isoelectronic units HC$^+$ or C:, as represented by the generic carbocation **A** or the carbene **B**.

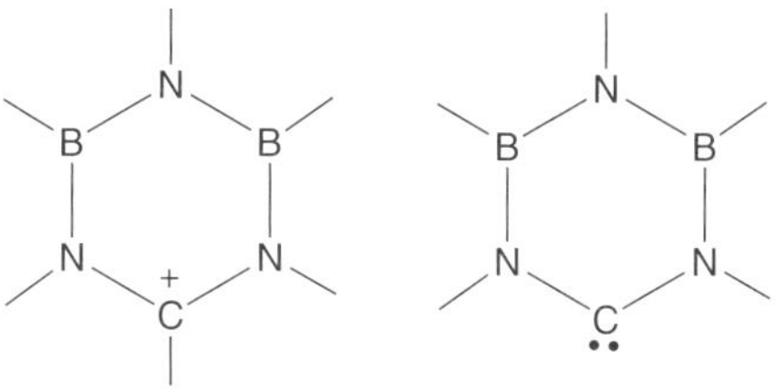

[Structures of cation **A** and carbene **B**]

As illustrated in Scheme 9.3, the synthesis of carbocations of type **A** is achieved by two different routes. The first (top) method is only applicable to derivatives with amino substituents on boron (R = R' = Me$_2$N), whereas the alternative (bottom) route may be used for the preparation of carbocations with the same or different alkyl or aryl groups attached to the two boron atoms.[32a] Subsequent deprotonation of these carbocations with lithium tetramethylpiperidine in THF produces the corresponding carbenes as pale-yellow crystals with high thermal stability. The endocyclic bond lengths for the carbocation indicate that the six π-electrons are localised on two pseudo-allyl fragments, a four π-electron cationic NCN unit and a two π-electron neutral BNB unit. By contrast, the bond lengths for the carbene ring system are consistent with a more delocalised π-system;[32a] however, analysis of the electronic

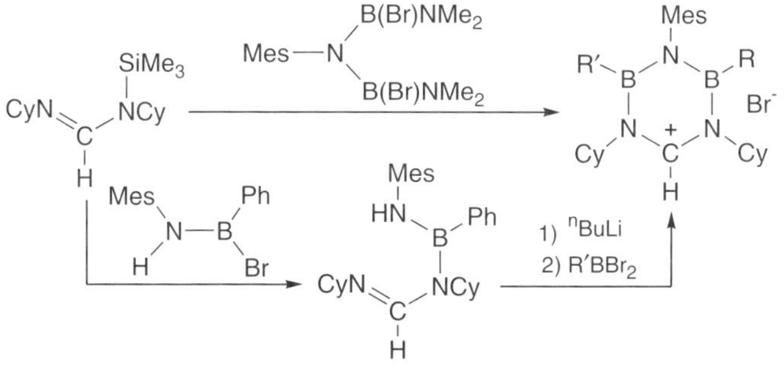

Scheme 9.3 Synthetic routes to carbocation-containing borazines.

structure on the basis of DFT calculations reveals only limited conjugation in the six-membered ring.[32b] The carbene acts as a two-electron ligand towards transition metals such as Rh.[32a]

9.2.2 Other Unsaturated Boron–Nitrogen Rings

The steric bulk of the substituents on B and N in an iminoborane RB≡NR′ has a major influence on the outcome of the oligomerisation process.[25] Small substituents favour the formation of borazines. For example, only the borazine $Me_3B_3N_3Me_3$ is formed for the combination R = R′ = Me. An increase in the steric bulk of R and/or R′ may lead to the formation of a Dewar borazine **C**, a four-membered ring **D** or an eight-membered ring **E**. Specific examples of each of these ring systems and the mechanisms of their interconversions are discussed below.

C   **D**   **E**

The prerequisite for making Dewar borazines more stable than borazines is steric strain as exemplified by the derivative $B,B′,B″$-triphenyl-$N,N′,N″$-tri-*tert*-butylborazine (Figure 9.3a).[33] The structure of $Ph_3B_3N_3{}^tBu_3$ is folded about the transannular B–N bond, which (at 1.75 Å) is significantly longer than a single bond. The four B–N bond distances of 1.55 Å involving the four-coordinate boron and nitrogen atoms are close to the single bond value, while the remaining two B–N bond lengths (1.38 Å) represent double bonds.

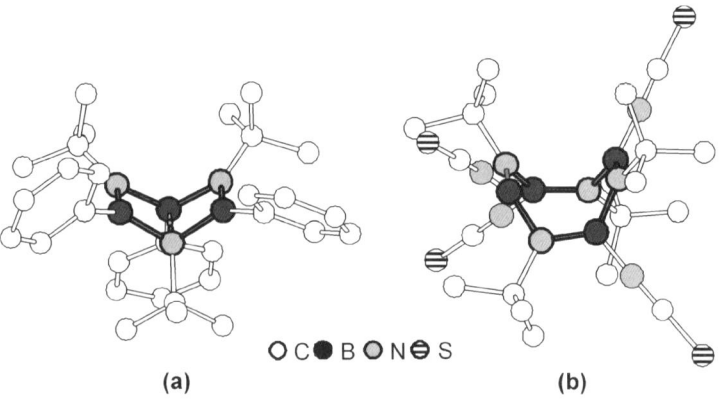

Figure 9.3 Molecular structures of (a) $Ph_3B_3N_3{}^tBu_3$ and (b) $(SCN)_4B_4N_4{}^tBu_4$.

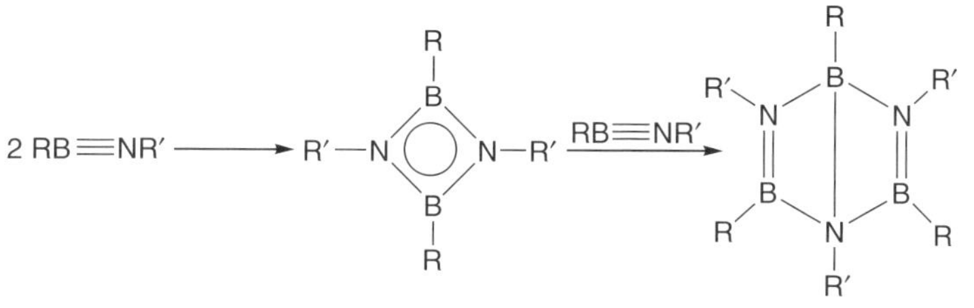

Scheme 9.4 Formation of a Dewar borazine from iminoboranes.

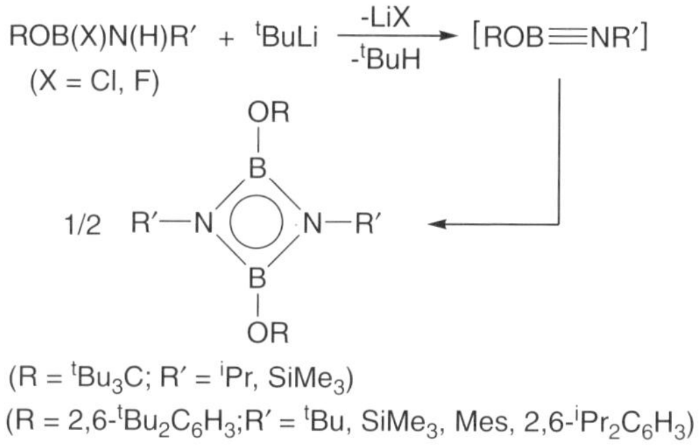

Scheme 9.5 Synthesis of B_2N_2 ring systems.

(R = $^tBu_3C$; R' = ^iPr, $SiMe_3$)
(R = 2,6-$^tBu_2C_6H_3$; R' = ^tBu, $SiMe_3$, Mes, 2,6-$^iPr_2C_6H_3$)

As indicated in Scheme 9.4, the formation of Dewar borazines involves the dimerisation of an iminoborane followed by the [4 + 2] cycloaddition of the cyclodimer thus formed with an iminoborane. For smaller R/R' groups, the isomerisation of the bicyclic Dewar borazine to a monocyclic borazine occurs readily.

On the other hand, extremely bulky R/R' substituents stabilise the four-membered ring.[25] The synthetic route to *B*-alkoxy or -aryloxy derivatives is illustrated in Scheme 9.5.[34] Other examples of R/R' combinations that give rise to stable four-membered rings include $^tBu/^tBu$, TMP/^tBu, $(Me_3Si)_2N/SiMe_3$ and $C_6F_5/^tBu$. The B_2N_2 rings are rhombic, rather than rectangular, with B–N bond lengths that are intermediate between single and double bonds; both planar and slightly folded rings have been reported.[34]

The most stable eight-membered B–N rings are of the type $X_4B_4N_4R_4$ (X = Cl, Br, NCS; R = ^tBu). They have a tub-shaped structure with S_4 symmetry (Figure 9.3b) and, like cyclooctatetraene, alternating single and double bond lengths.[25,35] These eight-membered rings do not participate in equilibria with the corresponding four-membered rings. However, when the substituent X on boron in $X_4B_4N_4R_4$ is an alkyl group, equilibria between the four- and eight-membered rings prevail. The mechanism for this interconversion may involve face-to-face dimerisation of two four-membered rings, followed by asynchronous rupture of the two transannular B–N bonds to form the monocyclic eight-membered ring.[25] This proposal is supported by computational studies.[36]

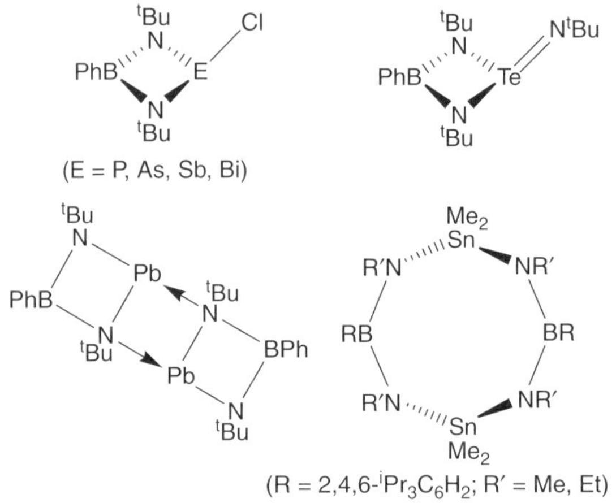

Figure 9.4 Heteroatom-substituted four- and eight-membered B–N rings.

Other p-block elements may be introduced into an unsaturated four-membered B–N ring by using metathetical reactions of reagents of the type $Li_2[RB(NR')_2]$ with main group element halides.[37] These hetero-substituted B–N rings form either monomers, ladder-like dimers via N→M dative bonding or, in a few cases, eight-membered rings (Figure 9.4). Two particularly interesting features of this class of ring system are (a) the stabilisation of a terminal Te=N^{t}Bu bond in PhB(N^{t}Bu)$_2$Te=N^{t}Bu[38a] and (b) the formation of deeply coloured, spirocyclic radicals of the type $[M\{PhB(N^tBu)_2\}_2]^{\bullet}$ (M = Al, Ga) that are stable in the solid state.[38b] The existence of these neutral radicals illustrates the ability of the group 13 metal centre to stabilise the acyclic anion radical $[PhB(N^tBu)_2]^{-\bullet}$.

A number of odd-membered unsaturated boron–nitrogen rings involving either B–B or N–N bonds are known. The best known examples are three-membered rings of the type $(R_2N)_2B_2N^tBu$ [R = ^{i}Pr, Me$_2$C(CH$_2$)$_3$CMe$_2$][39] or $^tBu_2B_2NMes$ (**9.12**),[40] which are isoelectronic with a cyclopropenium cation, *e.g.* $[C_3{}^tBu_3]^+$. The *B*-alkyl derivative **9.12** is prepared from an iminoborane by the route shown in Scheme 9.6. The B–B bond length of 1.56 Å in **9.12** is notably shorter (by 0.04–0.05 Å) than that in the *B*-dialkylamino derivatives $(R_2N)_2B_2N^tBu$, suggesting significant double bond character in **9.12**, as expected for a two π-electron system.

The three-membered $^tBu_2B_2N^tBu$ ring undergoes ring expansion with azides or iminoboranes via insertion into the B–B bond to give four-membered B_2N_2 or five-membered B_3N_2 rings, respectively, (*e.g.* **9.13**, Scheme 9.7).[41] The short B–N bond lengths of 1.40–1.41 Å in **9.13** indicate localised double bonds.

The nitrogen-rich B_2N_3 ring was first reported in the early 1960s.[42a,b] An improved synthesis involves the hydrazinolysis of boron thiolates.[42c] A recent extension of this synthetic approach using methylhydrazine and $PhB(NMe_2)_2$ produces a B_2N_3 ring with an exocyclic MeNH group, which undergoes

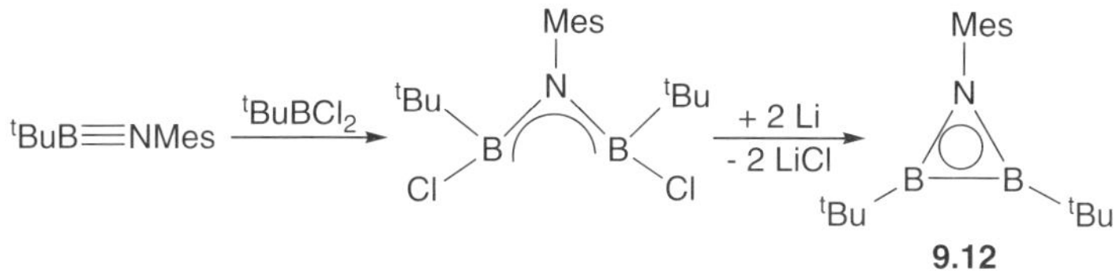

Scheme 9.6 Synthesis of the three-membered ring $^tBu_2B_2NMes$.

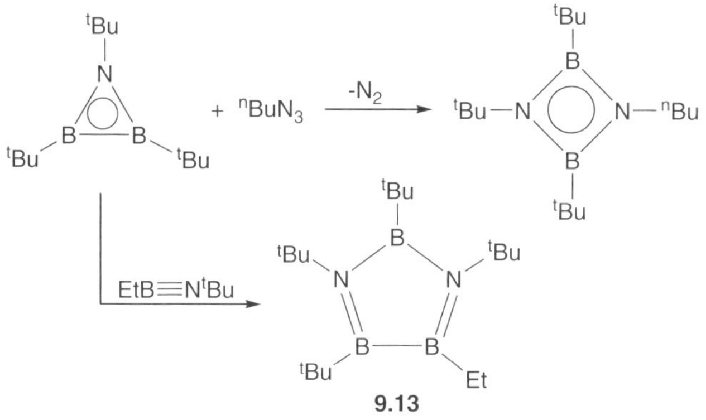

Scheme 9.7 Ring-expansion reactions of $^tBu_2B_2N^tBu$.

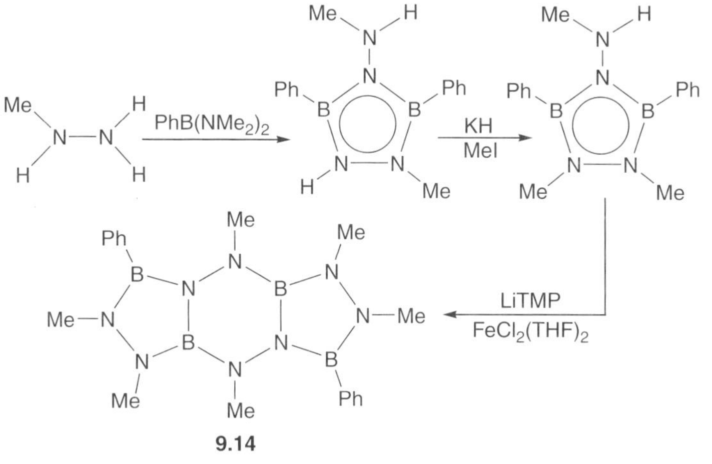

Scheme 9.8 Synthesis, deprotonation and oxidation of a B_2N_3 ring.

deprotonation of the ring NH group with strong bases to produce a six π-electron anionic B_2N_3 ligand.[42d] *N*-Methylation of this anion followed by deprotonation of the exocyclic MeNH group and oxidation with $FeCl_2$ generates the almost planar, 16 π-electron tricyclic B–N heterocycle **9.14** (Scheme 9.8).[42e]

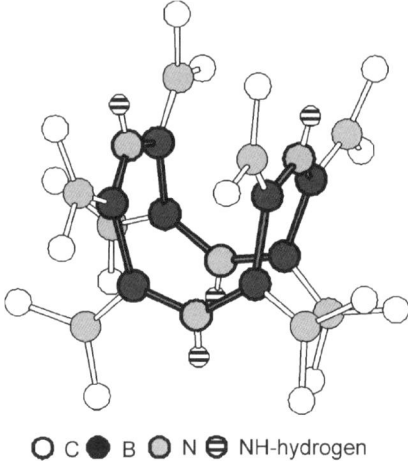

Figure 9.5 Molecular structure of the 12-membered ring $B_8(NH)_4(NMe_2)_8$.

A B_8N_4 ring that incorporates two B–B bonds is obtained in low yield from the reaction of $B_2(NMe_2)_4$ and two equivalents of $[NH_4][PF_6]$ in THF.[43] The unique 12-membered ring adopts a cyclooctatetraene-like boat structure (Figure 9.5).

9.2.3 Saturated Boron–Nitrogen Rings (Cycloborazanes)

Saturated boron–nitrogen ring systems of the type $(H_2BNH_2)_n$ ($n = 2, 3, 4$ or 5) are white crystalline solids, which are obtained from the reaction of sodium amide with the diammoniate of diborane (DADB) in liquid ammonia [eqn (9.6)].[44a] The individual cyclic oligomers are separated and purified, with difficulty, by careful fractional sublimation. The cyclotriborazane $(H_2BNH_2)_3$ adopts a chair-shaped structure,[44b] *cf.* cyclohexane. In the solid state the B–N bond distances are equal (*ca* 1.58 Å), *cf.* 1.43 Å in borazine (Section 4.1.2.1), consistent with B–N single bonding in the saturated ring system; short intermolecular BH···HN contacts (2.29 Å) are indicative of weak dihydrogen bonding.

$$[(NH_3)_2BH_2]^+[BH_4]^- + NaNH_2 \rightarrow 1/n(H_2BNH_2)_n + NaBH_4 + 2NH_3 \quad (9.6)$$
(DADB)

Ammonia–borane, $H_3B \cdot NH_3$, is a potential hydrogen storage material (see Section 9.2.4.2). The decomposition of this adduct in solution has been monitored by ^{11}B and ^{15}N NMR spectroscopy. The reaction is second order in $H_3B \cdot NH_3$ to give DADB, which rapidly cyclises to generate cyclodiborazane $(H_2BNH_2)_2$. This dimer then reacts with $H_3B \cdot NH_3$ to produce the derivative cyclo-$(NH_2BH_2NH_2BH)NH_2BH_3$ (**9.15**) (Scheme 9.9).[44c]

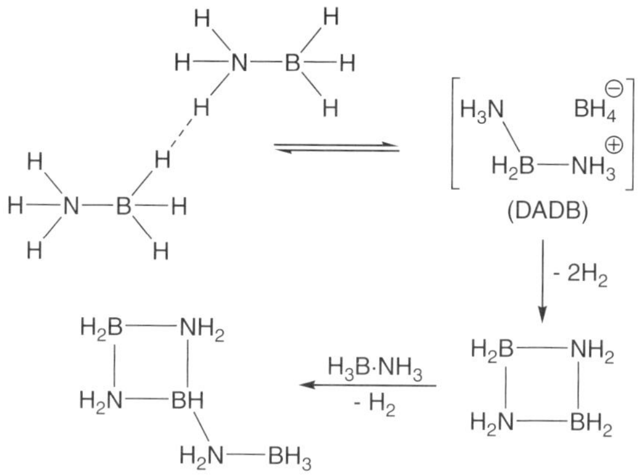

Scheme 9.9 Formation of cyclodiborazane from decomposition of $H_3B \cdot NH_3$ in solution.

The formation of individual cycloborazanes can be achieved, in some cases, by transition metal-catalysed dehydrocoupling reactions. By using this strategy, secondary amine–boranes are converted to the four-membered ring $(H_2BNMe_2)_2$ under mild conditions, whereas primary amine–boranes produce borazines. The cyclic pentamer is obtained as the exclusive product from ammonia–borane adduct $H_3B \cdot NH_3$ employing an iridium(III) catalyst [see eqn (2.13) in Section 2.3)].

9.2.4 Applications of Boron–Nitrogen Rings

9.2.4.1 Precursors for Ceramic Materials. Both unsaturated and saturated boron–nitrogen ring systems have attracted attention recently in the context of potential applications. Boron nitride may exist as hexagonal boron nitride (*h*-BN) or cubic boron nitride (*c*-BN). These structural analogues of the carbon allotropes, graphite and diamond, are of interest as ceramics, *i.e.* materials that can withstand very high temperatures. Applications include their use in the semiconductor industry (as crucibles), in the aerospace industry (for rocket combustion chambers) and as high-temperature abrasives. The traditional industrial synthesis of *h*-BN involves the reaction of diboron trioxide B_2O_3 with ammonia at temperatures above 1000 °C followed by treatment with nitrogen at 1800 °C. An alternative route that requires lower temperatures is the thermolysis of borazine-based molecular or polymeric precursors.[45] An interesting recent development of the molecular precursor approach involves the decomposition of $B_3N_3H_6$ at *ca* 1000 °C on a hot rhodium surface, which generates thin films of an *h*-boron nitride nanomesh on the Rh(111) surface; this nanomesh acts as a template to trap substrates such as C_{60} or naphthalocyanine.[46]

Ternary materials that incorporate carbon into a boron–nitrogen framework are expected to exhibit extreme hardness as a result of their diamond-like structures. Alkyl-, vinyl- and alkynyl-substituted borazines are potential molecular precursors to novel BCN materials. As an example, the pyrolysis of the cross-linked hydrocarbon-bridged polymer, prepared by the hydroboration of B,B',B''-triethynylborazine with $B_3N_3H_6$ in the absence of a catalyst, produces a homogeneous, amorphous boron carbonitride ceramic.[47] A different approach to these ternary materials makes use of a sol–gel process, which involves a condensation reaction to produce a gel that is subsequently converted into a ceramic at high temperature. For example, the reaction of B,B',B''-trichloroborazine with bis(trimethylsilyl)carbodiimide generates an amorphous xerogel comprised of a network of borazine rings linked by bridging –N=C=N– functionalities [eqn (9.7)].[48] Pyrolysis of the xerogel at 1200 °C generates an amorphous B_4CN_4 material, while the use of even higher temperatures (2000 °C) produces boron carbide (B_4C), the hardness of which is comparable to that of diamond.

$$n \, [\text{Cl}_3\text{B}_3\text{N}_3\text{H}_3] + 1.5n \, \text{Me}_3\text{SiNCNSiMe}_3 \longrightarrow \underbrace{[(B_3N_3H_3)(NCN)_{1.5}]_n + 3n \, \text{ClSiMe}_3}_{\text{gel}} \quad (9.7)$$

9.2.4.2 Intermediates in Hydrogen Storage Materials. The utility of hydrogen as a fuel in transportation is currently under intensive investigation. Hydrogen-powered cars and buses have been manufactured, but their use has been limited to short journeys owing to the problems associated with the storage of large volumes of hydrogen. Compounds that can be exploited to store hydrogen in the form of E–H bonds (where E is a light main group element) are possible candidates for this purpose. More specifically, ammonia–borane $H_3N \cdot BH_3$,[49a-c] ammonia–triborane $H_3N \cdot B_3H_7$[50] and amine–borane adducts[49d] are being investigated intensively in view of the combination of their low molecular weights and high gravimetric hydrogen capacity. The thermolysis of $H_3N \cdot BH_3$ is known to produce the cyclic species $(HBNH)_3$ and $(H_2BNH_2)_3$, in addition to polyamino-borane $(H_2BNH_2)_n$, the formation of which is an impediment to the full utilisation of hydrogen. In addition, volatile materials such as borazine are undesirable because they may poison a fuel cell. The employment of an ionic liquid, such as 1-butyl-3-methylimidazolium chloride, as the medium for ammonia–borane dehydrogenation increases the extent and rate of hydrogen release significantly and also minimises the production of borazine.[51] The acid-initiated decomposition is an alternative to the use of heat for dehydrogenation of $H_3N \cdot BH_3$.[49b]

9.2.5 Boron–Nitrogen Polymers

Several examples of well-characterised boron–nitrogen polymers are known. Polyborazylene **9.16** (with molecular weights M_w up to *ca* 7600 and M_n up to *ca*

3400) is a cyclolinear polymer based on six-membered borazine rings, which has been prepared via the thermally induced dehydropolymerisation of borazine [eqn (9.8)].[52]

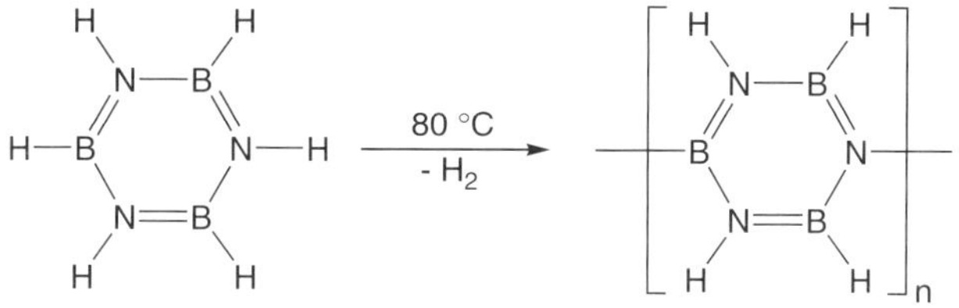

$$\text{(borazine)} \xrightarrow[-H_2]{80\ °C} \text{9.16} \qquad (9.8)$$

The polymers **9.16** were isolated as white solids and characterisation suggested the presence of a significantly branched structure. Pyrolysis at 1200 °C yielded white boron nitride in 85–93% yield.[53] An idealised representation of this process is shown in Scheme 9.10.

Processing the polymer **9.16** followed by pyrolysis provides an excellent method for preparing this valuable ceramic in shaped forms such as fibres and monoliths. Derivatisation of **9.16** by a thermal dehydrocoupling reaction with amines NHR_2 to replace –H substituents on boron by $-NR_2$ groups can be achieved.[53] Examples of void-free BN fibres prepared from polyaminoborazylene precursors formed by

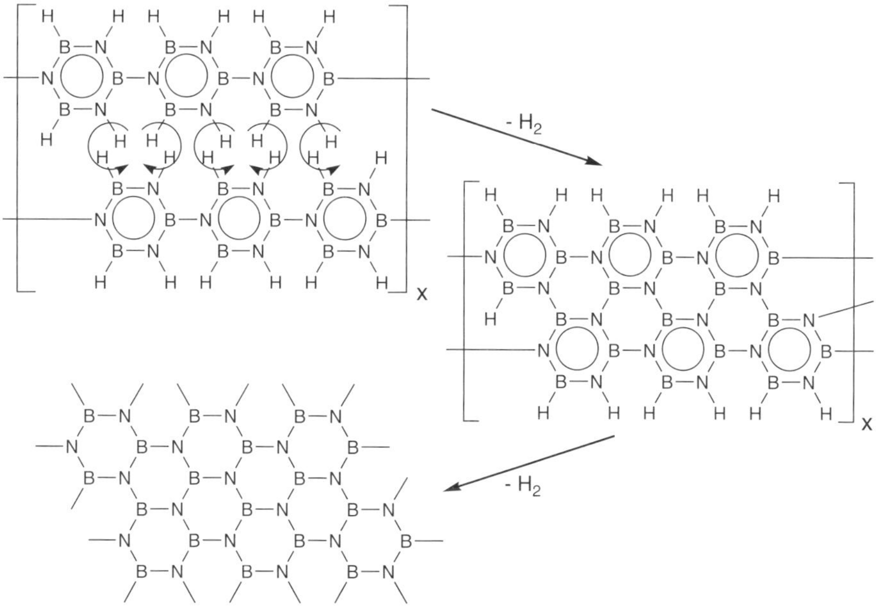

Scheme 9.10 Conversion of polyborazylene **9.16** into hexagonal boron nitride. Reproduced with permission from T. Wideman et al., Chem. Mater., 1998, **10**, 412.[53]

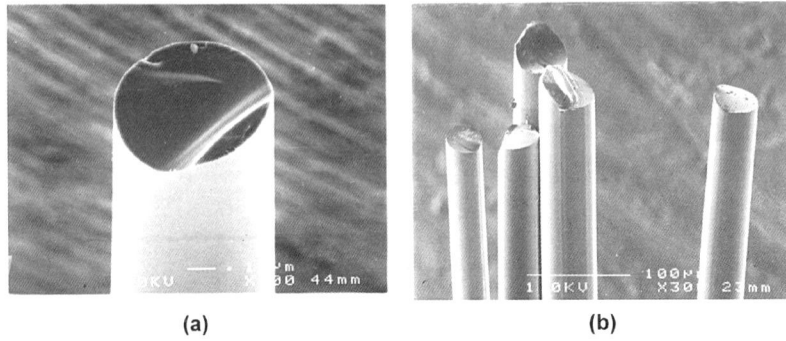

Figure 9.6 (a) High-resolution and (b) low-resolution scanning electron micrographs of BN fibres fabricated by pyrolysis of preceramic fibres of polyaminoborazylenes. Reproduced with permission from T. Wideman *et al.*, *Chem. Mater.*, 1998, **10**, 412.[53] Copyright (1998) American Chemical Society.

melt-extrusion followed by pyrolysis at 1200 °C are shown in Figure 9.6. Such fibres have technological applications in composite materials with enhanced oxidation resistance, thermal stability and electrical insulation properties.[53]

In addition to polymers based on borazine rings, materials containing boron–nitrogen four-membered rings have also been prepared.[54] Linear polymers based on boron–nitrogen skeletons have been much less studied. Examples include polyiminoboranes, $(RB=NR')_n$, boron–nitrogen analogues of polyacetylene $(RC=CR')_n$, which are prepared from the oligomerisation of sterically unencumbered iminoborane monomers $RB=NR'$ (R = *n*-alkyl). However, these materials are insoluble in common organic solvents and have not been studied in any detail.[55] On the other hand, soluble polyaminoboranes, $(BH_2-NHR)_n$ (**9.17**, R = Me, ^{n}Bu), boron–nitrogen analogues of polyolefins, have been prepared with molecular weights (M_w) up to 400 000 by iridium-catalysed dehydrocoupling of primary amine–borane adducts [eqn (9.9)].[56] Several examples of these materials also show promise as BN-based ceramic precursors.

$$H_3B \cdot NH_2R \xrightarrow[-H_2]{\text{[Ir] cat.}} \left[\begin{array}{c} H \quad R \\ | \quad | \\ B-N \\ | \quad | \\ H \quad H \end{array} \right]_n$$

9.17

(9.9)

9.3 BORON–PHOSPHORUS RINGS

9.3.1 Unsaturated Systems

By analogy with borazines, the initial interest in the analogous unsaturated boron–phosphorus rings focused on the fundamental issue of the extent of π-electron delocalisation. In practice, boron–phosphorus analogues of borazine are relatively scarce. However, stable six-membered rings of the type $(RPBR')_3$ are obtained by salt-elimination reactions between an organoboron dihalide and a lithium organophosphide, if the substituents around boron and phosphorus provide sufficient steric protection [eqn (9.10)].[57,58] For certain combinations of more bulky substituents, unsaturated four-membered $(RPBR')_2$ rings are formed in this cyclocondensation process.

$R'BX_2 + 2LiPHR \rightarrow 1/n(RPBR')_n + 2LiX + H_2PR$

$(n = 3; R/R' = Mes/Ph, Mes/Cy, Mes/Mes, Mes/^tBu, Ph/Mes)$ (9.10)

$(n = 2; R/R' = Mes/1\text{-}Ad, Thexyl/Mes, Mes/^tBu)$

The common structural features of these borazine analogues, *e.g.* $(MesBPPh)_3$ (Figure 9.7), are (a) a planar $B_3P_3C(ipso)_6$ arrangement and (b) equivalent B–P bond lengths that are *ca* 0.1 Å shorter than a B–P single bond. Although these structural data imply some degree of delocalisation, calculations employing modern computational methods are needed to address this issue in the context of the continuing debate over the extent of delocalisation in borazine (see Section 4.1.2.1). By contrast, the four-membered $(RBPR')_2$ rings, *e.g.* $(ThexylBPMes)_2$ (Figure 9.7), exhibit pyramidal phosphorus centres and B–P bond lengths that are closer to the single-bond value of 1.94 Å. Four-membered rings of the type $(R_2NBPR')_2$ are also known in which donation of electron density from the dialkylamino substituent to boron results in elongation of the B–P bonds.[59,60]

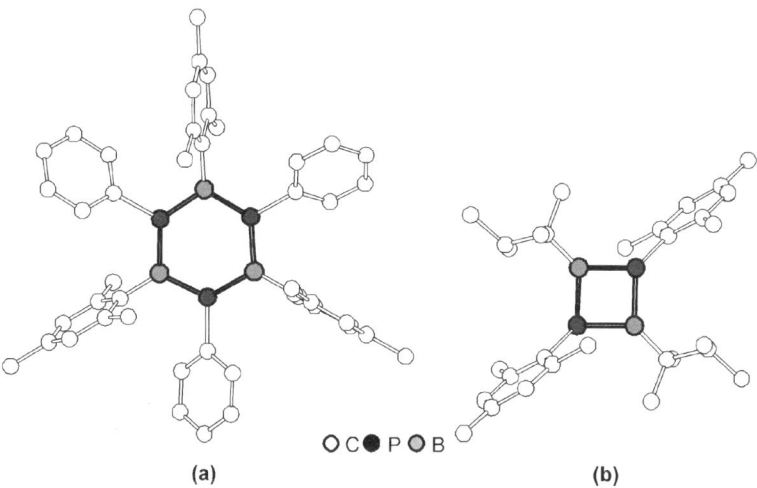

Figure 9.7 Structures of unsaturated (a) B_3P_3 and (b) B_2P_2 ring systems.

9.3.2 Cyclic Biradicals

A fascinating feature of the chemistry of boron–phosphorus heterocycles is the existence of stable singlet biradicals, e.g. (^{t}BuBP^{i}Pr$_2$)$_2$ (for a discussion of the electronic structure, see Section 5.4.2.2).[61] The synthesis of this four-membered ring is illustrated in Scheme 5.2. Related derivatives of the type (RBPR'$_2$)$_2$ are also prepared by a salt-elimination reaction of the appropriate lithium phosphide with a 1,2-dichlorodiborane or, in the case of the perphenylated derivative (PhBPPh$_2$)$_2$, by reduction of the cyclic dimer [Ph(Cl)BPPh$_2$]$_2$ with lithium naphthalenide (Scheme 9.11).[62]

The structures of the four-membered rings (RBPR'$_2$)$_2$ are markedly dependent on the nature of the substituents on boron and phosphorus. The yellow derivative (^{t}BuBP^{i}Pr$_2$)$_2$ is comprised of a planar four-membered B$_2$P$_2$ ring with equal B–P bond lengths, which are ca 0.05 Å shorter than expected for single bonds and a B–B distance that is ca 0.7 Å longer than the longest known B–B bond (Figure 9.8).[61] By contrast, the organic analogues, bicyclo[1.1.0]butanes, are bicyclic systems with a transannular C–C bond. Bicyclic (RBPR'$_2$)$_2$ rings are formed, however, upon changing the substituents on boron and/or phosphorus.[62]

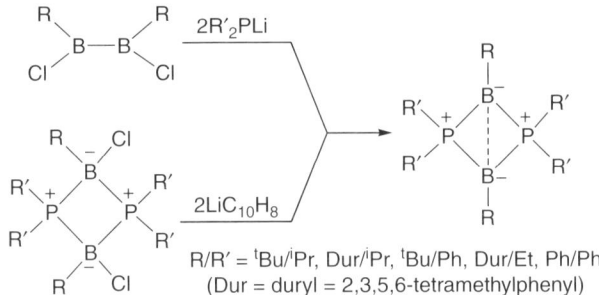

Scheme 9.11 Synthesis of four-membered (RBPR'$_2$)$_2$ rings.

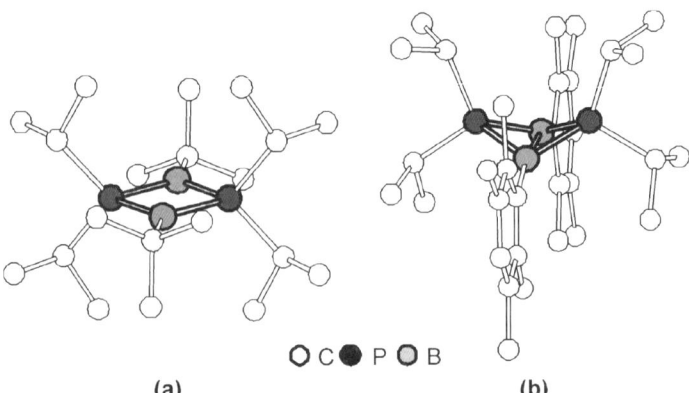

Figure 9.8 Molecular structures of (a) (^{t}BuBP^{i}Pr$_2$)$_2$ and (b) (DurBP^{i}Pr$_2$)$_2$.

Table 9.1 Structural parameters and NMR chemical shifts for $(RBPR'_2)_2$.

Parameter	$(^tBuBP^iPr_2)_2$	$(DurBP^iPr_2)_2$	$(^tBuBPPh_2)_2$	$(DurBPEt_2)_2$	$(PhBPPh_2)_2$
B–B (Å)	2.57	2.24	1.99	1.89	1.83
τ (°)^a	180	130	118	115	114
δ^{31}P (ppm)	+3	–26	–55	–55	–54
δ^{11}B (ppm)	+25	–13	–12	–19	–13

^{a}Dihedral angle.

Replacement of the *tert*-butyl group on boron by a 2,3,5,6-tetramethylphenyl (duryl) group generates the folded structure of $(DurBP^iPr_2)_2$ with an interplanar angle between PBB units of 130° and a B–B distance that is reduced to 2.24 Å (Figure 9.8). Changing the *iso*-propyl groups on phosphorus to phenyl groups produces an even more folded structure (interplanar angle = 118°) and a B–B distance of 1.99 Å in $(^tBuBPPh_2)_2$. The shortest transannular B–B bond (1.83 Å) is observed for the all-phenylated derivative $(PhBPPh_2)_2$.

The trends in structural parameters and NMR chemical shifts for $(RBPR'_2)_2$ derivatives are summarised in Table 9.1. The relatively low-field ^{11}B and ^{31}P NMR chemical shifts for $(^tBuBP^iPr_2)_2$ indicate the absence of four-coordinate boron and three-membered B_2P rings, respectively. The folding of the B_2P_2 rings is accompanied by substantial upfield shifts in both of these resonances, consistent with the presence of four-coordinate boron centres in a bicyclic structure. These trends in NMR chemical shifts provide diagnostic evidence for the presence of 'bond-stretch isomers' in solution for the purple derivative $(PhBP^iPr_2)_2$, which has a planar B_2P_2 ring with a transannular B···B distance of 2.57 Å in the solid state, consistent with biradical character; the phenyl rings are almost coplanar with the B_2P_2 ring.[63] In the solid state, $(PhBP^iPr_2)_2$ shows a single ^{31}P resonance at +5.9 ppm, whereas the solution-state NMR spectra at room temperature exhibit signals at –28 ppm (^{31}P) and –9 ppm (^{11}B) which, by comparison with the data in Table 9.1, indicate conversion to the folded structure. Variable-temperature NMR spectra showed the presence of the open structure at low temperature; a prominent ^{31}P resonance appears at +4.0 ppm.[58] The counterintuitive observation of the rupture of a B–B σ-bond on decreasing the temperature is strongly entropy driven. The diradical isomer with coplanar phenyl groups has fewer degrees of freedom than the bicyclic isomer in which free rotation of the phenyl groups and inversion at boron are both allowed. This type of temperature-dependent interconversion has potential applications in electrical switch devices.

The absence of an EPR signal in solution or in the solid state is indicative of a singlet ground state for the diradical $(^tBuBP^iPr_2)_2$. An indication of the radical character of this derivative is provided by a variety of facile oxidative addition reactions (Scheme 9.12).[64] For example, the treatment of $(^tBuBP^iPr_2)_2$ with diphenyl diselenide (or elemental selenium) produces a bicyclic compound in which a selenium atom bridges the two boron atoms. Trimethyltin hydride reacts rapidly with $(^tBuBP^iPr_2)_2$ to give the *trans* adduct. Finally, $(^tBuBP^iPr_2)_2$ is slowly oxidised by deuterated chloroform to produce a B,B'-dichloro adduct as a mixture of *cis* and *trans* isomers.

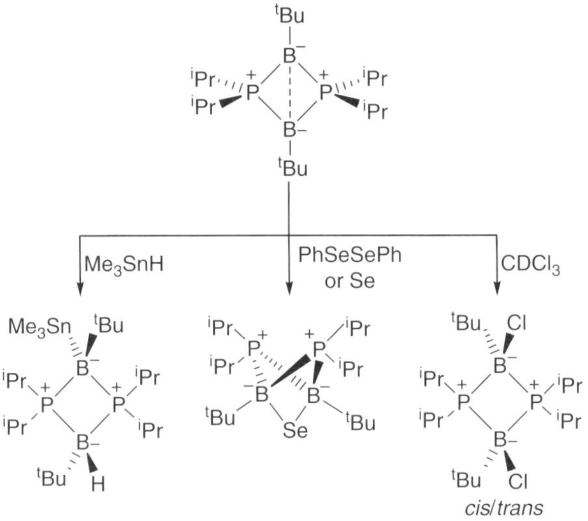

Scheme 9.12 Some oxidative-addition reactions of (^{t}BuBP^{i}Pr$_2$)$_2$.

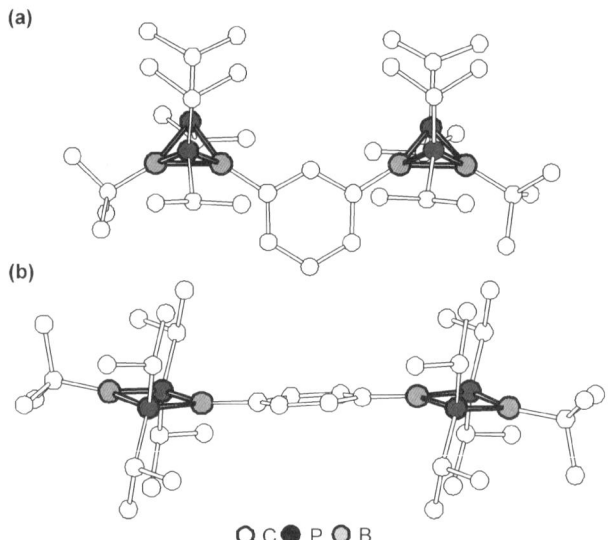

Figure 9.9 Molecular structures of (a) *meta*- and (b) *para*-[^{t}BuB(μ-P^{i}Pr$_2$)B]$_2$C$_6$H$_4$.

The synthetic methodology used to prepare diradicals involving four-membered (RBPR′$_2$)$_2$ rings (Scheme 9.11) has been adapted for the preparation of stable radicals in which two B$_2$P$_2$ diradicals are separated by *meta*- or *para*-C$_6$H$_4$ spacer groups.[65] Significantly, the *para* isomer is violet whereas the *meta* isomer is colourless. The reason for this difference is apparent from the molecular structures depicted in Figure 9.9. The B$_2$P$_2$ rings in the *para* isomer are almost planar with $d(B-B) = 2.57$ Å. In marked contrast, these four-membered rings are folded in the *meta* isomer with $d(B-B) = 1.88$ and 1.91 Å.

9.3.3 Saturated Systems

Saturated ring systems of the type $(R_2MER'_2)_2$, where M is a group 13 element and E is a group 15 element, are of interest as single-source precursors for binary metal pnictides (Section 9.4.2). Four-membered rings of the type $(X_2BPR_2)_2$ (X = halogen) are prepared by salt-elimination reactions.[66] For example, the reactions of (a) BCl_3 with $LiPEt_2$ or (b) boron trihalides with $LiP(SiMe_3)_2$ at room temperature produce the dimers $(Cl_2BPEt_2)_2$ and $[X_2BP(SiMe_3)_2]_2$ (X = Cl, Br), respectively [eqn (9.11)].[67a] The related *B*-dialkylated derivatives $[R_2BP(SiMe_3)_2]_2$ (R = Me, Et) are obtained by trimethylsilyl halide elimination between diorganoboron halides and tris(trimethylsilyl)phosphine $P(SiMe_3)_3$.[67b]

$$2BX_3 + 2LiP(SiMe_3)_2 \rightarrow [X_2BP(SiMe_3)_2]_2 + 2LiX \quad (9.11)$$
$$(X = Cl, Br)$$

The state of aggregation of phosphinoboranes $(R_2PBR'_2)_n$ is influenced markedly by the nature of the substituents on the group 13 and 15 elements. This is illustrated by the series $[R_2PB(C_6F_5)_2]_n$, which are monomeric ($n = 1$) for relatively bulky groups on phosphorus (R = Cy and ^tBu), but dimeric for R = Et, Ph (Scheme 9.13).[68] The state of aggregation in solution is readily determined from the ^{11}B NMR spectra, which exhibit triplet resonances for the dimers and doublets for the monomers as a result of coupling with ^{31}P centres. Interestingly, in the context of hydrogen storage devices (Section 9.2.4.2), the monomeric phosphinoboranes $R_2PB(C_6F_5)_2$ (R = Cy, ^tBu) undergo facile, but irreversible, addition of H_2 to give the corresponding phosphine–borane adducts $R_2PH \cdot HB(C_6F_5)_2$, whereas the four-membered rings $[R_2PB(C_6F_5)_2]_2$ (R = Et, Ph) are unreactive towards H_2.[68]

Six- and eight-membered rings of the type $(H_2BPR_2)_n$ ($n = 3$ or 4; R = alkyl or aryl) are made by dehydrogenation of the borane adducts $H_3B \cdot PHR_2$. Thus, the thermally very stable trimer $(H_2BPMe_2)_3$ and smaller amounts of the somewhat less stable tetramer $(H_2BPMe_2)_4$ are obtained by thermolysis of $H_3B \cdot PHMe_2$ at 150 °C.[69] More recently, the transition metal-catalysed dehydrogenation of $H_3B \cdot PHPh_2$ was shown to give a mixture of the related *P*-phenylated derivatives $(H_2BPPh_2)_n$ (**9.18**, $n = 3$; **9.19**, $n = 4$) [eqn (9.12)].[71]

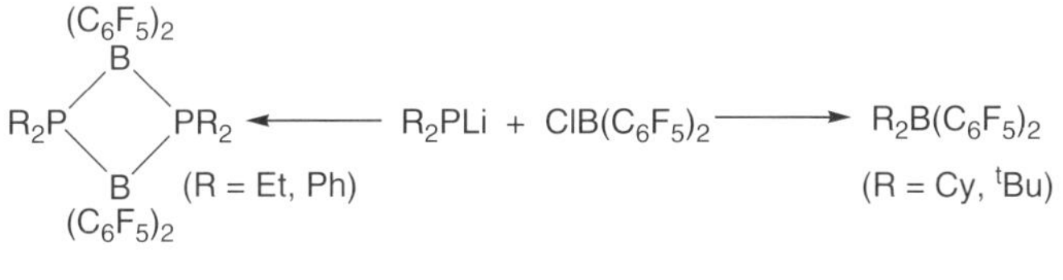

Scheme 9.13 Synthesis of monomeric and dimeric phosphinoboranes.

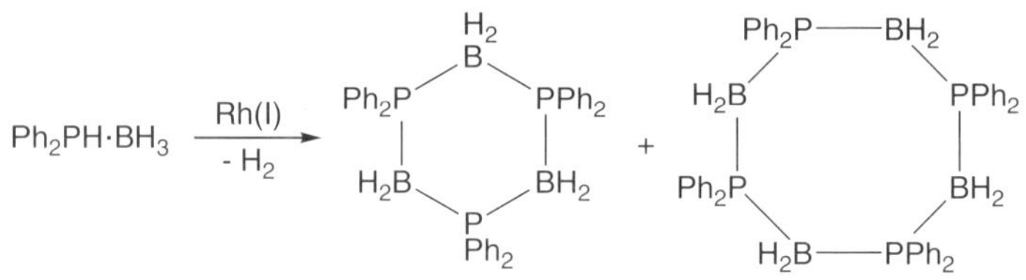

$$Ph_2PH \cdot BH_3 \xrightarrow[-H_2]{Rh(I)} \mathbf{9.18} + \mathbf{9.19} \qquad (9.12)$$

The four-membered rings in $(Me_2BP^tBu_2)_2$,[67b] $[X_2BP(SiMe_3)_2]_2$ (X = Cl, Br)[67a] and $[Et_2PB(C_6F_5)_2]_2$[68] are essentially planar with distorted tetrahedral geometry at the four-coordinate boron and phosphorus centres; the mean B–P bond lengths are 2.08, 2.02–2.03 and 2.06 Å, respectively, cf. 1.79 Å for the monomer $^tBu_2PB(C_6F_5)_2$.[68] The relatively weak B–P bond in $(Me_2BP^tBu_2)_2$ is manifested by dissociation of this dimer into two $Me_2BP^tBu_2$ monomers in solution.[67b] The six-membered ring $(H_2BPPh_2)_3$ has a chair conformation in the solid state with P–B bond lengths of 1.95 Å,[70] whereas the eight-membered ring $(H_2BPPh_2)_4$ adopts the boat–boat conformation with a mean P–B distance of 1.94 Å (Figure 9.10).[71]

9.3.4 Boron–Phosphorus Chains and Polymers

The transition metal-catalysed dehydrocoupling of phosphine–borane adducts has proved to be a useful route to phosphinoborane chains and polymers in addition to cyclic species. The dehydrocoupling of the secondary phosphine–borane adduct $H_3B \cdot PHPh_2$ in the presence of $[\{Rh(\mu\text{-Cl})(1,5\text{-cod})\}_2]$ or $[Rh(1,5\text{-cod})_2][OTf]$ (0.5–1 mol% Rh) gives the linear dimer $H_3B-PPh_2-BH_2-PPh_2H$ (**9.20**) at 90 °C [eqn (9.13)].[71] The catalytic potential of other complexes based on other transition metals (e.g. Ru, Ir, Pd, Pt) was also demonstrated. In the absence of any added catalyst, no significant conversion of $Ph_2PH \cdot BH_3$ was observed at this temperature.

$$2H_3B \cdot PHPh_2 \rightarrow \underset{\mathbf{9.20}}{H_3B-PPh_2-BH_2-PPh_2H} + H_2 \qquad (9.13)$$

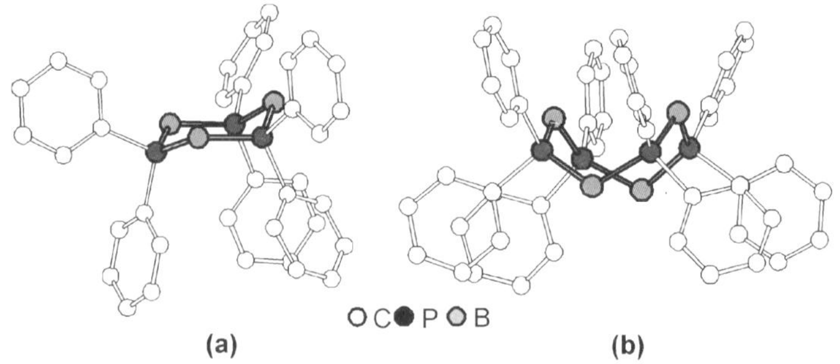

Figure 9.10 Molecular structures of (a) $(H_2BPPh_2)_3$ and (b) $(H_2BPPh_2)_4$.

Early work on the pyrolysis of the primary phosphine–borane adduct $H_3B \cdot PH_2Ph$ in the mid-1960s suggested that low molecular weight polymers of possible formula $(BH_2-PhPH)_n$ could exist. For instance, heating $H_3B \cdot PH_2Ph$ at temperatures between 100 and 250 °C was reported to give a benzene-soluble polymer with an approximate composition corresponding to $(PhPH-BH_2)_n$ (**9.21**) as determined by elemental analysis and a molecular weight (M_n) of up to 2600, together with substantial amounts of insoluble material.[72] When $H_3B \cdot PH_2Ph$ was heated in toluene at 110 °C with 0.5–1 mol% of [Rh(1,5-cod)$_2$][OTf], polyphenylphosphinoborane **9.21** was isolated as an off-white solid that is air- and moisture-stable in the solid state [eqn (9.14)].[71] This material is an analogue of polystyrene with a B–P instead of a C–C backbone. The molecular weight of **9.21** prepared by this solution method was found to be relatively low ($M_w = 5600$). Heating $H_3B \cdot PH_2Ph$ and a catalytic amount of [{Rh(μ-Cl)(1,5-cod)}$_2$] (*ca* 0.6 mol% rhodium) in the absence of solvent afforded much higher molecular weight material ($M_w = 33\,300$).

$$H_3B \cdot PH_2Ph \xrightarrow[-H_2]{[Rh]\,cat.} \left[\begin{array}{cc} H & Ph \\ | & | \\ B & -P \\ | & | \\ H & H \end{array} \right]_n$$
9.21

(9.14)

Several other examples of polyphosphinoboranes have been generated by using a similar catalytic dehydrogenation route.[73] The presence of electron-withdrawing groups at phosphorus lowers the temperature required for the catalytic dehydrogenation. For example, the introduction of *para*-CF$_3$ groups on the phenyl substituent allows temperatures as low as 60 °C to be used for the Rh-catalysed dehydrogenation reaction.[74]

Polyphosphinoboranes are of interest with respect to their potentially useful physical properties such as flame-retardant behaviour and an ability to function as precursors to BP-based ceramics. In addition, the electron-beam sensitivity of some polyphosphinoboranes has been demonstrated. This allows their use as lithographic resists for patterning applications when coated as thin films on substrates such as silicon (Figure 9.11).[74]

9.4 ALUMINIUM–, GALLIUM– AND INDIUM–NITROGEN RINGS

9.4.1 Unsaturated Systems

In contrast to iminoboranes (Section 9.2), iminoalanes usually oligomerise to give cage structures $(RAlNR')_n$ ($n = 4-16$) as a result of the larger size of aluminium and the high polarity of the Al–N bond.[75] Cubanes ($n = 4$) and hexagonal prisms ($n = 6$) are the most common architectures; cubane structures are also formed by iminogallanes. Aluminium analogues of the six-membered borazines $(RBNR')_3$ or the four-membered $(RBNR')_2$ rings can, however, be obtained by using the stabilising influence of suitably bulky substituents. Several examples of

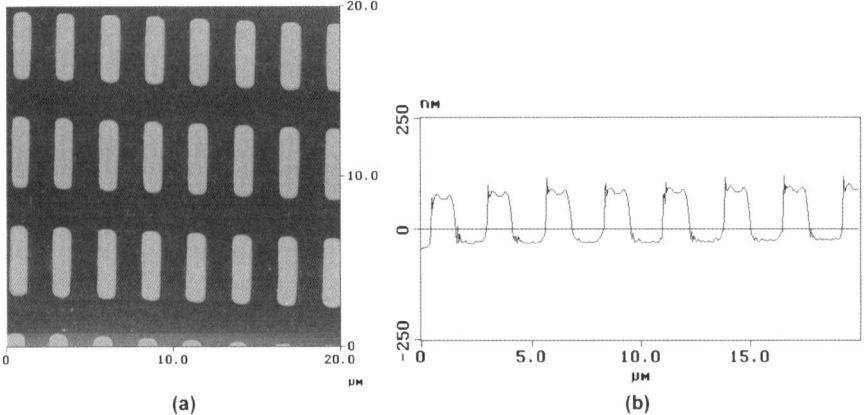

Figure 9.11 (a) Tapping-mode atomic force microscopy images and (b) cross-sectional analysis images of bars of $[BH_2-PH(p-CF_3C_6H_4)]_n$ lithographically patterned on silicon using electron-beam lithography. Reproduced with permission from T. J. Clark et al., Chem. Eur. J., 2005, **11**, 4526.[74]

unsaturated gallium–nitrogen rings, and also cyclic systems involving alternating group 13 elements and the heavier pnictogens, are known. The primary interest in these inorganic heterocycles is the extent of π-electron delocalisation.

The general approach to oligomers of the type $(RMNR')_n$ (M = Al, Ga, In) involves alkane elimination reactions between a trialkyl group 13 derivative and a primary alkyl- or arylamine. The first step proceeds under relatively mild conditions and gives aminometallanes, usually as dimeric four-membered rings $[R_2MN(H)R']_2$; however, the subsequent alkane elimination requires more vigorous conditions, e.g. boiling aromatic solvents [eqn (9.15)]. The subtle influence of the steric size of the group on nitrogen on the nature of the oligomer formed is strikingly illustrated by the reaction of trimethylaluminium with arylamines, $ArNH_2$.[76] When Ar = Dipp (2,6-di*iso*propylphenyl), a six-membered Al_3N_3 ring is formed, whereas an Al_4N_4 cubane is obtained for Ar = mesityl and an Al_6N_6 hexagonal pyramid is preferred when Ar = phenyl.

$$R_3M + H_2NR' \rightarrow \tfrac{1}{2}[R_2MN(H)R']_2 + RH \rightarrow 1/n(RMNR')_n + RH \quad (9.15)$$

The structure of the six-membered ring $(MeAlNdipp)_3$ (Figure 9.12b) consists of a planar $Al_3N_3C_6$ arrangement with Al–N bond lengths of ca 1.78 Å. The endocyclic bond angles deviate somewhat from the idealised value of 120° [NAlN (ca 115°) and AlNAl (ca 125°)].[76] Although the planar configuration of the Al_3N_3 ring allows efficient $Al(3p_\pi)$–$N(2p_\pi)$ overlap and the Al–N bonds are short compared with the predicted single bond value, calculations on the model system $(HAlNH)_3$ do not support significant resonance energy for this six-membered ring.[58b,77] The absence of significant delocalisation in the Al_3N_3 ring compared with the B_3P_3 ring (Section 9.3.1) is attributed to the higher polarity of the Al–N bond. The alumazine $(MeAlNDipp)_3$ behaves as a Lewis acid in

Group 13: Rings and Polymers 137

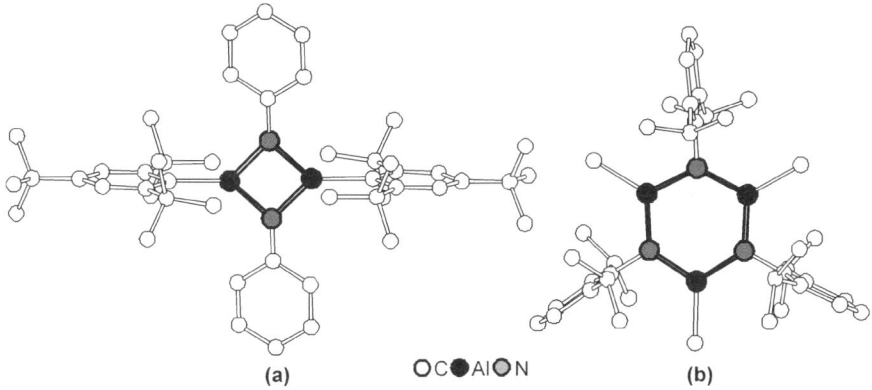

Figure 9.12 Molecular structures of (a) (Mes*AlNPh)$_2$ and (b) (MeAlNDipp)$_3$.

forming a variety of mono-, bis- and tris-adducts with Lewis bases such as pyridine and 4-dimethylaminopyridine.[78]

Dimeric iminoalanes, which contain an unsaturated Al$_2$N$_2$ ring, may also be prepared by alkane elimination. For example, the reaction of Cp$_3$Al with DippNH$_2$ produces (CpAlNDipp)$_2$ in hot toluene.[79] In a different approach, the treatment of the arylalane (Mes*AlH$_2$)$_2$ with aniline in ethylbenzene at 125 °C yields the dimer (Mes*AlNPh)$_2$ via the intermediate formation of the bis(amido)alane Mes*Al(NHPh)$_2$.[80] The Al$_2$N$_2$ rings in these two four-membered rings form almost perfect squares (Figure 9.12a) with Al–N bond lengths in the range 1.80–1.82 Å, slightly longer than the Al–N bonds in the six-membered ring (MeAlNDipp)$_3$. The reagent (Mes*AlH$_2$)$_2$ has also been used to generate a five-membered Al$_2$N$_3$ ring with three-coordinate Al and N atoms and one N–N bond by reaction with azobenzene PhN=NPh.[81]

The alkane elimination route is not suitable for the synthesis of unsaturated gallium–nitrogen rings because the electrophilic gallium centres engender activation of C–H groups on the alkyl substituents or the aryl groups attached to nitrogen at the high temperatures necessary for thermolysis.[76] The four-membered ring (η^1-Cp*GaNXyl)$_2$ (Xyl = 2,6-dimethylphenyl) is obtained, however, by oxidation of Cp*Ga(I) with xylyl azide.[82] The Ga$_2$N$_2$ ring in this derivative is a slightly distorted square with a mean Ga–N distance of 1.86 Å, *cf.* 1.99 Å in the gallium–nitrogen cages in which the gallium centre is three-coordinate. A value of 1.88 Å is found for a related 1,3-Ga$_2$N$_2$ ring [ArGaN(NCPh$_2$)]$_2$ (Ar = 2, 6-Dipp-C$_6$H$_3$) obtained by the oxidation ArGa(I) with diphenyldiazomethane Ph$_2$CN$_2$.[83] Although the isomeric 1,2-Ga$_2$N$_2$ ring is predicted to be substantially less stable than the 1,3-isomer, the cyclic derivative ArGaN(p-tolyl)N(p-tolyl)-GaAr is isolated from the reaction of solutions of ArGa(I) with (p-tolyl)N=N(p-tolyl). The trapezoidal Ga$_2$N$_2$ core in this blue–green compound is planar; all bond lengths are close to single-bond values (Figure 9.13).[83]

Monomeric imides of aluminium, gallium and indium have been stabilised by installing extremely bulky groups on both the group 13 element and the

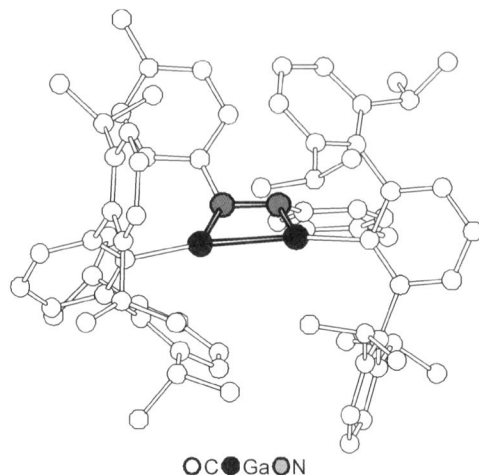

Figure 9.13 Molecular structure of ArGaN(p-tolyl)N(p-tolyl)GaAr.

nitrogen atoms;[84,85] however, their synthesis requires a different approach. A monomeric iminoalane or iminogallane is obtained by the oxidation of the metal(I) β-ketiminate monomers [{HC(CMeDippN)$_2$}M:] with the sterically encumbered azide N$_3$-2,6-Trip$_2$C$_6$H$_3$ (Trip = 2,4,6-^{i}Pr$_3$C$_6$H$_2$).[84] Alternatively, the treatment of the dimetallenes ArMMAr (M = Ga, In) with an azide N$_3$Ar′ (Ar and Ar′ are both extremely bulky aryl groups) produces iminogallanes or an iminoindane.[85] The iminogallanes and the iminoindane have been structurally characterised,[84,85] but the structure of the iminoalane is unknown.

9.4.2 Saturated Systems

The reaction of trialkylaluminium compounds with ammonia gas in hot hydrocarbon solvents proceeds with alkane elimination to give the saturated trimers (R$_2$AlNH$_2$)$_3$ (R = Me, ^{t}Bu) in which the conformation of the six-membered Al$_3$N$_3$ rings is markedly influenced by the substituent on aluminum.[86] The methyl derivative exhibits a twist-boat conformation whereas the *tert*-butyl compound is planar (Figure 9.14). By contrast, the Ga$_3$N$_3$ ring in (H$_2$GaNH$_2$)$_3$ adopts a chair conformation [*cf.* (H$_2$BNH$_2$)$_3$].[87] The potential application of this all-protonated cyclotrigallazane as a single-source precursor for gallium nitride, a light-emitting diode, is discussed in Section 9.5.3.

9.5 ALUMINIUM–, GALLIUM– AND INDIUM–PNICTOGEN RINGS

9.5.1 Unsaturated Systems

Several methods are available for the synthesis of ring systems comprised of the heavier group 13 and group 15 elements. Six-membered Al$_3$E$_3$ rings (E = P, As) are obtained from the reactions of (Mes*AlH$_2$)$_2$ with phenylphosphine or phenylarsine at elevated temperatures (*ca* 160 °C) [eqn (9.16)].[80] The six-membered rings in (Mes*AlEPh)$_3$ (E = P, As) adopt a boat conformation in the solid state (Figure 9.15b). The aluminium atoms exhibit planar coordination, whereas the phosphorus and arsenic atoms are distinctly pyramidal.[72] This geometry and the Al–E bond lengths, which are close to single-bond values, indicate a lack of delocalised π-bonding in these formally unsaturated ring systems.

$$(\text{Mes}^*\text{AlH}_2)_2 + 2\text{H}_2\text{EPh} \rightarrow 2/3(\text{Mes}^*\text{AlEPh})_3 + 4\text{H}_2$$
$$(\text{E} = \text{P, As})$$
(9.16)

The reactions of organogallium dihalides with lithium phosphide reagents provide a route to unsaturated four- or six-membered gallium–phosphorus ring systems. The four-membered ring (^{t}BuGaPMes*)$_2$ (**9.23**) is synthesised from *tert*-butylgallium dichloride and LiPHMes* in a two-step process (Scheme 9.14); the intermediate ^{t}BuGa(HPMes*)$_2$ (**9.22**) is isolated prior to thermolysis in boiling toluene.[88] The six-membered ring (TriphGaPCy)$_3$ (Triph = 2,4,6-Ph$_3$C$_6$H$_2$) is prepared by treatment of (Triph)gallium dichloride with Li$_2$PCy in a mixture of toluene and diethyl ether at room temperature.[89]

The mean Ga–P bond distances in (^{t}BuGaPMes*)$_2$ and (TriphGaPCy)$_3$ (2.27 and 2.30 Å, respectively) are close to single-bond values. The phosphorus

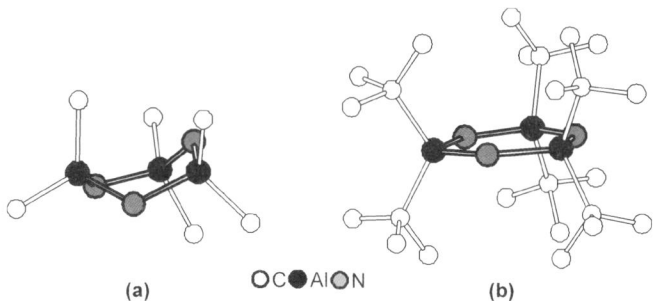

Figure 9.14 Conformations of the Al$_3$N$_3$ ring in (R$_2$AlNH$_2$)$_3$, (a) R = Me and (b) R = ^{t}Bu.

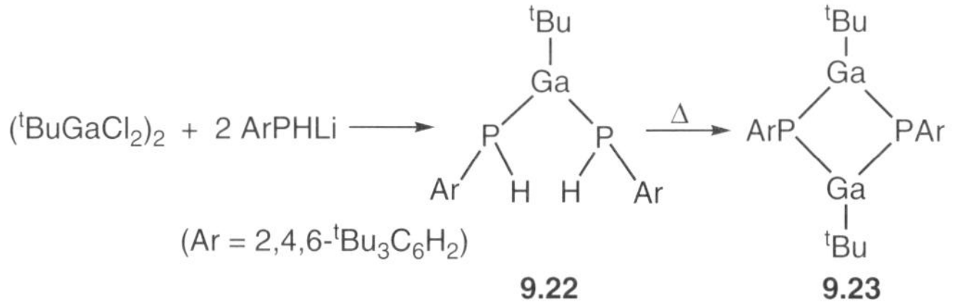

Scheme 9.14 Synthesis of a Ga$_2$P$_2$ ring.

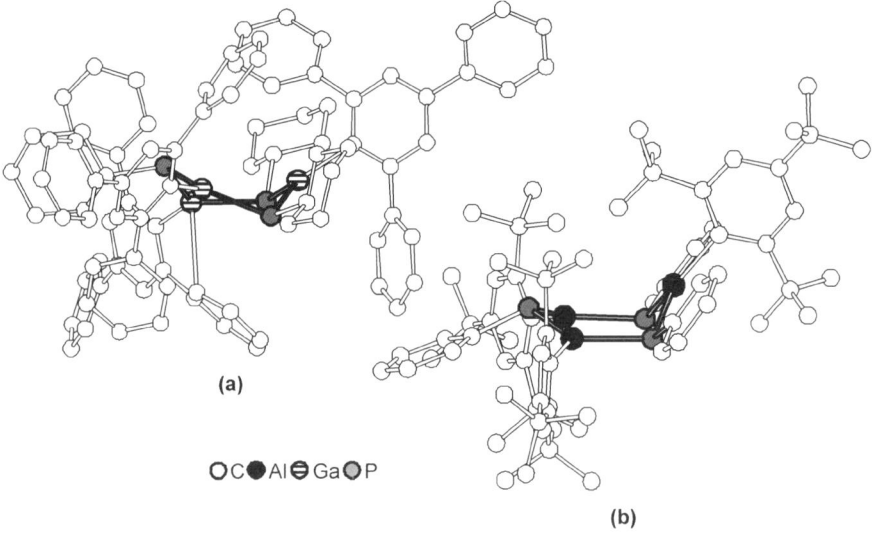

Figure 9.15 Molecular structures of (a) (TriphGaPCy)$_3$ and (b) (Mes*AlPPh)$_3$.

centres in both the four- and six-membered rings are pyramidal and the Ga$_3$P$_3$ ring is markedly puckered (Figure 9.15a), indicating minimal π-delocalisation in these unsaturated gallium–phosphorus rings.[89]

9.5.2 Saturated Systems

Pioneering work on the chemistry of saturated heavier group 13–group 15 ring systems outdates the discovery of the related unsaturated heterocycles by more than 25 years.[90] The saturated systems (R$_2$MER$'_2$)$_n$ form either four-membered ($n = 2$) or six-membered ($n = 3$) rings in which the bridging pnictogen group: ER$'_2$ donates a pair of electrons to the group 13 metal (Figure 9.16). Since the different resonance forms contribute equally to these structures, the M–E bond lengths in these ring systems are approximately equal.

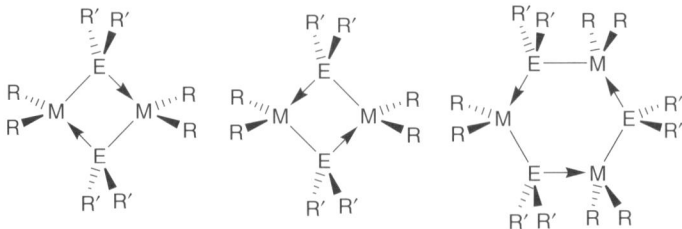

Figure 9.16 Schematic structures of saturated group 13–pnictogen ring systems (only one resonance form is shown for the six-membered ring).

Interest in this class of inorganic heterocycles was rekindled in the late 1980s because of the potential applications of binary semiconductors involving the heavier group 13 and 15 elements.[91] In the materials science community, these semiconductors are referred to as III–V compounds. For example, compound semiconductors such as gallium arsenide (GaAs) and indium phosphide (InP) are used in the fabrication of microelectronic and optoelectronic devices. Gallium nitride (GaN) is a very efficient light-emitting diode that emits green light; it is used in traffic signals in various parts of the world.

In many of these applications, it is necessary to generate these compound semiconductors as highly pure, thin films. The current industrial technology involves the transport of two precursors, often an organometallic compound, e.g. Me$_3$Ga or Me$_3$In, and a volatile pnictogen source, e.g. the hydrides EH$_3$ (E = P, As). The obvious disadvantages of this process include (a) the pyrophoric nature of the alkyl–metal reagents, (b) the toxicity of the pnictogen sources and (c) the high temperatures (ca 600 °C) required for the deposition of thin films.[92] In order to circumvent these problems, the concept of single-source precursors has been developed. These precursors contain both the group 13 and the group 15 element in the same molecule, preferably as nearest neighbours. Although these precursors should be sufficiently volatile for transport in the vapour phase to the heated substrate surface using the traditional chemical vapour deposition (CVD) technique, the requirement for high volatility has been overcome in recent years through the development of methods such as aerosol-assisted (AA) CVD or liquid-injection CVD, both of which use solutions of the precursor in organic solvents.[93]

There are several synthetic approaches to saturated group 13–group 15 rings.[91] Alkane or arene elimination reactions are effective in many cases [eqn (9.17)].[90] Dehydrosilation reactions between Sb(SiMe$_3$)$_3$ and R$_2$AlH (R = Et, ^{i}Bu), with elimination of Me$_3$SiH, have been employed for the preparation of Al$_2$Sb$_2$ rings.[94]

$$Me_3M + R_2EH \rightarrow 1/n(Me_2MER_2)_n + CH_4 \quad (9.17)$$
$$(M = Ga, In; E = P, As; R = Me, Ph)$$

As an alternative to the hydrocarbon-elimination method, salt-elimination reactions involving an organometallic halide, which can be generated *in situ*, and an alkali-metal pnictide have also been used (Scheme 9.15a).[95] The elimination of a volatile triorganosilyl halide is viable in cases where the steric bulk of substituents results in a very slow alkane eliminations[96a] or for the synthesis of antimony-containing rings (Scheme 9.15b).[97b]

The four-membered rings (R$_2$MER'$_2$)$_n$ are either planar or approximately planar with equal M–E bond lengths. The six-membered rings (n = 3) adopt a variety of conformations including planar [(^{t}Bu$_2$GaPH$_2$)$_3$],[97] and (Me$_2$-InAsMe$_2$)$_3$; the latter also exists in a puckered ring[95]], boat [(Me$_2$GaAs^{i}Pr$_2$)$_3$],[81] twist-boat [(Br$_2$GaAs(CH$_2$SiMe$_3$)$_2$)$_3$][98] and (Me$_2$MSb^{t}Bu$_2$)$_3$ (M = Ga, In)][98] and irregular boat [(Cl$_2$GaSb^{t}Bu$_2$)$_3$].[96] The Al$_3$P$_3$ ring in (Me$_2$AlPMe$_2$)$_3$ assumes a chair conformation in the vapour phase according to an electron diffraction

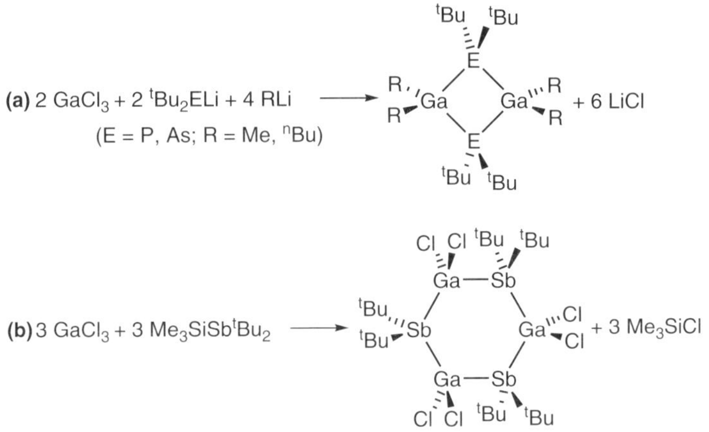

(a) 2 GaCl$_3$ + 2 ^{t}Bu$_2$ELi + 4 RLi ⟶ [ring product] + 6 LiCl

(E = P, As; R = Me, ^{n}Bu)

(b) 3 GaCl$_3$ + 3 Me$_3$SiSb^{t}Bu$_2$ ⟶ [ring product] + 3 Me$_3$SiCl

Scheme 9.15 Syntheses of saturated group 13–pnictogen rings by (a) salt elimination and (b) trimethylsilyl chloride elimination.

study.[99] In general, the M–E bond lengths in the six-membered rings are significantly shorter than those in the corresponding four-membered rings. This is attributable, in part, to the relief of ring strain, but the steric influence of bulky substituents on the group 13 and/or 15 atoms also plays a role.

9.5.3 Single-source Precursors for III–V Semiconductors

The four-membered rings (Me$_2$GaAs^{t}Bu$_2$)$_2$ and (Me$_2$InP^{t}Bu$_2$)$_2$ are suitable single-source precursors for GaAs and InP, respectively. These compound semiconductors are generated as carbon-free thin films by using the CVD technique. The decomposition process involves the elimination of isobutylene, implying the intermediate formation of E–H bonds.[100] The six-membered rings (^{t}Bu$_2$GaEH$_2$)$_3$ (E = P, As), with hydrogen substituents on the group 15 atoms, decompose to give GaE at *ca* 250 °C in the solid state and at *ca* 110 °C in toluene solution.[97]

The unique properties of gallium nitride, including superior robustness and power efficiency compared with silicon or gallium arsenide, combine to make this wide-bandgap (3.5 eV) semiconductor a promising candidate for reducing the demand for electricity.[101] The technologically significant semiconductors have two common structures, zinc blende (cubic) and wurtzite (hexagonal). Among the III–V semiconductors, only the nitrides of aluminium, gallium and indium exhibit the wurtzite structure. Since there is often a relationship between the structures of the molecular precursor and the solid-state product, various gallium–nitrogen ring systems have been investigated as a source of gallium nitride. Such ring systems should preferably exclude the direct attachment of organic substituents to gallium in order to avoid carbon contamination of the final product.

The cyclotrigallazane (H$_2$GaNH$_2$)$_3$ is readily prepared by passing gaseous ammonia over solid Me$_3$N · GaH$_3$ at room temperature.[87] Pyrolysis of

(H_2GaNH_2)$_3$ at 500 °C produces nanocrystalline GaN with a significant impurity of gallium metal.[102] The nanocrystalline material slowly converts to the wurtzite phase at 900 °C. Since dimethylhydrazine H_2NNMe_2 decomposes at much lower temperatures than ammonia, the dimeric four-membered ring ($H_2GaNHNMe_2$)$_2$ has also been considered as a precursor to GaN.[103]

9.6 BORON–CHALCOGEN RINGS

9.6.1 Boron–Oxygen Rings

The most common arrangement for boron–oxygen heterocycles, by far, is the six-membered B_3O_3 ring system, which occurs either in borates $[B_3O_6]^{3-}$ (**9.24**) or in the organic derivatives $(RBO)_3$ (**9.25**).[104a] Although the monocyclic B_3O_3 ring system occurs in metaborates such as $M[B_3O_6]$ (M = K, Cs) and $M[B_3O_6]_2$ (M = Ba), many borates involve bi- or tricyclic structures in which one (or more) of the boron centres is four-coordinate through addition of an OH$^-$ group; polymeric structures in which B_3O_3 rings are linked into chains are also known.[104b]

9.24 **9.25**

Cyclotriboroxanes $(RBO)_3$ (also known as boroxines) are formally anhydrides of the corresponding boronic acids $RB(OH)_2$; only mild heating is required to convert these dibasic acids into a wide range of cyclic trimeric anhydrides [eqn (9.18)]. This dehydration process can be adapted for the synthesis of B_3O_3 rings with different aryl groups on the boron atoms by using two or more arylboronic acids in the appropriate stoichiometric ratio.[105]

$$3RB(OH)_2 \rightarrow (RBO)_3 + 3H_2O \qquad (9.18)$$

The recent resurgence of interest in B_3O_3 rings emanates from a variety of potential applications in materials science,[106] and also from the use of arylboroxines as components of catalysts in organic synthesis.[107] In an interesting application of the methodology shown in eqn (9.18), three-dimensional covalent organic frameworks based on a central B_3O_3 core have been synthesised.[106a] These porous materials exhibit a combination of properties [high thermal stabilities (400–500 °C), high surface areas and low densities] that make them attractive candidates for applications in gas separation and storage technology. Other potential uses of molecules based on a cyclic B_3O_3 framework include their incorporation in non-linear optical or photoluminescent

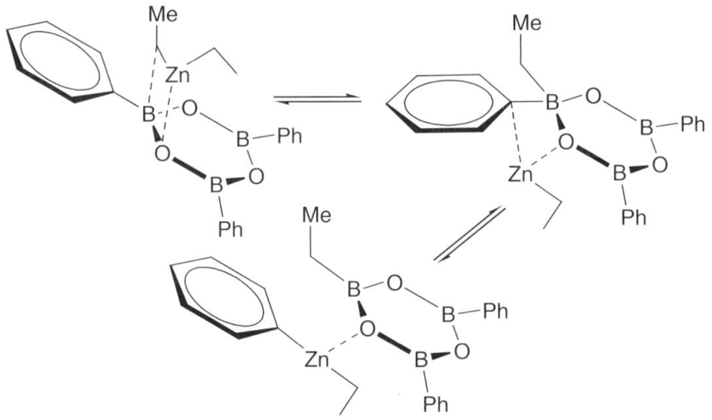

Scheme 9.16 Mechanism of transmetallation from (PhBO)₃ to Et₂Zn.

materials.[106b,106c] The reactions of [(MeO)BO]₃ with alkoxides of Zr or Al produce mixed oxide ceramic materials, *e.g.* $ZrO_2 \cdot B_2O_3$, in high yields and purity.[106d]

The transmetallation from (PhBO)₃ to diethylzinc has been shown to occur in a two-step process (Scheme 9.16). This catalytic system is highly effective for the enantioselective arylation of aldehydes to give diaryl carbinols.[107]

The structure and bonding in (HBO)₃ are of interest in comparison with those of the isoelectronic six-membered borazine ring (HBNH)₃, which has been studied in much more detail (see Section 4.1.2.1). The B_3O_3 ring in (HBO)₃ (and other derivatives of the B_3O_3 ring) is essentially planar (see structure **3.1**). The mean B–O bond length of 1.38 Å indicates significant multiple bond character. However, π-delocalisation in (HBO)₃ is less pronounced than that in borazine because of the higher electronegativity of oxygen compared with that of nitrogen. The planarity of the B_3O_3 ring is disrupted in the anions $[Ph_4B_3O_3]^-$ and $[(C_6F_5)_5B_3O_3]^{2-}$, which contain one or two four-coordinate boron centres, respectively.[108a,b]

9.6.2 Boron–Sulfur and –Selenium Rings

In marked contrast to the structure of boric oxide, B_2O_3, which consists of a three-dimensional network of BO_3 units linked by oxygen atoms, the binary boron sulfide, B_2S_3, has a fascinating layer structure comprised of planar B_3S_3 and B_2S_2 rings linked by sulfur bridges (Figure 9.17).[109a] All the boron atoms are trigonal planar with mean B–S distances of 1.81 Å and an interlayer separation of 3.55 Å. Another interesting binary boron sulfide has the empirical formula BS_2 (see **6.12** in Section 6.2.1). It is comprised of five-membered B_2S_3 rings linked by sulfur atoms to give a planar porphin-like molecule.[109b]

The most common organic derivatives of boron–sulfur rings are based on the three ring systems found in the binary boron sulfides, *viz.* the five-membered rings $(RB)_2S_3$ (**9.27**) and six-membered rings $(RBS)_3$ (**9.28**) and, less frequently,

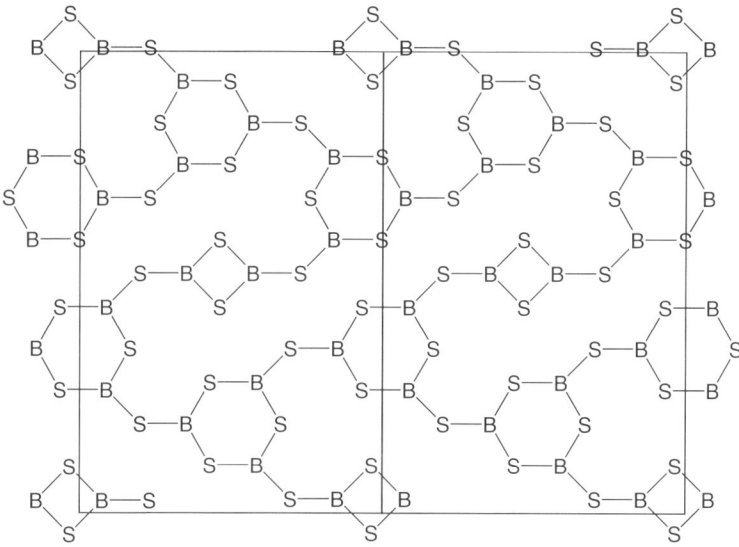

Figure 9.17 Schematic structure of B_2S_3 showing the B_2S_2 and B_3S_3 rings.

the four-membered rings $(RBS)_2$ (**9.26**). Selenium analogues of the five- and six-membered rings are also known.[110]

Although the synthesis of these ring systems can be achieved by the reactions of organoboron dihalides with a variety of sulfur sources, e.g. elemental sulfur, organic disulfides RSSR, H_2S_x ($x > 1$), $(Me_3Si)_2S_x$ ($x = 1-3$), HgS or alkali metal sulfides, the outcome of these reactions is not always predictable.[110] Nevertheless, a variety of triaryl derivatives of the six-membered rings $(ArBS)_3$ are reliably obtained in excellent yields from the reaction of $ArBBr_2$ with HgS in benzene at reflux.[111a] The five-membered rings $Ar_2B_2S_3$ are isolated in moderate yields by treatment of $ArBBr_2$ with $^tBuSS^tBu$ in boiling toluene.[111b] B-Halo derivatives are obtained when boron trihalides are employed as the boron source. For example, the reaction of BI_3 with elemental sulfur in CS_2 at room temperature gives the five-membered ring $(IB)_2S_3$ (**9.29**), which is converted to the six-membered ring $(IBS)_3$ (**9.30**) on treatment with additional BI_3 at 90 °C (Scheme 9.17).[112]

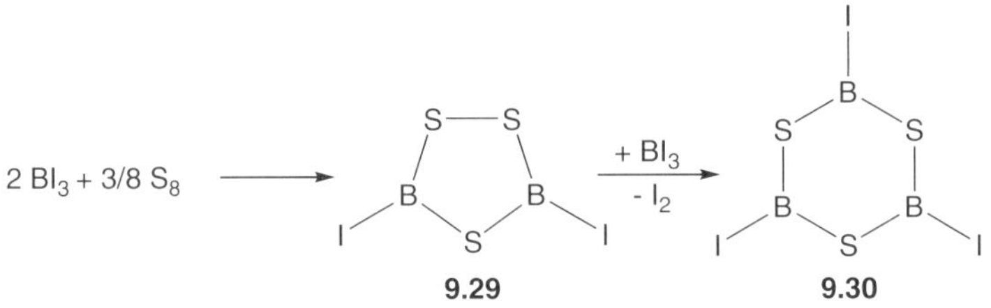

Scheme 9.17 Synthesis of iodinated B_2S_3 and B_3S_3 rings.

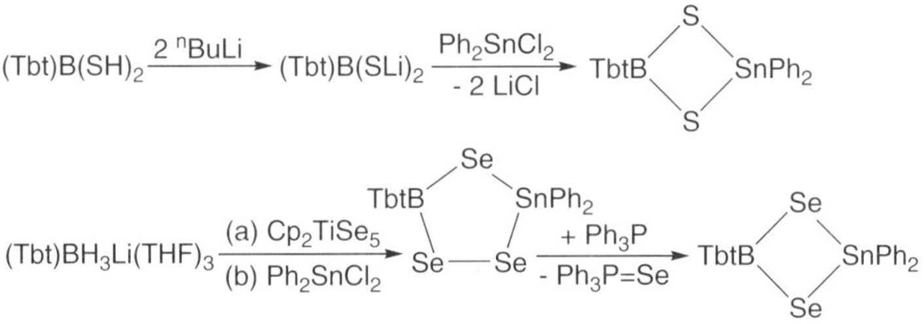

Scheme 9.18 Syntheses of four-membered BE_2Sn rings (E = S, Se).

The only structurally characterised four-membered boron–sulfur ring, $(Et_2NBS)_2$, has a planar structure;[113] the six-membered rings $(XBS)_3$ (X = Br,[114] SH[115]) are also essentially planar. The five-membered ring in $(PhB)_2S_3$ forms an envelope structure with the unique sulfur atom located 0.12 Å out of the BSSB plane.[116] The structures of the five-membered rings $(XB)_2S_3$ (X = Cl, Me)[117] in the gas phase have been determined by electron diffraction. The B–S and S–S distances in all of these boron–sulfur rings are close to single-bond values. Estimates of the delocalisation in the six-membered rings $(HBE)_3$ (E = O, S, Se) using the criterion of nucleus-independent chemical shift (see Section 4.1.2.1) indicate very small (ca 5 kcal mol^{-1}) aromatic stabilisation energies. On the other hand, magnetic susceptibility exaltation data are consistent with significant aromaticity.[118]

A fascinating aspect of the chemistry of boron–chalcogen ring systems is the use of four-membered rings containing tin, $(Tbt)B(\mu-E)_2SnPh_2$ {E = S, Se; Tbt = 2,4,6-tris[bis(trimethylsilyl)methyl]phenyl},[119,120] as a source of monomeric chalcogenaboranes RB=E (isovalent with iminoboranes RB=NR'). The synthetic approaches to the sulfur and selenium analogues of these heterocycles are different, as illustrated in Scheme 9.18.

The monomeric thia- or selenaboranes (Tbt)B=E (E = S, Se)[119a,120] are generated by heating the appropriate four-membered ring in toluene at 100–120 °C and trapped by cycloaddition with 2,3-dimethyl-1,3-butadiene (Scheme 9.19). The oxo analogue (Tbt)B=O is formed from the reaction of $(Tbt)B(\mu-S)_2SnPh_2$ with dimethyl sulfoxide.[121]

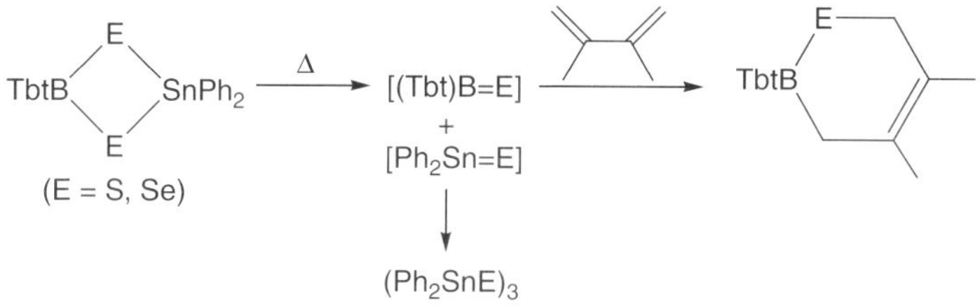

Scheme 9.19 Formation and trapping of monomeric (Tbt)B=E (E = S, Se) from cyclic precursors.

9.7 ALUMINIUM–, GALLIUM– AND INDIUM–OXYGEN RINGS

The tendency of aluminium to become four-coordinate, which was discussed for iminoalanes in Section 9.4, is also exhibited by organoalumoxanes $(RAlO)_n$. There are no aluminium analogues of the planar six-membered $(RBO)_3$ ring system. The most important class of alumoxanes are methylalumoxanes (MAOs), which are obtained by the partial hydrolysis of trimethylaluminium. The significance of MAOs is related to their use as effective co-catalysts for olefin polymerisation by metallocene catalysis. Since MAOs are comprised of a complex mixture of methylalumoxanes $(MeAlO)_n$, primarily involving cage structures with four-coordinate aluminium centres,[122,123a] the exact structure of this widely used reagent is unknown. Recent quantum chemical calculations show that nanotubular MAOs are readily formed from the hydrolysis of trimethylaluminium.[123b] The replacement of the methyl substituents in MAOs by bulkier *tert*-butyl groups has facilitated the structural characterisation of a series of organoalumoxane cages $[^tBuAl(\mu_3\text{-}O)]_n$ ($n = 6$–9) in which four-membered Al_2O_2 and six-membered Al_3O_3 rings serve as the building blocks.[124] Organogalloxanes form similar cage structures.

The attachment of bulky aryl groups to aluminium has allowed the synthesis of the cyclic tetramer $(Mes^*AlO)_4$ (**9.31**) from the reaction of $(Mes^*AlH_2)_2$ with the cyclic siloxane $(Me_2SiO)_3$; the eight-membered Al_4O_4 ring in **9.31** is almost planar.[125] A nine-membered dianionic Ga_3O_6 ring (**9.32**), in which the three four-coordinate gallium atoms are linked by three peroxo groups and bridged by a μ_3-oxo ligand, has been structurally characterised.[126]

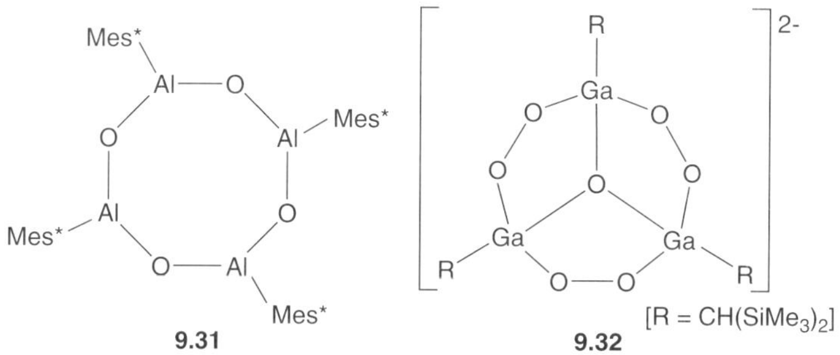

9.8 ALUMINIUM–, GALLIUM– AND INDIUM–CHALCOGEN RINGS

Interest in ring systems incorporating the heavier group 13 elements and chalcogens is driven by their potential use as single-source precursors to binary group 13–16 (III–VI) semiconductors. Like the corresponding oxygen-containing systems, compounds of the type $(RME)_n$ (M = Al, Ga, In; E = S, Se, Te) form cage structures, which are most commonly cubanes, *e.g.* [^{t}BuM(μ_3-E)]$_4$ (M = Al, Ga; E = S, Se Te)[127] and [Cp*M(μ_3-E)]$_4$ (M = Al; E = Se, Te[128] or M = Ga; E = S, Se).[129] The latter gallium complexes have been used to generate thin films of Ga_2E_3 (E = S, Se) using MOCVD techniques.[129] By contrast, the tetramer (^{t}Bu-GaS)$_4$, which is shown by electron diffraction studies to retain the cubane structure in the vapour phase, serves as a source of the cubic phase of GaS.[130]

The aggregation of group 13–chalcogen rings to form cages may be prevented or, alternatively, the cage structure is broken down by the formation of adducts with Lewis bases that coordinate to the group 13 centre. This strategy has been used to stabilise both four- and six-membered group 13–chalcogen rings. For example, the reaction of the adduct $Me_3N \cdot AlH_3$ with elemental selenium or tellurium generates bis(trimethylamine) adducts of the prototypical four-membered rings [HAl(μ-E)]$_2$ as the *trans* isomers [eqn (9.19)];[131] when E = Se, the latter react with diphenyl dichalcogenides to give *trans*-[{Me_3N(PhE)Al(μ-Se)}$_2$] (E = S, Se, Te).[132,133]

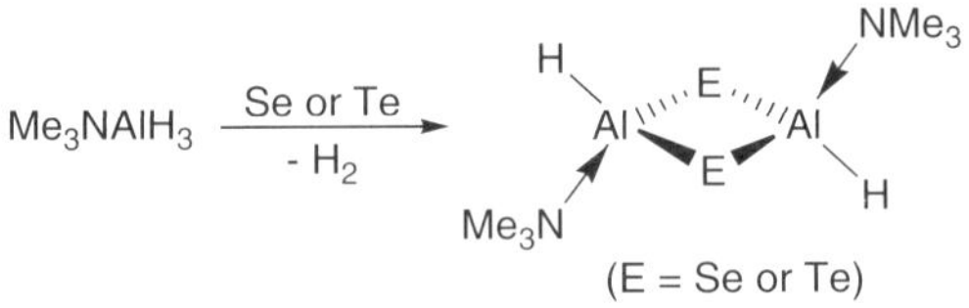

$$Me_3NAlH_3 \xrightarrow[-H_2]{\text{Se or Te}} \text{[four-membered ring]} \quad (E = Se \text{ or } Te) \tag{9.19}$$

The reaction of *trans*-[{Me_3N(H)Al(μ-Se)}$_2$] with N,N,N',N'',N''-pentamethyldiethylenetriamine (PMEDTA) affords an interesting bicyclic compound with an Al_4Se_5 (**9.33b**) skeleton; the sulfur analogue of this bicycle **9.33a** is obtained from the direct treatment of $Me_3N \cdot AlH_3$ with bis(trimethylsilyl) sulfide at 90 °C in toluene (Scheme 9.20).[134]

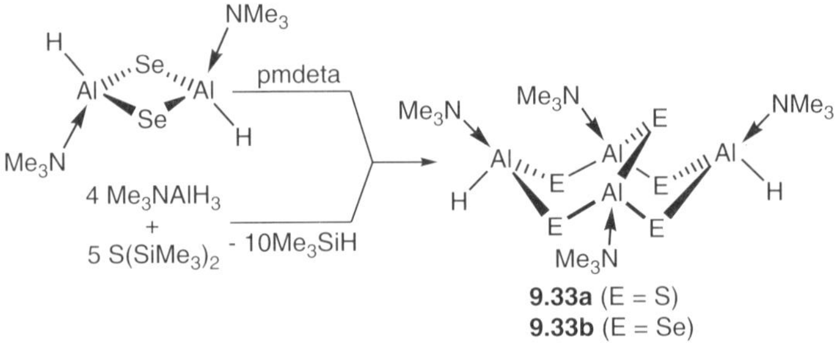

9.33a (E = S)
9.33b (E = Se)

Scheme 9.20 Formation of bicyclic Al_4E_5 (E = S, Se) frameworks.

The cage structure in the oligomers (^{t}BuGaS)$_n$ ($n = 4, 7, 8$) is broken down upon treatment with pyridine to give the trimer (^{t}BuGaS·py)$_3$ with a central Ga$_3$S$_3$ ring in a twist-boat conformation.[135] Similar soluble adducts with Ga$_3$E$_3$ (E = S, Se) rings are formed when pyridine donors are added to the insoluble ternary compounds GaEBr (E = S, Se) in acetonitrile (Scheme 9.21).[136]

An interesting approach to tellurium-containing six-membered rings of the group 13 elements **9.34a** and **9.34b** involves the *in situ* oxidation of GaI or InCl, respectively, with elemental tellurium in the presence of the sodium salt of the chelating ligand [TeP(^{i}Pr)$_2$N(^{i}Pr)$_2$PTe]$^-$ (Scheme 9.22).[137] The four-coordination of the group 13 metal centres in **9.34** prevents further aggregation of the six-membered M$_3$Te$_3$ (M = Ga, In) rings, which have a distorted boat conformation. The indium complex generates pure thin films of cubic In$_2$Te$_3$, a low-bandgap semiconductor, using aerosol-assisted chemical vapour deposition and substrate temperatures of 325–475 °C.[138]

An extensive series of chalcogenolates of group 13 metals of the type (R$_2$MER')$_n$ is known for all combinations of M and E. Most commonly these compounds are dimeric ($n = 2$) with a central, planar, four-membered M$_2$E$_2$ ring. For example, all three aluminium derivatives [^{t}Bu$_2$Al(μ-E^{t}Bu)]$_2$ are

Scheme 9.21 Formation of pyridine adducts of Ga$_3$E$_3$ (E = S, Se) rings.

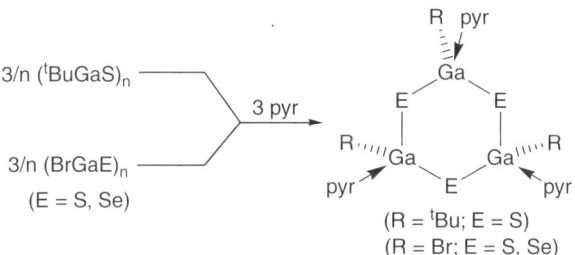

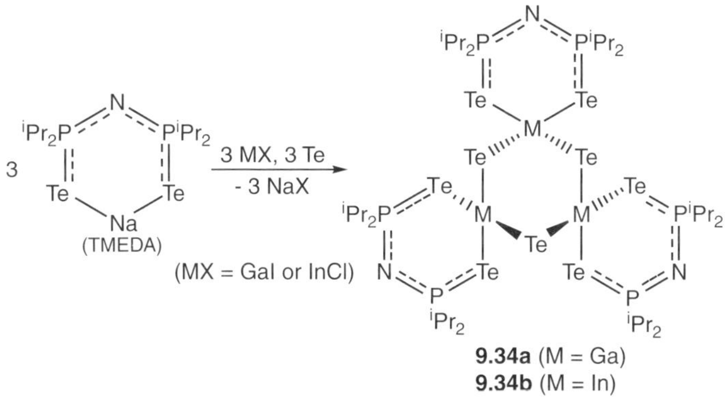

9.34a (M = Ga)
9.34b (M = In)

Scheme 9.22 Formation of six-membered M$_3$Te$_3$ rings (M = Ga, In).

obtained by insertion of the chalcogen into an Al–C bond of $^tBu_3Al$ at 25 °C (E = S, Se) or 70 °C (E = Te) [eqn (9.20)].[127]

$$2\,^tBu_3Al + 2\,E \longrightarrow \begin{array}{c} ^tBu\!-\!E\!-\!^tBu \\ ^tBu\!-\!Al\!\diagdown\!\!\diagup\!Al\!-\!^tBu \\ ^tBuE^tBu \end{array}$$

(E = S, Se, Te) (9.20)

A variety of other approaches are available for the synthesis of this class of group 13–chalcogen ring system, including (a) redistribution reactions,[139] (b) salt-elimination reactions[140] and (c) the use of dichalcogenide reagents REER.[141] Examples are given in eqns (9.21a–c). Several of these four-membered rings have been investigated as single-source precursors to group 13–16 (III–VI) semiconductors.

$$2Me_3In + In(SePh)_3 \longrightarrow 3/2[Me_2In(\mu\text{-}SePh)]_2 \quad (9.21a)$$

$$(Me_3CCH_2)_2GaCl + LiTePh \longrightarrow {}^1\!/_2[(Me_3CCH_2)_2Ga(\mu\text{-}TePh)]_2 \\ + LiCl \quad (9.21b)$$

$$Mes_3In + PhTeTePh \longrightarrow [Mes_2In(\mu\text{-}TePh)]_2 \quad (9.21c)$$

REFERENCES

1. C. E. Housecroft and A. G. Sharpe, *Inorganic Chemistry*, 2nd edn., Pearson Education, Harlow, 2005.
2. T. Davan and J. A. Morrison, *Inorg. Chem.*, 1979, **18**, 3194.
3. (a) T. Davan and J. A. Morrison, *J. Chem. Soc., Chem. Commun.*, 1981, 250; (b) T. Mennekes, P. Paetzold, R. Boese and D. Bläser, *Angew. Chem. Int. Ed. Engl. Engl.*, 1991, **30**, 172.
4. H. Klusik and A. Berndt, *J. Organomet. Chem.*, 1982, **234**, C17.
5. M. L. McKee, *Inorg. Chem.*, 1999, **38**, 321.
6. C.-J. Maier, H. Pritzkow and W. Siebert, *Angew. Chem. Int. Ed.*, 1999, **38**, 1666.
7. H. Nöth and H. Pommerening, *Angew. Chem. Int. Ed. Engl.*, 1980, **19**, 482.
8. M. Baudler, K. Rockstein and W. Oehlert, *Chem. Ber.*, 1991, **124**, 1149.
9. A. Maier, M. Hofmann, H. Pritzkow and W. Siebert, *Angew. Chem. Int. Ed.*, 2002, **41**, 1529.
10. C. Präsang, M. Hofmann, G. Geiseler, W. Massa and A. Berndt, *Angew. Chem. Int. Ed.*, 2002, **41**, 1526.
11. G. Linti, H. Schnöckel, W. Uhl and N. Wiberg, in *Molecular Clusters of the Main Group Elements*, ed. M. Driess and H. Nöth, Wiley-VCH, Weinheim, 2004, pp. 126–168.

12. C. Elschenbroich, *Organometallics*, Wiley-VCH, Weinheim, 2006, p. 133.
13. A. N. Alexandrova, A. I. Boldyrev, H.-J. Zhai and L.-S. Wang, *Coord. Chem. Rev.*, 2006, **250**, 2811.
14. A. N. Alexandrova, K. A. Birch and A. I. Boldyrev, *J. Am. Chem. Soc.*, 2003, **125**, 10786.
15. (a) J. Namatsu, N. Nakagawa, T. Muranaka, Y. Zenitani and J. Akimitsu, *Nature*, 2001, **410**, 63; (b) M. Monteverde, M. Nuñez-Regueiro, N. Rogado, K. A. Regan, M. A. Hayward, T. He, S. M. Loureiro and R. J. Cava, *Science*, 2001, **292**, 75.
16. B. Le Guennie, H. Jiao, S. Kahlal, J.-Y. Saillard, J.-F. Halet, S. Ghosh, M. Shang, A. M. Beatty, A. L. Rheingold and T. P. Fehlner, *J. Am. Chem. Soc.*, 2004, **126**, 3203.
17. (a) X. Li, A. E. Kuznetsov, H. F. Zhang, A. I. Boldyrev and L.-S. Wang, *Science*, 2001, **291**, 859; (b) A. E. Kuznetsov, A. I. Boldyrev, X. Li and L.-S. Wang, *J. Am. Chem. Soc.*, 2001, **123**, 8825.
18. J. M. Mercero, E. Formoso, J. M. Matxain, L. A. Eriksson and J. M. Ugalde, *Chem. Eur. J.*, 2006, **12**, 4496.
19. A. E. Kuznetsov, K. A. Birch, A. I. Boldyrev, X. Li, H.-J. Zhai and L.-S. Wang, *Science*, 2003, **300**, 622.
20. A. Datta and S. K. Pati, *J. Am. Chem. Soc.*, 2005, **127**, 3496.
21. X.-W. Li, Y. Xie, P. R. Schreiner, K. D. Gripper, R. C. Crittendon, C. F. Campana, H. F. Schaefer and G. H. Robinson, *Organometallics*, 1996, **15**, 3798.
22. R. J. Wright, M. Brynda and P. P. Power, *Angew. Chem. Int. Ed.*, 2006, **45**, 5953.
23. (a) A. Purath, R. Köppe and H. Schnöckel, *Angew. Chem. Int. Ed.*, 1999, **38**, 2926; (b) P. Yang, R. Köppe, T. Duan, J. Hartig, G. Hadiprono, B. Pilawa, I. Keilhauer and H. Schnöckel, *Angew. Chem. Int. Ed.*, 2007, **46**, 3579.
24. B. Twamley and P. P. Power, *Angew. Chem. Int. Ed.*, 2000, **39**, 3500.
25. (a) P. Paetzold, *Phosphorus Sulfur Silicon*, 1994, **93/94**, 39; (b) P. Paetzold, *Adv. Inorg. Chem.*, 1987, **31**, 123.
26. (a) T. Wideman and L. G. Sneddon, *Inorg. Chem.*, 1996, **34**, 1002; (b) P. J. Fazen and L. G. Sneddon, *Organometallics*, 1994, **13**, 2867.
27. (a) W. Maringgele, in *The Chemistry of Inorganic Homo- and Heterocycles*, ed. I. Haiduc and D. B. Sowerby, Academic Press, London, Vol. 1, 1987, Chapter 2, pp. 17–101; (b) B. Anand, H. Nöth, H. Schwenk-Kircher and A. Troll, *Eur. J. Inorg. Chem.*, 2008, 3186.
28. (a) A. Wakamiya, T. Ide and S. Yamaguchi, *J. Am. Chem. Soc.*, 2005, **127**, 14859; (b) I. H. T. Sham, C.-C. Kwok, C.-M. Che and N. Zhu, *Chem. Commun.*, 2005, 3547.
29. (a) H. Nöth, S. Rojas-Lima and A. Troll, *Eur. J. Inorg. Chem.*, 2005, 1895; (b) H. Nöth, B. Gemünd and A. Troll, *Eur. J. Inorg. Chem.*, 2007, 4282.
30. (a) H. Braunschweig, C. Kollann and M. Müller, *Eur. J. Inorg. Chem.*, 1998, 291; (b) H. Braunschweig, H. Green, K. Radacki and K. Uttinger, *Dalton Trans.*, 2008, 3531.

31. H.-J. Koch, H. W. Roesky, S. Besser and R. Herbst-Irmer, *Chem. Ber.*, 1993, **126**, 571.
32. (a) C. Prasang, B. Donnadieu and G. Bertrand, *J. Am. Chem. Soc.*, 2005, **127**, 10182; (b) A. Kausamo, H. M. Tuononen, K. E. Krahulic and R. Roesler, *Inorg. Chem.*, 2008, **47**, 1145.
33. P. Paetzold, J. Kiesgen, K. Krahé, H.-U. Meier and R. Boese, *Z. Naturforsch., Teil B*, 1991, **46**, 853.
34. E. v. Steuber, G. Elter, M. Noltemeyer, H.-G. Schmidt and A. Meller, *Organometallics*, 2000, **19**, 5083.
35. P. T. Clarke and H. M. Powell, *J. Chem. Soc. B*, 1966, 1172.
36. T. M. Gilbert and B. D. Gailbreath, *Organometallics*, 2001, **20**, 4727.
37. C. Fedorchuk, M. Copsey and T. Chivers, *Coord. Chem. Rev.*, 2007, **251**, 897.
38. (a) T. Chivers, X. Gao and M. Parvez, *Angew. Chem. Int. Ed. Engl.*, 1995, **34**, 2549; (b) T. Chivers, D. J. Eisler, C. Fedorchuk, G. Schatte, H. M. Tuononen and R. T. Boeré, *Chem. Commun.*, 2005, 3930.
39. (a) K.-H. van Bonn, P. Schreyer, P. Paetzold and R. Boese, *Chem. Ber.*, 1988, **121**, 1045; (b) F. Dirschl, E. Hanecker, H. Nöth, W. Rattay and W. Wagner, *Z. Naturforsch., Teil B*, 1986, **41**, 32.
40. E. Eversheim, U. Englert, R. Boese and P. Paetzold, *Angew. Chem. Int. Ed. Engl.*, 1994, **33**, 201.
41. (a) P. Paetzold, B. Redenz-Stormanns and R. Boese, *Chem. Ber.*, 1991, **124**, 2435; (b) P. Paetzold, B. Redenz-Stormanns and R. Boese, *Angew. Chem. Int. Ed. Engl.*, 1990, **29**, 900.
42. (a) H. Nöth and Regnet, *Z. Naturforsch., Teil B*, 1963, **18**, 1138; (b) K. Niedenzu, P. Fritz and H. Jenne, *Angew. Chem. Int. Ed. Engl.*, 1964, **3**, 506; (c) D. Nölle and H. Nöth, *Angew. Chem. Int. Ed. Engl.*, 1971, **10**, 126; (d) H. V. Ly, J. H. Chow, M. Parvez, R. McDonald and R. Roesler, *Inorg. Chem.*, 2007, **46**, 9303; (e) H. V. Ly, H. M. Tuononen, M. Parvez and R. Roesler, *Chem. Commun.*, 2007, 4522.
43. G. Bramham, J. P. H. Charmant, A. J. R. Cook, N. C. Norman, C. A. Russell and S. Saithong, *Chem. Commun.*, 2007, 4605.
44. (a) K. W. Böddeker, S. G. Shore and R. K. Bunting, *J. Am. Chem. Soc.*, 1966, **88**, 4396; (b) P. W. R. Corfield and S. G. Shore, *J. Am. Chem. Soc.*, 1973, **95**, 1480; (c) W. J. Shaw, J. C. Linehan, N. K. Szymczak, D. J. Heldebrandt, C. Yonker, D. M. Camaioni, R. T. Baker and T. Autrey, *Angew. Chem. Int. Ed.*, 2008, **47**, 7493.
45. R. T. Paine and C. K. Narula, *Chem. Rev.*, 1990, **90**, 73.
46. (a) M. Corso, W. Auwarter, M. Muntwiler, A. Tamai, T. Greber and J. Osterwalder, *Science*, 2004, **303**, 217; (b) S. Berner, M. Corso, R. Widmer, O. Groening, R. Laskowski, P. Blaha, K. Schwarz, A. Goriachko, S. Over, S. Gsell, M. Schreck, H. Sachdev, T. Greber and J. Osterwalder, *Angew. Chem. Int. Ed.*, 2007, **46**, 5115.
47. J.-K. Jeon, Y. Uchimaru and D.-P. Kim, *Inorg. Chem.*, 2004, **43**, 4796.
48. E. Kroke, K. W. Volger, A. Klonczynski and R. Riedel, *Angew. Chem. Int. Ed.*, 2001, **40**, 1698.

49. (a) F. H. Stephens, V. Pons and R. T. Baker, *Dalton Trans.*, 2007, 2613; (b) F. H. Stephens, R. T. Baker, M. H. Matus, D. J. Grant and D. A. Dixon, *Angew. Chem. Int. Ed.*, 2007, **46**, 746; (c) W. R. Nutt and M. L. McKee, *Inorg. Chem.*, 2007, **46**, 7633; (d) A. Staubitz, M. Besora, J. N. Harvey and I. Manners, *Inorg. Chem.*, 2008, **47**, 5910.
50. C. W. Yoon and L. G. Sneddon, *J. Am. Chem. Soc.*, 2006, **128**, 13992.
51. M. E. Bluhm, M. G. Bradley, R. Butterick III, U. Kusari and L. G. Sneddon, *J. Am. Chem. Soc.*, 2006, **128**, 7748.
52. P. J. Fazan, J. S. Beck, A. T. Lynch, E. E. Remsen and L. G. Sneddon, *Chem. Mater.*, 1990, **2**, 96.
53. T. Wideman, E. E. Remsen, E. Cortez, V. L. Chlanda and L. G. Sneddon, *Chem. Mater.*, 1998, **10**, 412.
54. Y. Chujo, I. Tomita and T. Saegusa, *Macromolecules*, 1992, **25**, 3005.
55. (a) P. Paetzold and Y. von Bennigsen-Mackiewicz, *Chem. Ber.*, 1981, **114**, 298; (b) P. Paetzold, *Adv. Inorg. Chem.*, 1987, **31**, 123.
56. A. Staubitz, A. Presa Soto and I. Manners, *Angew. Chem. Int. Ed.*, 2008, **47**, 6212.
57. H. V. R. Dias and P. P. Power, *J. Am. Chem. Soc.*, 1989, **111**, 144.
58. (a) P. P. Power, *J. Organomet. Chem.*, 1990, **400**, 49; (b) P. P. Power, A. Moezzi, D. C. Pestana, M. A. Petrie, S. C. Shoner and K. M. Waggoner, *Pure Appl. Chem.*, 1991, **63**, 859; (c) P. P. Power, in *The Chemistry of Inorganic Ring Systems*, ed. R. Steudel, Elsevier, Amsterdam, 1992, Chapter 2.
59. A. M. Arif, A. H. Cowley, M. Pakulski and J. M. Power, *J. Chem. Soc., Chem. Commun.*, 1996, 889.
60. P. Kölle, H. Nöth and R. T. Paine, *Chem. Ber.*, 1986, **119**, 2681.
61. D. Scheschkewitz, H. Amii, H. Gornitzka, W. W. Schoeller, D. Bourissou and G. Bertrand, *Science*, 2002, **295**, 1880.
62. D. Scheschkewitz, H. Amii, H. Gornitzka, W. W. Schoeller, D. Bourissou and G. Bertrand, *Angew. Chem. Int. Ed.*, 2004, **43**, 585.
63. A. Rodriquez, R. A. Olsen, N. Ghaderi, D. Scheschkewitz, F. S. Tham, L. J. Mueller and G. Bertrand, *Angew. Chem. Int. Ed.*, 2004, **43**, 4880.
64. H. Amii, L. Vranicar, H. Gornitzka, D. Bourissou and G. Bertrand, *J. Am. Chem. Soc.*, 2004, **126**, 1344.
65. A. Rodriquez, F. S. Tham, W. W. Schoeller and G. Bertrand, *Angew. Chem. Int. Ed.*, 2004, **43**, 4876.
66. R. T. Paine and H. Nöth, *Chem. Rev.*, 1995, **95**, 343.
67. (a) M. S. Lube, R. L. Wells and P. S. White, *Inorg. Chem.*, 1996, **35**, 5007; (b) T. J. Groshens, K. T. Higa, R. Nissan, R. J. Butcher and A. J. Freyer, *Organometallics*, 1993, **12**, 2904.
68. S. J. Geier, T. M. Gilbert and D. W. Stephan, *J. Am Chem. Soc.*, 2008, **130**, 12638.
69. A. B. Burg and R. I. Wagner, *J. Am. Chem. Soc.*, 1953, **75**, 3872.
70. G. J. Bullen and P. R. Mallinson, *J. Chem. Soc., Dalton Trans.*, 1973, 1295.
71. H. Dorn, R. A. Singh, J. A. Massey, J. M. Nelson, C. A. Jaska, A. J. Lough and I. Manners, *J. Am. Chem. Soc.*, 2000, **122**, 6669.

72. V. V. Korshak, V. A. Zamyatina and A. I. Solomatina, *Izv. Akad. Nauk SSSR Ser. Khim.*, 1961, 1541.
73. H. Dorn, J. M. Rodezno, B. Brunnhofer, E. Rivard, J. A. Massey and I. Manners, *Macromolecules*, 2003, **36**, 291.
74. T. J. Clark, J. M. Rodezno, S. B. Clendenning, S. Aouba, P. M. Brodersen, A. J. Lough, H. E. Ruda and I. Manners, *Chem. Eur. J.*, 2005, **11**, 4526.
75. (a) C. E. Housecroft, *Cluster Molecules of the p-Block Elements*, Oxford University Press, Oxford, 1994, pp. 24–25; (b) A. Y. Timoshkin, *Coord. Chem. Rev.*, 2005, **249**, 2094.
76. (a) K. M. Waggoner and P. P. Power, *J. Am. Chem. Soc.*, 1991, **113**, 3385; (b) K. M. Waggoner, H. Hope and P. P. Power, *Angew. Chem. Int. Ed. Engl.*, 1988, **27**, 1699.
77. N. Matsuanga and M. S. Gordon, *J. Am. Chem. Soc.*, 1994, **116**, 11407.
78. J. Lobl, A. Y. Timoshkin, T. Cong, M. Necas, H. W. Roesky and J. Pinkas, *Inorg. Chem.*, 2007, **46**, 5678.
79. J. D. Fisher, P. J. Shapiro, G. P. A. Yap and A. L. Rheingold, *Inorg. Chem.*, 1996, **35**, 271.
80. R. J. Wehmschulte and P. P. Power, *J. Am. Chem. Soc.*, 1996, **118**, 791.
81. R. J. Wehmschulte and P. P. Power, *Inorg. Chem.*, 1996, **35**, 2717.
82. P. Jutzi, B. Neumann, G. Reumann and H.-G. Stammler, *Organometallics*, 1999, **18**, 2037.
83. R. J. Wright, M. Brynda, J. C. Fettinger, A. R. Betzer and P. P. Power, *J. Am. Chem. Soc.*, 2006, **128**, 12498.
84. N. J. Hardman, C. Cui, H. W. Roesky, W. H. Fink and P. P. Power, *Angew. Chem. Int. Ed.*, 2001, **40**, 2172.
85. R. J. Wright, A. D. Phillips, T. L. Allen, W. H. Fink and P. P. Power, *J. Am. Chem. Soc.*, 2003, **125**, 1694.
86. L. V. Interrante, G. A. Sigel, M. Garbauskas, C. Hejna and G. A. Slack, *Inorg. Chem.*, 1989, **28**, 252.
87. J.-W. Hwang, S. A. Hanson, D. Britton, J. F. Evans, K. F. Jensen and W. L. Gladfelter, *Chem. Mater.*, 1990, **2**, 342.
88. D. A. Atwood, A. H. Cowley, R. A. Jones and M. A. Mardones, *J. Am. Chem. Soc.*, 1991, **113**, 7050.
89. H. Hope, D. C. Pestana and P. P. Power, *Angew. Chem. Int. Ed. Engl.*, 1991, **30**, 691.
90. O. T. Beachley and G. E. Coates, *J. Chem. Soc.*, 1965, 3241.
91. A. H. Cowley and R. A. Jones, *Angew. Chem. Int. Ed. Engl.*, 1989, **28**, 1208.
92. P. D. Gurney and R. J. Seymour, *Inorganic chemicals and metals in the electronic industries*, in *Insights into Speciality Inorganic Chemicals*, ed. D. Thompson, Royal Society of Chemistry, Cambridge, 1995, Chapter 9.
93. (a) A. C. Jones and P. O'Brien, *CVD of Compound Semiconductors: Precursor Synthesis, Development and Applications*, VCH, Weinheim, 1997; (b) J. S. Ritch, T. Chivers, M. Afzaal and P. O'Brien, *Chem. Soc. Rev.*, 2007, **36**, 1622.

94. S. Schulz and M. Nieger, *Organometallics*, 1998, **17**, 3398.
95. A. M. Arif, B. L. Benac, A. H. Cowley, R. L. Geerts, R. A. Jones, K. B. Kidd, J. M. Power and S. T. Schwab, *J. Chem. Soc., Chem. Commun.*, 1986, 1534.
96. (a) C. G. Pitt, A. P. Purdy, K. T. Higa and R. L. Wells, *Organometallics*, 1986, **5**, 1266; (b) A. H. Cowley, R. A. Jones, K. B. Kidd, C. M. Nunn and D. L. Westmoreland, *J. Organomet. Chem.*, 1988, **341**, C1.
97. A. H. Cowley, P. R. Harris, R. A. Jones and C. M. Nunn, *Organometallics*, 1991, **10**, 2099.
98. A. P. Purdy, R. L. Wells, A. T. McPhail and C. G. Pitt, *Organometallics*, 1986, **6**, 1266.
99. A. Haaland, J. Hougen, H. V. Volden, G. Hanika and H. H. Karsch, *J. Organomet. Chem.*, 1987, **322**, C24.
100. A. H. Cowley, B. L. Benac, J. G. Ekerdt, R. A. Jones, K. B. Kidd, J. Y. Lee and J. E. Miller, *J. Am. Chem. Soc.*, 1988, **110**, 6248.
101. J. Emsley, *Chem. World*, 2004, March, 30.
102. J.-W. Hwang, J. P. Campbell, J. Kozubowski, S. A. Hanson, J. F. Evans and W. L. Gladfelter, *Chem. Mater.*, 1995, **7**, 517.
103. B. Luo, S. Y. Lee and J. M. White, *Chem. Mater.*, 2004, **16**, 629.
104. (a) I. Haiduc, *Boron–oxygen heterocycles*, in *The Chemistry of Inorganic Homo- and Heterocycles*, ed. I. Haiduc and D. B. Sowerby, Academic Press, London, 1987, pp. 109–141; (b) For a discussion of borates, see N. N. Greenwood and A. Earnshaw, *Chemistry of the Elements*, 2nd edn., Butterworth-Heinemann, Oxford, 1998, pp. 205–207.
105. P. M. Iovine, C. R. Gyselbrecht, E. K. Perttu, C. Klick, A. Neuwelt, J. Loera, A. G. DiPasquale, A. L. Rheingold and J. Kua, *Dalton Trans.*, 2008, 3791.
106. (a) H. M. El-Kaderi, J. R. Hunt, J. L. Mendoza-Cortés, A. P. Côté, R. E. Taylor, M. O'Keefe and O. M. Yaghi, *Science*, 2007, **316**, 268; (b) G. Alcaraz, L. Euzenat, O. Mongin, C. Katan, I. Ledoux, J. Zyss, M. Blancard-Desce and M. Vaultier, *Chem. Commun.*, 2003, 2766; (c) Y. Li, J. Ding, M. Day, Y. Tao, J. Lu and M. D'Iorio, *Chem. Mater.*, 2003, **15**, 4936; (d) M. A. Beckett, M. P. Rugen-Hankey, J. L. Timmis and K. S. Varma, *Dalton Trans.*, 2008, 1503.
107. C. Jimeno, S. Sayalero, T. Fjermestad, G. Colet, F. Maseras and M. A. Pericàs, *Angew. Chem. Int. Ed.*, 2008, **47**, 1098.
108. (a) W. Kliegel, H.-W. Motzkus, S. J. Rettig and J. Trotter, *Can. J. Chem.*, 1985, **63**, 3516; (b) J. L. Priego, L. H. Doerrer, L. H. Rees and M. L. H. Green, *Chem. Commun.*, 2000, **1**, 779.
109. (a) H. Diercks and B. Krebs, *Angew. Chem. Int. Ed. Engl.*, 1977, **16**, 313; (b) B. Krebs and H.-W. Hürter, *Angew. Chem. Int. Ed. Engl.*, 1980, **19**, 481.
110. (a) For details of the synthesis of boron–sulfur rings, see W. Siebert, Boron–sulfur and boron–selenium heterocycles, in *The Chemistry of Inorganic Homo- and Heterocycles*, ed. I. Haiduc and D. B. Sowerby, Academic Press, London, 1987, pp. 143–165; (b) M. A. Beckett,

Compounds containing the boron–chalcogen bond, in *Handbook of Chalcogen Chemistry; New Perspectives in Sulfur, Selenium and Tellurium*, ed. F. A. Devillanova, Royal Society of Chemistry, Cambridge, 2007, pp. 17–19.
111. (a) M. A. Beckett, P. R. Minton and B. Werschkun, *J. Organomet. Chem.*, 1994, **468**, 37; (b) M. A. Beckett and P. R. Minton, *J. Organomet. Chem.*, 1995, **487**, 209.
112. M. Schmidt and W. Siebert, *Angew. Chem. Int. Ed. Engl.*, 1966, **5**, 597.
113. G. W. Bushnell and G. A. Rivett, *Can. J. Chem.*, 1977, **55**, 3294.
114. W. Schwarz, H. D. Hausen and H. Hess, *Z. Naturforsch., Teil B*, 1974, **29**, 596.
115. W. Schwarz, H. D. Hausen, H. Hess, M. Mandt, W. Schmelzer and B. Krebs, *Acta Crystallogr., Sect. B*, 1973, **29**, 2029.
116. B. Krebs, *Angew Chem. Int. Ed. Engl.*, 1983, **22**, 113.
117. (a) H. M. Seip, R. Seip and W. Siebert, *Acta Chem. Scand.*, 1973, **27**, 15; (b) A. Almenningen, H. M. Seip and P. Vassbotn, *Acta Chem. Scand.*, 1973, **27**, 21.
118. E. D. Jemmis and B. Kiran, *Inorg. Chem.*, 1998, **37**, 2110.
119. (a) M. Ito, N. Tokitoh and R. Okazaki, *Organometallics*, 1997, **16**, 4314; (b) N. Tokitoh, M. Ito and R. Okazaki, *Tetrahedron Lett.*, 1996, **37**, 5145.
120. M. Ito, N. Tokitoh and R. Okazaki, *Chem. Commun.*, 1998, 2495.
121. M. Ito, N. Tokitoh and R. Okazaki, *Tetrahedron Lett.*, 1997, **38**, 4451.
122. E. Zurek, T. K. Woo, T. K. Forman and T. Ziegler, *Inorg. Chem.*, 2001, **40**, 361.
123. (a) M. Linnolahti, T. N. P. Luhtanen and T. A. Pakkanen, *Chem. Eur. J.*, 2004, **10**, 5977; (b) M. Linnolahti, J. R. Severn and T. A. Pakkanen, *Angew. Chem. Int. Ed.*, 2008, **47**, 9279.
124. (a) M. R. Mason, J. M. Smith, S. G. Bott and A. R. Barron, *J. Am. Chem. Soc.*, 1993, **115**, 4971; (b) C. F. Harlan, M. R. Mason and A. R. Barron, *Organometallics*, 1994, **13**, 2957.
125. R. J. Wehmschulte and P. P. Power, *J. Am. Chem. Soc.*, 1997, **119**, 8387.
126. W. Uhl and M. R. Halvagar, *Angew. Chem. Int. Ed.*, 2008, **47**, 1955.
127. A. H. Cowley, R. A. Jones, P. R. Harris, D. A. Atwood, L. Contreras and C. J. Burek, *Angew. Chem. Int. Ed. Engl.*, 1991, **30**, 1143.
128. S. Schulz, H. W. Roesky, H. J. Koch, G. M. Sheldrick, D. Stalke and A. Kuhn, *Angew. Chem. Int. Ed. Engl.*, 1993, **32**, 1729.
129. S. Schulz, E. G. Gillan, J. L. Ross, L. M. Rogers, R. D. Rogers and A. R. Barron, *Organometallics*, 1996, **15**, 4880.
130. W. M. Cleaver, M. Späth, D. Hnyk, G. McMurdo, M. P. Power, M. Stuke, D. W. H. Rankin and A. R. Barron, *Organometallics*, 1995, **14**, 690.
131. M. G. Gardiner, C. L. Raston and V.-A. Tolhurst, *J. Chem. Soc., Chem. Commun.*, 1995, 2501.
132. P. D. Godfrey, C. L. Raston, B. W. Skelton, V.-A. Tolhurst and A. H. White, *Chem. Commun.*, 1997, 2235.

133. W. J. Grigsby, C. L. Raston, V.-A. Tolhurst, B. W. Skelton and A. H. White, *J. Chem. Soc., Dalton Trans.*, 1998, 2547.
134. R. J. Wehmschulte and P. P. Power, *J. Am. Chem. Soc.*, 1997, **119**, 9566.
135. M. B. Power, J. W. Ziller and A. R. Barron, *Organometallics*, 1993, **11**, 2783.
136. S. D. Nogai and H. Schmidbaur, *Dalton Trans.*, 2003, 2488.
137. M. C. Copsey and T. Chivers, *Chem. Commun.*, 2005, 4938.
138. S. S. Garje, M. C. Copsey, M. Afzall, P. O'Brien and T. Chivers, *J. Mater. Chem.*, 2006, **16**, 4542.
139. H. J. Gysling, A. A. Wernberg and T. N. Blanton, *Chem. Mater.*, 1992, **4**, 900.
140. M. A. Banks, O. T. Beachley Jr, H. J. Gysling and H. R. Luss, *Organometallics*, 1990, **9**, 1979.
141. H. Rahbarnoohi, R. Kumar, J. Heeg and J. P. Oliver, *Organometallics*, 1995, **14**, 502.

CHAPTER 10
Group 14: Rings and Polymers

The interest in ring systems involving group 14 elements originates from the fact that they may be considered as analogues of cyclic organic molecules. An extensive chemistry of homocyclic organopolysilanes $(R_2Si)_n$ and, to a lesser extent, their heavier group 14 congeners has been developed over the past 60 years. Significant advances have also been made in our knowledge of the structures and reactions of analogues of unsaturated organic rings such as cyclopropene and cyclobutadiene in the last 15 years. The photolysis of cyclotrisilanes and -trigermanes is used to generate multiply bonded group 14 compounds of the type $R_2M=MR_2$ (M = Si, Ge), *i.e.* alkene analogues. From a more applied perspective, polymers based on a heavy group 14 element (Si, Ge or Sn) backbone have also attracted intense attention because of their unusual electronic delocalisation that involves σ-electrons in the main chain. The photosensitivity of these materials also allows photopatterning and several derivatives act as efficient, easily processed precursors to SiC-based ceramic fibres and monoliths on pyrolysis. Some polysilanes can be prepared from cyclic precursors *via* ring-opening polymerisation. In this chapter, particular emphasis is accorded to a comparison of the chemistry and physical properties of these inorganic ring systems and polymers with those of their organic counterparts.

Heterocycles involving group 14 elements and oxygen (or the heavier chalcogens) are equally important, especially cyclic siloxanes $(R_2SiO)_n$ and the polymers derived from them. For example, cyclic siloxanes such as $(Me_2SiO)_n$ ($n = 4$–6) are common ingredients in beauty and personal-care products such as antiperspirants and deodorants. Silicone polymers are ubiquitous in everyday life as a result of their widespread use in a variety of materials, including oils, greases, polishes, water repellents, rubbers and gaskets. Ring systems involving group 14 elements and the heavier chalcogens have been shown to be excellent precursors of ketone analogues of the type $R_2M=E$ (M = Si, Ge, Sn, Pb; E = S, Se, Te).

Inorganic Rings and Polymers of the p-Block Elements: From Fundamentals to Applications
By Tristram Chivers and Ian Manners
© Tristram Chivers and Ian Manners 2009
Published by the Royal Society of Chemistry, www.rsc.org

10.1 HOMOATOMIC RINGS AND POLYMERS: SATURATED SYSTEMS

10.1.1 Cyclopolysilanes

An extensive homologous series of cyclic polysilanes $(R_2Si)_n$ ($n = 4$–35) is known;[1a] the first example, the hexamer $(Me_2Si)_6$, was reported in 1949 by the General Electric Co. Laboratories.[1b] Derivatives containing Si–H bonds, *e.g.* $(H_2Si)_n$ ($n = 5$, 6),[2] *cf.* cycloalkanes, are known, but they are difficult to manipulate owing to their pyrophoric nature. Consequently, the methyl derivatives (R = Me) have been most widely studied because of their desirable physical properties, including air and moisture stability, with the exception of the smaller homologues, *e.g.* $(Me_2Si)_n$ ($n = 4$, 5). The standard synthesis of permethylated cyclopolysilanes $(Me_2Si)_n$ ($n = 5$–35) involves the reductive coupling of dimethyldichlorosilane, Me_2SiCl_2, with sodium-potassium alloy in THF, which, under thermodynamically controlled conditions, gives primarily the hexamer $(Me_2Si)_6$ (90%), with smaller amounts of pentamer $(Me_2Si)_5$ (9%) and a trace of heptamer $(Me_2Si)_7$ (1%).[3] If the addition of Me_2SiCl_2 is carried out very slowly, a mixture of cyclic oligomers $(Me_2Si)_n$ ($n = 8$–19) is obtained that can be separated by HPLC; the composition of this mixture is kinetically controlled to some extent. The larger rings $(Me_2Si)_n$ ($n = 20$–35) have only been detected by HPLC (see Section 3.2). Larger amounts of the medium-sized rings $(Me_2Si)_n$ ($n = 7$–9) are formed when lithium is used as the reducing agent.

The cyclic tetramer $(Me_2Si)_4$ is not obtained by using the reductive-coupling method. However, this highly reactive four-membered ring is formed in low yields upon photolysis of $(Me_2Si)_6$ in cyclohexane at 45 °C. The pentamer $(Me_2Si)_5$ is the major product of this ring-contraction process [eqn (10.1)].[4]

$$(Me_2Si)_6 \xrightarrow{h\nu} (Me_2Si)_5 + (Me_2Si)_4 \tag{10.1}$$

Replacement of one of the methyl groups on each silicon in $(Me_2Si)_4$ by a *tert*-butyl group provides sufficient steric protection to render this cyclic tetramer air stable. All four isomers of $[Me(^tBu)Si]_4$ are obtained by the reductive coupling of $Me(^tBu)SiCl_2$ with lithium in THF. The major product is the isomer in which the *tert*-butyl groups alternate above and below the Si_4 ring (Figure 10.1a).[5]

The permethylated cyclopolysilanes are colourless, crystalline solids. The hexamer $(Me_2Si)_6$ adopts a chair conformation with negligible methyl–methyl interactions in the solid state,[6] whereas the larger rings $(Me_2Si)_n$ ($n = 13$ or 16) are distorted in response to endocyclic interactions between methyl groups (see Figure 6.1). In solution, the permethylated cyclopolysilanes show only sharp

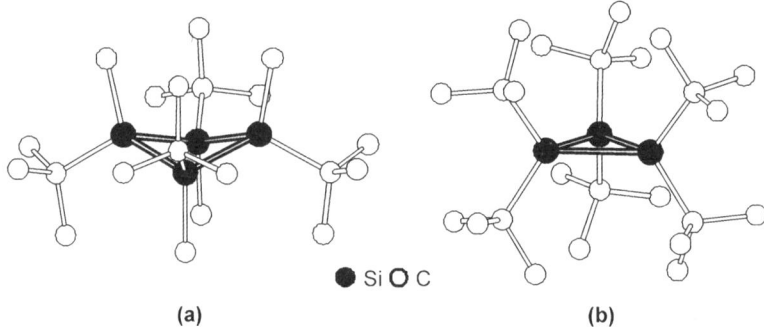

Figure 10.1 Molecular structures of (a) [Me(^{t}Bu)Si]$_4$ (all-*trans* isomer) and (b) (^{t}Bu$_2$Si)$_3$.

singlets in their ^{1}H and ^{13}C NMR spectra, indicating rapid interconversion of equatorial and axial methyl groups on the NMR time scale.

The most intriguing difference between the chemical properties of cyclopolysilanes and those of cycloalkanes is the ability of the former to form either anion or cation radicals upon one-electron reduction or oxidation, respectively. For example, the cyclic pentamer (Me$_2$Si)$_5$ is reduced to the corresponding radical anion by sodium–potassium alloy in diethyl ether [see eqn (4.1) in Section 4.1.3], whereas the hexamer (Me$_2$Si)$_6$ is oxidised by aluminium trichloride in dichloromethane to the corresponding cation radical.[7] In both cases the EPR spectra of the radical ions can be interpreted in terms of σ-electron delocalisation over the entire polysilane ring (see Section 10.1.4.1). In this respect, the cyclosilanes resemble aromatic hydrocarbons rather than their aliphatic analogues.

Perphenylated cyclopolysilanes were first isolated in the 1920s from the reaction of sodium with diphenyldichlorosilane in toluene, although their cyclic structures were not recognised at the time. Subsequently, the three products of this reaction were shown to be the cyclic oligomers (Ph$_2$Si)$_n$ (n = 4, 5, 6). The tetramer (Ph$_2$Si)$_4$ is formed in better yields if lithium is used as the reducing agent.[8] One of the silicon–silicon bonds in this ring system is easily cleaved by halogens or lithium metal in THF to give bifunctional reagents[9a] that can be used to introduce various heteroatoms into a polysilane ring system as illustrated in Scheme 10.1.[9b,c]

The Si–phenyl bonds in the cyclopolysilanes (Ph$_2$Si)$_n$ (n = 5, 6) are cleaved by treatment with hydrogen halides in the presence of aluminium trihalides to give perhalogenated cyclopolysilanes such as (Br$_2$Si)$_5$ and (Cl$_2$Si)$_6$, which are subsequently converted into the prototypical cyclopolysilanes (H$_2$Si)$_n$ (n = 5, 6) by treatment with LiAlH$_4$ in diethyl ether.[2] The parent cyclopentasilane (H$_2$Si)$_5$ is a suitable precursor for the solution processing of polycrystalline silicon films by both spin-coating and ink-jet printing techniques.[10] An improved preparation of the cyclic hexamer (H$_2$Si)$_6$ (in 95% yield) involves the reaction of the tetradecachlorocyclohexasilane dianion [Si$_6$Cl$_{14}$]$^{2-}$ with LiAlH$_4$ in diethyl ether.[11] The structure of this fascinating dianion is comprised of a planar cyclohexasilane ring with the additional two chlorine atoms located on the six-fold axis of the ring (Figure 10.2).[11]

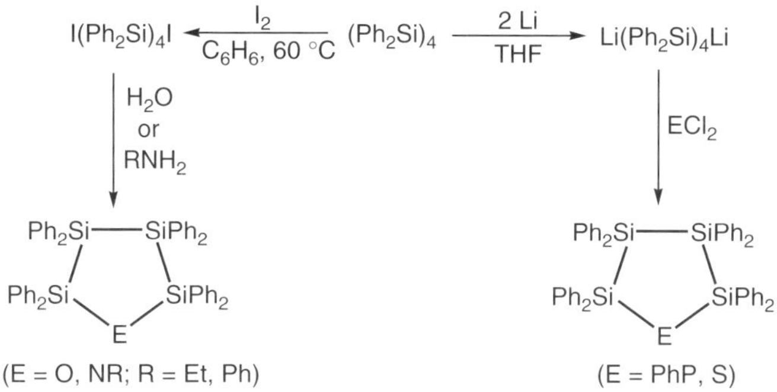

Scheme 10.1 Reactions of $(Ph_2Si)_4$.

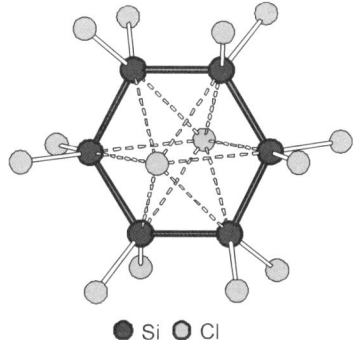

Figure 10.2 Structure of the $[Si_6Cl_{14}]^{2-}$ dianion.

Cyclotrisilanes $(Ar_2Si)_3$ are obtained as air-stable, high-melting solids *via* reductive coupling if moderately bulky substituents, *e.g.* Ar = 2,6-dimethylphenyl, are attached to the silicon atoms [eqn (10.2)].[12a] The germanium and tin analogues $(Ar_2M)_3$ (M = Ge, Ar = 2,6-dimethylphenyl; M = Sn, Ar = 2,6-diethylphenyl) are prepared in a similar manner.[12b,c] All three of these cyclic trimers form equilateral triangles in the solid state with M–M bond lengths that are elongated by *ca* 0.08 Å compared with single-bond values. Bulky alkyl substituents also provide sufficient steric protection to stabilise the Si_3 ring. For example, the air-stable *tert*-butyl derivative $(^tBu_2Si)_3$ is obtained in 70% yield from the reaction of $^tBu_2SiI_2$ with lithium naphthalenide; the Si–Si bond lengths in this cyclic trimer are 0.1 Å longer than those in the aryl derivative (Figure 10.1b).[13]

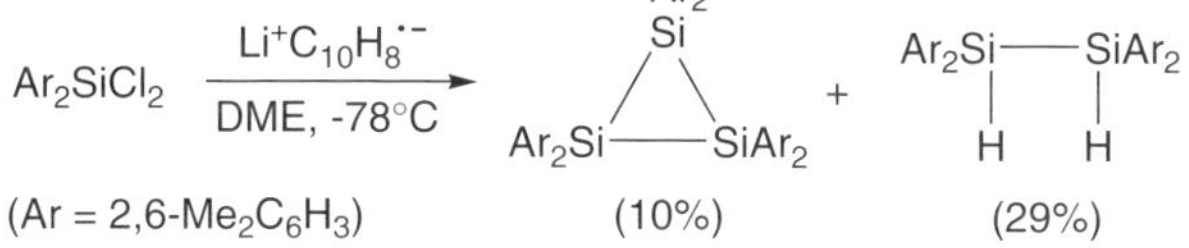

(10.2)

Although $(Ar_2Si)_3$ ($Ar = 2,6$-dimethylphenyl) is thermally stable up to 275 °C, the ring system is labile to ultraviolet light. The progress of this irradiation, which is complete in 5 min at room temperature, is readily monitored by ^1H NMR spectroscopy. The colourless solution becomes intensely yellow due to the formation of the disilene $Ar_2Si=SiAr_2$ [eqn (10.3)].[12a] This transformation was one of the first syntheses of a disilene; the congeneric digermenes are obtained in a similar manner.[12b] An alternative, but similar, approach to the synthesis of disilenes involves the photolysis of an open-chain trisilane $Mes_2Si(SiMe_3)_2$.[14]

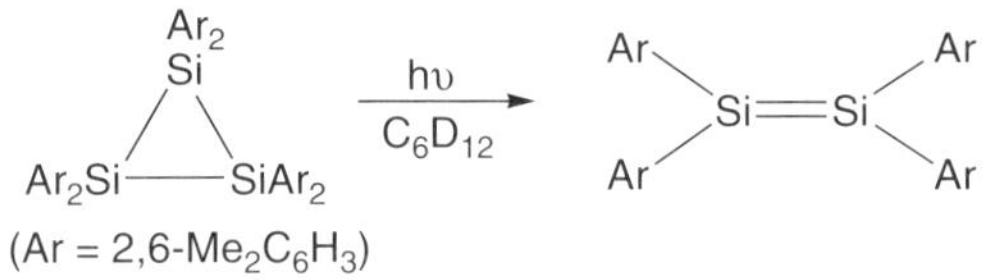

$$\underset{(Ar\,=\,2,6\text{-}Me_2C_6H_3)}{\underset{Ar_2Si\text{——}SiAr_2}{\overset{Ar_2}{\overset{Si}{\triangle}}}} \xrightarrow[C_6D_{12}]{h\upsilon} \underset{Ar}{\overset{Ar}{>}}Si=Si\underset{Ar}{\overset{Ar}{<}} \qquad (10.3)$$

Three-membered rings containing two different group 14 elements are also accessible by using different synthetic approaches, as illustrated for Mes_6Ge_2Si (**10.1**)[15a] and $R_2Ge(SiR_2)_2$ (**10.2**, $R = SiMe_3$)[15b] in Scheme 10.2.

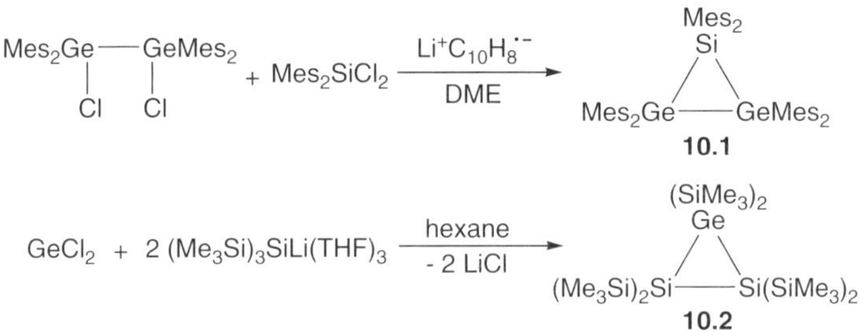

Scheme 10.2 Syntheses of cyclotrimetallanes containing two different group 14 atoms.

10.1.2 Cyclopolygermanes, -stannanes and -plumbanes

The structures and reaction chemistry of cyclopolygermanes and -stannanes resemble those of cyclopolysilanes, except that the homologous series of characterised cyclic oligomers $(R_2M)_n$ is more restricted in the case of the heavier group 14 elements ($M = Ge$, Sn). The ring sizes of cyclic polygermanes ($M = Ge$) are limited to $n = 4$–6 when $R = Ph$.[16] However, the cyclic heptamer $(Me_2Ge)_7$ is obtained in small amounts by reductive coupling of Me_2GeCl_2 with lithium in THF; the hexamer $(Me_2Ge)_6$ is the major product.[17] Both the hexamer and heptamer are air-stable crystalline compounds, whereas the pentamer $(Me_2Ge)_5$ is readily oxidised in air. Reductive coupling of dichlorogermanes R_2GeCl_2 with a mixture of magnesium and magnesium bromide produces either cyclotrigermanes $(R_2Ge)_3$ or cyclotetragermanes $(R_2Ge)_4$, depending on the bulk of the substituents R.[18] All four phenyl groups on germanium in the tetramer [Ph(^{t}Bu)Ge]$_4$ are cleaved by treatment with HCl in the presence of aluminium trichloride to give [^{t}Bu(Cl)Ge]$_4$ (**10.3**) as the all-*trans* isomer [eqn (10.4)].[19] The Ge–Ge bonds in (Ph$_2$Ge)$_4$ are readily cleaved by I$_2$ to give the bifunctional linear oligomer I(Ph$_2$Ge)$_4$I, which is useful for the synthesis of heterocyclic polygermanes,[16b] *cf.* I(Ph$_2$Si)$_4$I (Scheme 10.1).

$$\text{[}^t\text{Bu(Ph)Ge]}_4 \xrightarrow{\text{HCl, AlCl}_3} \text{[}^t\text{Bu(Cl)Ge]}_4 \quad \textbf{10.3}$$

(10.4)

The cyclic trimer (^{t}Bu$_2$Ge)$_3$ is obtained from ^{t}Bu$_2$GeCl$_2$ by reductive coupling;[20] the Ge–Ge distance in this three-membered ring is *ca* 0.17 Å longer than that in the aryl derivative, (Ar$_2$Ge)$_3$ (Ar = 2,6-dimethylphenyl). The four-membered ring in (Ph$_2$Ge)$_4$ is planar with single Ge–Ge bond lengths,[21] whereas the Ge$_4$ ring in the tetramer **10.3** has a fold angle of 21°.[19]

Cyclic polystannanes (R$_2$Sn)$_n$ are known for ring sizes from $n = 3$ to 9;[22] however, the identification of the larger homocycles ($n = 7$–9) depends on a combination of HPLC, ^{119}Sn NMR spectroscopy and FAB mass spectrometry.[23] The preparation of cyclotristannanes by reductive coupling is shown in eqn (2.5) in Section 2.2. The method of choice for the synthesis of other cyclic polystannanes is the dehydrocoupling of diorganotin dihydrides R$_2$SnH$_2$ [see eqn 2.10 in Section 2.3].[24] This process is catalysed by amines, but the outcome is markedly dependent on the nature of the amine. For example, in the presence of DMF, diphenyltin dihydride yields the cyclic pentamer (Ph$_2$Sn)$_5$, whereas the hexamer (Ph$_2$Sn)$_6$ is the preferred product when the decomposition is carried out in pyridine.[25a] By contrast, the pale yellow cyclic tetramer (Bz$_2$Sn)$_4$ is obtained almost quantitatively from dibenzyltin dihydride in DMF at 50 °C.[25b] In the presence of R$_2$SnCl$_2$, the dehydrocoupling of diorganotin dihydrides, R$_2$SnH$_2$, produces the cyclic nonamers (R$_2$Sn)$_9$ (R = Et, ^{i}Bu);[26] surprisingly, since medium-sized cyclic polystannanes are unreactive in air, crystals of the diethyl derivative (Et$_2$Sn)$_9$ are pyrophoric. The use of early and late transition metal catalysts for this dehydropolymerisation process produces high molecular weight linear and branched polystannanes, respectively (see Section 10.1.4.2)

The condensation of diorganotin dihydrides with diorganotin diamides is another route to cyclic polystannanes. This method is especially useful for the high-yield synthesis of the dimethyl derivative (Me$_2$Sn)$_6$ [eqn (10.5)],[27] which cannot be obtained by the dehydrocoupling process.

$$3\text{Me}_2\text{SnH}_2 + 3\text{Me}_2\text{Sn(NEt}_2)_2 \rightarrow (\text{Me}_2\text{Sn})_6 + 6\text{Et}_2\text{NH} \quad (10.5)$$

Unexpectedly, the cyclic tetramer ($^{t}Bu_2Sn)_4$ (rather than $^{t}Bu_4Sn$) is obtained as bright-yellow crystals from the reaction of $^{t}Bu_2SnCl_2$ with an excess of $^{t}BuMgCl$ in boiling THF.[28a] Evidence for the cyclic structure of ($^{t}Bu_2Sn)_4$ in this early work was based on the reaction with iodine to give the diodotetratin reagent $I(^{t}Bu_2Sn)_4I$, which has subsequently been used to create linear tin oligomers **10.4** (Scheme 10.3).[29] The cyclic structure of ($^{t}Bu_2Sn)_4$ was later confirmed by an X-ray structure; the strained four-membered ring is planar and the Sn–Sn bonds are longer than those in other cyclostannanes.[28b]

In the early literature, the reactions of Grignard or organolithium reagents with tin(II) dihalides were reported to give products with the composition 'R_2Sn', which have subsequently been shown to be mixtures of both straight- and branched-chain polymers with Sn–Sn backbones.[30] A fascinating outcome of this synthetic approach to diorganotin compounds occurs in the reaction of 1-lithio-2,6-diethylbenzene (prepared *in situ* from the corresponding aryl bromide and $^{n}BuLi$) with $SnCl_2$. As illustrated in Scheme 10.4, this reaction produces the cyclotristannane ($Ar_2Sn)_3$ in *ca* 50% yield with THF as solvent, whereas only a dark-red viscous oil is obtained in diethyl ether.[31] Orange–red crystals of a bicyclo[2.2.0]hexastannane **10.5**, which was the first example of a polycyclic polystannane, can be isolated from a pentane solution of this red oil.

The successful synthesis of a cyclic oligoplumbane was not reported until 2003. The outcome of reactions of lead(II) halides with aryl Grignard reagents ArMgBr depends on both the steric requirements of the ligand and the reaction conditions. When $Ar = 2$-^{t}Bu-$4,5,6$-Me_3C_6H a monomeric plumbylene

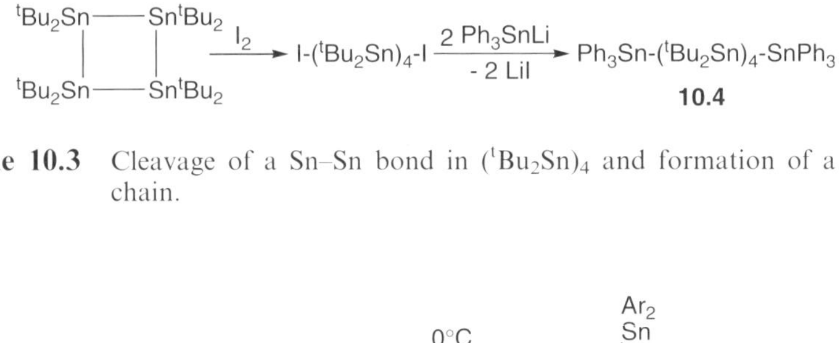

Scheme 10.3 Cleavage of a Sn–Sn bond in ($^{t}Bu_2Sn)_4$ and formation of a hexatin chain.

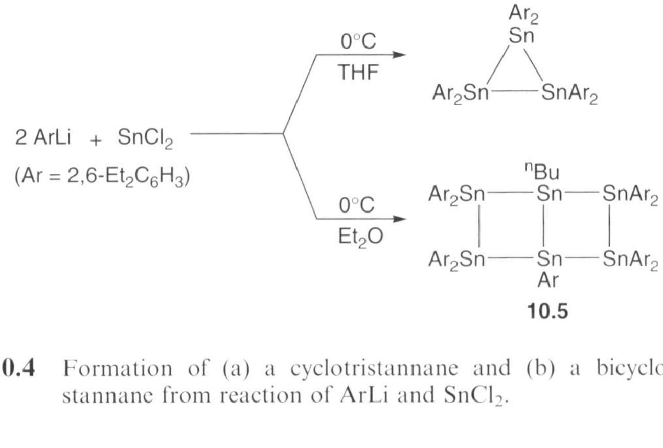

Scheme 10.4 Formation of (a) a cyclotristannane and (b) a bicyclo[2.2.0]hexastannane from reaction of ArLi and $SnCl_2$.

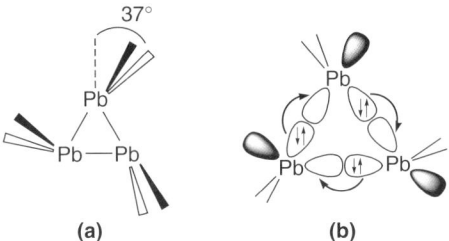

Figure 10.3 (a) Structure and (b) bonding in a cyclotriplumbane.

Ar$_2$Pb: is obtained,[32a] whereas the 2,4,6-triisopropylphenyl ligand affords the diplumbene Ar$_2$Pb=PbAr$_2$ (Ar = 2,4,6-^{i}Pr$_3$C$_6$H$_2$).[32b] The reaction between PbBr$_2$ and a Grignard reagent with the slightly less bulky 2,4,6-triethylphenyl group results in the isolation of black crystals of the cyclotriplumbane (Ar$_2$Pb)$_3$ (Ar = 2,4,6-Et$_3$C$_6$H$_2$), which is thermally stable up to 80 °C.[33] The X-ray structure reveals an approximately equilateral triangle with Pb–Pb bond lengths that are substantially longer (by 0.34 Å) than those of a typical diplumbane, e.g. Ph$_3$Pb–PbPh$_3$; in addition, the substituents at the lead atoms are twisted by ca 37° out of their ideal positions (Figure 10.3a). These structural parameters suggest that, in contrast to the M–M single bonds in other cyclotrimetallanes (Ar$_2$M)$_3$ (M = Si, Ge, Sn), the cyclic structure in the Pb$_3$ ring system is held together by weak interactions between three singlet plumbylenes Ar$_2$Pb: (Figure 10.3b).

Interest in saturated ring systems containing two different group 14 elements, e.g. silicon and tin, arises from the possibility that they may serve as precursors to the corresponding polymers *via* ring-opening polymerisation. A variety of approaches are available for the synthesis of hybrid Si–Sn systems with ring sizes from four to eight group 14 atoms.[34] However, the formation of polymers from these heteroatomic systems has not been successful, apparently because the bulky groups necessary to stabilise the smaller rings decrease the reactivity of these cyclic systems.

10.1.3 Cyclic Polyanions of Silicon, Germanium, Tin and Lead

The best known examples of polyanions of group 14 elements are the so-called Zintl anions, e.g. M$_4^{4-}$, (tetrahedral), M$_5^{2-}$ (trigonal bipyramidal) and M$_9^{4-}$ (mono-capped square antiprismatic), all of which have electron-deficient polyhedral structures that can be predicted by using Wade's rules.[35] The formation of these classic inorganic clusters was first investigated in detail by Zintl in the 1930s, but the solid-state structures were not established until the mid-1970s when it was recognised that the use of crypt or crown ether ligands to encapsulate the alkali metal counterions prevented the occurrence of internal electron transfer from the strongly reducing polyanions.

More recently, a number of polyanions of the group 14 elements that have cyclic structures have been synthesised and structurally characterised. These include the triangular dianion M_3^{2-} (M = Si, Sn),[36] the pentagonal hexaanions M_5^{6-} (M = Sn, Pb),[37] and the hexagonal Sn_6^{6-} (ref. 38) and M_6^{10-} (M = Si, Ge)[39] polyanions, all of which exist as solid-state materials in which the polyanions are stabilised by coordination to alkali or alkaline earth metal cations. From the perspective of inorganic rings, the interest in these highly charged species is their isovalent relationship with cyclic organic systems such as the cyclopropenium cation, $[C_3H_3]^+$, the cyclopentadienide anion, $[C_5H_5]^-$, and benzene, C_6H_6.

The cyclic group 14 polyanions are usually found in ternary materials that contain two types of cations which, in addition to providing charge balance, perform different roles. The structures typically consist of cyclic polyanions that are stacked in an eclipsed fashion to form columns. One type of cation coordinates to the anions and separates them within the column, whereas the second type segregates the columns. These ternary materials are typically prepared by heating stoichiometric amounts of the three elements in an unreactive metal tube, *e.g.* niobium, at very high temperatures (>800 °C) followed by slow annealing at lower temperatures.

As a specific example of these general principles, we will consider the structure of the ternary materials Na_8BaM_6 (M = Sn, Pb) prepared by direct reaction of the elements at 750 °C.[37b] These isostructural compounds contain two types of anions, *viz.* pentagonal M_5^{6-} anions and monoatomic M^{4-} anions, arranged as depicted in Figure 10.4. The pentagonal rings are

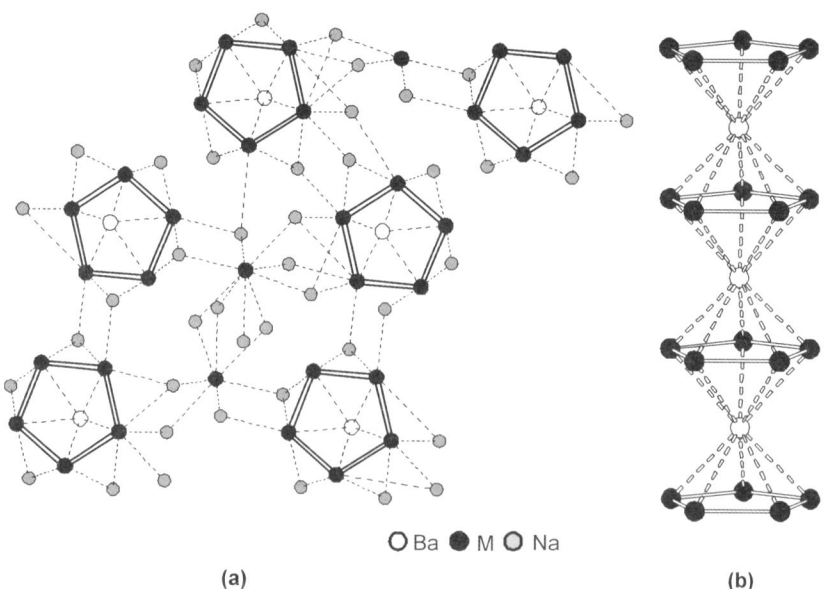

Figure 10.4 (a) Structure of Na_8BaM_6 (M = Sn, Pb) viewed along the *b* axis; (b) columns of M_5^{6-} ions separated by Ba^{2+} cations.

stacked on top of each other in an eclipsed fashion and form columns in which the Ba^{2+} ions are sandwiched symmetrically between the five-membered rings (Figure 10.4b). The monoatomic M^{4-} anions are located between the pentagons and the Na^+ cations coordinate these ions and bridge bonds in the pentagons (Figure 10.4a).

According to Wade's rules, one non-bonding electron pair is allocated to each metal atom in polyionic metal clusters of the p-block elements. Thus the planar five-membered M_5^{6-} anions are 16-electron species comprised of two valence electrons contributed by each metal atom and the 6– charge. Since there are eight electron pairs for bonding five cluster atoms, Wade's rules predict an *arachno* structure ($n+3$ electron pairs for n cluster atoms), *i.e.* a pentagonal bipyramid with two vacant sites.

10.1.4 Polymers with Backbones of Heavier Group 14 Elements

10.1.4.1 Polysilanes. Following the first reports of soluble and processable polysilanes in the late 1970s, these macromolecules have attracted substantial interest from both fundamental and applied perspectives.[40–42] The backbone of silicon atoms gives rise to unique electronic and optical properties as a result of the delocalisation of σ-electrons. Several polysilanes have also been found to function as useful thermal precursors to silicon carbide fibres and these materials have also attracted attention with respect to microlithographic applications and as polymerisation initiators.[40–42]

The first report of a soluble polysilane appeared in 1978 and the material was prepared by the treatment of a mixture of organodichlorosilanes with sodium metal. Instead of only the expected cyclic oligomers, a polymeric product, termed 'polysilastyrene' (**10.6**), was formed [eqn (10.6)]. Polydimethylsilane had been previously prepared as a highly crystalline insoluble material; the introduction of phenyl groups in the random copolymer reduces the crystallinity and allows the material to be soluble in common organic solvents. Although yields are usually low, the use of this Wurtz coupling route subsequently allowed the preparation of a wide range of polymers with alkyl or aryl substituents at silicon that are of high molecular weight with $M_n > 10^5$.

$$Me_2SiCl_2 + PhMeSiCl_2 \xrightarrow{Na, toluene, 110\,°C} {-}{\left[\begin{array}{c}Me\\|\\Si\\|\\Me\end{array}\right.}\Big/{\left.\begin{array}{c}Ph\\|\\Si\\|\\Me\end{array}\right]}_n$$

10.6 (10.6)

One of the most remarkable features of the all-silicon backbone is that it leads to the delocalisation of σ-electrons, a phenomenon which is essentially unknown in carbon chemistry.[42,43] This can be understood in terms of the nature of the molecular orbitals associated with the Si–Si σ-bonds. These are more diffuse than those associated with C–C σ-bonds as they are constructed from higher energy 3s and 3p atomic orbitals. This leads to significant interactions between the adjacent Si–Si σ-bonds along a polysilane chain, a situation analogous to that for the π-bonds in π-delocalised polymers such as polyacetylene. Thus a band model rather than a localised model is more appropriate. As a consequence of the delocalisation of σ-electrons, the σ–σ* transition which occurs at 220 nm in Me_3Si–$SiMe_3$, moves to lower energy as the number of silicon atoms in the chain increases. In the high polymers the σ–σ* bandgap transitions occur in the near-UV region at ca 300–400 nm.[41–43] The electron delocalisation also leads to appreciable electrical conductivity following doping with oxidants such as iodine. In addition, many of the polymers are thermochromic as the conformations adopted by the polymer change with temperature, which alters the degree of σ-delocalisation along the main chain. Due to their low energy σ–σ* transitions, polysilanes are photosensitive and have attracted considerable attention as photoresist materials in microlithography.[41,42]

The main method used to synthesise polysilanes **10.7** involves the thermally induced Wurtz coupling reaction of organodichlorosilanes with alkali metals [eqn (10.7)]. Although improvements in this process have been reported (*e.g.* by the use of ultrasound), the harsh conditions for this reaction tend to limit the side groups that can be successfully introduced to non-functionalised alkyl and aryl units and makes scale-up unattractive.

$$RR'SiCl_2 \xrightarrow{\text{Na, toluene}}_{110\ °C} \left[\begin{array}{c} R \\ | \\ -Si- \\ | \\ R' \end{array}\right]_n$$

10.7

(10.7)

Because of these limitations, considerable effort has been focused on the development of new synthetic routes to polysilanes. The early transition metal-catalysed dehydrocoupling process discovered in 1985 [eqn (10.8)] is potentially very attractive; however, the molecular weights of the polysilanes formed to date are generally fairly low ($M_n < 8000$).[44] The catalysts used for these coupling reactions are usually titanocene or zirconocene derivatives.[44,45]

$$RSiH_3 \xrightarrow{\text{catalyst}} H\left[\begin{array}{c} R \\ | \\ -Si- \\ | \\ H \end{array}\right]_n H + H_2$$

(10.8)

A novel ROP route to polysilanes [eqn (10.9)] was reported in 1991.[46] The key to this approach is to take readily accessible octaphenylcyclotetrasilane, which is too sterically crowded to undergo ROP, and to replace some of the phenyl groups by smaller methyl substituents (via a two-step process) to give [Ph(Me)Si]$_4$ (**10.8**), which is polymerisable.

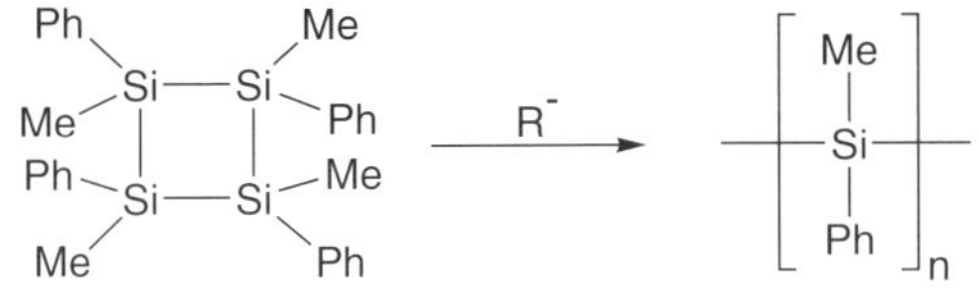

Group 14: Rings and Polymers

[Structure 10.8: cyclotetrasilane with Ph/Me substituents] →(R⁻)→ [−Si(Me)(Ph)−]ₙ

(10.9)

Another innovative route to polysilanes involves the anionic polymerisation of disilabicyclooctadienes **10.9**, which function as sources of masked disilenes, *e.g.* $Me_2Si=SiMe^nBu$ [eqn (10.10)].[47]

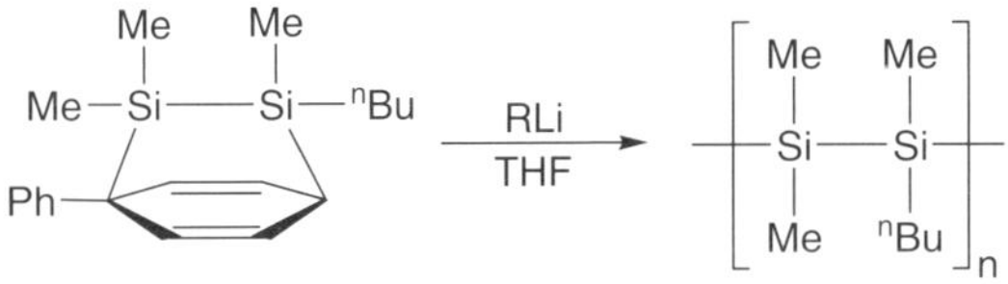

[Structure 10.9] →(RLi, THF)→ [−Si(Me)(Me)−Si(Me)(^{n}Bu)−]ₙ

(10.10)

An important application of polydimethylsilane is as a source of silicon carbide (SiC) fibres, which are manufactured under the trade-name Nicalon by Nippon Carbon in Japan. Heating in an autoclave under pressure converts polydimethylsilane to spinnable polycarbosilane $(-Me_2Si-CH_2-)_n$ with elimination of methane. The spun fibres are then subjected to temperatures of 1200–1400 °C to produce silicon carbide fibres with very high tensile strengths and elastic moduli.[48]

As a result of their conductivity, polysilanes have also been used as hole transport layers in electroluminescent devices.[49] In addition, the photoconductivity of polymethylphenylsilane doped with C_{60} has been found to be particularly impressive.[50]

10.1.4.2 Polygermanes and Polystannanes. The development and remarkable properties of polysilanes led to interest in polymer chains based on the heavier group 14 elements germanium and tin. Polygermanes **10.10** were prepared in the mid-1980s by Wurtz coupling techniques similar to those used to prepare the silicon polymers [eqn (10.11)]. Studies of these materials indicated that the σ-delocalisation is even more extensive than for polysilanes and that the σ–σ* bandgap transition for the high polymers was red shifted by *ca* 20 nm in comparison with the silicon analogues.[51,52]

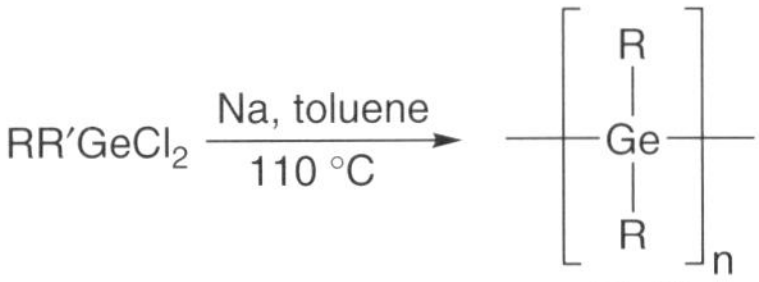

$RR'GeCl_2$ →(Na, toluene, 110 °C)→ [−GeR R−]ₙ **10.10**

(10.11)

Polystannanes would be expected to possess even more σ-delocalised structures, as suggested by studies of linear oligostannanes with up to six tin atoms, where the σ–σ* transition associated with the Sn–Sn bonds moves dramatically to lower energy as the chain length is increased. This observation led to the proposed term 'molecular metals' for prospective high molecular weight analogues.[53] However, attempts to generate high polymers by Wurtz coupling of organodichlorostannanes initially yielded only low molecular weight oligomers and reduction products. The key breakthrough in this area involved the use of transition metal-catalysed dehydrocoupling reactions, which had been extensively applied to the synthesis of polysilanes (Section 10.1.4.1), for the polymerisation of secondary stannanes R_2SnH_2 [eqn (10.12)].[54] Yellow polystannanes $(-R_2Sn-)_n$ (**10.11**, R = n-butyl, n-hexyl or n-octyl) of substantial molecular weight (up to $M_w \approx 96\,000$, $M_n \approx 22\,000$) were prepared by using various zirconocene catalysts. These materials indeed possess σ-electrons that are extensively delocalised, as illustrated by the bandgap transition, which occurs at 384–388 nm (in THF). In addition, exposure of thin films of the polymers to the oxidant AsF_5 leads to significant electronic conductivities of ca 0.01–0.3 S cm^{-1}.[54]

$$R_2SnH_2 \xrightarrow[25\,°C]{Zr\ catalyst} \left[\begin{array}{c} R \\ | \\ Sn \\ | \\ R \end{array}\right]_n$$
10.11

(10.12)

Polystannanes have been shown to display very interesting properties. They are highly photosensitive and exhibit photobleaching behaviour and on UV irradiation depolymerise to yield cyclic oligomers. The materials are thermally stable to 200–270 °C in air and at more elevated temperatures function as precursors to SnO_2.[54] By using rhodium catalysts, branched polystannanes have been prepared.[55]

10.2 HOMOATOMIC RINGS: UNSATURATED SYSTEMS

10.2.1 Three-membered Rings

The discovery of cyclotrimetallanes $(R_2M)_3$ (M = Si, Ge and Pb) and, concomitantly, disilenes $R_2Si=SiR_2$ in the early 1980s provided the incentive for the synthesis of cyclotrimetallenes, *i.e.* heavy group 14 analogues of cyclopropenes.[56] However, it was not until 1995 that a cyclotrigermene, the first example of this interesting class of unsaturated inorganic homocycle, was discovered.[57a] The successful synthesis of cyclotrigermenes requires the use of bulky $^tBu_3M'$ (M' = Si, Ge) groups on germanium; these electropositive substituents provide crucial relief of the strain in three-membered rings comprised of group 14 elements. The initial synthesis involved the reaction of $^tBu_3SiNa$ with $GeCl_2$–dioxane in a 2:1 molar ratio.[57a] When this reaction was carried out in a 1:1 molar ratio, the cyclotrigermane *cis,trans*-($^tBuSiGeCl$)$_3$ (**10.12**) was isolated in quantitative yield and shown to be an intermediate in the formation of the cyclotrigermene (Scheme 10.5). Several other cyclotrigermenes have subsequently been reported; they are all dark red.[56]

A few years later, synthetic routes to cyclotrisilenes were reported.[58,59a] In the first method, the dark-red compound $R_3(R_3Si)Si_3$ (**10.13**, $R = {}^tBuMe_2Si$) was obtained by reduction of $(R_3Si)SiBr_2Cl$ with potassium graphite (KC_8) and characterised by spectroscopic methods and oxidative-addition of Cl_2 (with CCl_4 as the source of Cl_2) (Scheme 10.6a).[58] An alternative approach involves the reductive coupling of an R_2Si unit and two RSi units (Scheme 10.6b). A cleaner route to cyclotrisilenes utilises the salt-elimination reaction illustrated in Scheme 10.6c.[59b]

Only one example of a cyclotristannene is known. The dark-red–brown compound R_4Sn_3 (**10.14**, $R = {}^tBu_3Si$) is prepared by the reaction of the tin(II) reagents $Sn(O^tBu)_2$ or $Sn[N(SiMe_3)_2]_2$ with ${}^tBu_3SiNa$ in pentane at $25\,°C$.[60] At lower temperatures this reaction produces the dark-blue tristannaallene $R_2Sn=Sn=SnR_2$, which may be isolated; however, this thermally unstable,

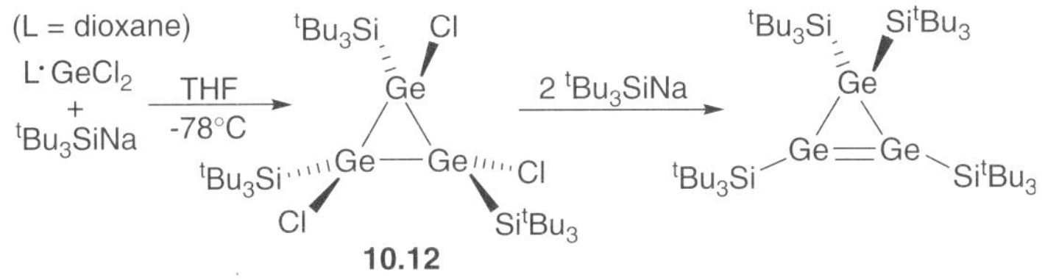

Scheme 10.5 Synthesis of a cyclotrigermene *via* a cyclotrigermane.

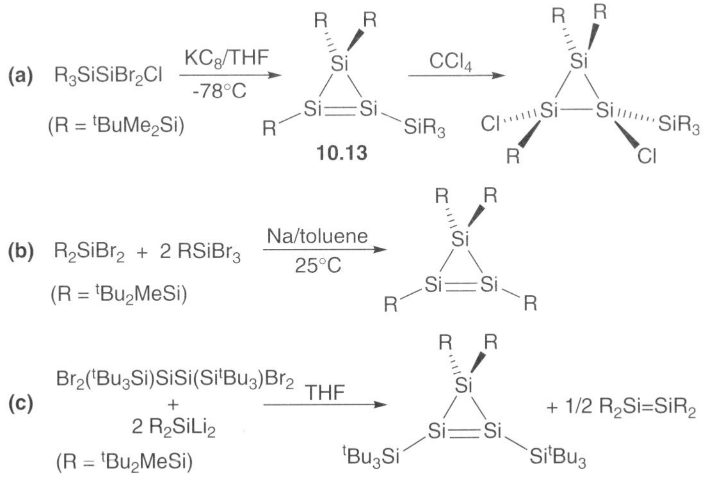

Scheme 10.6 Synthetic routes to cyclotrisilenes.

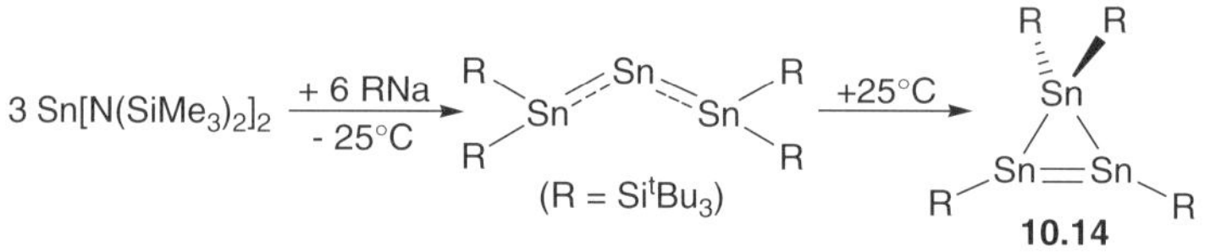

Scheme 10.7 Synthesis of a cyclotristannene.

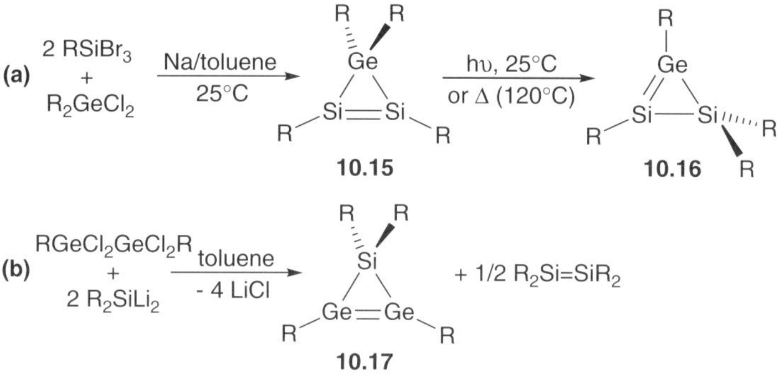

Scheme 10.8 (a) Synthesis and isomerisation of a disilagermirene and (b) synthesis of a siladigermirene (R = SiMe^{t}Bu$_2$).

acyclic compound isomerises to the corresponding cyclotristannene at room temperature (Scheme 10.7).

A reductive-coupling strategy similar to that used for the synthesis of a cyclotrisilene (Scheme 10.6b) can also be applied to the preparation of a mixed cyclotrimetallene. As shown in Scheme 10.8a, the disilagermirene **10.15** (unsaturated Si$_2$Ge ring) prepared in this manner has an >Si=Si< double bond; however, it readily isomerises to **10.16**, the thermodynamically more stable ring containing an >Si=Ge< double bond, upon photolysis or heating.[61] A different approach was necessary for the synthesis of the related SiGe$_2$ ring containing a >Ge=Ge< double bond (**10.17**, Scheme 10.8b).[62]

The solid-state structures of cyclotrimetallenes of the type R$_4$M$_3$ (M = Si, Ge) consist of isosceles triangles with metrical parameters that are consistent with two M–M single bonds and one M=M double bond.[56] For the silicon-containing rings ^{29}Si NMR spectroscopy is a useful diagnostic tool because the resonances for the sp^2 silicon centres occur significantly downfield of those for sp^3 silicon centres (*cf.* ^{13}C NMR chemical shifts of alkenes and alkanes).[58,59] Similarly, the ^{119}Sn NMR chemical shifts of the different tin environments in the cyclotristannene (Scheme 10.7) differ by *ca* 1100 ppm, with the low-field signal being attributable to the three-coordinate tin centres.[60]

As illustrated in Scheme 10.6a, cyclometallenes react readily with chlorocarbons, *e.g.* CCl$_4$, to give *trans*-1,2-dichloro derivatives.[56] In an interesting extension of this type of reactivity, the treatment of a cyclotrigermene or siladigermirene with an excess of dichloromethane produces four-membered Ge_3C or $SiGe_2C$ rings as a result of ring expansion following the initial oxidative-addition process [eqn (10.13)].[62]

$$\text{(E = Si, Ge; R = Si}^t\text{Bu}_2\text{Me)} \quad (10.13)$$

In the context of inorganic ring systems, the most significant reaction of cyclometallenes involves the conversion of the neutral $(^tBu_3Si)_4Ge_3$ ring to the corresponding cyclotrigermenium cation $[(^tBu_3SiGe)_3]^+$ (**10.18**) by reaction with trityl tetraphenylborate [eqn (10.14)].[63a] In the solid state there are no close interactions between this cation and the $[BPh_4]^-$ counterion; nevertheless, highly fluorinated borate anions, e.g. $[B(C_6F_5)_4]^-$, enhance the stability of the $[(^tBu_3SiGe)_3]^+$ cation in solution.[63b] Similar methodology has been used for the synthesis of the unsymmetrically substituted cyclotrisilenium cation **10.19** [eqn (10.15)].[59b] The three-membered rings in these cyclic cations form an equilateral triangle in which the M–M (M = Si, Ge) bond lengths are intermediate between single- and double-bond values, consistent with the description of these species as two π-electron systems [i.e. heavy congeners of the cyclopropenium cation $(C_3H_3)^+$]. The $[(^tBu_3SiGe)_3]^+$ cation is isoelectronic with the gallium-containing dianion $[(ArGa)_3]^{2-}$ (Ar = 2,6-$Mes_2C_6H_3$) (see Section 9.1.2). The synthesis and EPR spectrum of the related neutral three π-electron germanium-containing radical $[(ArGe)_3]^\bullet$ are discussed in Section 5.1.1 [see eqn (5.2)].[64]

$$(R = Si^tBu_3) \quad \textbf{10.18} \quad (10.14)$$

$$\textbf{10.19} \quad (10.15)$$

A disilacyclopropenylium cation **10.20** has also been prepared in a similar manner from the corresponding disilacyclopropene (Scheme 10.9).[64] The Si–C bond distance in the neutral, unsaturated Si_2C ring is typical for an Si=C double bond.

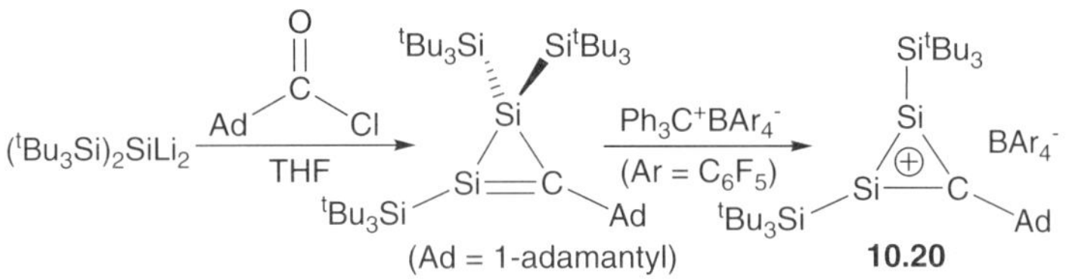

Scheme 10.9 Synthesis of neutral and cationic unsaturated Si_2C rings.

10.2.2 Four-membered and Larger Rings

Significant progress has been made in the past 12 years in the synthesis and characterisation of heavy group 14 analogues of cyclobutene, cyclobutadiene and related charged species.[65] The first tetrasila analogue of cyclobutene R_6Si_4 (**10.21**, R = ^{t}BuMe$_2$Si) was obtained as a bright-orange solid by reductive coupling [eqn (10.16)].[66] The silicon–silicon double bond length in the folded four-membered Si_4 ring is significantly shorter that those in related acyclic disilenes, whereas the opposite silicon–silicon single bond is unusually long. The unsaturated Si_4 ring undergoes isomerisation to the corresponding tetrasilabicyclo[1.1.0]butane upon photolysis. The photolysis product reverts to the thermodynamically more stable cyclotetrasilene in the dark at room temperature.

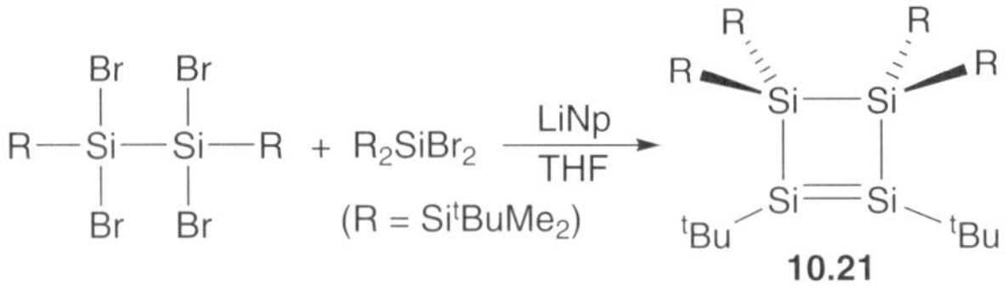

$$(10.16)$$

An alternative route to an unsaturated Si_4 ring involves the oxidative addition of iodine to the tetrahedral tetrasilane R_4Si_4 (R = Si^{t}Bu$_3$) (Scheme 10.10).[67]

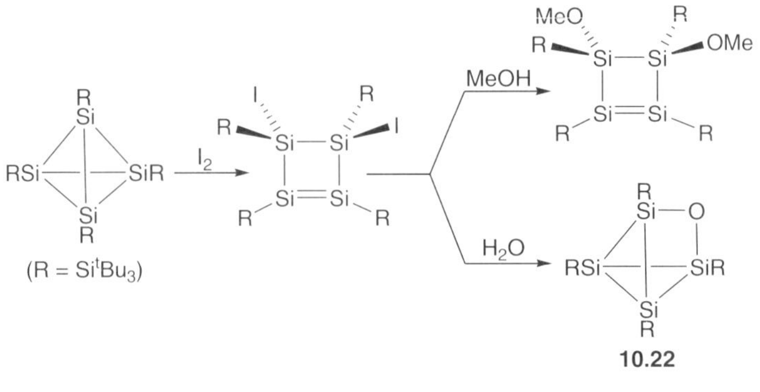

Scheme 10.10 Synthesis and reactions of a diiodocyclotetrasilene.

The four-membered ring in the orange–red diiodo derivative *trans*-$R_4Si_4I_2$ exhibits a folded structure with a silicon–silicon double bond length similar to that in the related acyclic disilene $R_2Si=SiR_2$ (R = Si^iPr_3) and long silicon–silicon single bonds. Interestingly, the silicon–silicon double bond is retained upon reaction of *trans*-$R_4Si_4I_2$ with methanol, whereas treatment with water results in the replacement of the two iodide substituents by a bridging oxo group in a bicyclotetrasilane structure **10.22** (Scheme 10.10).

The cationic four-membered ring $[R_3R'_2Si_4]^+$ (**10.23**, R = $^tBu_2MeSi$, R' = ^tBu) is unexpectedly obtained from the reaction of a cyclotrisilene with $[Et_3Si(C_6H_6)][B(C_6F_5)_4]$ [eqn (10.17)].[68] The yellow tetrakis(pentafluorophenyl)-borate salt of **10.23** is comprised of non-interacting $[R_3R'_2Si_4]^+$ cations and $[B(C_6F_5)_4]^-$ anions. The Si–Si bond lengths in the cation are consistent with localisation of the positive charge on three silicon atoms of the folded Si_4 ring; the fold angle is *ca* 47°. A further interesting feature of the structure is a cross-ring Si–Si distance that is only *ca* 15% longer than an Si–Si single bond, suggesting a 1,3-orbital interaction to give the homoaromatic cyclotrisilenylium cation **10.23** [eqn (10.17)]. The ^{29}Si NMR spectrum exhibits three resonances for the cyclic Si atoms in the ratio 1:2:1; the central Si atom of the cationic part of the ring is strongly deshielded.

$$\text{(10.17)}$$

Investigations of the reduction of the cyclotetrasilenylium cation **10.23** have led to the synthesis and structural characterisation of the corresponding neutral radical and monoanion.[69,70] The red–purple, thermally stable cyclotetrasilenyl radical $[R_3R'_2Si_4]^•$ is obtained upon reduction of the cation with $^tBu_3SiNa$ or potassium graphite,[69] whereas the use of an excess of lithium metal produces green crystals of the lithium salt of the cyclotetrasilenide anion $[R_3R'_2Si_4]^-$.[70] In marked contrast to the precursor cation, the four-membered ring in the radical is almost planar and the 1,3-orbital interaction observed in the cation is absent in the radical (Figure 10.5a). The lithium salt $(THF)Li[R_3R'_2Si_4]$ exists as a contact-ion pair in the solid state with the lithium cation coordinated to three Si atoms of the folded four-membered ring; the fold angle is *ca* 27° (Figure 10.5b).

Investigations of the redox behaviour of cyclotetrasilenyl rings have revealed a reversible redox system for the cation, radical and anion (Scheme 10.11).[70]

The anionic four-membered ring $[Tip_5Ge_4]^-$ (**10.24**, Tip = 2,4,6-$^iPr_3C_6H_2$) is obtained as the dark-red $[Li(DME)_3]^+$ salt by reduction of the corresponding digermene with lithium metal [eqn (10.18)].[71] The disparate Ge–Ge bond distances (2.37 and 2.51 Å) indicate that the planar Ge_4 ring incorporates an allyl-like Ge_3

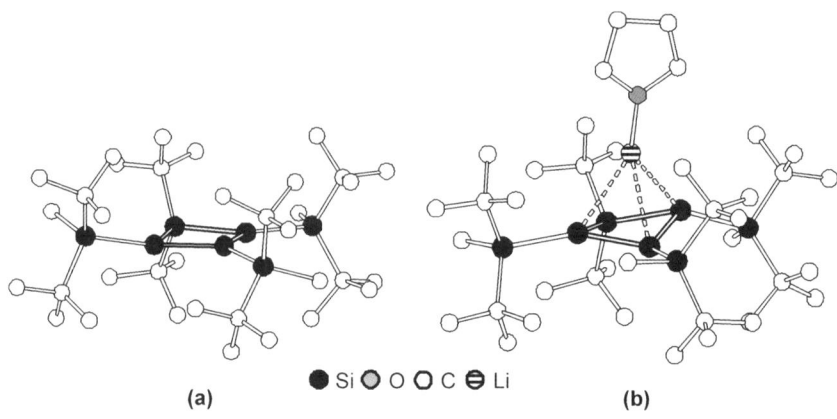

Figure 10.5 Molecular structures of (a) [R₃R′₂Si₄]• and (b) (THF)Li[R₃R′Si₄] (R = ᵗBu₂MeSi, R′ = ᵗBu).

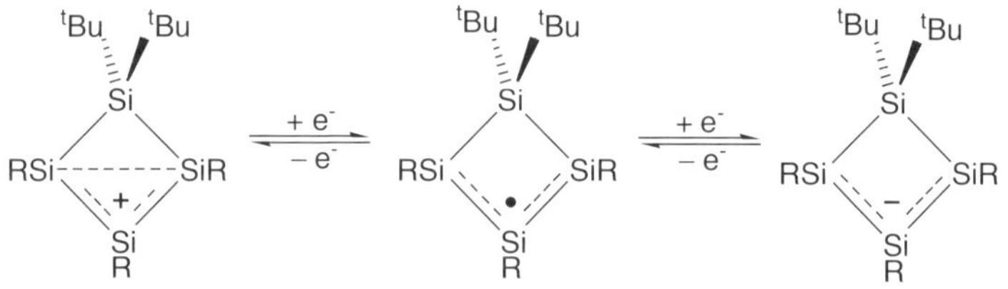

Scheme 10.11 Interconversion of cyclotetrasilenyl cation, radical and anion (R = ᵗBu₂MeSi).

anion. Shorter reaction times produce the hexaaryltetragermabuta-1,3-diene Ar₂Ge=Ge(Ar)–Ge(Ar)=GeAr₂ (Ar = Tip) with conjugated germanium–germanium double bonds.

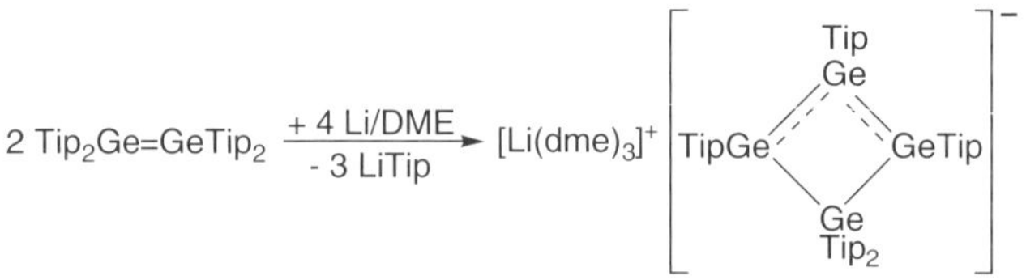

$$2\ \text{Tip}_2\text{Ge}=\text{GeTip}_2 \xrightarrow[-3\ \text{LiTip}]{+4\ \text{Li/DME}} [\text{Li(dme)}_3]^+ \begin{bmatrix} \text{TipGe} \begin{array}{c} \text{Tip} \\ \text{Ge} \\ \end{array} \text{GeTip} \\ \begin{array}{c} \text{Ge} \\ \text{Tip}_2 \end{array} \end{bmatrix}^-$$

10.24

(10.18)

Several examples of unsaturated four-membered rings containing two different heavy group 14 elements are known. For example, a neutral Si₂Ge₂ analogue of cyclobutene, **10.25a**, is obtained by a ring expansion reaction of a disilagermirene with germanium(II) dichloride [eqn (10.19)].[72] The use of SnCl₂ instead of GeCl₂ produces a four-membered Si₂GeSn ring **10.25b** with a germanium–tin double bond.

A heteroatomic (Si_2Ge_2) analogue of the cyclobutadiene dianion **10.26** is prepared by reductive dehalogenation of appropriate dihalogenated precursor with potassium graphite (Scheme 10.12a); a homoatomic (Si_4) dianion is obtained in a similar manner.[65a,73] The four-membered rings in the dipotassium salts of these dianions are puckered with the two potassium ions located above and below the ring and η^2-coordinated at the 1,3- and 2,4-positions (Figure 10.6). Reduction of the same dihalogenated precursor with an alkaline earth metal produces a bicyclic dianion **10.27**, which is chelated through the two germanium atoms to the solvated metal dication (Scheme 10.12b).[74]

The formation of η^4-metal complexes of cyclobutadiene is a classic example of the stabilisation of a highly reactive organic ligand by coordination to a metal. In this context, the preparation of metal complexes of the tetrasilacyclobutadiene ligand represents an intriguing target. The first examples of such complexes were reported for cobalt [eqn (10.20)].[75] The Si_4 ring in the 18 π-electron $[(R_4Si_4)Co(CO)_2]^-$ anion (**10.28**, R = $^tBuMe_2Si$) is a planar rectangle with opposite Si–Si distances of ca 2.26 and 2.31 Å. The synthesis, structure and bonding in the related neutral complex $[(R_4Si_4)Fe(CO)_3]$ are discussed in Section 7.2.1 (see Scheme 7.4).

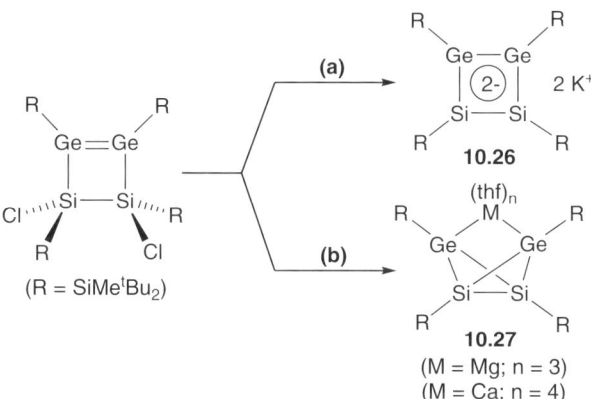

Scheme 10.12 Reduction of the unsaturated Si_2Ge_2 ring with (a) KC_8 and (b) M–THF (M = Mg, Ca).

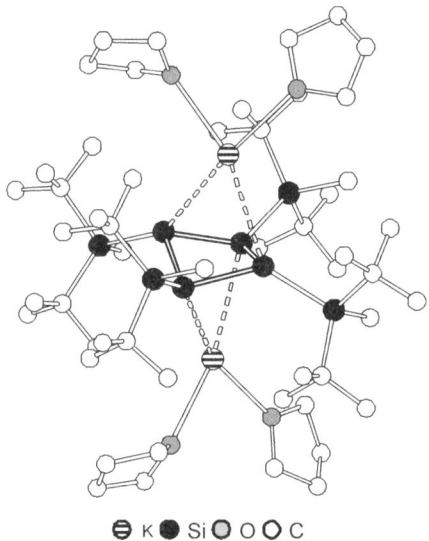

Figure 10.6 Molecular structure of [K(THF)$_2$]$_2$[R$_4$Si$_4$] (R = SiMe^{t}Bu$_2$).

$$\text{(10.20)}$$

The synthesis of analogues of the cyclopentadienide anion [C$_5$H$_5$]$^-$ or benzene C$_6$H$_6$ in which all the carbon atoms are replaced by heavier group 14 elements is a formidable, but intriguing, synthetic challenge. Progress towards making analogues of the five-membered cyclopentadienide anion has so far been limited to ring systems that incorporate only three silicon or germanium atoms;[76] these carbon-containing rings are outside the scope of this book.

A fascinating bicyclic compound **10.29**, in which two five-membered Si$_5$ rings share a common Si=Si double bond, is obtained by reductive coupling of a 1,1-dichlorocyclotetrasilane [eqn (10.21)].[77]

$$\text{(10.21)}$$

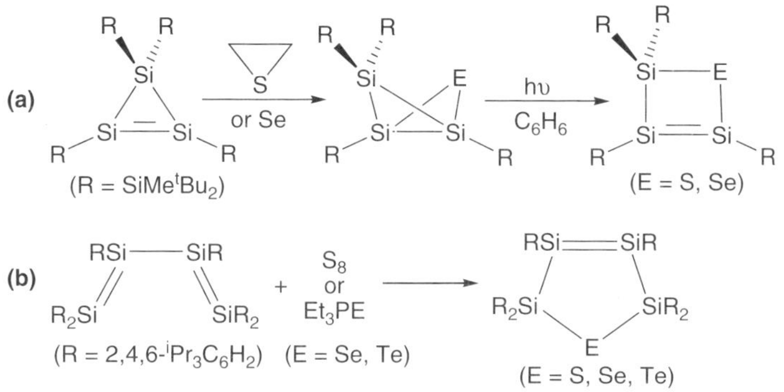

Scheme 10.13 Synthesis of unsaturated (a) Si$_3$E (E=S, Se) and (b) Si$_4$E (E=S, Se, Te) rings.

Unsaturated silicon analogues of cyclobutene and cyclopentene containing a single chalcogen atom are accessible *via* (a) the reaction of a cyclotrisilene with propylene sulfide (or elemental chalcogen) followed by photolysis of the initially formed bicyclic product (Scheme 10.13a)[78a] and (b) a formal (4 + 1) cycloaddition of an acyclic tetrasilabutadiene with a source of chalcogen (Scheme 10.13b), respectively.[78b]

10.3 SILICON–, GERMANIUM– AND TIN–NITROGEN RINGS

10.3.1 Saturated Systems

Saturated silicon–nitrogen rings of the type (R$_2$SiNR')$_n$ (cyclosilazanes) are isoelectronic with the corresponding silicon–oxygen rings (R$_2$SiO)$_n$ (cyclosiloxanes, see Section 10.5). In contrast to cyclosiloxanes, however, ring sizes are limited to four-, six- and eight-membered rings for the cyclosilazanes. There is also a wide diversity of silicon–nitrogen rings that contain either Si–Si or N–N bonds.[79]

The most versatile synthesis of rings with alternating silicon and nitrogen atoms involves cyclocondensation of primary amines R'NH$_2$ (or ammonia) with dichlorodiorganosilanes R$_2$SiCl$_2$.[79] This method is readily adapted for the preparation of heterocycles with different substituents on the same silicon atom by using dichlorosilanes of the type RR'SiCl$_2$ (R ≠ R'). Unsymmetrical systems are made by salt-elimination reactions, *e.g.* between R$_2$Si(NLiR'')$_2$ and R'$_2$SiCl$_2$; this approach is also used to incorporate other p-block elements into Si–N rings by the reaction of R$_2$Si(NLiR'')$_2$ with main group element halides. Low temperatures and short reaction times are necessary in order to optimise the yields of the kinetically favoured four-membered rings relative to those of six- and eight-membered rings, which are the thermodynamically preferred products, as exemplified in Scheme 10.14.[80] In addition, a low concentration of reagents is desirable to suppress polymer formation *via* intermolecular processes.

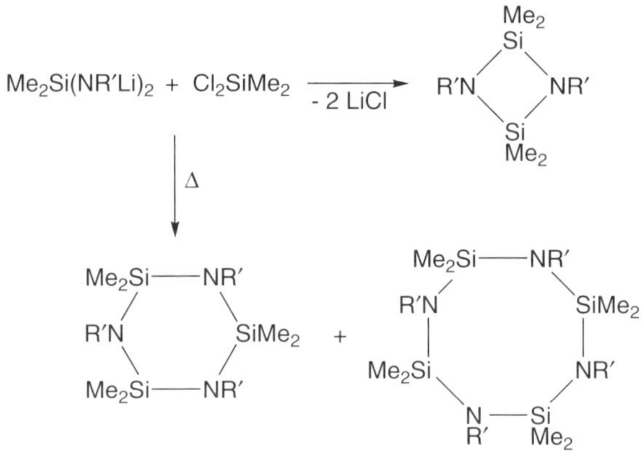

Scheme 10.14 Formation of cyclosilazanes *via* salt elimination.

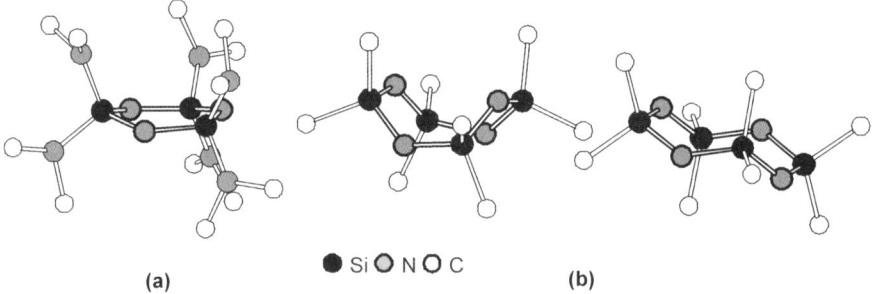

Figure 10.7 Molecular structures (a) [(Me$_2$N)$_2$SiNH]$_3$ and (b) the boat and chair forms of (Me$_2$SiNH)$_4$.

Cyclodisilazanes, *e.g.* (Me$_2$SiNSiMe$_3$)$_2$,[81] form planar four-membered rings in which the substituent on nitrogen is coplanar with the Si$_2$N$_2$ ring and the Si–N–Si bond angle is slightly larger than 90°.[79b] According to an electron diffraction study, hexamethyltrisilazane (Me$_2$SiNH)$_3$ has a puckered structure in the gas phase, but the deviation from planarity is small.[82] By contrast, the hexachloro derivative (Cl$_2$SiNH)$_3$ is planar in the solid state with Si–N bond lengths that are *ca* 0.05 Å shorter than those in (Me$_2$SiNH)$_3$.[83] Replacement of the Cl substituents by dimethylamino groups in [(Me$_2$N)$_2$SiNH]$_3$ leads to a slight puckering of the Si$_3$N$_3$ ring; the mean deviation of the Si and N atoms from the best plane is *ca* 0.10 Å (Figure 10.7a).[84] The solid-state structure of the tetramer (Me$_2$SiNH)$_4$ consists of two different molecules, one in a boat and the other in a chair conformation (Figure 10.7b).[85] Remarkably, experimental charge density studies of (Me$_2$SiNH)$_4$ indicate almost purely ionic Si–N bonds.[85]

A significant difference between cyclosilazanes, *e.g.* (Me$_2$SiNH)$_3$ and the isoelectronic cyclosiloxanes (Me$_2$SiO)$_3$, is the presence of the reactive N–H functionality in the former, which is readily lithiated with ^{n}BuLi to produce N-lithiated reagents that are reactive towards main group element halides. For example, ring coupling is achieved by reaction of an N-lithiated cyclotrisilazane with chloropentamethylborazine.[86] In some cases, the reactions of N-lithiated cyclosilazanes result in ring transformations, e.g. contraction of six-membered into four-membered rings. However, the tetramer $(Me_2SiNH)_4$ is completely deprotonated by reaction with $LiAlH_4$ in THF to give the tetraanion $[(Me_2SiN)_4]^{4-}$, which retains the eight-membered Si_4N_4 ring.[85]

The ability of cyclosilazanes to act as macrocyclic ligands is limited by (a) the cleavage of the Si–N bond by many metal halides and (b) the unavailability of ring sizes larger than eight atoms (cf. cyclosiloxanes, Section 10.5). However, the eight-membered ring $(Me_2SiNH)_4$ forms bis-adducts with certain early transition metal halides, e.g. $TiCl_4$, in which only two of the Lewis basic nitrogen centres are coordinated to a metal centre.[87a] In the mono-adducts $(Me_2SiNH)_3 \cdot MCl_3$ (M = Ti, V), the six-membered Si_3N_3 ring acts as a tridentate nitrogen donor on the basis of spectroscopic data.[87b]

Both germanium and tin form saturated ring systems of the type $(R_2MNR')_n$ that are synthesised in a similar manner to cyclosilazanes; however, they are limited to four- and six-membered rings (n = 2 or 3, respectively).[88,89] Transamination reactions with the elimination of a volatile amine are also used to prepare cyclostannazanes [eqn (10.22)].[90]

$$R_2Sn(NR''_2)_2 + H_2NR' \rightarrow (R_2SnNR')_n + 2HNR''_2 \qquad (10.22)$$

The reactivity of the tin–nitrogen bond in cyclostannazanes towards organoelement halides may be used to prepare other inorganic heterocycles, e.g. B_2N_2 rings [eqn (10.23)].[91]

(10.23)

Current interest in saturated Si–N rings stems primarily from the industrial uses of silicon nitride (and related materials), which is a hard, chemically resistant insulator,[92] and as precursors to silicon–nitrogen polymers (polysilazanes). The formation and precursor chemistry of these polymers are discussed in Section 10.3.3.

10.3.2 Unsaturated Systems and Biradicals

Investigations of unsaturated group 14 element–nitrogen ring systems are primarily focused on cyclic oligomers of the type $(:MNR)_n$ in which M is in the divalent state. The six-membered rings (n = 3) are potentially Hückel six π-electron systems in which each of the nitrogens contribute two π-electrons.

Examples of this class of inorganic heterocycle are limited to germanium. Both four- and six-membered rings have been characterised with the ring size being determined by the nature of the substituent on the nitrogen atoms.[93–95]

The four-membered rings [Ge(μ-NAr)]$_2$ [**10.30a**, Ar = 2,4,6-^{t}Bu$_3$C$_6$H$_2$;[93] **10.30b**, 2,4,6-(CF$_3$)$_3$C$_6$H$_2$[95]] are obtained as yellow solids by the salt-elimination reaction of germanium(II) chloride with an alkali metal derivative of the arylamine [eqn (10.24)]. The germanium–nitrogen bond lengths within each Ge$_2$N$_2$ ring are equal with mean values of 1.85 and 1.88 Å, respectively. The aryl groups are oriented at 90° with respect to the planar Ge$_2$N$_2$ ring. The dimer **10.30a** forms an adduct [(CO)$_4$Fe{Ge(μ-NAr)}$_2$] upon treatment with Fe$_2$(CO)$_9$; it also undergoes oxidative addition at both germanium(II) centres with two equivalents of MeI.[93]

$$4 \text{ ArNHM} + 2 \text{ GeCl}_2 \cdot \text{dioxane} \xrightarrow{-4 \text{ MCl}} \begin{array}{c} \text{Ge} \\ \text{ArN} \diagup \diagdown \text{NAr} \\ \diagdown \diagup \\ \text{Ge} \end{array}$$

(M = Li; Ar = 2,4,6-^{t}Bu$_3$C$_6$H$_2$)
(M = K; Ar = 2,4,6-(CF$_3$)$_3$C$_6$H$_2$)

10.30

(10.24)

The use of a slightly less bulky aryl group (Ar = 2,6-^{i}Pr$_2$C$_6$H$_3$) produces the trimeric species [Ge(μ-NAr)]$_3$, which is prepared by the transamination reaction between Ge[N(SiMe$_3$)$_2$]$_2$ and the arylamine ArNH$_2$.[94] The Ge$_3$N$_3$ ring in the trimer is essentially planar with mean Ge–N bond lengths of 1.86 Å (Figure 10.8). Although the planarity of this six-membered ring suggests some degree of delocalisation, the calculated stabilisation energy indicates very little aromatic character in germanazenes.[96]

With less bulky groups on the nitrogen atoms, the most common oligomer formed by the hypothetical monomeric group 14 imides, :M=NR (M = Ge, Sn, Pb), is the cubic tetramer, *e.g.* [M(μ$_3$-N^{t}Bu)]$_4$.[97] This is especially true for the heavier (larger) elements tin and lead for which, in contrast to germanium, cubanes are observed even when R = 2,6-^{i}Pr$_2$C$_6$H$_3$.[94]

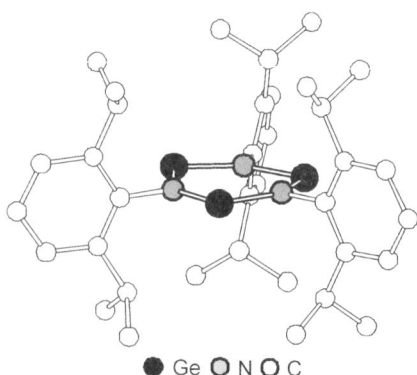

● Ge ○ N ○ C

Figure 10.8 Molecular structure of [Ge(μ-NAr)]$_3$ (Ar = 2,6-^{i}Pr$_2$C$_6$H$_3$).

Group 14: Rings and Polymers

An intriguing germanium–nitrogen ring system, ArGe(μ-NSiMe$_3$)$_2$GeAr (**10.31**, Ar = 2,6-Dipp$_2$C$_6$H$_3$, Dipp = 2,6-^{i}Pr$_2$C$_6$H$_3$), is obtained as dark-violet crystals by the reaction of a 'digermyne' ArGeGeAr with trimethylsilyl azide (Scheme 10.15a).[98] The related four-membered tin–nitrogen ring ClSn(μ-NSiMe$_3$)$_2$SnCl (**10.32**) is a colourless compound, which is formed unexpectedly in the reaction of (Me$_3$Si)$_2$NSn(μ-Cl)$_2$SnN(SiMe$_3$)$_2$ with silver isocyanate (Scheme 10.15b).[99]

The M$_2$N$_2$ rings in **10.31** and **10.32** are both planar with the groups attached to the group 14 metal centres (aryl and Cl, respectively) in a *trans* arrangement

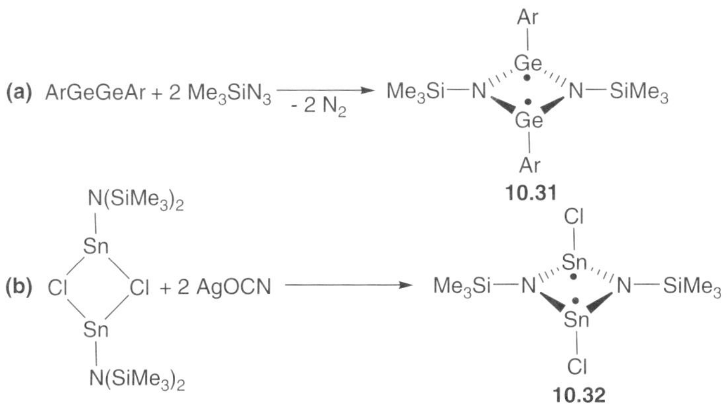

Scheme 10.15 Syntheses of M$_2$N$_2$ (M = Ge, Sn) rings with biradicaloid character.

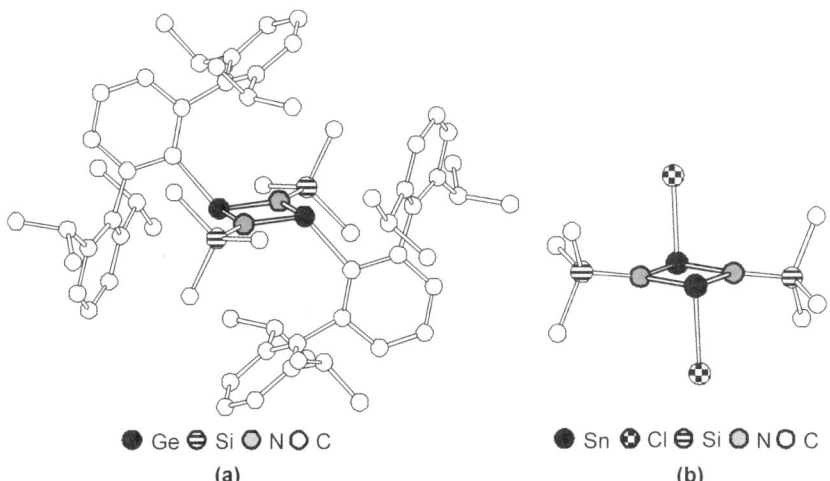

Figure 10.9 Molecular structures of (a) ArGe(μ-NSiMe$_3$)$_2$GeAr (Ar = 2,6-Dipp$_2$C$_6$H$_3$) and (b) ClSn(μ-NSiMe$_3$)$_2$SnCl.

(Figure 10.9). The transannular distance between the two metal atoms in these four-membered rings is substantially longer than the corresponding M–M single bond (by *ca* 0.3 and 0.6 Å, respectively). DFT calculations support the lack of a bonding interaction between the two metal atoms and indicate a singlet ground state for both of these biradical species.[98,99] The singlet–triplet energy gap is estimated to be *ca* 17 kcal mol^{-1} for both heterocyclic systems. The biradical character of **10.31** is also indicated by facile addition of molecular hydrogen at room temperature.[98a]

10.3.3 Silicon–Nitrogen Polymers: Ceramic Precursors

Polysilazanes possess an Si–N backbone and have attracted significant attention as polymeric precursors to Si_3N_4-based ceramics. These materials can be used to fabricate fibres, coatings or 3D continuous fibre-reinforced ceramic matrix composites, which usually cannot be made through traditional ceramic processing methods.[100,101] In many cases the polysilazane coating is converted into silica. This commercially important class of polymers includes many relatively poorly characterised low molecular weight materials that contain both cyclic and linear structures; they are prepared by the condensation of dichlorosilanes or organodichlorosilanes with primary amines or ammonia. More well-defined, higher molecular weight polysilazanes **10.33** are prepared by the ROP approach using the four-membered rings as a precursor [eqn (10.25)].[102]

(10.25)

10.4 SILICON–, GERMANIUM– AND TIN–PHOSPHORUS RINGS

This section focuses on rings in which the group 14 element and phosphorus alternate. In the case of silicon, saturated heterocycles of the type $(R_2SiPR')_n$ (*i.e.* cyclosilaphosphanes) are limited to four- and six-membered rings.[103] A series of cyclophosphapolysilanes containing Si–Si bonds is also known, *e.g.* $PhP(SiMe_2)_n$ ($n = 4$–6) and the six-membered ring $[PhP(SiMe_2)_2]_2$.[104] The formation of polycyclic structures, *e.g.* the adamantane-like $P_4(SiMe_2)_6$, is also a common feature of the silicon–phosphorus(III) system.[103]

Cyclosilaphosphanes are normally prepared by salt-elimination reactions between dichlorosilanes R_2SiCl_2 and an alkali metal phosphide, $LiPHR'$ or Li_2PR' For example, the four-membered ring $(^tBu_2SiPH)_2$ is obtained either by (a) reaction of $^tBu_2SiF_2$ with $LiPH_2$ followed by treatment of the intermediate $^tBu_2Si(F)PH_2$ with one equivalent of ^nBuLi[105] or (b) by reaction of $^tBu_2SiCl_2$

with Li$_2$PH.[106] As indicated by this example, bulky substituents such as *tert*-butyl favour the formation of the smaller (four-membered) ring. Interestingly, however, the ring size that is obtained from the reaction of dimethyldichlorosilane with dilithium phenylphosphide is temperature dependent. The four-membered ring (Me$_2$SiPPh)$_2$ (**10.34**) is isolated when the reaction is conducted at low temperature, whereas the six-membered ring (Me$_2$SiPPh)$_3$ (**10.35**) is the preferred product at higher temperatures (Scheme 10.16).[104] NMR studies show that, in solution, these two ring sizes exist in equilibrium with the trimer favoured at higher temperatures. The reaction of Ph$_2$SiCl$_2$ with KHPPh in ether solvents also gives both the dimeric and trimeric all-phenyl ring systems (Ph$_2$SiPPh)$_n$ ($n = 2, 3$).[107]

Information on the solid-state structures of cyclosilaphosphanes is lacking and there have been limited studies of their chemical behaviour. As might be expected, these ring systems may act as *P*-mono, -di- or -tridentate ligands towards transition metal centres, as illustrated by the examples given in Figure 10.10.[104,108,109]

A silicon–phosphorus bond in (Me$_2$PSiMe)$_3$ is readily cleaved by methyllithium to generate the acyclic dilithiated derivative Me$_2$Si(PLiMe)$_2$ [**10.36**, eqn (10.26)].[106] The related dilithiated dianion Me$_2$Si(PLi^{t}Bu)$_2$ has been used to

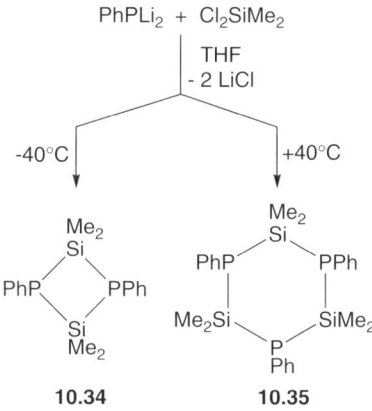

Scheme 10.16 Formation of Si$_2$P$_2$ and Si$_3$P$_3$ rings *via* salt elimination.

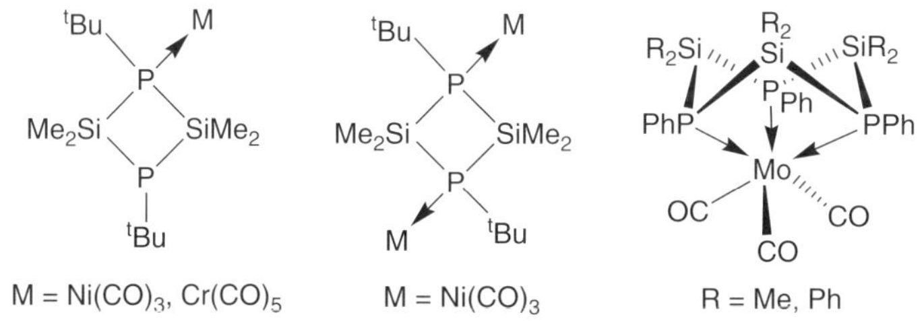

Figure 10.10 Metal complexes of cyclosilaphosphanes.

prepare the unsymmetrical four-membered ring $Me_2Si(\mu\text{-}P^tBu)_2SiPh_2$ or the silicon-centred spirocycle $[Me_2Si(\mu\text{-}P^tBu)_2]_2Si$ by reactions with Ph_2SiCl_2 or $SiCl_4$, respectively.[106]

$$\begin{array}{c}\text{MeP——SiMe}_2\\ /\quad\quad\quad\backslash\\ \text{Me}_2\text{Si}\quad\quad\text{PMe}\\ \backslash\quad\quad\quad /\\ \text{MeP——SiMe}_2\end{array} + 2\text{ MeLi} \longrightarrow \begin{array}{c}\text{Me}\\ \text{LiP——SiMe}_2\\ /\quad\quad\quad\backslash\\ \quad\quad\quad\text{PMe}\\ \backslash\quad\quad\quad /\\ \text{LiP——SiMe}_2\\ \text{Me}\end{array}$$

10.36 (10.26)

Saturated germanium– and tin–phosphorus rings of the type $(R_2MPR')_n$ (M = Ge, Sn; n = 2,3) have been the subject of a limited number of studies.[110,111] They are prepared in a manner similar to cyclosilaphosphanes. For example, the reactions of Ph_2GeCl_2 with KHPPh or K_2PPh produce the four-membered cyclogermaphosphanes $(Ph_2GePPh)_2$ or the six-membered $(Ph_2GePPh)_3$, respectively.[107] The P–P bonded isomer of the four-membered Ge_2P_2 ring is obtained from Ph_2GeCl_2 and $K_2[^tBuPP^tBu]$.[112] The generation of the transient monomer $[Me_2Ge=PPh]$ by the method illustrated in Scheme 10.17 produces the dimer $(Me_2GePPh)_2$ (**10.37**) and the trimer $(Me_2GePPh)_3$ (**10.38**) in a 3:1 molar ratio.[113a] The tin-containing six-membered ring $(Me_2SnPPh)_3$ is obtained, *via* the monomer $[Me_2Sn=PPh]$, by using similar synthetic methodology.[113b] Trimeric cyclostannaphosphanes $(R_2SnPPh)_3$ (R = Me, ^{n}Bu, Ph) are more simply prepared in good yields by the cyclocondensation reaction of phenylphosphine with diorganotin dihalides in the presence of triethylamine; molecular masses were determined cryoscopically.[107,114]

A different class of four-membered Sn_2P_2 ring **10.39** is obtained as orange–red crystals from the salt-elimination reaction of potassium di-*tert*-butyl phosphide with the triethylphosphine adduct of tin(II) dichloride [eqn (10.27)]; the dimeric structure was deduced from multinuclear NMR spectra and solution molecular weight measurements. The ^{119}Sn NMR spectrum exhibits a doublet-of-triplets

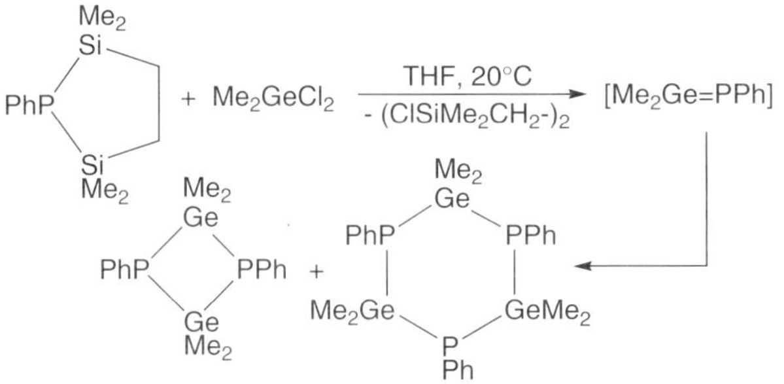

Scheme 10.17 Formation of Ge_2P_2 and Ge_3P_3 rings *via* the monomer $[Me_2Ge=PPh]$.

pattern arising from coupling to the two types of phosphorus centres, P_a and P_b, respectively. The formation of the dimer **10.39** involves a Lewis base–Lewis acid interaction.[115]

$$2\ KP^tBu_2 + Et_3P\cdot SnCl_2 \xrightarrow[-PEt_3]{-2\ KCl} \mathbf{10.39} \tag{10.27}$$

The reaction of $K_2[^tBuPP^tBu]$ with R_2SnCl_2 gives either the three-membered P_2Sn ring $[R_2Sn(^tBuPP^tBu)]$ ($R = {}^tBu$) or the six-membered P_4Sn_2 ring $[R_2Sn(\mu\text{-}^tBuPP^tBu)]_2$ ($R = Et$) as the major product.[116]

10.5 SILICON–, GERMANIUM– AND TIN–OXYGEN RINGS

Silicon–oxygen rings of the type $(R_2SiO)_n$ (cyclosiloxanes) are compounds of considerable consequence as precursors to silicone polymers (Section 10.6). In addition to their commercial importance, these inorganic heterocycles are of historical significance. They are prepared by the hydrolysis of disubstituted dichlorosilanes R_2SiCl_2 or depolymerisation of dimethylsiloxane polymer $(Me_2SiO)_n$.[117] Early studies on the phenyl derivative Ph_2SiCl_2 were carried out in an attempt to generate the monomer $Ph_2Si=O$, the silicon analogue of a ketone; hence they were referred to as 'silico(keto)nes'.[118] This work led to the characterisation of both six- and eight-membered rings $(Ph_2SiO)_n$ ($n = 3, 4$), which are high-melting white solids. Subsequently, investigations of the hydrolysis of Me_2SiCl_2 established that an extensive series of cyclic oligomers exists for the all-methylated derivatives $(Me_2SiO)_n$ ($n = 3$–8) with the tetramer and pentamer as the major products.[119a,b] The trimer $(Me_2SiO)_3$ and the octamer $(Me_2SiO)_8$ are white crystalline solids, whereas the other members of the series are colourless liquids. Milligram quantities of the individual rings up to $n = 25$, formed by the KOH-catalysed equilibration of the tetramer, have been separated by chromatographic methods and identified spectroscopically.[119c] A more recent analysis of the mixtures obtained by hydrolysis of H_2SiCl_2, $Me(H)SiCl_2$ and Me_2SiCl_2 under comparable reaction conditions provided evidence for the oligomeric series $(H_2SiO)_n$ ($n = 3$–23), $[Me(H)SiO]_n$ ($n = 3$–19) and $(Me_2SiO)_n$ ($n = 3$–17).[120]

Although the isolation of a silanone $R_2Si=O$ remains elusive, evidence for the existence of a heteroatom-stabilised monomer has been obtained by trapping experiments. The precursor for this species is a silyl formate **10.40** formed by CO_2 insertion into a Si–H bond (Scheme 10.18). Decomposition of this ester at 85 °C generates the transient silanone, which can be trapped by insertion into $(Me_2SiO)_3$ to give the eight-membered Si_4O_4 ring **10.41**.[121a] The stabilisation

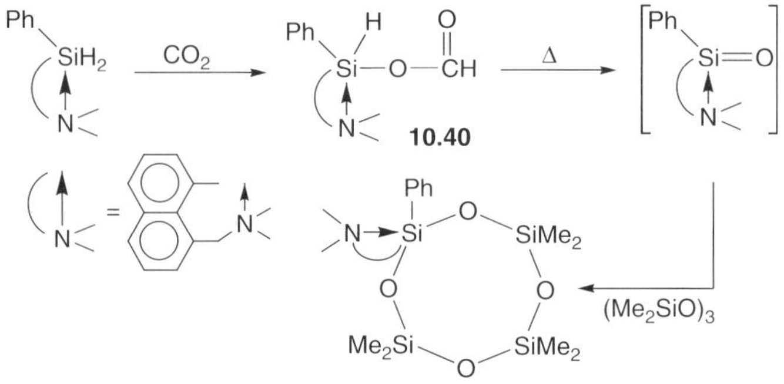

Scheme 10.18 Trapping of a silanone *via* insertion into an Si_3O_3 ring.

of a silaformamide R(H)Si=O by complexation to BH_3 has been reported recently.[121b]

There is an interesting analogy between cyclosiloxanes and the cyclic metasilicates $[Si_nO_{3n}]^{2n-}$ ($n = 3$–6), *e.g.* $(Me_2SiO)_3$ (**10.42**) and $[Si_3O_9]^{6-}$ (**10.43**); the methyl groups in the dimethylsiloxanes are replaced by the formally isoelectronic O^- in the silicates. Cyclic metasilicates occur naturally in certain minerals. For example, the trimer ($n = 3$) is found in $Ca_3Si_3O_9$ (α-wollastonite), whereas the hexamer ($n = 6$) is a constituent of $Be_3Al_2Si_6O_{18}$ (beryl).

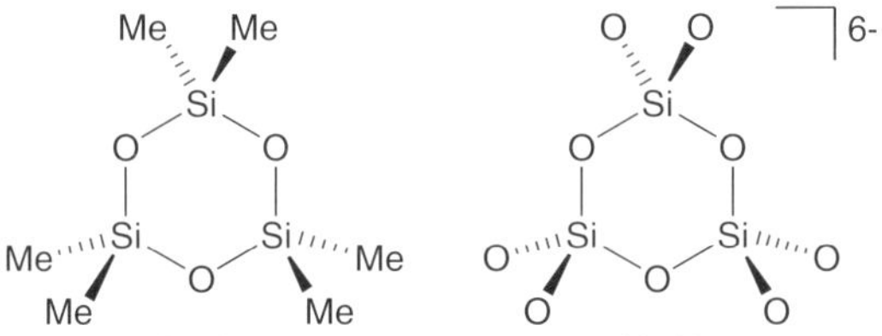

Four-membered rings of the type $(R_2SiO)_2$ are not obtained by the hydrolysis of R_2SiCl_2. The first cyclodisiloxane $(Mes_2SiO)_2$ was prepared in 1981 by the oxidation of tetramesityldisilene with atmospheric oxygen.[122] It is a white powder of high thermal stability, but it undergoes hydrolysis to give the corresponding diol $(HO)Mes_2SiOSiMes_2(OH)$.[123] Cyclodisiloxanes of the type **10.44** with different bulky substituents on each silicon atom are also be obtained by this route [eqn (10.28)].[123] The four-membered Si_2O_2 ring in **10.44** is either planar or slightly puckered with mean Si–O bond lengths that are somewhat longer than those in other cyclosiloxanes.[122b] The narrow Si–O–Si angles of 86–91° are indicative of a strained ring system. The transannular Si–Si distance in **10.44** is close to the silicon–silicon single bond value, raising the question of a possible bonding interaction.[123] Calculations of the nature of this interaction indicate that, although there is not a σ-type bond between the silicon atoms, there may be weak π-bonding.[124]

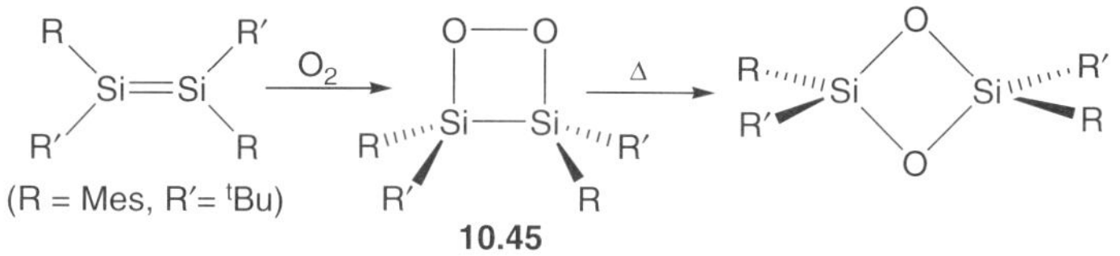

Group 14: Rings and Polymers

$$\underset{\substack{R'\\(R = Mes, R' = {}^tBu)}}{\overset{R}{\underset{R'}{\sum}}Si{=}Si\overset{R'}{\underset{R}{\sum}}} \xrightarrow{O_2} \underset{\substack{R'\\\\10.45}}{\overset{R}{\underset{R}{\sum}}}\overset{O{-}O}{\underset{|}{Si{-}Si}}\overset{R'}{\underset{\substack{R\\}}{\sum}} \xrightarrow{\Delta} \underset{R'}{\overset{R}{\sum}}Si\overset{O}{\underset{O}{\sum}}Si\overset{R'}{\underset{R}{\sum}}$$

Scheme 10.19 Formation of a 1,2-Si$_2$O$_2$ ring by oxidation of disilenes with oxygen.

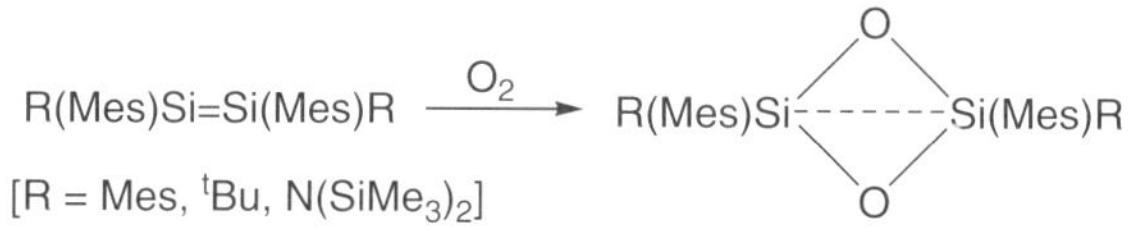

$$R(Mes)Si{=}Si(Mes)R \xrightarrow{O_2} R(Mes)Si\overset{O}{\underset{O}{\diamond}}Si(Mes)R$$

[R = Mes, ^{t}Bu, N(SiMe$_3$)$_2$]

10.44

(10.28)

The mechanism of the oxidation process depicted in Scheme 10.19 has been investigated in detail for 1,2-dimesityl-1,2-di-*tert*-butyldisilene by using isotopically labelled $^{18}O_2$.[125] The intermediate 1,2-disiladioxetane **10.45** was isolated and structurally characterised. The four-membered ring is almost planar with a normal oxygen–oxygen single bond length. Both the oxidation process and the transformation of the 1,2-Si$_2$O$_2$ ring into the 1,3-isomer occur with retention of configuration. The isomerisation, which also takes place in the solid state, involves a first-order (intramolecular) process ($E_a = +21.7$ kcal mol^{-1}), as demonstrated by the lack of mass spectrometric evidence for the crossover product when equimolar quantities of 1,2-Si$_2^{16}$O$_2$ (unlabelled) and 1,2-Si$_2^{18}$O$_2$ (labelled) rings were allowed to rearrange at room temperature.[125] The four-membered 1,2-Si$_2$O$_2$ ring is deoxygenated by phosphines or sulfides to form the corresponding three-membered Si$_2$O ring (disilaoxirane).

The primary interest in six- and eight-membered rings of the type (R$_2$SiO)$_n$ ($n=3, 4$) is the cationic polymerisation of the methylated derivatives by protonic acids as a source of high molecular weight polydimethylsiloxane (Section 10.6).[126] An alternative approach to functional materials based on Si$_n$O$_n$ ($n=3, 4$) ring systems involves the installation of exocyclic hydroxyl groups on the silicon atoms. This is achieved by hydrolysis of ^{i}Pr(Ar)SiCl$_2$ to give a mixture of [^{i}Pr(Ar)SiO]$_4$ isomers followed by dearylchlorination and hydrolysis to give [^{i}Pr(OH)SiO]$_4$. This procedure has been used to synthesise all four stereoisomers of 1,3,5,7-tetrahydroxy-1,3,5,7-tetraisopropylcyclotetrasiloxane [all-*trans*, *cis*,*cis*,*trans*, *cis*,*trans*,*cis* and all-*cis* (Figure 10.11)].[127a] These hydroxy-functionalised eight-membered rings are versatile precursors for siloxanes with cage or ladder structures, and also nanosized aggregates. The same approach has been used to generate *cis*,*trans*-1,3,5-trihydroxy-1,3,5-triisopropylcyclotrisiloxane.[127b]

A wide range of p-block elements and transition metals have been incorporated into silicon–oxygen ring systems (heterocyclosiloxanes), primarily with a view to their use as precursors to Si–O polymers incorporating another element.[128] The most common synthetic approaches to six-membered heterocyclosiloxanes containing another p-block element involve cyclocondensation reactions between 1,3-dichloro- or 1,3-dihydroxytetraalkyl/aryldisiloxane

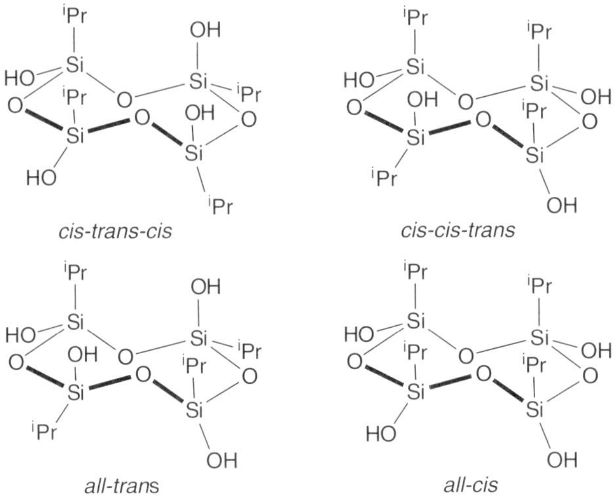

Figure 10.11 Four stereoisomers of the cyclotetrasiloxanetetraol [^{i}Pr(OH)SiO]$_4$.

XSiR$_2$SiOSiR$_2$X (X=Cl, OH) and a bifunctional reagent containing the p-block element. For example, boracyclotrisiloxanes **10.46** are prepared by the [3+3] cyclocondensation reaction of 1,3-dichlorotetramethyldisiloxane and phenylboronic acid. This method can also be used to make eight- and 10-membered cyclosiloxanes containing only one boron atom [**10.47** and **10.48** in eqn (10.29)].[129] The structural parameters for (PhBO)(Ph$_2$SiO)$_2$ compared with those of (Ph$_2$SiO)$_3$ indicate that the replacement of one silicon (covalent radius 1.17 Å) by a smaller boron atom (covalent radius 0.80 Å) increases the ring strain in the former significantly. As a consequence, the BSi$_2$O$_3$ rings undergo extensive ring-ring transformations upon heating or in the presence of KOSiMe$_3$ to give larger rings containing a single boron atom, in addition to cyclic and polymeric siloxanes and (PhBO)$_3$.[129]

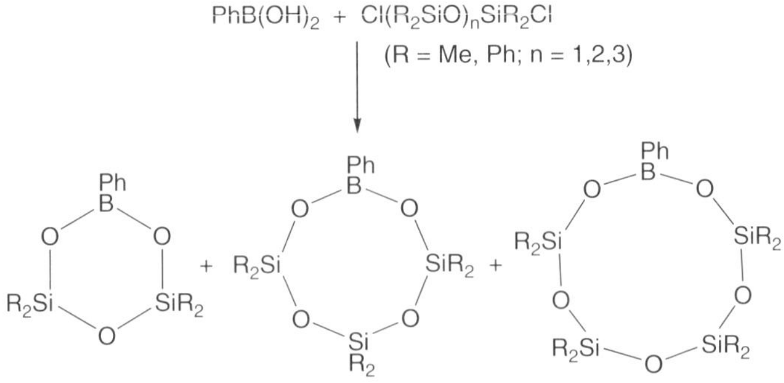

(10.29)

Eight- and 12-membered hybrid silicon–oxygen–nitrogen rings, $Si_4N_2O_2$ (**10.49**) and $Si_6N_2O_4$ (**10.50**), respectively, are obtained by the cyclocondensation reactions of $C_6F_5NLi_2$ and α,ω-dichlorosiloxanes (Scheme 10.20).[130] The 12-membered ring **10.50** adopts a chair-like conformation.[130b]

Larger ring systems of the type $(R_2SiO)_n$ ($n > 4$) may be considered as silicon analogues of crown ethers (for a detailed discussion, see Section 7.1.3).[131a] Although potassium complexes of the 14-membered ring $(Me_2SiO)_7$ have been known for some time, as a result of serendipitous reactions of highly reactive potassium reagents with silicon grease,[131b] the direct synthesis of stable alkali metal complexes of cyclosiloxanes was achieved only recently. The stability of these complexes is determined by the nature of the counterion, which must be a weakly coordinating anion such as $[Al\{OC(CF_3)_3\}_4]^-$ (Al_F).[132] The reason for this requirement becomes evident from a consideration of a Born–Haber cycle for the overall reaction (Scheme 10.21). For instance, when $X = I^-$, the lattice energy terms are $\Delta U_1 = 175$ and $\Delta U_2 = 89$ kcal mol^{-1}, whereas for $X = [Al_F]^-$, the corresponding contributions are $\Delta U_1 = 88$ and $\Delta U_2 = 76$ kcal mol^{-1}. The lattice energy of LiI is much larger than that of the host–guest complex and so the overall enthalpy change is positive ($\Delta H = +16$ kcal mol^{-1}). The lattice energy of Li[Al_F], however, is comparable to that of [LiD$_6$][Al_F] and the overall enthalpy change is negative ($\Delta H = -58$ kcal mol^{-1}). In summary, weakly coordinating anions minimise changes in lattice energy; they also reduce the possibility of unfavourable cation–anion interactions that would perturb the complexation of the metal ion.

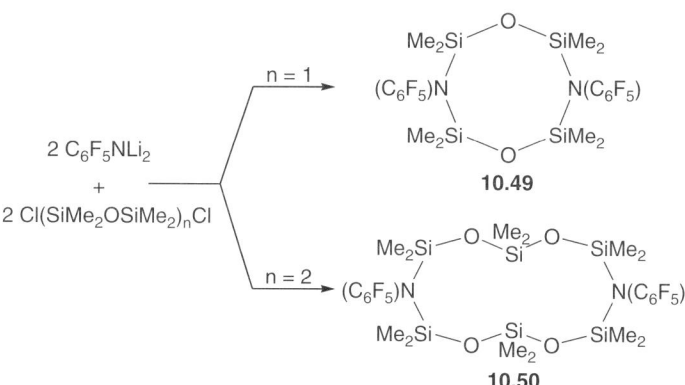

Scheme 10.20 Synthesis of $Si_4N_2O_2$ and $Si_6N_2O_4$ rings.

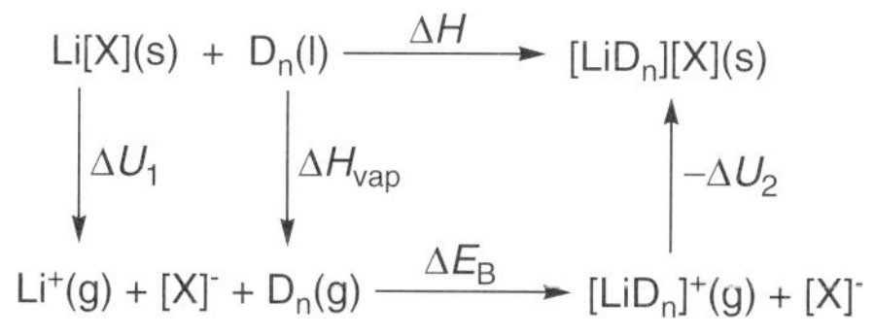

Scheme 10.21 Born–Haber cycle for formation of Li$^+$ complexes of cyclosiloxanes (D$_n$).

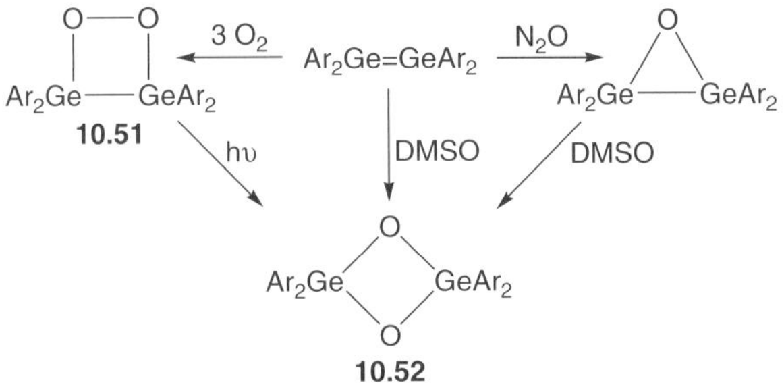

Scheme 10.22 Formation of 1,2-Ge$_2$O$_2$, 1,3-Ge$_2$O$_2$ and Ge$_2$O rings by oxidation of a digermene [Ar = 2,6-R$_2$C$_6$H$_3$ (R = Et, ^{i}Pr)].

Only four-, six- and eight-membered germanium–oxygen rings of the type (R$_2$GeO)$_n$ (n = 2–4) are known. Both isomers of the Ge$_2$O$_2$ ring are prepared in a manner similar to the silicon analogues, *i.e.* by oxidation of digermene with bulky aryl groups attached to the germanium atoms. Details of the products of oxidation are given in Scheme 10.22.[133] The trapezoidal Ge$_2$O$_2$ ring in the 1,2-isomer **10.51** is significantly more puckered than the almost square ring in the 1,3-isomer **10.52**. The Ge–Ge distance in the 1,2-isomer is close to the single-bond value, while the ring strain is reflected in somewhat elongated Ge–O bonds. In contrast to the silicon analogue, the 1,3-Ge$_2$O$_2$ ring can be described with four equivalent localised Ge–O bonds, since the transannular Ge–Ge distance is *ca* 0.16 Å longer than a typical single bond. A planar 1,3-Ge$_2$O$_2$ ring is also formed by the oxidation of the germylene :Ge[N(SiMe$_3$)$_2$] with molecular oxygen.[134] The hydrolysis of diorganogermanium dihalides gives six-membered rings (R$_2$GeO)$_3$ (R = ^{n}Bu, Ph, C$_6$F$_5$) or eight-membered rings (R$_2$GeO)$_4$ (R = Ph, ^{i}Pr).[135] The Ge$_3$O$_3$ ring in the all-phenylated derivatives is slightly puckered,[136a] whereas the Ge$_4$O$_4$ ring in (Ph$_2$GeO)$_4$ is non-planar with S_4 symmetry.[136b]

In contrast to cyclosiloxanes and cyclogermoxanes, stannoxanes of the type (R$_2$SnO)$_n$, which are prepared by hydrolysis of diorganotin dihalides, are often insoluble polymers. However, crystalline, lipophilic dimers or trimers may be formed and structurally characterised when bulky alkyl or aryl groups are attached to tin. The orange, sparingly soluble cyclodistannoxane (R$_2$SnO)$_2$ [**10.53**, R = CH(SiMe$_3$)$_2$] is obtained by oxidation of the corresponding tin(II) dimer Sn$_2$R$_4$ with trimethylamine *N*-oxide [eqn (10.30)].[137] The transannular Sn–Sn distance in the almost planar Sn$_2$O$_2$ rings is *ca* 0.13 Å longer than the single-bond value, whereas the Sn–O distances are comparable to those in Sn$_3$O$_3$ rings.

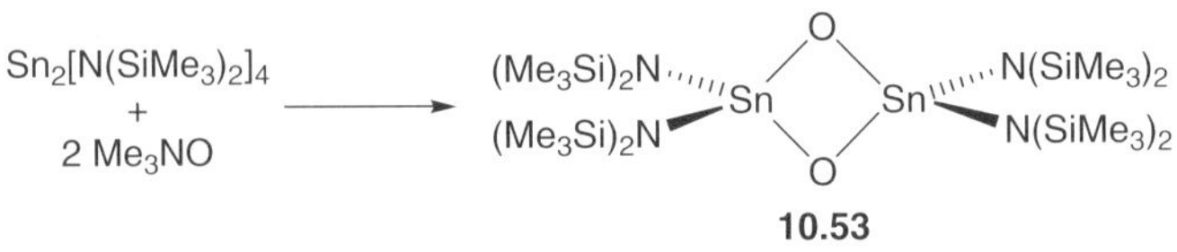

(10.30)

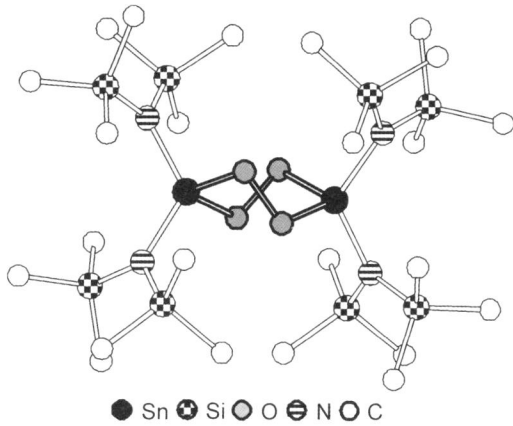

Figure 10.12 Molecular structure of the Sn_2O_4 ring in $[Sn\{N(SiMe_3)_2\}_2(\mu\text{-}O_2)]_2$.

The six-membered Sn_3O_3 ring in the three derivatives $(R_2SnO)_3$ (R = ^tBu,[138a] Am,[138b] 2,6-$Et_2C_6H_3$[12c]) is essentially planar. These tin–oxygen heterocycles are prepared by hydrolysis of the corresponding diorganotin dihalides, which, in the case of the aryl derivative, is formed by the reaction of (2,4-pentanedionato)tin dichloride with the aryl Grignard reagent.[12c]

The unique six-membered Sn_2O_4 ring is the central feature of the complex $[Sn\{N(SiMe_3)_2\}_2(\mu\text{-}O_2)]_2$, which is obtained as colourless crystals by the atmospheric oxidation of $Sn[N(SiMe_3)_2]_2$.[139] The two tin atoms are linked by two 1,2-peroxo bridges to give a puckered Sn_2O_4 ring in a twist-boat conformation (Figure 10.12). The kinetic stability of this unusual ring system is attributed to the protection of the two peroxy linkages by the four bulky amido groups.

10.6 SILICON–OXYGEN POLYMERS: POLYSILOXANES (SILICONES)

Polysiloxanes were first developed in the 1930s and 1940s and currently represent the most commercially important inorganic polymer system. These materials and their products comprise a billion dollar global industry.[42,140–143] Indeed, these materials are now often regarded as commodity polymers rather than speciality products. The main methods of synthesis involve condensation and anion- or cation-induced ring-opening routes, and these are illustrated for the synthesis of the most common siloxane polymer, poly(dimethylsiloxane), in eqns (10.31) and (10.32). The starting materials are organodichlorosilanes of which the most common, Me_2SiCl_2, is readily available *via* the Rochow–Müller process. The cyclics used in the ROP route are available by controlled hydrolysis of the corresponding organodichlorosilane.

$$\text{Me}_2\text{SiCl}_2 \xrightarrow{\text{H}_2\text{O}} \text{HO}\left[\begin{array}{c}\text{Me}\\|\\\text{Si}\\|\\\text{Me}\end{array}-\text{O}\right]_n\text{H} \quad (10.31)$$

$$\text{cyclic siloxanes} \xrightarrow{\text{heat and/or anionic or cationic initiators}} \left[\begin{array}{c}\text{Me}\\|\\\text{Si}\\|\\\text{Me}\end{array}-\text{O}\right]_n \quad (10.32)$$

The exceptional properties of polysiloxanes are a direct result of their inorganic backbone of silicon and oxygen atoms and have resulted in their widespread use as high-performance elastomers and fluids, surface modifiers, adhesives and biomedical materials.[42,140–143] The siloxane backbone, with the long Si–O bond (1.64 Å compared with 1.54 Å for a C–C bond), the absence of substituents on alternate skeletal atoms (oxygen) and the wide bond angle at oxygen (Si–O–Si 143° compared with C–C–C 109°), possesses unique dynamic flexibility. This leads to materials which retain elasticity and do not become brittle even at very low temperatures. For example, the most common polymer, poly(dimethylsiloxane), has a glass transition temperature (T_g, see Section 8.2.3) of –123 °C and for poly(methylhydrosiloxane) the T_g is even lower (–137 °C). In addition, Si–O bonds are stronger than C–C bonds (bond energies: Si–O ca 450 kJ mol^{-1}, C–C ca 348 kJ mol^{-1}) and are more stable to oxidation and UV radiation. This results in higher thermooxidative stability and leads, for example, to the use of polysiloxanes in oil baths found in most chemical laboratories. Poly(dimethylsiloxane) will crystallise ($T_m = -40$ °C) and for many elastomer applications the introduction of a small percentage of phenyl substituents by random copolymerisation is necessary to ensure amorphous character. Cross-linking for elastomer applications can be achieved by a variety of techniques, including heating with peroxides or room temperature vulcanisation where cross-linking can be induced by the transition metal-catalysed hydrosilylation reactions between vinyl substituents and Si–H groups introduced along the polymer chain. In general, the addition of reinforcing agents or fillers such as high surface area silica is also necessary to improve mechanical properties.[42]

Polysiloxanes also possess a variety of other useful properties such as hydrophobicity and exceptionally high permeability to gases. Indeed, snails can live submerged beneath the surface of low molecular weight poly(dimethylsiloxane) fluids for up to 72 h by breathing oxygen that diffuses through the material.[143] Other applications in the biomedical field, which also take advantage of the high permeability of polysiloxanes, include uses as soft contact lenses and artificial skin. However, the leakage of silicone-based materials through polysiloxane membranes in breast implants has raised considerable public concern. Nevertheless, claims of health problems arising from the consequential presence of silicones in the body appear unproven.

Although polydimethylsiloxane is the best-studied example, a wide variety of other polysiloxanes with aliphatic or aromatic side groups is known and these are accessible either by condensation or by cationic- or anionically-induced ROP routes. These materials also possess very interesting properties and many have been the subject of recent studies. For example, the morphology of simple high molecular weight ($M_w = ca.\ 10^5$–10^6) n-alkyl-substituted polysiloxanes, which are available by ROP, has been studied in detail.[144] Polysiloxanes with fluoroalkyl substituents (*e.g.* trifluoropropyl) or 'fluorosilicones' possess novel surface properties and are technologically important with a range of applications. The living anionic ROP of strained cyclic siloxanes such as $(Me_2SiO)_3$ (see Section 8.2.2) is also important as this permits access to block copolymers with organic monomers.[145,146]

A very versatile and important methodology for controlling the properties of polysiloxanes involves hydrosilylation [eqn (10.33)]. Reaction of poly(methylhydrosiloxane) or poly(methylhydrosiloxane)–poly(dimethylsiloxane) copolymers with vinyl-capped species allows the introduction of side groups which give rise to a variety of interesting properties.[147]

$$\left[\begin{array}{c}Me\\|\\-Si-O-\\|\\H\end{array}\right]_n + \diagup\!\!\diagup\!\!\diagdown_R \xrightarrow{\text{Pt catalyst}} \left[\begin{array}{c}Me\\|\\-Si-O-\\|\\\\\end{array}\right]_n$$
$$R$$

(10.33)

Polysiloxanes are attracting attention for a range of 'high-tech' materials science applications. For example, liquid crystalline polysiloxanes with short switching times have been reported.[148a] Polysiloxanes functionalised with polar cyanopropyl and cross-linkable methacryloxypropyl groups have been used as the matrices of Na^+ sensing membranes for chemically modified field-effect transistors (CHEMFETs).[148b] The advantages of polysiloxanes for this type of application revolve around their high diffusion and permeability coefficients. Polysiloxanes have also been used to fabricate materials than can bend or move on the surface of fluids in response to exposure to light.[149]

A silicone polymer with intriguing properties is the material known as Silly (or Bouncing) Putty. This polymer was discovered accidentally in the laboratories of General Electric, as a result of attempts to make synthetic rubber during World War II. It was obtained by mixing silicone oil $Me_3SiO(Me_2SiO)_nSiMe_3$ with boric acid. In addition to bouncing like a rubber ball, Silly Putty exhibits flow properties. These unusual characteristics are attributed to a small amount of cross-linking between neighbouring polymer chains involving weak coordinate bonds between Lewis basic oxygen centres and Lewis acidic boron centres. The laboratory synthesis now involves heating a mixture of silicone oil with *ca* 7% of its weight of boric acid at 200 °C, so the incorporation of boron in the polymer chain is limited. Despite its unique properties, no practical use has been found for this inorganic backbone polymer. It was originally sold as a novelty item in a toy store but dropped after a year. Subsequently, the name Silly Putty was adopted and the curiosity was packaged in plastic eggs. Marketing of this product was given a tremendous boost in 1950 by an article in the *New Yorker* magazine. In 2000, the total annual sales of this popular children's toy were estimated to be about 4500 tonnes (equivalent to 300×10^6 eggs).[150]

10.7 SILICON–, GERMANIUM–, TIN– AND LEAD–CHALCOGEN RINGS

10.7.1 $(R_2ME)_n$ (M = Si, Ge, Sn, Pb; E = S, Se, Te) and Metal-rich Rings

The majority of the investigations of group 14–group 16 heterocycles have focused on silicon or tin derivatives. Consequently, the emphasis in this section is on these two group 14 elements.

In contrast to the extensive homologous series known for cyclosiloxanes, monocyclic ring systems of the type $(R_2SiS)_n$ are limited to four- and six-membered rings ($n = 2, 3$). The most common synthetic method involves the reaction of a diorganodichlorosilane with H_2S in the presence of a base such as triethylamine or pyridine [eqn (10.34)].[151]

$$R_2SiCl_2 + H_2S \xrightarrow{NEt_3 \text{ or pyridine}} (R_2SiS)_n \qquad (10.34)$$

Other approaches include the reactions of (a) R_2SiCl_2 with alkali metal sulfides, (b) R_2SiH_2 with elemental sulfur and (c) disilene $R_2Si=SiR_2$ with elemental sulfur. Six-membered rings are formed preferentially when R is a non-bulky alkyl or aryl group. In the case of R=Me, Ph the four-membered rings $(R_2SiS)_2$ are produced by the thermally induced ring contraction of the corresponding six-membered ring $(R_2SiS)_3$.[152,153] For example, the all-phenylated derivative $(Ph_2SiS)_3$ is obtained in 55% yield as a white moisture-sensitive solid by treatment of Ph_2SiCl_2 in the presence of two equivalents of pyridine with H_2S in benzene.[153] Alternatively, the use of freshly prepared anhydrous Na_2S (instead of toxic H_2S) produces $(Ph_2SiS)_3$ in 63% yield and the selenium analogue $(Ph_2SiSe)_3$ is obtained in a similar manner.[154]

Similar methods are used for making monocyclic tin–chalcogen ring systems of the type $(R_2SnE)_n$ (E=S, Se, Te; $n=2, 3$). The all-methylated derivative $(Me_2SnS)_3$ is prepared by the reaction of dimethyltin dichloride with anhydrous sodium sulfide in benzene,[155] whereas the selenium analogue $(Me_2SnSe)_3$ is formed by passing hydrogen selenide gas into an aqueous solution of dimethyltin dichloride.[156] The all-phenylated derivatives (Ph$_2$SnE)$_3$ (E = S, Se) and the all-benzylated compounds (Bz$_2$SnE)$_3$ (E = S, Se) are prepared by the former method.[154,157] The reaction of dialkyltin oxides (R$_2$SnO)$_n$ with carbon disulfide has also been used to make (R$_2$SnS)$_3$ (R = Et, ^{i}Pr) or (^{t}Bu$_2$SnS)$_2$.[158]

In view of their high toxicity, however, it is desirable to avoid the use of the gaseous reagents H$_2$Se and H$_2$Te for the synthesis of Se- and Te-containing ring systems. Thus, the six-membered tin–tellurium ring (Me$_2$SnTe)$_3$ is prepared by the reaction of (a) Me$_2$SnH$_2$ with elemental tellurium or (b) dimethyltin dichloride with NaHTe in aqueous solution [eqn (10.35)].[159] The latter synthesis is preferred, because the former method also produces the five-membered ring (Me$_2$Sn)$_3$Te$_2$. The phenyl derivative (Ph$_2$SnTe)$_3$ and the germanium–tellurium ring (Me$_2$GeTe)$_3$ are also prepared by method (b).[160] In a minor modification of this method, the benzyl-substituted six-membered ring (Bz$_2$SnTe)$_3$ is made by treatment of Bz$_2$SnCl$_2$ with aqueous ammonium telluride (NH$_4$)$_2$Te.[161] As implied by the use of an aqueous medium for synthesis, the tin–chalcogen ring systems exhibit higher stability towards hydrolysis than their silicon analogues. However, tin–tellurium compounds are photolytically sensitive.

$$3R_2SnCl_2 + 3NaHTe \rightarrow (R_2SnTe)_3 + 3NaCl + 3HCl \quad (10.35)$$

$$(R = Me, Ph)$$

The installation of bulky *tert*-butyl substituents on tin results in the formation of four-membered Sn$_2$E$_2$ rings (E = S, Se, Te), which are obtained by simple metathesis [eqn (10.36)].[162] The dimerisation of the heavy ketone analogues of the type R$_2$M=E to give the corresponding four-membered (R$_2$ME)$_2$ rings (R = bulky aryl group) is discussed in Section 10.7.2. Four-membered M$_2$E$_2$ rings are also a central feature of tetraanions of the type [M$_2$S$_6$]$^{4-}$ (M = Ge, Sn), which are prepared by reaction of the corresponding metal sulfide MS$_2$ with an alkali metal sulfide in aqueous solution.[163]

$$2^tBu_2SnCl_2 + 2Na_2E \rightarrow (^tBu_2SnE)_2 + 4NaCl \quad (10.36)$$

$$(E =, S, Se, Te)$$

Four-membered M$_2$E$_2$ rings are also obtained from the reactions of divalent compounds of the type R$_2$M: (M = Si, Ge, Sn) with elemental chalcogens [or a labile source of the chalcogen, *e.g.* R$_3$P=E (E = Se, Te)]. The formation of smaller rings in these reactions is favoured because the isolation of these reactive reagents requires bulky substituents on the group 14 element centre. An interesting example of this type of transformation involves the oxidation of decamethylsilicocene Cp*$_2$Si with chalcogens, which gives either four-membered Si$_2$E$_2$ rings (**10.54**, E = S, Se) or a dark-red five-membered Si$_2$Te$_3$ ring (**10.55**, Scheme 10.23).[164]

Cyclic compounds containing Si$_2$E$_2$ rings (**10.56**, E = S, Se) are also formed from the reaction of a stable silylene with sulfur or selenium [eqn (10.37)].[165] The reaction of the divalent germanium compound :Ge[N(SiMe$_3$)$_2$]$_2$ with

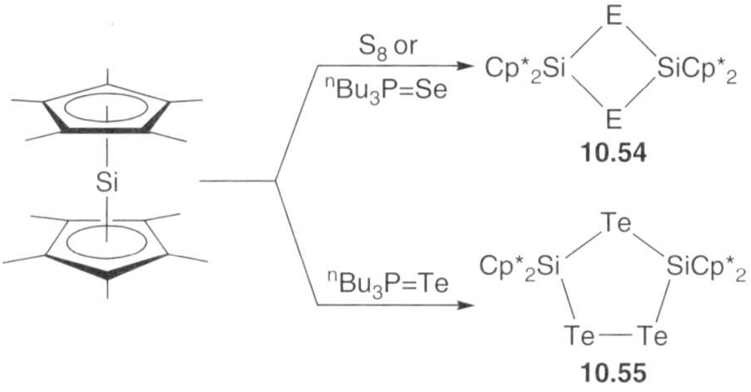

Scheme 10.23 Formation of Si_2E_2 (E = S, Se) or Si_2Te_3 rings from $Cp*_2Si$ and chalcogens.

tellurium in THF produces a four-membered Ge_2Te_2 ring.[166]

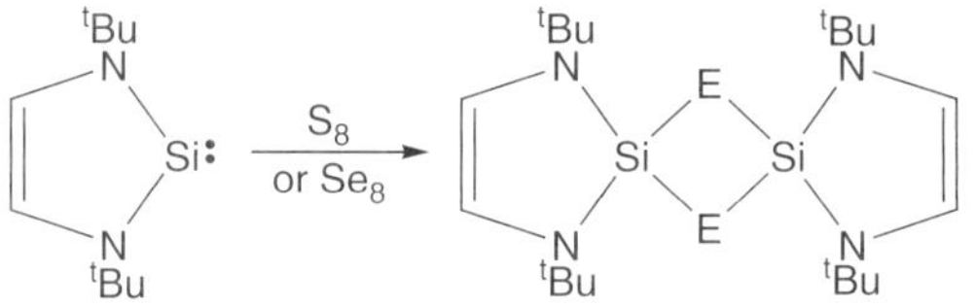

10.56 (E = S, Se) (10.37)

The oxidation of dimetallenes $R_2M=MR_2$ with chalcogens provides a general route to three-membered M_2E rings [eqn (10.38)]. Thus, tetramesityldisilene reacts with sulfur, selenium or tellurium in benzene at room temperature to give Si_2E rings (E = S, Se, Te), which exhibit remarkably high thermal stability.[167] For example, the colourless disilaselenirane $(Mes_2Si)_2Se$ (**10.57**, E = Se) does not decompose at the melting point (272–274 °C). The corresponding germanium-containing three-membered Ge_2E rings (E = S, Se, Te) are obtained in a similar manner.[168] The tellurium-containing compound is thermochromic, being colourless at low temperature, pale yellow at 25 °C and orange at *ca* 140 °C.

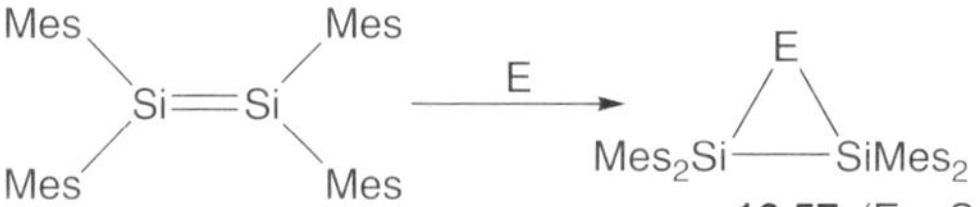

10.57 (E = S, Se, Te) (10.38)

The oxidation of the distannene $Ar_2Sn=SnAr_2$ (Ar = 2,4,6-^{i}Pr$_3$C$_6$H$_2$) with chalcogens produces either the three-membered Sn_2E ring or the four-membered 1,3-Sn_2E_2 rings (E = S, Se, Te), depending on the amount of chalcogen that is used.[169] The reaction with sulfur is particularly interesting, since it produces the yellow 1,2-isomer *via* the formal cycloaddition of an S_2 unit and $Ar_2Sn=SnAr_2$ (Figure 10.13).[169b] The structure of 1,2-$(Ar_2Sn)_2S_2$ is comprised

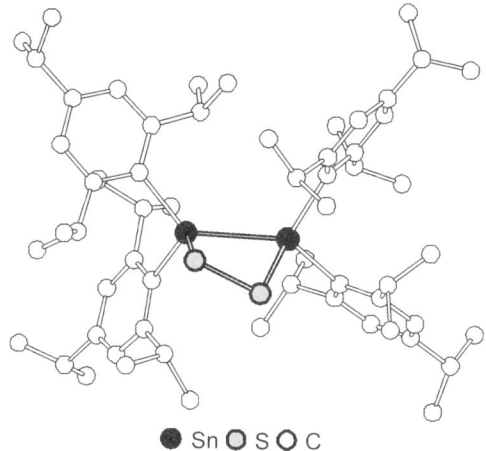

Figure 10.13 Molecular structure of 1,2-$(Ar_2Sn)_2S_2$ $(Ar = 2,4,6\text{-}^iPr_3C_6H_2)$.

of a folded four-membered ring with an Sn–Sn bond length of 2.89 Å and a slightly elongated S–S bond length of 2.09 Å.

The metathesis reaction of $ClMe_2SnSnMe_2Cl$ with NaTeH produces the photolytically sensitive, six-membered ring $[(Me_2Sn)_4Te_2]$,[170a] which is a dimer of the three-membered Sn_2Te ring formed when very bulky aryl groups are attached to tin. The analogous Sn_4Se_2 ring is obtained from $ClMe_2SnSnMe_2Cl$ and Na_2Se.[170b]

The cyclic structures of $(Me_2SiS)_n$ ($n = 2, 3$), which are volatile liquids, have been ascertained by electron diffraction; the Si–S bond lengths are typical of single bonds.[152] A large number of X-ray structural investigations of ring systems of the type $(R_2ME)_n$ (M = Si, Ge, Sn; E = S, Se, Te; $n = 2, 3$) have been reported. In general, the four-membered rings are planar, whereas the six-membered systems form puckered rings in a twisted boat conformation with normal M–E single bond lengths. Representative examples are shown in Figure 10.14. The three-membered M_2E rings are close to planar and exhibit approximately equal M–E bond lengths. However, when E = S or Se, the M–M bond lengths are significantly shorter than typical single bonds, leading to the suggestion that the three-membered rings have retained some of the π-bond character of the corresponding dimetallene.[167,168]

^{119}Sn NMR spectroscopy provides a convenient probe for monitoring the behaviour of tin–chalcogen rings in solution, especially for those of selenium and tellurium, which give rise to characteristic satellites due to the spin-active isotopes (^{77}Se, $I = \frac{1}{2}$, 7.6%; ^{125}Te, $I = \frac{1}{2}$, 7.0%). By using this technique it can be shown that mixed chalcogen rings are formed when a mixture of any two of the series of six-membered rings $(Me_2SnE)_3$ (E = S, Se, Te) is allowed to equilibrate in benzene solution.[158] A similar exchange process has been demonstrated for the pair of phenylated rings $(Ph_2SnSe)_3$ and $(Ph_2SnTe)_3$.[160] The redistribution of R_2Sn groups between six-membered rings may also occur. For example, a

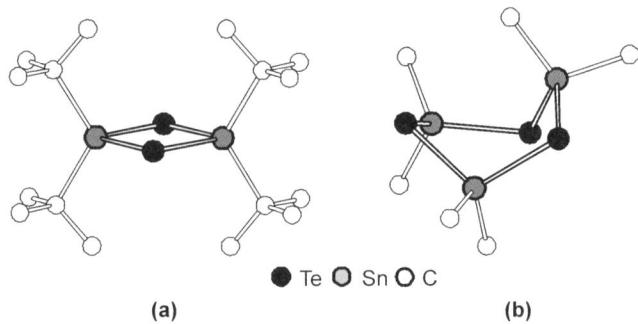

Figure 10.14 Molecular structures of (a) (^{t}Bu$_2$SnTe)$_2$[162] and (b) (Me$_2$SnTe)$_3$.[159]

mixture of (Me$_2$SnS)$_3$ and (Ph$_2$SnS)$_3$ produces an equilibrium mixture with the mixed systems [(Me$_2$SnS)$_2$(Ph$_2$SnS)] and [(Ph$_2$SnS)$_2$(Me$_2$SnS)]; this process is catalysed by a platinum(II) complex.[160]

Tin chalcogenides have a variety of applications as narrow-bandgap semiconductors. Tin sulfide (SnS) exhibits high conversion efficiency in photovoltaic devices and polycrystalline films of SnSe are used in memory-switching devices. Tin telluride (SnTe) is a component of infrared detectors and thermoelectric devices. For these reasons, cyclic tin–chalcogen compounds (R$_2$SnE)$_3$ (R = Ph, Bz; E = S, Se, Te) have been investigated as single-source precursors to pure thin films of tin chalcogenides.[124,157,161] Pyrolysis of (Ph$_2$SnE)$_3$ at 450 °C produces the corresponding tin chalcogenides SnE (E=S, Se) as black powders.[154] The benzyl derivatives (Bz$_2$SnE)$_3$ (E=S, Se, Te) are promising alternatives for this purpose in view of the relatively low dissociation energy of the Sn–benzyl bond.[157,161] Pyrolysis of a mixture of (Bz$_2$SnS)$_3$ and (Bz$_2$SnSe)$_3$ produces solid solutions of formula Sn(S$_x$Se$_{1-x}$), the composition of which can be controlled by varying the ratio of the two precursors.[157] The tellurium derivative (Bz$_2$SnTe)$_3$ is an efficient precursor to pure cubic SnTe upon heating at 200–275 °C [eqn (10.39)].[161]

$$[(C_6H_5CH_2)_2SnTe]_3 \rightarrow 3SnTe + 3C_6H_5CH_2CH_2C_6H_5 \qquad (10.39)$$

Acyclic silicon–chalcogen reagents (Me$_3$Si)$_2$E (E = S, Se, Te) are widely used as a source of chalcogens for the construction of metal chalcogenide clusters. The use of cyclic silicon–, germanium– or tin–chalcogen compounds for this purpose leads to the generation of novel chelating ligands. For example, the reaction of (Me$_2$SiS)$_3$ with copper acetate in the presence of PEt$_3$ produces a binuclear complex **10.58** in which the dithiolato ligand [S(SiMe$_2$S)$_2$]$^{2-}$ bridges two Cu(I) centres.[171] The oxidative addition of the tin–chalcogen rings (R$_2$SnE)$_3$ (R = Me, Ph; E = S, Se, Te) to Pt(II) generates octahedral Pt(IV) complexes **10.59** containing the chelating [ESnR$_2$ESnR$_2$]$^{2-}$ ligand in a five-membered metallacyclic ring.[160] The cyclotrithiasilane (Me$_2$SiS)$_3$ may also serve as a source of the chelating [S$_2$SiMe$_2$]$^{2-}$ ligand in cyclic Pd(II) or Ti(IV) complexes **10.60** via reactions with palladium acetate or (MeCp)$_2$TiCl$_2$ and Li$_2$S, respectively.[172,173]

10.7.2 Chalcogen-rich Rings

Monomers of the type $R_2M=E$ (M = Si, Ge, Sn, Pb, E = S, Se, Ge, Pb) involving the heavier group 14 and group 16 elements are of fundamental interest as congeners of ketones $R_2C=O$. As such, they are referred to as 'heavy ketones'.[174] In order to provide kinetic stabilisation of these monomeric species and, hence, prevent oligomerisation to the corresponding cyclic system, it is necessary to attach extremely sterically demanding aryl substituents to the group 14 metal centre. Several approaches for the synthesis of heavy ketones have been investigated, the most effective and versatile of which is the dechalcogenation of five-membered Ar_2ME_4 rings with trivalent phosphorus reagents.[174] This procedure is effective for the generation of monomeric $Ar_2M=E$ provided that sufficiently bulky aryl groups, *e.g.* 2,4,6-[(Me$_3$Si)$_2$CH]$_3$C$_6$H$_2$ (Tbt) and 2,4,6-^{i}Pr$_3$C$_6$H$_2$ (Tip), are attached to the group 14 metal centre [eqn (10.40)].[175,176] If the steric protection is insufficient, oligomerisation to the corresponding four-membered 1,3-M_2E_2 rings occurs. For example, it is necessary to replace the Tip substituent with the even bulkier terphenyl ligand, Ditp [eqn (10.40)] in order to stabilise monomeric $R_2Sn=E$ (E = S, Se).[177] In some cases, chalcogen-rich heterocycles, *e.g.* SnSe$_2$[178] or Ge$_2$S$_4$,[176] are isolated as minor products. This synthetic procedure cannot be used to generate tellurium-containing heavy ketones because the Ar_2MTe_4 precursors are not known. Consequently, $R_2M=Te$ monomers (M = Ge, Sn) are generated by direct oxidation of highly sterically hindered metallylenes :MR$_2$ with tellurium or ^{n}Bu$_3$PTe.[179]

(10.40)

The desulfurisation of a tetrathioplumbolane Ar(Ar')PbS$_4$ with different substituents on the lead centre unexpectedly produces the four-membered

[Ar'Pb(μ-S)]$_2$ ring **10.61** with the same two groups attached to each lead atom, together with the plumbylene :PbAr(SAr). The observed symmetrisation is rationalised by invoking the rearrangement of the initially formed plumbanethione [eqn (10.41)].[180]

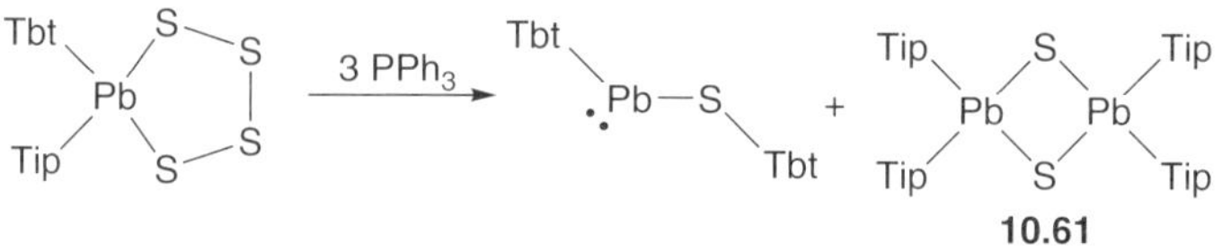

(10.41)

The most versatile synthesis of tetrachalcogenametallolanes R$_2$ME$_4$ (M = Si, Ge, Sn) involves the thermal chalcogenation of the corresponding dihydrides with elemental chalcogen (Scheme 10.24a).[181] Alternatively, the dihalides R$_2$SnCl$_2$ are treated with two equivalents of ^{t}BuLi followed by the chalcogen.[182] Lead-containing tetrasulfides are prepared by oxidation of the corresponding plumbylene with sulfur (Scheme 10.24b).[183] The MS$_4$ ring in these group 14 polysulfides is highly puckered with typical S–S single bond lengths of ca 2.05 Å.[181,182]

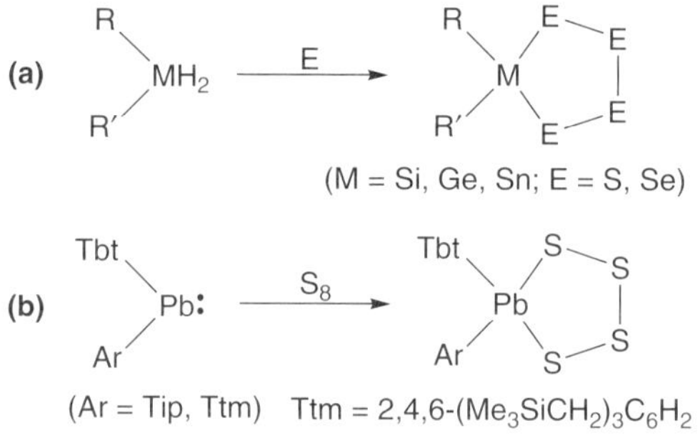

Scheme 10.24 Synthetic approaches to R$_2$ME$_4$ rings (M = Si, Ge, Sn, Pb; E = S, Se).

REFERENCES

1. (a) E. West and E. Carberry, *Science*, 1975, **189**, 179; (b) C. A. Burkhard, *J. Am. Chem. Soc.*, 1949, **71**, 963.
2. (a) E. Hengge and G. Bauer, *Angew. Chem. Int. Ed. Engl.*, 1973, **12**, 316; (b) E. Hengge and D. Kovar, *Angew. Chem. Int. Ed. Engl.*, 1977, **16**, 403.
3. L. F. Brough and R. West, *J. Am. Chem. Soc.*, 1981, **103**, 3049.
4. M. Ishikawa and M. Kumada, *J. Organomet. Chem.*, 1972, **42**, 325.
5. C. J. Hurt, J. C. Calabrese and R. West, *J. Organomet. Chem.*, 1975, **91**, 273.
6. H. L. Carrell and J. Donohue, *Acta Crystallogr., Sect. B*, 1972, **28**, 1566.
7. R. West, *Pure Appl. Chem.*, 1982, **54**, 1041.

8. (a) A. W. P. Jarvie, H. J. S. Winkler, D. J. Peterson and H. Gilman, *J. Am. Chem. Soc.*, 1961, **83**, 1924; (b) H. Gilman, W. H. Atwell and F. K. Cartledge, *Adv. Organomet. Chem.*, 1964, **4**, 1.
9. (a) D. A. Armitage, *Inorganic Rings and Cages*, Edward Arnold, London, 1972, p. 157; (b) E. Hengge and D. Wolfer, *Angew. Chem. Int. Ed. Engl.*, 1973, **12**, 315.
10. T. Shimoda, Y. Matsuki, M. Furusawa, T. Aoki, I. Yudasaka, H. Tanaka, H. Iwasawa, D. Wang, M. Miyasaka and Y. Takeuchi, *Nature*, 2006, **440**, 783.
11. S.-B. Choi, B.-K. Kim, P. Boudjouk and D. G. Grier, *J. Am. Chem. Soc.*, 2001, **123**, 8117.
12. (a) S. Masumune, Y. Hanzawa, S. Murukami, T. Bally and J. F. Blount, *J. Am. Chem. Soc.*, 1982, **104**, 1150; (b) S. Masumune, Y. Hanzawa and D. J. Williams, *J. Am. Chem. Soc.*, 1982, **104**, 6136; (c) S. Masumune, L. Sita and D. J. Williams, *J. Am. Chem. Soc.*, 1983, **105**, 630.
13. A. Schäfer, M. Weidenbruch, K. Peters and H. G. von Schnering, *Angew. Chem. Int. Ed. Engl.*, 1984, **4**, 302.
14. R. West, M. J. Fink and J. Michl, *Science*, 1980, **214**, 1343.
15. (a) K. M. Baines and J. A. Cooke, *Organometallics*, 1991, **10**, 3419; (b) A. Heine and D. Stalke, *Angew. Chem. Int. Ed. Engl.*, 1994, **33**, 113.
16. (a) K. M. Baines and W. G. Stibbs, *Coord. Chem. Rev.*, 1995, **145**, 157; (b) I. Haiduc and M. Dräger, in *The Chemistry of Inorganic Homo- and Heterocycles*, ed. I. Haiduc and D. B. Sowerby, Academic Press, London, 1987, Vol. **1**, pp. 361–365.
17. E. Carberry, B. D. Dombek and S. C. Cohen, *J. Organomet. Chem.*, 1972, **36**, 61.
18. W. Ando and T. Tsumuraya, *J. Chem. Soc., Chem. Commun.*, 1987, 1514.
19. A. Sekiguchi, T. Yartabe, H. Naito, C. Kabuto and H. Sakurai, *Chem. Lett.*, 1992, 1697.
20. M. Weidenbruch, F.-T. Grimm, M. Herrndorf, A. Schäfer, K. Peters and H. G. von Schnering, *J. Organomet. Chem.*, 1988, **341**, 335.
21. L. Ross, L. Ross and M. Dräger, *J. Organomet. Chem.*, 1980, **199**, 195.
22. L. Sita, *Adv. Organomet. Chem.*, 1995, **38**, 189.
23. B. Jousseaume, N. Noiret, M. Pereyre, A. Saux and J.-M. Francès, *Organometallics*, 1994, **13**, 1034.
24. A. G. Davies, *Organotin Chemistry*, 2nd edn., Wiley, Chichester, 2004, Chapter 18.
25. (a) W. P. Neumann and K. König, *Angew. Chem. Int. Ed. Engl.*, 1962, **1**, 212; (b) W. P. Neumann and K. König, *Angew. Chem. Int. Ed. Engl.*, 1964, **3**, 751.
26. W. P. Neumann, *The Organic Chemistry of Tin*, Wiley, Chichester, 1970, pp. 140–146.
27. B. Watta, W. P. Neumann and J. Sauer, *Organometallics*, 1985, **4**, 1954.
28. (a) W. V. Farrar and H. A. Skinner, *J. Organomet. Chem.*, 1964, **1**, 434; (b) H. Puff, C. Bach, W. Schuh and R. Zimmer, *J. Organomet. Chem.*, 1986, **312**, 313.

29. S. Adams and M. Dräger, *Angew. Chem. Int. Ed. Engl.*, 1987, **26**, 1255.
30. W. P. Neumann, *Angew. Chem. Int. Ed. Engl.*, 1963, **2**, 164.
31. L. R. Sita, *J. Am. Chem. Soc.*, 1989, **111**, 3769.
32. (a) M. Stürmann, M. Weidenbruch, K. W. Klinkhammer, F. Lissner and H. Marsmann, *Organometallics*, 1998, **17**, 4425; (b) M. Stürmann, W. Saak, H. Marsmann and M. Weidenbruch, *Angew. Chem. Int. Ed.*, 1999, **38**, 187.
33. F. Stabenow, W. Saak, H. Marsmann and M. Weidenbruch, *J. Am. Chem. Soc.*, 2003, **125**, 10172.
34. R. Fischer and F. Uhlig, *Coord. Chem. Rev.*, 2005, **249**, 2075.
35. C. E. Housecroft and A. G. Sharpe, *Inorganic Chemistry*, Pearson Education, Harlow, 2005, pp. 358–360.
36. (a) J. Goodey, J. Mao and A. Guloy, *J. Am. Chem. Soc.*, 2000, **122**, 10478; (b) T. F. Fassler and C. Kronseder, *Angew. Chem. Int. Ed. Engl.*, 1997, **36**, 2683.
37. (a) R. Nesper, J. Curda and H.-G. von Schnering, *J. Solid State Chem.*, 1986, **62**, 199; (b) I. Todorov and S. C. Sevov, *Inorg. Chem.*, 2004, **43**, 6490; (c) I. Todorov and S. C. Sevov, *Inorg. Chem.*, 2005, **44**, 5361.
38. I. Todorov and S. C. Sevov, *Inorg. Chem.*, 2006, **45**, 4478.
39. H.-G. von Schnering, U. Bolle, J. Curda, K. Peters, W. Carillo-Cabrera, M. Somer, M. Schultheiss and U. Wedig, *Angew. Chem. Int. Ed. Engl.*, 1996, **35**, 984.
40. R. West, *J. Organomet. Chem.*, 1986, **300**, 327.
41. (a) R. D. Miller and J. Michl, *Chem. Rev.*, 1989, **89**, 1359; (b) S. Hayase, *Prog. Polym. Sci.*, 2003, **28**, 359.
42. J. E. Mark, H. R. Allcock and R. West, *Inorganic Polymers*, 2nd edn., Oxford University Press, Oxford, 2004.
43. (a) J. Nelson and W. J. Pietro, *J. Phys. Chem.*, 1988, **92**, 1365; (b) A. Savin, O. Jepsen, J. Flad, O. K. Andersen, H. Preuss and H.-G. von Schnering, *Angew. Chem. Int. Ed. Engl.*, 1992, **31**, 187.
44. C. T. Aitken, J. F. Harrod and E. Samuel, *J. Am. Chem. Soc.*, 1986, **108**, 4059.
45. T. D. Tilley, *Acc. Chem. Res.*, 1993, **26**, 22.
46. (a) M. Cypryk, Y. Gupta and K. Matyjaszewski, *J. Am. Chem. Soc.*, 1991, **113**, 1046; (b) E. Fossum and K. Matyjaszewski, *Macromolecules*, 1995, **28**, 1618.
47. K. Sakamoto, K. Obata, H. Hirata, M. Nakajima and H. Sakurai, *J. Am. Chem. Soc.*, 1989, **111**, 7641.
48. W. Büchner, R. Schliebs, G. Winter and K. H. Büchel, *Industrial Inorganic Chemistry*, VCH, Weinheim, pp. 365–367.
49. H. Suzuki, H. Meyer, J. Simmerer, J. Yang and D. Haarer, *Adv. Mater.*, 1993, **5**, 743.
50. Y. Wang, R. West and C. H. Yuan, *J. Am. Chem. Soc.*, 1993, **115**, 3844.
51. P. Trefonas III and R. West, *J. Polym. Sci., Polym. Chem. Ed.*, 1985, **23**, 2099.

52. R. D. Miller and R. Sooriyakumaran, *J. Polym. Sci., Polym. Chem. Ed.*, 1987, **25**, 111.
53. S. Adams and M. Dräger, *Angew. Chem. Int. Ed. Engl.*, 1987, **26**, 1255.
54. T. Imori, V. Lu, H. Cai and T. D. Tilley, *J. Am. Chem. Soc.*, 1995, **117**, 9931.
55. L. R. Sita, K. W. Terry and K. Shibata, *J. Am. Chem. Soc.*, 1995, **117**, 8049.
56. A. Sekiguchi and V. Y. Lee, *Chem. Rev.*, 2003, **101**, 1429.
57. (a) A. Sekiguchi, H. Yamazaki, C. Kabuto and H. Sakurai, *J. Am. Chem. Soc.*, 1995, **117**, 8025; (b) M. Ichinohe, H. Sekiyama, N. Fukaya and A. Sekiguchi, *J. Am. Chem. Soc.*, 2000, **122**, 6781.
58. T. Iwamoto, C. Kabuto and M. Kira, *J. Am. Chem. Soc.*, 1999, **121**, 886.
59. (a) M. Ichinohe, T. Matsuno and A. Sekiguchi, *Angew. Chem. Int. Ed.*, 1999, **38**, 2194; (b) M. Ichinohe, M. Igarashi, K. Sanuki and A. Sekiguchi, *J. Am. Chem. Soc.*, 2005, **127**, 9978.
60. N. Wiberg, H.-W. Lerner, S.-K. Vashist, S. Wagner, K. Karaghiosoff, H. Nöth and W. Ponikwar, *Eur. J. Inorg. Chem.*, 1999, 1211.
61. V. Ya. Lee, M. Ichinohe and A. Sekiguchi, *J. Am. Chem. Soc.*, 2000, **122**, 9034.
62. V. Ya. Lee, H. Yasuda, M. Ichinohe and A. Sekiguchi, *Angew. Chem. Int. Ed.*, 2005, **44**, 2.
63. (a) A. Sekiguchi, M. Tsukamoto and M. Ichinohe, *Science*, 1997, **275**, 60; (b) A. Sekiguchi, N. Fukaya, M. Ichinohe and Y. Ishida, *Eur. J. Inorg. Chem.*, 2000, 1155.
64. M. Igarashi, M. Ichinohe and A. Sekiguchi, *J. Am. Chem. Soc.*, 2007, **129**, 12660.
65. (a) V. Ya. Lee and A. Sekiguchi, *Acc. Chem. Res.*, 2007, **40**, 410; (b) V. Ya. Lee and A. Sekiguchi, *Angew. Chem. Int. Ed.*, 2007, **46**, 6596.
66. M. Kira, T. Iwamoto and C. Kabuto, *J. Am. Chem. Soc.*, 1996, **118**, 10303.
67. N. Wiberg, H. Auer, H. Nöth, J. Knoizek and K. Polborn, *Angew. Chem. Int. Ed.*, 1998, **37**, 2869.
68. A. Sekiguchi, T. Matsuno and M. Ichinohe, *J. Am. Chem. Soc.*, 2000, **122**, 11250.
69. A. Sekiguchi, T. Matsuno and M. Ichinohe, *J. Am. Chem. Soc.*, 2001, **123**, 12436.
70. T. Matsuno, M. Ichinohe and A. Sekiguchi, *Angew. Chem. Int. Ed.*, 2002, **41**, 1575.
71. H. Schafer, W. Saak and M. Weidenbruch, *Angew. Chem. Int. Ed.*, 2000, **39**, 3703.
72. V. Ya. Lee, K. Takanashi, M. Ichinohe and A. Sekiguchi, *J. Am. Chem. Soc.*, 2003, **125**, 6012.
73. V. Ya. Lee, K. Takanashi, T. Matsuno, M. Ichinohe and A. Sekiguchi, *J. Am. Chem. Soc.*, 2004, **126**, 4758.
74. V. Ya. Lee, K. Takanashi, M. Ichinohe and A. Sekiguchi, *Angew. Chem. Int. Ed.*, 2004, **43**, 6703.

75. K. Takanashi, V. Ya. Lee, T. Matsuno, M. Ichinohe and A. Sekiguchi, *J. Am. Chem. Soc.*, 2005, **127**, 5768.
76. (a) V. Ya. Lee, K. Takanashi, R. Kato, T. Matsuno, M. Ichinohe and A. Sekiguchi, *J. Organomet. Chem.*, 2007, **692**, 2800; (b) V. Ya. Lee and A. Sekiguchi, *Chem. Soc. Rev.*, 2008, **37**, 1652.
77. H. Kobayashi, T. Iwamoto and M. Kira, *J. Am. Chem. Soc.*, 2005, **127**, 15376.
78. (a) V. Ya. Lee, S. Miyazaki, H. Yasuda and A. Sekiguchi, *J. Am. Chem. Soc.*, 2008, **128**, 2758; (b) A. Grybat, S. Boomgaarden, W. Saak, H. Marsmann and M. Weidenbruch, *Angew. Chem. Int. Ed.*, 1999, **38**, 2010.
79. (a) W. Fink, *Angew. Chem. Int. Ed. Engl.*, 1966, **5**, 760; (b) U. Klingebiel, in *The Chemistry of Inorganic Homo- and Heterocycles*, ed. I. Haiduc and D. B. Sowerby, Academic Press, London, 1987, Vol. **1**, pp. 221–275.
80. M. Bouquey, C. Brochon, S. Bruzaud, A.-F. Mingotaud, M. Schappacher and A. Soum, *J. Organomet. Chem.*, 1996, **521**, 21.
81. P. J. Wheatley, *J. Chem. Soc.*, 1962, 1721.
82. B. Rozsondai, J. Hargittai, A. V. Golubinskii, L. V. Vilkov and V. S. Mastryukov, *J. Mol. Struct.*, 1975, **28**, 339.
83. D. Mootz, J. Fayos and A. Zinnius, *Angew. Chem. Int. Ed. Engl.*, 1972, **11**, 58.
84. R. Rovai, C. W. Lehmann and J. S. Bradley, *Angew. Chem. Int. Ed.*, 1999, **38**, 2036.
85. N. Kocher, C. Selinka, D. Leusser, D. Kost, I. Kalikhman and D. Stalke, *Z. Anorg. Allg. Chem.*, 2004, **630**, 1777.
86. D. Enterling, U. Klingebiel and A. Meller, *Z. Naturforsch., Teil B*, 1978, **33**, 527.
87. (a) J. Hughes and G. R. Willey, *J. Am Chem. Soc.*, 1973, **95**, 8758; (b) G. R. Willey, *J. Am Chem. Soc.*, 1968, **90**, 3362.
88. M. Schmidt and I. Ruidisch, *Angew. Chem. Int. Ed. Engl.*, 1964, **3**, 637.
89. M. Veith, in *The Chemistry of Inorganic Homo- and Heterocycles*, ed. I. Haiduc and D. B. Sowerby, Academic Press, London, 1987, Vol. **1**, pp. 382–387.
90. D. Haenssgen and I. Pohl, *Angew. Chem. Int. Ed. Engl.*, 1974, **13**, 607.
91. W. Storch, W. Jackstiess, H. Nöth and G. Winter, *Angew. Chem. Int. Ed. Engl.*, 1977, **16**, 478.
92. U. Schubert and N. Hüsing, *Synthesis of Inorganic Materials*, Wiley-VCH, Weinheim, 2000.
93. P. B. Hitchcock, M. F. Lappert and A. J. Thorne, *J. Chem. Soc., Chem. Commun.*, 1990, 1587.
94. (a) R. A. Bartlett and P. P. Power, *J. Am. Chem. Soc.*, 1990, **112**, 3660; (b) H. Chen, R. A. Bartlett, H. V. R. Dias, M. M. Olmstead and P. P. Power, *Inorg. Chem.*, 1991, **112**, 3660.
95. J.-T. Ahlemann, H. W. Roesky, R. Murugavel, E. Parisini, M. Noltemeyer, H.-G. Schmidt, O. Müller, R. Herbst-Irmer, L. N. Markovskii and Y. G. Shermolovich, *Chem. Ber.*, 1997, **130**, 1113.

96. S. Boughdiri, K. Hussein, B. Tangour, M. Dahrouch, M. Rivière-Baudet and J.-C. Barthelat, *J. Organomet. Chem.*, 2004, **689**, 3279.
97. T. Chivers and D. J. Eisler, in *Tin Chemistry: Fundamentals, Frontiers and Applications*, ed. M. Gielen, A. G. Davies, K. Pannell and E. Tiekink, Wiley, Chichester, 2008, pp. 53–58.
98. (a) C. Cui, M. Brynda, M. M. Olmstead and P. P. Power, *J. Am. Chem. Soc.*, 2004, **126**, 6510; (b) C. Cui, M. M. Olmstead, J. C. Fettinger, G. H. Spikes and P. P. Power, *J. Am Chem. Soc.*, 2005, **127**, 17530.
99. H. Cox, P. B. Hitchcock, M. F. Lappert and L. J.-M. Pierssens, *Angew. Chem. Int. Ed.*, 2004, **43**, 4500.
100. H. N. Han, D. A. Lindquist, J. S. Haggerty and D. Seyferth, *Chem. Mater.*, 1992, **4**, 705.
101. M. Hörz, A. Zern, F. Berger, J. Haug, K. Müller, F. Aldinger and M. Weinmann, *J. Eur. Ceram. Soc.*, 2005, **25**, 99.
102. E. Duguet, M. Schappacher and A. Soum, *Macromolecules*, 1992, **25**, 4835.
103. G. Fritz and J. Härer, in *The Chemistry of Inorganic Homo- and Heterocycles*, ed. I. Haiduc and D. B. Sowerby, Academic Press, London, 1987, Vol. **1**, pp. 277–286.
104. R. T. Oakley, D. A. Stanislawski and R. West, *J. Organomet. Chem.*, 1978, **157**, 389.
105. U. Klingebiel and N. Vater, *Angew. Chem. Int. Ed. Engl.*, 1982, **21**, 875.
106. G. Fritz and R. Biastoch, *Z. Anorg. Allg. Chem.*, 1986, **535**, 95.
107. H. Schumann and H. Benda, *Chem. Ber.*, 1971, **104**, 333.
108. H. Schumann and H. Benda, *Angew. Chem. Int. Ed. Engl.*, 1970, **9**, 76.
109. G. Fritz and R. Uhlmann, *Z. Anorg. Allg. Chem.*, 1980, **463**, 149.
110. I. Haiduc, in *The Chemistry of Inorganic Homo- and Heterocycles*, ed. I. Haiduc and D. B. Sowerby, Academic Press, London, 1987, Vol. **1**, pp. 370–371.
111. M. Veith, in *The Chemistry of Inorganic Homo- and Heterocycles*, ed. I. Haiduc and D. B. Sowerby, Academic Press, London, 1987, Vol. **1**, pp. 395–399.
112. R. Fröhlich and K. F. Tebbe, *Z. Anorg. Allg. Chem.*, 1983, **506**, 27.
113. (a) C. Couret, J. Satgé, J. D. Andriamizaka and J. Escudié, *J. Organomet. Chem.*, 1978, **157**, C35; (b) C. Couret, J. D. Andriamizaka, J. Escudié and J. Satgé, *J. Organomet. Chem.*, 1981, **208**, C3.
114. H. Schumann and H. Benda, *Angew. Chem. Int. Ed. Engl.*, 1968, **7**, 812.
115. W.-W. du Mont and H.-J. Kroth, *Angew. Chem. Int. Ed. Engl.*, 1977, **16**, 792.
116. M. Baudler and H. Suchomel, *Z. Anorg. Allg. Chem.*, 1983, **505**, 39.
117. V. Chavlovsky, in *The Chemistry of Inorganic Homo- and Heterocycles*, ed. I. Haiduc and D. B. Sowerby, Academic Press, London, 1987, Vol. **1**, pp. 287–348.
118. F. S. Kipping and R. Robinson, *J. Chem. Soc.*, 1914, **105**, 484.
119. (a) W. Patnode and D. F. Wilcock, *J. Am. Chem. Soc.*, 1946, **68**, 358; (b) M. J. Hunter, J. F. Hyde, E. L. Warrick and H. J. Fletcher, *J. Am. Chem.*

Soc., 1946, **68**, 667; (c) J. F. Brown Jr. and G. M. J. Sluscarzuk, *J. Am. Chem. Soc.*, 1965, **87**, 932.
120. D. Seyferth, C. Prud'homme and G. H. Wiseman, *Inorg. Chem.*, 1983, **22**, 2163.
121. (a) P. Arya, J. Boyer, F. Carré, R. Corriu, G. Lanneau. J. Lapasset, M. Perrot and C. Priou, *Angew. Chem. Int. Ed. Engl.*, 1989, **28**, 1016; (b) S. Yao, M. Brym, C. van Wüllen and M. Driess, *Angew. Chem. Int. Ed.*, 2007, **46**, 4159.
122. R. West, M. J. Fink and J. Michl, *Science*, 1981, **214**, 1343.
123. M. J. Michalczyk, M. J. Fink, K. J. Haller, R. West and J. Michl, *Organometallics*, 1986, **5**, 531.
124. R. S. Grev and H. F. Schaeffer III, *J. Am. Chem. Soc.*, 1987, **109**, 6577.
125. K. L. McKillop, G. R. Gillette, D. R. Powell and R. West, *J. Am. Chem. Soc.*, 1992, **114**, 5203.
126. G. Toskas, M. Moreau, M. Masure and P. Sigwalt, *Macromolecules*, 2001, **34**, 4730.
127. (a) M. Unno, Y. Kawagachi, Y. Kishimoto and H. Matsumoto, *J. Am. Chem. Soc.*, 2005, **127**, 2256; (b) M. Unno, Y. Kishimoto and H. Matsumoto, *Organometallics*, 2004, **23**, 6221.
128. A. C. Sullivan, *Coord. Chem. Rev.*, 1999, **189**, 19.
129. D. A. Foucher, A. J. Lough and I. Manners, *Inorg. Chem.*, 1992, **31**, 3034.
130. (a) I. Haiduc and H. Gilman, *J. Organomet. Chem.*, 1969, **18**, P5; (b) S. Blaurock, F. T. Edelmann, A. Fischer and I. Haiduc, *Z. Anorg. Allg. Chem.*, 2008, **634**, 34.
131. (a) J. S. Ritch and T. Chivers, *Angew. Chem. Int. Ed.*, 2007, **46**, 4610; (b) I. Haiduc, *Organometallics*, 2004, **23**, 3.
132. A. Decken, J. Passmore and X. Wang, *Angew. Chem. Int. Ed.*, 2006, **45**, 2773.
133. A. Masumune, S. A. Batcheller, J. Park and W. M. Davis, *J. Am. Chem. Soc.*, 1989, **111**, 1888.
134. D. Ellis, P. B. Hitchcock and M. F. Lappert, *J. Chem. Soc., Dalton Trans.*, 1992, **1**, 3397.
135. I. Haiduc and M. Dräger, in *The Chemistry of Inorganic Homo- and Heterocycles*, ed. I. Haiduc and D. B. Sowerby, Academic Press, London, 1987, Vol. **1**, pp. 371–372.
136. (a) L. Ross and M. Dräger, *Chem. Ber.*, 1982, **115**, 615; (b) L. Ross and M. Dräger, *Z. Naturforsch., Teil B*, 1984, **39**, 868.
137. M. A. Edelman, P. B. Hitchcock and M. F. Lappert, *J. Chem. Soc., Chem. Commun.*, 1990, 1116.
138. (a) H. Puff, W. Schuh, R. Sievers and R. Zimmer, *Angew. Chem. Int. Ed. Engl.*, 1981, **20**, 591; (b) H. Puff, W. Schuh, R. Sievers, W. Wald and R. Zimmer, *J. Organomet. Chem.*, 1984, **260**, 271.
139. R. W. Chorley, P. B. Hitchcock and M. F. Lappert, *J. Chem. Soc., Chem. Commun.*, 1992, 525.
140. J. M. Ziegler and F. W. G. Fearon (eds), *Silicon-based Polymer Science*, Advances in Chemistry Series, Vol. **224**, American Chemical Society, Washington, D.C., 1990.

141. J. A. Semlyen and S. J. Clarson (eds), *Siloxane Polymers*, Prentice Hall, Englewood Cliffs, NJ, 1991.
142. E. G. Rochow, *Silicon and Silicones*, Springer-Verlag, Heidelberg, 1987.
143. B. Arkles, *Chemtech*, 1983, **13**, 542.
144. G. J. J. Out, A. A. Teretskii, M. Möller and D. Oelfin, *Macromolecules*, 1994, **27**, 3310.
145. J. Chojnowski, *J. Inorg. Organomet. Polym.*, 1991, **1**, 299.
146. J. Stein, L. N. Lewis, K. A. Smith and K. X. Lettko, *J. Inorg. Organomet. Polym.*, 1991, **1**, 325.
147. S. Boileau and D. Teyssie, *J. Inorg. Organomet. Polym.*, 1991, **1**, 247.
148. (a) H. Poths and R. Zentel, *Macromol. Rapid Commun.*, 1994, **15**, 433; (b) H. Gankema, R. J. W. Lugtenberg, J. F. J. Engbersen, D. N. Reinhoudt and M. Möller, *Adv. Mater.*, 1994, **6**, 944.
149. M. Camacho-Lopez, H. Finkelmann, P. Palffy-Muhoray and M. Shelley, *Nat. Mater.*, 2004, **3**, 307.
150. A. Thayer, *Chem. Eng. News*, 2000, **78**(48), 27.
151. I. Haiduc, in *The Chemistry of Inorganic Homo- and Heterocycles*, ed. I. Haiduc and D. B. Sowerby, Academic Press, London, 1987, Vol. **1**, pp. 349–359.
152. M. Yokoi, T. Nomura and K. Yamasaki, *J. Am. Chem. Soc.*, 1955, **77**, 4484.
153. D. L. Mayfield, R. A. Flath and L. R. Best, *J. Am. Chem. Soc.*, 1964, **29**, 2444.
154. S. R. Bahr, P. Boudjouk and G. J. McCarthy, *Chem. Mater.*, 1992, **4**, 383.
155. (a) B. Menzebach and P. Bleckman, *J. Organomet.Chem.*, 1975, **91**, 291; (b) H.-J. Jacobsen and B. Krebs, *J. Organomet. Chem.*, 1977, **136**, 333.
156. M. Dräger, A. Blecher, H.-J. Jacobsen and B. Krebs, *J. Organomet. Chem.*, 1978, **161**, 319.
157. P. Boudjouk, D. J. Seidler, D. Grier and G. J. McCarthy, *Chem. Mater.*, 1996, **8**, 1189.
158. A. Blecher, B. Mathiasch and T. N. Mitchell, *J. Organomet. Chem.*, 1980, **184**, 175.
159. A. Blecher and M. Dräger, *Angew. Chem. Int. Ed. Engl.*, 1979, **18**, 677.
160. M. C. Janzen, H. A. Jenkins, L. M. Rendina, J. J. Vittal and R. J. Puddephatt, *Inorg. Chem.*, 1999, **38**, 2123.
161. P. Boudjouk, M. P. Remington, D. G. Grier, W. Triebold and B. R. Jarabek, *Organometallics*, 1999, **18**, 4534.
162. H. Puff, R. Gattermayer, R. Hundt and R. Zimmer, *Angew. Chem. Int. Ed. Engl.*, 1977, **16**, 547.
163. B. Krebs, S. Pohl and W. Schiwy, *Angew. Chem. Int. Ed. Engl.*, 1970, **9**, 897.
164. P. Jutzi, A. Möhrke, A. Müller and H. Bögge, *Angew. Chem. Int. Ed. Engl.*, 1989, **28**, 1518.
165. M. Haaf, A. Schmiedl, T. A. Schmedake, D. R. Powell, A. J. Millevolte, M. Denk and R. West, *J. Am. Chem. Soc.*, 1998, **120**, 12714.

166. P. B. Hitchcock, H. A. Jasim, M. F. Lappert, W.-P. Leung, A. K. Rai and R. E. Taylor, *Polyhedron*, 1991, **10**, 1203.
167. (a) R. West, D. J. De Young and K. J. Haller, *J. Am. Chem. Soc.*, 1985, **107**, 4942; (b) R. P.-K. Tan, G. R. Gillette, D. R. Powell and R. West, *Organometallics*, 1991, **10**, 546.
168. (a) T. Tsumuraya, S. Sato and W. Ando, *Organometallics*, 1988, **7**, 2015; (b) T. Tsumuraya, Y. Kabe and W. Ando, *J. Chem. Soc., Chem. Commun.*, 1990, 1159.
169. (a) A. Schäfer, M. Weidenbruch, W. Saak, S. Pohl and H. Marsmann, *Angew. Chem. Int. Ed. Engl.*, 1991, **30**, 834; (b) A. Schäfer, M. Weidenbruch, W. Saak, S. Pohl and H. Marsmann, *Angew. Chem. Int. Ed. Engl.*, 1991, **30**, 962.
170. (a) B. Mathiasch, *J. Organomet. Chem.*, 1980, **194**, 37; (b) B. Mathiasch, *J. Organomet. Chem.*, 1977, **141**, 189.
171. T. Komuru, T. Matsuo, H. Kawaguchi and K. Tatsumi, *Angew. Chem. Int. Ed.*, 2003, **42**, 465.
172. T. Komuru, T. Matsuo, H. Kawaguchi and K. Tatsumi, *Chem. Commun.*, 2002, **1**, 988.
173. D. M. Giolando, T. B. Rauchfuss and G. M. Clark, *Inorg. Chem.*, 1987, **26**, 3082.
174. (a) R. Okazaki and N. Tokitoh, *Acc. Chem. Res.*, 2000, **33**, 625; (b) R. Okazaki and N. Tokitoh, *Adv. Organomet. Chem.*, 2001, **47**, 121; (c) N. Tokitoh and R. Okazaki, in *The Chemistry of Organic Germanium Tin and Lead Compounds*, ed. Z. Rappoport, Wiley, Chichester, 2002, pp. 843–901.
175. H. Suzuki, N. Tokitoh, R. Okazaki, S. Nagase and M. Goto, *J. Am. Chem. Soc.*, 1998, **120**, 11096.
176. T. Matsumoto, N. Tokitoh and R. Okazaki, *J. Am. Chem. Soc.*, 1999, **121**, 8811.
177. M. Saito, N. Tokito and R. Okazaki, *J. Am. Chem. Soc.*, 2004, **126**, 15572.
178. M. Saito, N. Tokito and R. Okazaki, *J. Am. Chem. Soc.*, 1997, **119**, 11124.
179. (a) M. Saito, N. Tokito and R. Okazaki, *J. Am. Chem. Soc.*, 1997, **119**, 2337; (b) T. Tajima, N. Takeda, T. Sasamori and N. Tokitoh, *Organometallics*, 2006, **25**, 3552.
180. N. Kano, N. Tokitoh and R. Okazaki, *Organometallics*, 1997, **16**, 4237.
181. N. Tokitoh, H. Suzuki, T. Matsumoto, T. Matsuhashi and R. Okazaki, *J. Am. Chem. Soc.*, 1991, **113**, 7047.
182. Y. Matsuhashi, N. Tokitoh, R. Okazaki, M. Goto and S. Nagase, *Organometallics*, 1993, **12**, 1351.
183. N. Tokitoh, N. Kano, K. Shibata and R. Okazaki, *Organometallics*, 1995, **12**, 1351.

CHAPTER 11
Group 15: Rings and Polymers

The group 15 elements, primarily phosphorus, form a wide variety of ring systems that are of both fundamental and practical interest. The properties of homocyclic systems vary enormously. Polynitrogen rings are unsaturated homocycles isoelectronic with aromatic carbon systems, *e.g.* N_5^- and $[C_5H_5]^-$. These nitrogen-rich rings are of interest as high-energy materials that may serve, for example, as environmentally benign propellants. The phosphorus and arsenic analogues of naked homocycles such as N_5^- are stabilised as π-coordinating ligands in transition metal complexes. Homocycles of the group 15 elements of the type $(RE)_n$ (R = alkyl, aryl; E = P, As, Sb, Bi) are known for a variety of ring sizes ($n = 3-6$). They are saturated systems involving single E–E bonds. The phosphorus-containing homocycles are formally isoelectronic with cyclosilanes $(R_2Si)_n$ (see Section 10.1.1) and with cyclic sulfur allotropes (see Section 12.1.1).

Phosphorus and, to a lesser extent, the heavier group elements form both saturated and unsaturated heterocycles in combination with nitrogen. The saturated systems, *e.g.* cyclophosphazanes, $(RPNR')_n$ ($n = 2, 3$), are based on pnictogen(III) centres with a lone pair available for coordination to metal sites. The unsaturated rings, which involve phosphorus(V), *e.g.* cyclophosphazenes, $(R_2PN)_n$ ($n = 3-17$), have been known since the 1830s, but an accurate description of the electronic structure of these classic inorganic heterocycles continues to present a challenge to theoretical chemists (see Section 4.1.2.2). The six-membered cyclophosphazene $(NPCl_2)_3$ is an important precursor, *via* ring-opening polymerisation, to the inorganic polymer $(NPCl_2)_n$, which is sometimes referred to as 'inorganic rubber' because of its physical properties. The replacement of the Cl substituents on phosphorus in the polymer $(NPCl_2)_n$ by groups such as alkylamido or aryloxy produces processable polymers whose properties rival or, in some cases, outperform those of the isoelectronic silicone polymers $(OSiR_2)_n$ (see Section 10.6). The commercial development of polyphosphazenes has been hampered, however, by the relatively high production costs.

Inorganic Rings and Polymers of the p-Block Elements: From Fundamentals to Applications
By Tristram Chivers and Ian Manners
© Tristram Chivers and Ian Manners 2009
Published by the Royal Society of Chemistry, www.rsc.org

11.1 NITROGEN-RICH RINGS

11.1.1 Nitrogen Homocycles: Energetic Materials

The reason for the interest in polynitrogen compounds as fuels or explosives lies in the nitrogen bond energies. The average single bond (N–N) and double bond (N=N) energies of 39 and 100 kcal mol^{-1} are significantly less than one-third and two-thirds, respectively, of the N≡N triple bond energy of 228 kcal mol^{-1}. Consequently, decomposition of any polynitrogen compound to N_2 will result in the release of a large amount of energy. The formation of environmentally innocuous N_2 gas is an attractive feature of the use of polynitrogen compounds as an energy source.

The best known nitrogen homocycles are the pentazoles RN_5. Phenylpentazole PhN_5 is obtained as a thermally unstable compound by the reaction of benzenediazonium chloride with lithium azide in methanol at $-40\,^\circ$C [eqn (11.1)].[1]

$$PhN_2^+Cl^- + LiN_3 \rightarrow PhN_5 + LiCl \qquad (11.1)$$

The introduction of electron-donating substituents in the 4-position of the benzene ring stabilises the N_5 homocycle. The X-ray structures of 4-dimethylaminophenylpentazole[2a] and phenylpentazole (**11.1**)[2b] reveal planar N_5 rings with N–N bond lengths in the narrow range 1.31–1.35 Å, consistent with a six π-electron aromatic ring system (*cf.* single and double bond values of 1.48 and 1.20 Å, respectively). The ^{15}N NMR spectrum of 4-dimethylaminophenylpentazole at $-35\,^\circ$C shows the expected three resonances in the ratio 2:2:1, but some decomposition to the corresponding azide (loss of N_2) is evident even at this low temperature after 40 h.[3] A ^{15}N NMR study of the tetrazolylpentazole 2-HN_4CN_5, prepared from tetrazolediazonium chloride and LiN_3, reveals that decomposition occurs rapidly above $-50\,^\circ$C.[4]

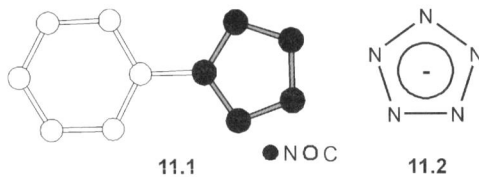

11.1 ● N ○ C **11.2**

The pentazolate anion, N_5^- (**11.2**), is estimated to have a half-life of 2.2 days, whereas that of the parent pentazole HN_5 is predicted to be only *ca* 10 min in methanol at 0 $^\circ$C.[5a] Although HN_5 is unknown, the cyclic anion N_5^- has been detected by tandem mass spectrometric studies of 4-hydroxyphenylpentazole.[5b] Similarly to its congener P_5^- (Section 11.2), N_5^- (isoelectronic with cyclopentadienide $[C_5H_5]^-$) has the potential to form metallocene-like complexes. The acyclic (V-shaped) cation N_5^+ has been isolated as a hexafluoroantimonate salt, which decomposes at *ca* 70 $^\circ$C.[6] The estimated energy density of $[N_5]^+[N_5]^-$ is approximately twice that of hydrazine, a well-known rocket propellant, suggesting that this ionic polynitrogen allotrope would be an excellent monopropellant for rocket propulsion or explosives if it can be stabilised.[7] However, attempts to make the related azide salt $[N_5]^+[N_3]^-$ resulted in violent explosions, a common occurrence in the synthesis of polynitrogen compounds.[7b]

11.1.2 Tetrazoles

Tetrazoles RCN_4 are a well-known class of heterocyclic compounds that are used extensively in medicinal chemistry research; pentamethylene tetrazole is a central nervous system stimulant.[8a] Some tetrazoles, *e.g.* 5-aminotetrazole H_2NCN_4 and its derivatives, are high-energy materials that show promise in applications as propellants or primers.[8b] Recently, tetrazoles in which the carbon atom has been replaced by either phosphorus or arsenic have been synthesised and structurally characterised.[9] These five-membered rings are obtained, as colourless $GaCl_3$ adducts, by application of the $GaCl_3$-assisted elimination of trimethylsilyl chloride to the reaction between Mes*N=ECl (E = P,[9a] As[9b]) and trimethylsilyl azide (Scheme 11.1). These tetrazoles can be formally regarded as the [2+3] cycloaddition products of $[Mes*E=N]^+$ and N_3^- ions. The use of an alkyl azide RN_3 (R = ^{t}Bu, CEt_3) rather than Me_3SiN_3 produces tetraazaphosphonium salts.[9c] The monomer Mes*N=PCl is well known (Section 11.4.1), but the arsenic analogue Mes*N=AsCl is generated *in situ* either by dissociation of the dimer ClAs(μ-NMes*)$_2$AsCl in solution or *via* elimination of Me_3SiCl from $Mes*N(SiMe_3)AsCl_2$.[9b]

The five-membered rings $Mes*N_4E$ (E = P, As) are planar with two N–N bond lengths in the range 1.35–1.37 Å and one shorter N–N bond (*ca.* 1.29 Å). The structural parameters and electronic structure calculations are consistent with six π-electron aromatic systems involving strongly polarised σ and π E–N bonds.[9] Interestingly, the As derivative exhibits significantly higher thermal stability than the P analogue, which is heat and shock sensitive and loses N_2 gas readily at room temperature.

11.2 CYCLIC ANIONS OF PHOSPHORUS, ARSENIC, ANTIMONY AND BISMUTH

There are three common allotropes of phosphorus. White phosphorus consists of tetrahedral P_4 molecules, whereas red phosphorus may be described as interlocking chains of phosphorus atoms linked by P–P bonds.[10] The most

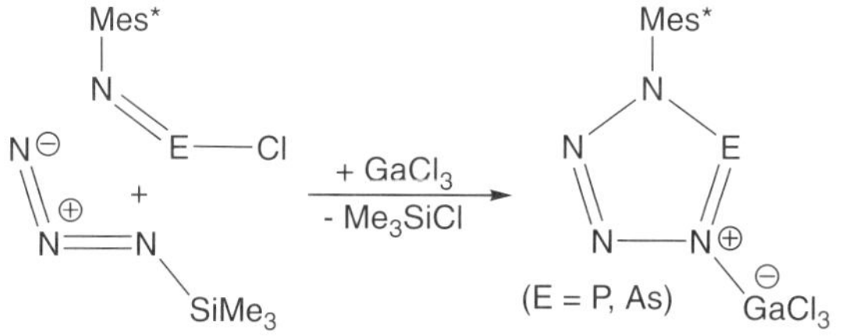

Scheme 11.1 Synthesis of a tetraazaphosphole and a tetraazaarsole.

stable allotrope, black phosphorus, has a double-layer lattice of chair-like six-membered rings.

A number of polyanions of phosphorus are obtained by reactions of phosphorus allotropes with various reducing agents, *e.g.* alkali metals. The identity of the reducing agent and the reaction conditions have a subtle, and unpredictable, influence on the nature of the anion formed; the polycyclic species P_7^{3-} and P_{11}^{3-} are the most common.[10] Monocyclic P_n anions (and their arsenic analogues) are stabilised by coordination to a transition metal. The synthetic route to these complexes usually involves thermal or photochemical reactions of P_4 (or As_4) with an organometallic reagent. The reaction conditions and the nature of the metal centre have an important influence on the nature of the products, which may include complexes involving E_n units ($E = P$, As; $n = 3$–6) and also smaller (E or E_2) fragments or larger polypnictogen ligands E_n ($n = 7, 8, 10, 12$).[11] In the case of antimony, the four-membered ring ($^tBuSb)_4$ has been used as a source of Sb_n units.[12] Some typical examples of complexes containing cyclic polypnictogen ligands, which include both sandwich and triple decker arrangements, are shown in Figure 11.1. Recent developments in this area have focused on the use of P_n and As_n ligand complexes, *e.g.* [Cp*Fe(η^5-P_5)][13a–c] and [Cp*Mo(CO)$_2$(η^3-As_3)][13d] as building blocks in supramolecular assemblies and also one- and two-dimensional polymers. Details of this exciting chemistry are given in Section 6.4.

The most widespread complexes of cyclic pnictogen anions are those containing the planar six π-electron ligand cyclo-P_5^- (isovalent with $[C_5H_5]^-$). The potassium salt of this anion is obtained *in solution* by the reaction of red phosphorus with KPH_2 in boiling DMF;[14a] this reagent can be used for the direct synthesis of η^5-P_5 complexes of metal carbonyls [eqns (11.2a) (M = Cr, Mo, W) and (11.2b)].[14b] An intriguing example of

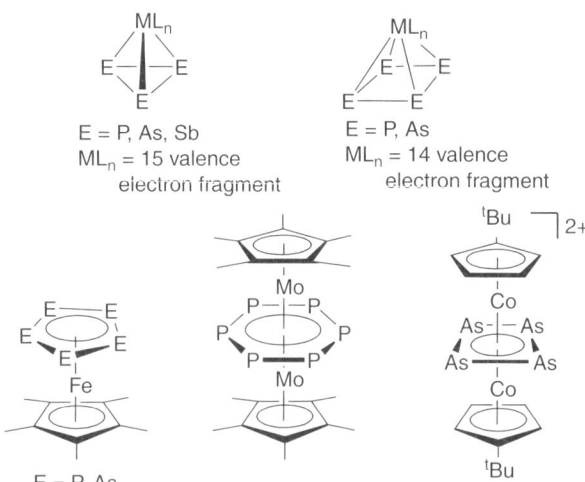

Figure 11.1 Transition metal complexes of cyclic P_n, As_n or Sb_n ligands.

this type of complex is the ferrocene-like carbon-free titanium compound $[(\eta^5\text{-}P_5)_2\text{Ti}]^-$ (see Section 7.2.1), which is prepared from TiCl_4 and white phosphorus in the presence of potassium naphthalenide as a reducing agent [eqn (11.2c)].[15]

$$\text{KP}_5 + [\text{M(CO)}_3(\text{MeCN})_3] \xrightarrow[155°\text{C}]{\text{DMF}} \text{K}[(\eta^5\text{-}P_5)\text{M(CO)}_3] \quad (11.2\text{a})$$

$$\text{KP}_5 + [\text{Mn(CO)}_5\text{Br}] \xrightarrow[155°\text{C}]{\text{DMF}} [(\eta^5\text{-}P_5)\text{Mn(CO)}_3] + \text{KBr} \quad (11.2\text{b})$$

$$\text{TiCl}_4(\text{THF})_2 + 6\text{K}[\text{C}_{10}\text{H}_8] \xrightarrow[\text{b) 2.5 P}_4, \text{THF}, 60°\text{C}]{\text{a) 18-crown-6}} [(18\text{-c-}6)_2][(\eta^5\text{-}P_5)_2\text{Ti}] \quad (11.2\text{c})$$

Although no salt-like compounds of the isolated cyclo-P_5^- anion are known in the solid state, the entire series of polypnictogen dianions of the type cyclo-E_4^{2-} has been isolated as solvated alkali metal salts and structurally characterised. Interestingly, the discovery of these Zintl anions in chronological order was as follows: Bi_4^{2-},[16a] Sb_4^{2-},[16b] As_4^{2-},[16c] P_4^{2-}.[16d] The key to the isolation of the stable black salts of the heavier congeners cyclo-E_4^{2-} (E = Sb, Bi) involves the use of 2,2,2-crypt to encapsulate the alkali metal cation and, hence, minimise internal electron transfer from the strongly reducing dianion.[16a,16b] The arsenic analogue cyclo-As_4^{2-} is obtained as yellow crystals of $[\text{Na(NH}_3)_5]_2\text{As}_4 \cdot 2\text{NH}_3$ upon reduction of arsenic with sodium in liquid ammonia.[16c] Cyclo-P_4^{2-} cannot be prepared in this manner from red phosphorus; however, the reaction of P_2H_4 with caesium in liquid ammonia produces yellow crystals of $\text{Cs}_2P_4 \cdot 2\text{NH}_3$ [eqn (11.3)].[16d] The crown ether-stabilised complexes $[\text{K(18-crown-6)}]_2E_4$ (E = P, As) have also been isolated and structurally characterised.[17]

$$6P_2H_4 + 10\,\text{Cs} \rightarrow \text{Cs}_2P_4 + 8\text{CsPH}_2 + 4H_2 \quad (11.3)$$

All members of the series E_4^{2-} exhibit planar four-membered rings with D_{4h} symmetry (Figure 11.2); they have approximately equal E–E bond lengths and endocyclic bond angles of ca 90°. Although the average bond lengths are intermediate between single and double bond values, this structural parameter alone cannot be used reliably as a criterion of aromaticity. Quantum chemical calculations show that cyclo-P_4^{2-} has a predicted bond order of ca 1.5, consistent with a six π-electron system, cf. the isoelectronic cyclo-S_4^{2+} dication (see Section 12.3.1).[16d] However, investigations of the bonding in polyphosphides using the electron localisation function (ELF) indicate that the aromaticity in homocyclic inorganic rings such as cyclo-P_4^{2-} and cyclo-P_5^- differs from that in aromatic hydrocarbons in being based on lone pairs rather than the E–E bonds.[18] The term 'lone pair aromaticity' has been suggested to describe this phenomenon. A similar description has been ascribed to the heavier congener cyclo-As_4^{2-}.[17]

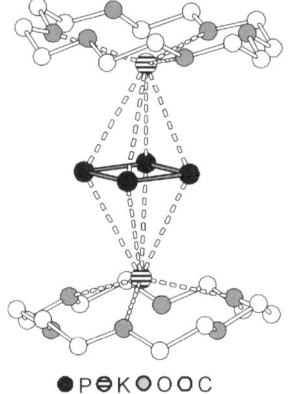

Figure 11.2 Molecular structure of [K(18-crown-6)]$_2$P$_4$ · 2NH$_3$ (NH$_3$ molecules omitted).

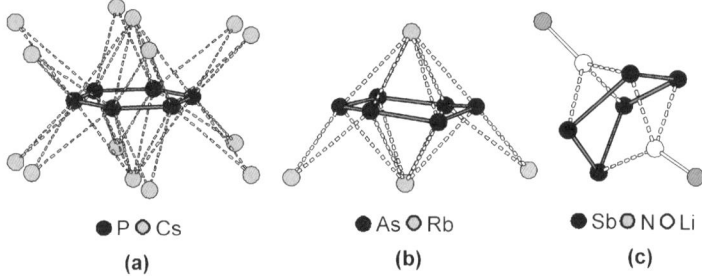

Figure 11.3 Coordination environment of (a) P$_6^{4-}$ in Cs$_4$P$_6$, (b) As$_6^{4-}$ in [Rb(18-crown-6)$_2$]Rb$_2$As$_6$ · 6NH$_3$ and (c) Sb$_5^{5-}$ in [Li$_2$(NH$_3$)$_2$Sb$_5$]$^{3-}$.

In addition to acting as an η^6 ligand in organometallic sandwich and triple-decker complexes (Figure 11.1),[9,19] the cyclohexaphosphide anion is encountered in binary compounds with alkali metals, M$_4$P$_6$ (M = K, Rb, Cs).[20] These extremely air- and moisture-sensitive black solids are made by heating the alkali metal with red phosphorus at temperatures above 400 °C. The arsenic analogues M$_4$As$_6$ (M = Rb, Cs) are obtained in a similar manner.[21a] In contrast to cyclo-P$_6^{4-}$, which disproportionates to give a variety of other polyphosphides,[18] the cyclo-As$_6^{4-}$ tetraanion is sufficiently stable in liquid ammonia solution to allow the isolation of green crystals of [Rb(18-crown-6)]$_2$Rb$_2$As$_6$ · 6NH$_3$.[21b]

The P$_6^{4-}$ ion in Cs$_4$P$_6$ forms a planar six-membered ring with P–P bond lengths of *ca* 2.15 Å (Figure 11.3a), whereas the As$_6^{4-}$ in [Rb(18-crown-6)$_2$]Rb$_2$As$_6$ · 6NH$_3$ adopts a chair conformation with As–As bond lengths in the range 2.40–2.42 Å. The coordination of the tetraanion to the four Rb$^+$ cations is depicted in Figure 11.3b. The earlier suggestion, based on structural evidence, that cyclo-P$_6^{4-}$ is a 10 π-electron system is not borne out by recent ELF calculations[21] or by solid-state NMR data for M$_4$P$_6$ (M = Rb, Cs).[22] The calculations indicate that the species cyclo-E$_6^{4-}$ (E = P, As) are not planar in the

Group 15: Rings and Polymers   217

absence of a highly coordinating environment. The ^{31}P NMR spectra for cyclo-P_6^{4-} do not show the low-field chemical shifts that are observed for the aromatic species cyclo-P_5^- and cyclo-P_4^{2-}.

Polypnictogen polyanions of the type E_n^{n-} (E = P, As, Sb, Bi) are isoelectronic with the corresponding neutral group 16 species E_n (E = S, Se, Te). Two examples are known: cyclo-Sb_5^{5-} and cyclo-E_8^{8-} (E = As, Sb). The cyclo-Sb_5^{5-} ion is isolated as the ionic complex $[Li(NH_3)_4]_3[Li_2(NH_3)_2Sb_5] \cdot 2NH_3$ in the form of dark-red crystals from the reaction of lithium with antimony in a 1:1 molar ratio in liquid ammonia.[23] The five-membered Sb_5^{5-} ring adopts an envelope conformation, which is capped by $[Li(NH_3)]^+$ cations on both sides of the ring, thus reducing the high negative charge to –3 (Figure 11.3c). The other three Li^+ ions, which are present as $[Li(NH_3)_4]^+$, separate the trianions from each other. The use of potassium instead of lithium as a reducing agent for antimony in liquid ammonia produces black needles of $[K_{17}(Sb_8)_2(NH_2)] \cdot 17.5NH_3$, which contains the crown-shaped Sb_8^{8-} ring,[24] cf. cyclo-S_8. The high negative charge of this octaanion is neutralized by extensive coordination to the potassium counterions. The arsenic analogue As_8^{8-} has been characterised in the anions $[NbAs_8]^{3-}$ and $[MoAs_8]^{2-}$ in which a highly charged transition metal cation is sequestered within the crown-shaped As_8 ring.[25]

11.3 HOMOATOMIC RINGS AND POLYMERS

11.3.1 Cyclophosphines

11.3.1.1 Neutral Systems. An analogy between phosphorus(III) and carbon chemistry based on the formation of ring and chain compounds was recognised more than 25 years ago.[26] The isolobal relationship between PR and CR_2 units implies that an extensive chemistry of catenated phosphorus(III) compounds, both cyclic and linear, will exist by comparison with cycloalkanes and linear alkanes. Indeed phosphorus has been referred to as 'the carbon copy' in one monograph.[27] In practice, monocyclic ring systems of the type $(RP)_n$ (cyclophosphines) form a much more limited homologous series ($n = 3$–6) (**11.3**–**11.6**)[28] than their isoelectronic counterparts in group 14 and group 16, viz. cyclosilanes $(R_2Si)_n$ ($n = 3$–35) (see Section 10.1.1) and the cyclic sulfur allotropes S_n ($n = 6$–25) (see Section 12.1.1).

11.3   **11.4**   **11.5**   **11.6**

The numerous methods available for the synthesis of cyclophosphines may be divided into two different types: (a) those that are not specific with respect to ring size and usually give rise to the thermodynamically more stable ring and (b) those that are designed to produce a specific ring size. In general, method (a) involves reagents that have a single RP unit. The most versatile examples of this approach are the cyclocondensation of a primary phosphine and a dichloroorganophosphine, preferably in the presence of a base to remove HCl [eqn (11.4)] and reductive coupling using a dichloroorganophosphine and either an alkali metal, magnesium or mercury [eqn (11.5)].[28] Metal hydrides, *e.g.* LiH or LiAlH$_4$, are also used in reductive coupling processes that, presumably, involve the intermediate formation of RPH$_2$. In this connection, the method of dehydrocoupling of RPH$_2$ under the influence of an early transition metal catalyst to give five-membered rings cyclo-(RP)$_5$ (R = Ph, Cy, 2,4,6-Me$_3$C$_6$H$_2$) has been noted in Section 2.3.[29]

$$RPH_2 + RPCl_2 \rightarrow 2/n(PR)_n + 2HCl \qquad (11.4)$$

$$RPX_2 + xM \rightarrow 1/n(PR)_n + xMX_y \qquad (11.5)$$

(X = Cl, Br)   ($x = 2$, $y = 1$; M = alkali metal
   $x = 1$, $y = 2$; M = Mg)

The salt-elimination approach can be designed to generate four-membered rings in excellent yields, including homocycles with different substituents on the phosphorus atoms [eqn (11.6)].[30]

$$2^iPr_2NPCl_2 + LiP(SiMe_3)_2 \cdot 2THF \xrightarrow[-2\,Me_3SiCl]{-2\,LiCl} {}^iPr_2N-P\begin{array}{c} SiMe_3 \\ | \\ P \\ \diagup \quad \diagdown \\ \quad \quad P-N^iPr_2 \\ P \\ | \\ SiMe_3 \end{array} \qquad (11.6)$$

The second approach to cyclophosphines [method (b)] makes use of reagents containing a preformed P–P bond. For example, cyclocondensation reactions are used to prepare cyclotriphosphines with either identical or mixed substituents by treating a dichloroorganophosphine with either bistrimethylsilyl derivatives (Me$_3$Si)RPPR(SiMe$_3$) or the dipotassium salts K(R)PP(R)K [eqn (11.7)]. The four- and five-membered rings cyclo-(EtP)$_n$ ($n = 4$, 5) have been prepared in a similar manner [eqn (11.8)].

$$K(R)PP(R)K + R'PCl_2 \rightarrow (RP)_2(PR') + 2KCl \qquad (11.7)$$

$$K(PEt)_nK + EtPCl_2 \rightarrow (EtP)_{n+1} + 2KCl$$
$$(n = 3, 4) \qquad (11.8)$$

This synthetic approach is readily adapted to the preparation of a wide variety of three-membered rings containing two phosphorus atoms and one other p-block element from groups 13, 14, 15 or 16 [eqn (11.9)].[26]

$$X(R)PP(R)X + R'_nEY_2 \xrightarrow{-2\,XY} \underset{RP-PR}{\overset{E-R'_n}{\triangle}}$$

X = K, Y = Cl; E = Si, Ge, Sn, B, As, Sb
X = Cl, Y = Me₃Sn ; E = S, Se

(11.9)

The solid-state structures of all four ring sizes of cyclophosphines cyclo-$(RP)_n$ ($n = 3-6$) have been established by X-ray crystallography for a variety of derivatives.[28] The P–P bond lengths fall within the range 2.19–2.24 Å, indicative of single bonds. Each phosphorus atom accommodates a stereochemically active lone pair giving rise to the possibility of isomers for each ring size depending on the relative orientation of the exocyclic substituents.

The P_6 ring in the hexamer cyclo-$(PhP)_6$ exists in the chair form with the phenyl groups in equatorial positions,[31a] *cf.* cyclohexane. The P_5 rings in pentamers exhibit a twisted envelope conformation, *e.g.* cyclo-$(PhP)_5$.[31b] The tetrameric systems cyclo-$(RP)_4$ (R = CF₃, C₆F₅, Cy, ^tBu) all have puckered four-membered rings; the extent of puckering is determined by steric interactions between the exocyclic substituents. For example, in cyclo-$(^tBuP)_4$ the *tert*-butyl groups alternate between opposite sides of the P_4 ring (Figure 11.4b);[32] in the trimer cyclo-$(^tBuP)_3$, two of the *tert*-butyl groups are located above the P_3 ring and the other is below it (Figure 11.4a).[33]

From the foregoing discussion, it is evident that a variety of isomers are possible for unsymmetrically substituted cyclophosphines; ³¹P NMR spectra are informative for distinguishing between these isomers. For example, two isomers of the cyclic trimers cyclo-$(PR)_2(PR')$ are observed. The symmetrical isomer with identical substituents on the same side of the ring exhibits an A_2B spin system, whereas the asymmetric isomer gives rise to an ABC spin system (Figure 11.5) (see Section 3.4.3).

Reactions of cyclophosphines may be divided into three types: (a) those that involve retention of the initial ring size, (b) those in which a P–P bond is cleaved and (c) those that result in a change in ring size, *i.e.* ring expansion or ring contraction. The former type (a) is observed for reactions with electrophilic reagents in which the lone pairs on the phosphorus(III) centres provide the nucleophilic site. These may involve simple adduct formation or oxidation to phosphorus(V); in some cases this oxidation may be accompanied by ring

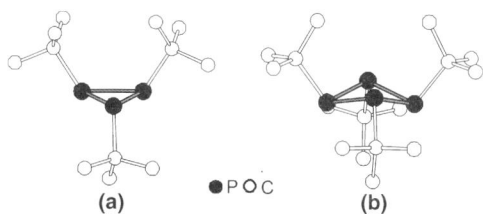

Figure 11.4 Molecular structures of (a) ($^tBuP)_3$ and (b) ($^tBuP)_4$.

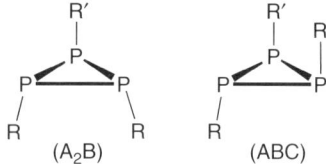

(A₂B)   (ABC)

Figure 11.5 NMR spin systems for two isomers of the three-membered rings $(PR)_2(PR')$.

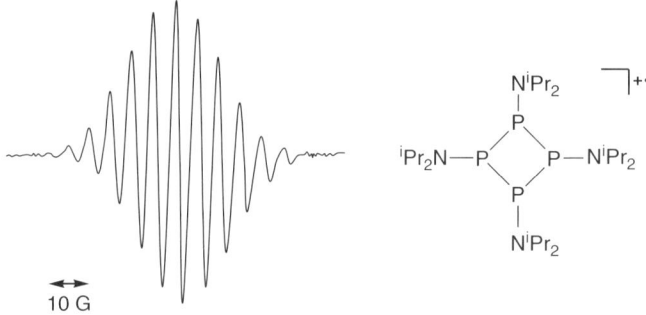

Figure 11.6 EPR spectrum of the cyclic cation radical cyclo-$[(^iPr_2NP)_4]^{+\bullet}$. Reproduced with permission from H.-G. Schäfer et al., J. Am. Chem. Soc., 1986, **108**, 7481.[34]

expansion [type (c)]. The second reaction type (b) is promoted by reducing or nucleophilic reagents and results in polyphosphorus chains.

Cyclotetraphosphines undergo electrochemical oxidation via a reversible one-electron process to form radical cations, which have been identified by EPR spectroscopy.[34] The violet radical cations are also produced by oxidation with AlCl₃ in dichloromethane. For example, the cation radical cyclo-$[(^iPr_2NP)_4]^{+\bullet}$ produced in this manner gives rise to a 13-line EPR spectrum (Figure 11.6), which is explained by coupling to four equivalent phosphorus nuclei and two equivalent nitrogen nuclei (the coupling to the other two nitrogen nuclei is not resolved).

Cyclophosphines $(PR)_n$ are readily oxidised by atmospheric oxygen. For example, the alkyl-substituted derivatives cyclo-$(P^tBu)_4$ and cyclo-$(PR)_5$ (R = Me, Et) react with a deficiency of O_2 to give the corresponding monoxides [eqn (11.10)].[28] Monosulfides are obtained in a similar manner by using elemental sulfur as the oxidising agent. The chalcogen atom in these monochalcogenides occupies an exocyclic position on the P_n skeleton of the parent cyclophosphine.

$$2\ (PR)_n + O_2 \xrightarrow[20°C]{C_6H_6} 2\ (PR)_nO$$

n = 4, R = ^{t}Bu
n = 5, R = Me, Et

(11.10)

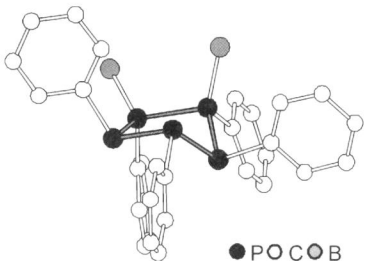

Figure 11.7 Molecular structure of cyclo-1,2-$(BH_3)_2(P_5Ph_5)$.

The presence of multiple donor sites enables cyclophosphines cyclo-$(PR)_n$ ($n = 3$–6) to function as versatile two-electron ligands in mono- di- or trinuclear metal complexes as discussed in Section 7.1.1 (see structures **7.1–7.3**); in some case the reaction of cyclo-$(PR)_5$ with transition metal complexes results in P–P bond activation, which may occur either with retention or fragmentation of the oligophosphine (for examples, see structures **7.4–7.5**).[35]

An interesting recent illustration of the Lewis base behaviour of these cyclic ligands is the formation of a bis-adduct of the pentamer cyclo-$(PPh)_5$ with borane monomer [eqn (11.11)].[36]

$$\text{cyclo-(PPh)}_5 \xrightarrow[- 2 \text{ SMe}_2]{2 \text{ BH}_3(\text{SMe}_2)} \text{cyclo-1,2-(BH}_3)_2(\text{cyclo-P}_5\text{Ph}_5)$$

(11.11)

The two BH_3 groups are coordinated to two adjacent phosphorus atoms of the P_5 ring in cyclo-1,2-$(BH_3)_2$(cyclo-P_5Ph_5) (Figure 11.7).[36] Only one diastereomer is present in the unit cell, corresponding to two enantiomers related by the crystallographic centre of symmetry. In solution, however, a complex mixture of isomers is evident from ^{31}P NMR spectra.

11.3.1.2 Cationic Systems. An intriguing development in the reactions of cyclophosphines with electrophilic reagents is the formation of cyclic monocations of the type (a) Me$(PR)_n^+$ ($n = 3$, 4, 5) by the alkylation of the neutral homocyclic rings with methyl trifluoromethanesulfonate (MeOTf) [eqn. (11.12)] or (b) R$'_2$P$(PR)_n^+$ ($n = 3$, 4) by insertion of a phosphenium cation R$_2$P$^+$ generated *in situ* from R$_2$PCl and trimethylsilyl trifluoromethanesulfonate into a P–P bond [eqn (11.13)].[37,38] Protonated ring systems are prepared in a similar manner by using trifluoromethanesulfonic acid [eqn (11.12)].

$$(RP)_n + R'OTf \rightarrow [(RP)_n R'][OTf] \quad (11.12)$$
$$(R' = Me, H)$$

$$(RP)_n + R'_2PCl + Me_3SiOTf \xrightarrow[-Me_3SiCl]{} [(RP)_n PR'_2][OTf] \quad (11.13)$$

For example, the five-membered rings $[R_2P(PPh)_4]^+$ (R = Ph, Me) are obtained as $[CF_3SO_3]^-$ salts from the reaction of a mixture of R_2PCl and trimethylsilyl trifluoromethanesulfonate with the tetramer cyclo-$(PhP)_4$ [eqn (11.13)].[38a] The same products are formed from the pentamer cyclo-$(PhP)_5$, demonstrating the thermodynamic preference for the formation of the five-membered ring over other alternatives for the phenyl-substituted rings. The solid-state structure of the perphenylated cations $[Ph_6P_5]^+$ consists of a highly puckered five-membered ring with P–P bond lengths in the narrow range 2.21–2.24 Å (Figure 11.8).[38a]

Related cations based on three- or four-membered polyphosphorus rings are obtained when sterically strained alkyl-substituted cyclo-tri- or -tetraphosphines are treated with methyl trifluoromethanesulfonate (MeOTf) or an *in situ*-generated phosphenium cation R_2P^+ (Scheme 11.2).[38b]

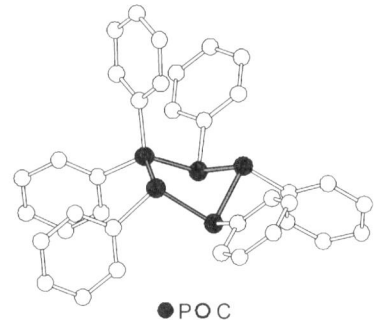

Figure 11.8 Molecular structure of the $[Ph_6P_5]^+$ cation (OTf$^-$ salt).

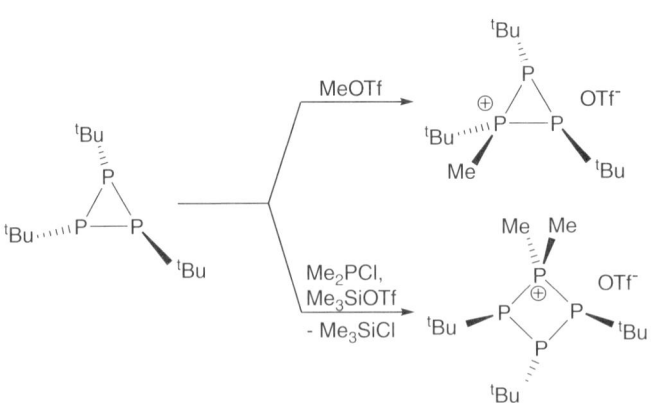

Scheme 11.2 Synthesis of cyclic $[Me^tBu_3P_3]^+$ and $[Me_2^tBu_3P_4]^+$ monocations.

By using appropriate reaction conditions, this synthetic approach can be adapted to the synthesis of dications of the type $[P_4R_4Me_2]^{2+}$ or $[P_6Ph_4R_4]^{2+}$.[39,40] For example, heating a mixture of the pentamer cyclo-$(PhP)_5$ with diphenyl- or dimethylchlorophosphine and gallium trichloride to *ca* 165 °C, in the absence of a solvent, produces the dications $[P_6Ph_4R_4]^{2+}$ (Ph, Me) as colourless tetrachlorogallate salts;[39a] the all-phenyl system $[P_6Ph_8]^{2+}$ is produced from cyclo-$(PhP)_5$ by using the melt formed by Ph_2PCl and $GaCl_3$ at room temperature.[39b] The six-membered P_6 ring in the $[P_6Ph_8]^{2+}$ dication adopts a twist-boat conformation (Figure 11.9a). This homocycle consists of two phosphonium centres in the 1,4-positions and four phosphine centres; the +2 charge has no significant effect on P–P bond lengths (2.23–2.24 Å).

In a similar manner, the methylation of cyclophosphines with MeOTf *in the absence of solvent* produces dications of the type $[Me_2(PR)_4]^{2+}$.[40] For example, the treatment of the cyclic tetramer cyclo-$(CyP)_4$ with an excess of MeOTf produces the dication $[Me_2(PCy)_4]^{2+}$ in excellent yield (Scheme 11.3). The observation of two 1:2:1 triplets in the ^{31}P NMR spectrum (an A_2X_2 spin system) suggests a symmetrical structure for this dication and this is confirmed by the X-ray structure, which shows that dimethylation occurs at the 1,3-positions of the four-membered ring (Figure 11.9b). Monitoring of the methylation process by ^{31}P NMR spectroscopy reveals the intermediate formation of the monomethylated cation $[Me(PCy)_4]^+$ (an A_2MX spin system) (Scheme 11.3)

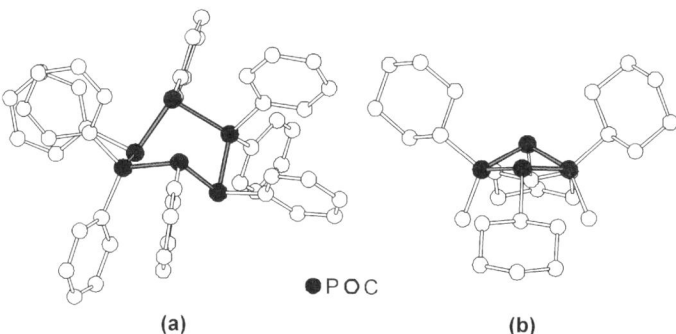

● P ○ C

(a)   (b)

Figure 11.9 Molecular structures of (a) $[P_6Ph_8]^{2+}$ ($GaCl_4^-$ salt) and (b) $[Me_2(PCy)_4]^{2+}$ (OTf^- salt).

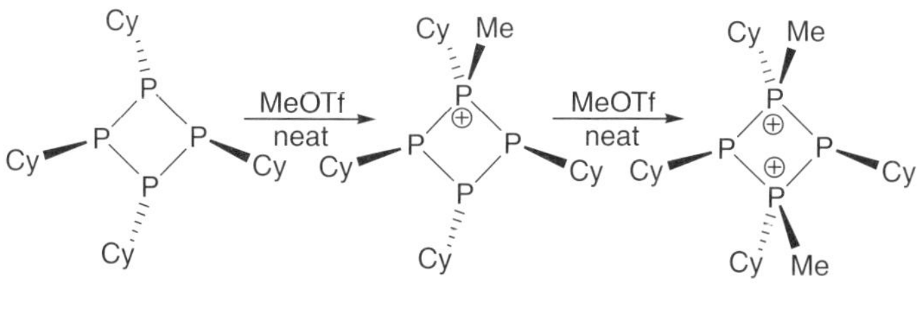

Scheme 11.3 Methylation of $(CyP)_4$ by methyl trifluoromethanesulfonate.

(see Section 3.4.3) after 1 h. Dicationic five-membered P_5 rings are prepared by methylation of cyclo-$(MeP)_5$ in a similar manner; the reaction proceeds quantitatively within 1 min and the monomethylated intermediate is not detected in this case.[40]

11.3.1.3 Anionic Systems. The monocyclic anion $[P_5H_4]^-$ is thermally unstable, but the thermal stability is enhanced by installing aryl or bulky alkyl groups on phosphorus. Cyclic monoanions of the type $[P_nR_{n-1}]^-$ are formally generated by cleavage of one of the P–R bonds in cyclopolyphosphines $(RP)_n$; under certain conditions this transformation is achieved by treatment with alkali metals. An early example of this process was the formation of $K[P_3{}^tBu_2]$ by the reaction of the cyclo-$({}^tBuP)_3$ with potassium.[41] The homocyclic species $[P_nR_{n-1}]^-$ have a tendency to undergo ring transformations in solution to give a mixture of cyclic oligomers. In some cases it is possible to identify the individual components of such mixtures, without isolation, by ^{31}P NMR spectroscopy because of their characteristic spin systems, *e.g.* $K[P_5Ph_4]$[42] and $Li[P_n{}^tBu_{n-1}]$ ($n = 3$–5).[43] An alternative route to this class of cycloorganophosphide anion involves the reaction of ${}^tBuPCl_2$ and PCl_3 with sodium in THF.[44] Under appropriate reaction conditions, this synthetic protocol can be adapted to the preparation of pure $Na[cyclo\text{-}P_5{}^tBu_4]$ [eqn (11.14)]. The reaction occurs *via* the intermediate formation of cyclo-$({}^tBuP)_4$.

$$4\,{}^tBuPCl_2 + PCl_3 + 12\,Na \xrightarrow[-11\,NaCl]{THF} [Na(THF)_4][P_5{}^tBu_4] \qquad (11.14)$$

The X-ray structure of $[Na(THF)_4][cyclo\text{-}P_5{}^tBu_4]$ shows an all-*trans* configuration of the *tert*-butyl groups on the phosphorus atoms in the anion with an envelope conformation of the chiral P_5 ring system (Figure 11.10).[44]

The anionic $[P_5{}^tBu_4]^-$ homocycle is a potentially versatile ligand in main group or transition metal complexes.[45–47] The reaction of $[Na(THF)_4][P_5{}^tBu_4]$ with Et_2AlCl produces the η^1 complex $Et_2Al(P_5{}^tBu_4)(THF)$ (**11.7**).[45] However,

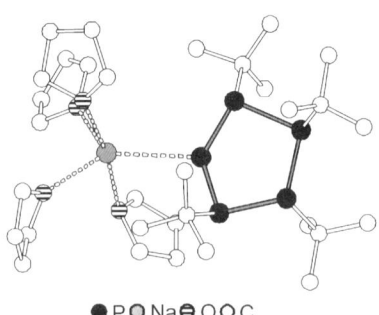

● P ○ Na ⊖ O ○ C

Figure 11.10 Structure of the cyclic $[P_5{}^tBu_4]^-$ anion {$[Na(THF)_4]^+$ salt}.

Group 15: Rings and Polymers　　　　　　　　　　　　　　　　　　　　225

reactions of this reagent with other main group metal halides, *e.g.* MCl$_2$ (M = Sn, Pb), result in reduction to the metal and the formation of the structural isomers (cyclo-P$_5^t$Bu$_4$)$_2$ (**11.8a**) and [(cyclo-P$_4^t$Bu$_3$)P^{t}Bu]$_2$ (**11.8b**) in which the monomer units are linked by P–P bonds.

A variety of behaviour is observed in the reactions of [Na(THF)$_4$][cyclo-P$_5^t$Bu$_4$] with phosphine halide complexes of the nickel triad.[46] Of particular interest is the formation of the η^2 complex Ni(cyclo-P$_5^t$Bu$_3$)(PEt$_3$)$_2$ in which the neutral cyclic diphosphene ligand P$_5^t$Bu$_3$ is formed by cleavage of one of the P–^{t}Bu bonds in the cyclo- [P$_5^t$Bu$_4$]$^-$ anion (see Scheme 7.3 in Section 7.1.1).[44,46a] Treatment of [Na(THF)$_4$][cyclo-P$_5^t$Bu$_4$] with [RhCl(PPh$_3$)$_3$] produces a low yield of the η^2 complex [Rh(cyclo-P$_5^t$Bu$_4$)(PPh$_3$)$_2$] in addition to redox-coupled products **11.8a** and **11.8b**.[46b] The cyclo-[P$_5^t$Bu$_4$]$^-$ anion forms a variety of oligomeric complexes with coinage metals, including the intriguing trinuclear gold(I) complex [Au$_3${cyclo-(P$_5^t$Bu$_4$)}$_3$] (**11.9**) in which intact P$_5$ rings bridge two gold centres.[47] By contrast, the reactions of [Na(THF)$_4$][P$_5^t$Bu$_4$] with Cu(I) or Ag(I) halides results in ring contraction to give the cyclo-(P$_4^t$Bu$_3$)P^{t}Bu ligand.[47]

Dianions of the type [P$_n$R$_n$]$^{2-}$, which were first generated by reduction of cyclophosphines with two equivalents of alkali metals in THF, have chain structures.[48,49] As such, they are potential synthons for the introduction of other elements into polyphosphine rings *via* metathesis reactions. More

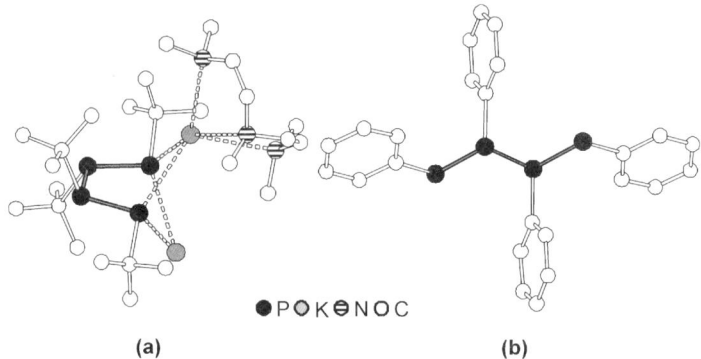

Figure 11.11 Molecular structures of (a) [K$_2$(PMDETA)(P$_4$^{t}Bu$_4$)] and (b) *meso*-[P$_4$Ph$_4$]$^{2-}$ in the [Na(2,2,2-cryptand)]$^+$ salt.

recently, optimised conditions for the selective synthesis of the TMEDA-solvated lithium salts of *catena*-[P$_n$Ph$_n$]$^{2-}$ (n = 2, 3, 4) from the reaction of the cyclic pentamer (PPh)$_5$ and lithium metal in THF (n = 2), DME (n = 3) or diethyl ether (n = 4) have been reported.[50] The linear catenated dianions [P$_n$R$_n$]$^{2-}$ (R = Ph, n = 2–4; R = Cy, ^{t}Bu, n = 4) have also been prepared by the direct reaction of sodium or potassium with RPCl$_2$ in boiling toluene or THF in the presence of TMEDA or PMDETA (pentamethyldiethylenetriamine).[51–53] In the solid state, the alkali metal salts form contact ion pairs, *e.g.* [K$_2$(PMDETA)(*catena*-P$_4$^{t}Bu$_4$)] (Figure 11.11a).[53] In the presence of 2,2,2-cryptand, however, the four-atom chains *catena*-[P$_4$R$_4$]$^{2-}$ (R = Ph, Cy) exist as solvent-separated ion pairs (Figure 11.11b).[51] The absence of the stabilising influence of the alkali metal counterions in the 2,2,2-cryptand-solvated salts is evident from the facile dissociation of *catena*-[P$_4$R$_4$]$^{2-}$ (R = Ph, Cy) dianions into the corresponding diphosphene radical monoanions [P$_2$R$_2$]$^{-•}$, which give rise to 1:2:1 triplets in the EPR spectrum.[51]

11.3.2 Cycloarsines, -stibines and -bismuthines

Arsenic forms a homologous series of ring systems cyclo-(RAs)$_n$, which, like the phosphorus congeners **11.3**–**11.6**, are limited to ring systems where n = 3–6.[54] The most common synthetic route involves the reductive coupling of RAsCl$_2$ with alkali metals or magnesium. In the case of ^{t}BuAsCl$_2$ the dehalogenation process produces the tetramer cyclo-(^{t}BuAs)$_4$ predominantly, with smaller amounts of the pentamer cyclo-(^{t}BuAs)$_5$. The trimer cyclo-(^{t}BuAs)$_3$ is obtained by the [2 + 1] cyclocondensation process shown in eqn (11.15).[55] The degree of oligomerisation of this thermally unstable, pyrophoric arsenic homocycle was determined by field ionisation mass spectrometry. The ^{1}H NMR spectrum shows two resonances in a 2:1 intensity ratio, consistent with the presence of the *cis,-trans*-isomer as established in the solid state for cyclo-(^{t}BuP)$_3$ (Figure 11.4a) The weakness of the As–As bonds (compared with P–P bonds) is evident from the facile conversion of cyclo-(^{t}BuAs)$_3$ into cyclo-(^{t}BuAs)$_4$ at room temperature.[55]

$$K_2[^tBuAsAs^tBu] + {}^tBuAsCl_2 \xrightarrow[-2\ KCl]{\text{pentane, } -78°C} \text{cyclo-}(^tBuAs)_3$$

(11.15)

Ring expansion also occurs for cyclo-(MeAs)$_5$ upon reactions with metal carbonyls to form binuclear products containing nine- or 10-membered rings, *e.g.* [cyclo-(MeAs)$_9$Cr$_2$(CO)$_6$] and [cyclo-(MeAs)$_{10}$Mo$_2$(CO)$_6$].[56] The 10-membered ring in [cyclo-(MeAs)$_{10}$Mo$_2$(CO)$_6$] has the boat–chair–boat conformation of cyclodecane (Figure 11.12).[57] With late transition metal reagents, *e.g.* Pd(PPh$_3$)$_4$, fragmentation of the cyclic tetramers (RAs)$_4$ (R = CF$_3$, C$_6$F$_5$, Ph) occurs readily to give η^2 complexes of the corresponding diarsene RAs=AsR.[58]

The best known cycloarsines are those that are present in the chemotherapeutic drug Salvarsan, which was used in the early 20th century for the treatment of syphilis. The synthesis of Salvarsan involves the reduction of an arylarsonic acid with aqueous sodium dithionite, which produces a mixture of cyclic oligomers (ArAs)$_n$ ($n = 3$–8, Ar = 3-NH$_2$-4-HOC$_6$H$_3$) as determined by electrospray ionisation mass spectrometry (see Figure 3.5 in Section 3.3). The major components of the mixture are the cyclic trimer and pentamer [eqn (11.16)].[59]

(11.16)

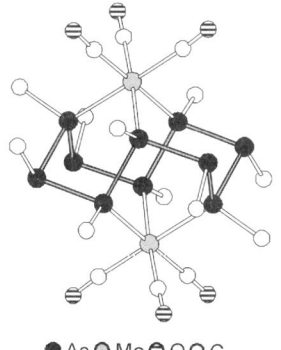

● As ○ Mo ⊖ O ○ C

Figure 11.12 Molecular structure of [cyclo-(MeAs)$_{10}$Mo$_2$(CO)$_6$].

Cyclic oligomers of the type $(RE)_n$ are also known for antimony ($E = Sb$, $n = 3$–6) and bismuth ($E = Bi$, $n = 3$–5).[60] A cyclic hexamer has not yet been characterised for the heaviest pnictogen. The first example of a cyclostibine, ($^tBuSb)_4$, was reported in 1965.[61] There is an enhanced tendency for the homocycles of the heavier pnictogens to engage in ring transformations because of their weaker E–E bond energies (P–P 48, As–As 35, Sb–Sb 30, Bi–Bi 25 kcal mol^{-1}) and this behaviour is prevalent for bismuth.[62] The most versatile substituent for the stabilisation of cyclostibines and cyclobismuthines is the bis(trimethylsilyl)methyl [$(Me_3Si)_2CH$] group. The orange tetramer $(RSb)_4$ [$R = (Me_3Si)_2CH$] is obtained in 60% yield by reductive coupling of $RSbCl_2$ with magnesium in THF.[63] A different synthetic approach, viz. the reaction of $RSbCl_2$ with lithium antimonide at low temperatures, produces the orange trimer (ca 60% yield). This three-membered ring adopts a cis, trans arrangement of substituents both in the solid state and in solution [eqn (11.17)],[64] cf. $(RE)_3$ ($E = P$, As). Although the cyclic tetramers are thermally stable, the UV irradiation of cyclo-$(RSb)_4$ [$R = (Me_3Si)_2CH$] produces the corresponding trimer cyclo-$(RSb)_3$ quantitatively.

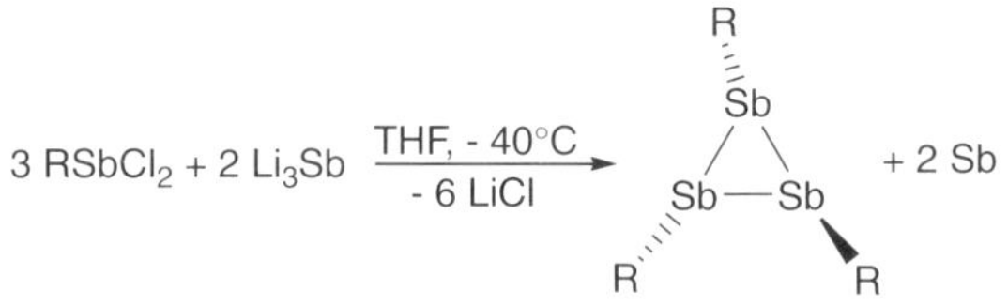

$$3\ RSbCl_2 + 2\ Li_3Sb \xrightarrow[-6\ LiCl]{THF,\ -40°C} \underset{R}{\underset{|}{Sb}}\diagdown\underset{R}{Sb}\diagup\underset{R}{Sb} + 2\ Sb \quad (11.17)$$

Cyclotetrastibines $(ArSb)_4$ are obtained in very high yields by metal-catalysed dehydrocoupling of arylstibines $ArSbH_2$, when Ar is a bulky aryl group [see eqn (2.11) in Section 2.3].[65a] The reaction is fast when the zirconium(IV) catalyst $Cp_2Zr(H)Cl$ is used, but sufficiently slow with the hafnium(IV) analogue $CpCp*Hf(H)Cl$ to allow characterisation of the intermediate Hf–Sb complex (Scheme 11.4). In addition to the formation of the tetramer cyclo-$(ArSb)_4$, thermal decomposition of $ArSbH_2$, generated in situ by hydride reduction of $ArSbCl_2$ [where Ar is the N, C, N chelating ligand C_6H_3-2,

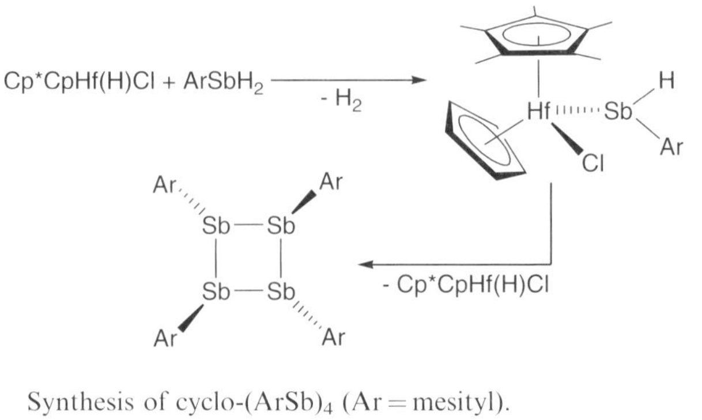

Scheme 11.4 Synthesis of cyclo-$(ArSb)_4$ (Ar = mesityl).

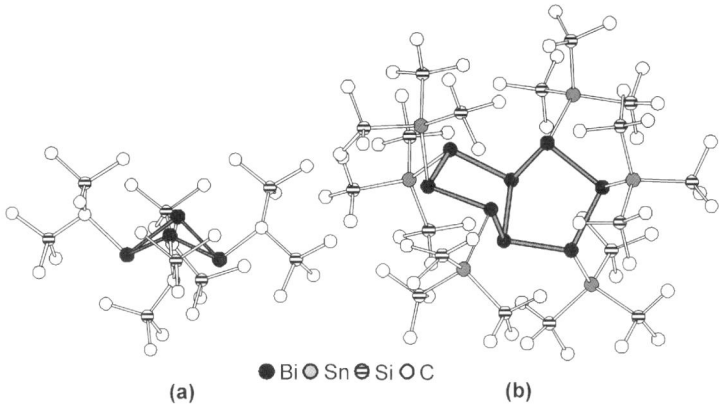

Figure 11.13 Molecular structures of (a) cyclo-[(Me$_3$Si)$_2$CHBi]$_4$ and (b) Bi$_8$[Sn(SiMe$_3$)$_3$]$_6$.

6-(CH$_2$NMe$_2$)$_2$], initially produces the tricyclic compound Ar$_3$Sb$_5$ with a trigonal bipyramidal arrangement of five Sb atoms.[65b]

The first cyclobismuthines (RBi)$_n$ [n = 3, 4; R = (Me$_3$Si)$_2$CH] were not characterised until 1998.[66] The red trimer (n = 3) and dark-green tetramer (n = 4) are obtained by reductive coupling of RBiCl$_2$ with Mg in THF at -35 °C; in solution these cyclic oligomers exist in a dynamic equilibrium that is shifted in favour of the tetramer at low temperatures. The tetramer cyclo-[(Me$_3$Si)$_2$CHBi]$_4$ incorporates a folded four-membered ring with an all-*trans* configuration of the exocyclic substituents; the Bi–Bi bond lengths are in the range 2.97–3.04 Å (Figure 11.13a). The structurally similar four-membered rings cyclo-[(Me$_3$E)$_3$SiBi]$_4$ (E = C, Si) are prepared by the reaction of BiBr$_3$ with the silyllithium reagents Li(THF)$_3$Si(EMe$_3$)$_3$,[67] whereas the stannyllithium reagent Li(THF)$_3$Sn(SiMe$_3$)$_3$ gives rise to Bi$_8$[Sn(SiMe$_3$)$_3$]$_6$ with a bicyclo [3.3.0]octane-like core of eight bismuth atoms (Figure 11.13b).[67a] The central Bi–Bi bond length in this novel bicyclic compound falls within the range of the other Bi–Bi bonds (2.97–3.02 Å).

Similarly to the antimony analogue, the ^{1}H NMR spectrum of the trimer cyclo-[(Me$_3$Si)$_2$CHBi]$_3$ reveals the presence of only the *cis,trans*-isomer in solution.[66] The major product of the thermal decomposition of the dihydride Me$_3$SiCH$_2$BiH$_2$ is the red trimer cyclo-(Me$_3$SiCH$_2$Bi)$_3$; a small amount of the pentamer cyclo-(Me$_3$SiCH$_2$Bi)$_5$ is also formed in this process.[68] The cyclic oligomers (Me$_3$SiCH$_2$Bi)$_n$ (n = 3, 5) are also formed by reduction of Me$_3$SiCH$_2$-BiCl$_2$ with LiAlH$_4$.[69]

The lability of the E–E and E–C bonds in cyclostibines and cyclobismuthines is readily evident in reactions of these homocycles with metal carbonyls.[60] Under mild conditions, complexes in which the antimony ligand acts as a two-electron donor towards the metal centres may be isolated, *e.g.* cyclo-(^{t}BuSb)$_4$[W(CO)$_5$]$_n$ (n = 1, 2) (*cf*. metal complexes of cyclophosphines, see Section 7.1.1). In boiling solvents, however, rupture of the inorganic homocycle and/or the E–C bonds occurs to give metal complexes that incorporate monomeric RSb units or naked Sb_n ($n = 2, 3$) fragments [eqn (11.18)].[70] The reaction of cyclo-($^tBuSb)_4$ with potassium in hot THF also results in Sb–C bond rupture, in addition to Sb–Sb cleavage, to give the linear anion [$^tBu_2SbSbSb^tBu_2$]$^-$.[71]

$$(^tBuSb)_4 \xrightarrow{[CpMo(CO)_3]_2} \underset{Sb}{Sb\!-\!\!\!|\!\!-\!MoCp(CO)_2} + \underset{Sb}{Sb\!-\!\!\!|\!\!-\!Sb} \overset{MoCp(CO)_2}{} \overset{MoCp(CO)_2}{}$$

(11.18)

In contrast to the simple adduct formation observed in the reaction of cyclostibines with [W(CO)$_5$(THF)], the treatment of cyclobismuthines (RBi)$_n$ ($n = 3, 5$; R = Me$_3$SiCH$_2$) with this labile reagent produces a binuclear complex of the dibismuthene RBi=BiR [eqn (11.19)].[68]

$$2/n \; cyclo\text{-}(RBi)_n + 2 \, [W(CO)_5(THF)] \xrightarrow[0°C]{THF} \underset{R}{\overset{(OC)_5W}{Bi}}\!\!=\!\!\underset{R}{\overset{W(CO)_5}{Bi}}$$

(11.19)

11.3.3 Ladder Polymers (RE)$_n$ (E = As, Sb)

High molecular weight polymers (MeAs)$_n$ are obtained as purple–black crystals from the reaction of methylarsine MeAsH$_2$ and methyldihaloarsines MeAsX$_2$ (X = Cl, Br, I) in either the presence or absence of pyridine.[72a] The polymer is produced more rapidly (1 day) as well-formed crystals from a 10% solution of MeAsH$_2$ in CCl$_4$ at room temperature *via* the intermediate MeAs(H)Cl. The treatment of cyclo-(MeAs)$_5$ with a trace amount of MeAsCl$_2$ in benzene also produces (MeAs)$_n$. This polymer has an unusual ladder structure in which the rungs of the ladder have a normal As–As single bond distance (*ca.* 2.4 Å) while the separation between the rungs is ca 2.9 Å (Figure 11.14).[72b] The distance

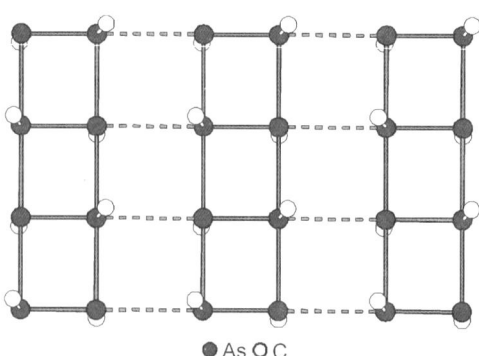

Figure 11.14 The ladder structure of (MeAs)$_n$.

between the neighbouring ladders is *ca* 3.4 Å (*cf.* sum of the van der Waals radii 4.0 Å). The polymer is a semiconductor with an energy gap of 1.2 eV.[72a] Antimony forms similar pale-green macromolecular species $(RSb)_n$ (R = Me, Et, ^{n}Bu) with a metallic sheen from the reaction of the stibines $RSbH_2$ with HCl or organosilicon chlorides.[72c] These stibines react with iodine-containing reagents to produce purple–black species of composition $(RSbI_{0.4})_n$.

11.4 PHOSPHORUS–NITROGEN RINGS AND POLYMERS

The combination of phosphorus and nitrogen generates an extraordinarily rich variety of ring systems. The best known of these inorganic heterocycles are (a) saturated systems involving P(III)–nitrogen linkages (cyclophosphazanes) and (b) unsaturated rings with P(V)–nitrogen bonds (cyclophosphazenes). The latter have attracted more attention in view of their role as precursors to phosphazene polymers and hybrid (organic–inorganic) materials (Section 11.4.3). However, cyclophosphazanes have attracted interest recently as ligands for metal complexes to be used in homogenous catalysis and as P_2N_2 building blocks in the construction of macrocycles that behave as anion receptors. The phosphorus(III) centres in cyclophosph(III)azanes are readily oxidized. *e.g.* by chalcogens, but the phosphorus–nitrogen rings in the resulting cyclophosph(V)azanes remain unsaturated.

11.4.1 Cyclophosphazanes

Cyclophosphazanes are obtained as four-membered ring systems, *i.e.* cyclodiphosphazanes, by the cyclocondensation reaction of PCl_3 with a primary amine, *e.g.* $^tBuNH_2$ or $PhNH_2$ [eqn (11.20)].[73] The exocyclic Cl substituents in the dichloro derivative $ClP(\mu-N^tBu)_2PCl$ are readily replaced by amido groups, *e.g.* by reaction with $LiNH^tBu$; however, the bisamido derivative $^tBu(H)NP(\mu-N^tBu)_2PN(H)^tBu$ may also be prepared directly from PCl_3 by using a large excess of the primary amine [eqn (11.21)].

$$2\, PCl_3 + 3\, H_2N^tBu + 3\, NEt_3 \xrightarrow[- [H_3N^tBu]Cl]{-3\,[HNEt_3]Cl} ClP(\mu-N^tBu)_2PCl \quad (11.20)$$

$$2\, PCl_3 + 10\, H_2N^tBu \xrightarrow{-6\,[H_3N^tBu]Cl} {}^tBu(H)NP(\mu-N^tBu)_2PN(H)^tBu \quad (11.21)$$

In the case of arylamines, however, the steric bulk of the aryl group influences the outcome of the dehydrodechlorination process in reactions with PCl_3. For example, the acyclic aminodiphosphine $DippN(PCl_2)_2$ is the major product of the reaction of $DippNH_2$ (Dipp = 2,6-diisopropylphenyl) and PCl_3; only a low yield of the four-membered ring $ClP(\mu-NDipp)_2PCl$ is obtained even with an excess of amine.[74a] With the more bulky arylamine Mes^*NH_2 (Mes^* = 2,4,6-tri-*tert*-butylphenyl) the reaction with PCl_3 produces the monomeric iminophosphine.[74b] Stabilisation of the analogous monomer DippN=PCl is achieved by adduct formation with an *N*-heterocyclic carbene, which causes dissociation of the cyclic dimer $ClP(\mu-NDipp)_2PCl$ [eqn (11.22)]; this process is reversed upon addition of a strong Lewis acid.[75]

(11.22)

The cyclic structure of $ClP(\mu-N^tBu)_2PCl$ was not established until 1971 (Figure 11.15a);[76] the bisamido derivative $^tBu(H)NP(\mu-N^tBu)_2PN(H)^tBu$ has also been structurally characterised (Figure 11.15b).[77] The P_2N_2 rings in both these cyclodiphosphazanes adopt an almost planar conformation and the exocyclic substituents on phosphorus are in the *cis* configuration. The exocyclic P–N bond lengths in $^tBu(H)NP(\mu-N^tBu)_2PN(H)^tBu$ are *ca* 0.06 Å shorter than the endocyclic P–N bonds, which are close to the single-bond value. Although only the *cis* isomers of cyclodiphosphazanes are observed in the solid state, ^{31}P NMR studies reveal that the *cis* and *trans* isomers co-exist in solution in approximately equal amounts indicating that their energy differences are small.[73a] The *trans* isomer is often the initial (kinetic) product in the synthesis of cyclodiphosphazanes, but it transforms into the thermodynamically more stable *cis* isomer. The mechanism of this interconversion is still a matter of debate, but pyramidal inversion at the phosphorus centre is the most likely alternative.[73b]

The chemistry of cyclophosphazanes is dominated by four-membered rings; six-membered P_3N_3 rings are far less common. The first examples of cyclotriphosphazanes $(MeNPX)_3$ (X = Cl, Br) were prepared by the cyclocondensation

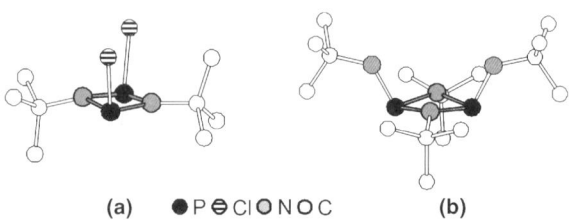

Figure 11.15 Molecular structures of $XP(\mu-N^tBu)_2PX$ (X = Cl, ^tBuNH).

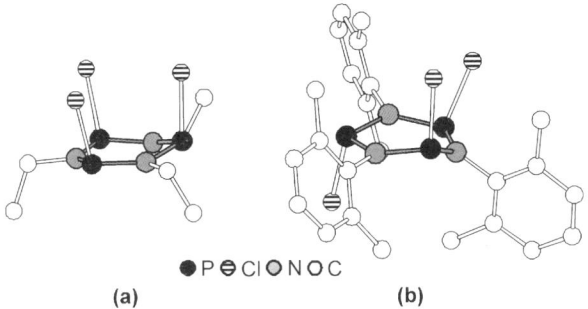

Figure 11.16 Molecular structures of (a) *cis*-(EtNPCl)$_3$ and (b) *trans*-(DmpNPCl)$_3$.

of PX$_3$ with MeN(SiMe$_3$)$_2$.[78] The ethyl derivative (EtNPCl)$_3$ is obtained by the reaction of PCl$_3$ with ethylamine hydrochloride in boiling 1,1,2,2-tetrachloroethane.[79] The ^{31}P NMR spectrum reveals the presence of both *cis* and *trans* isomers in solution; the former gives rise to a single resonance whereas the latter consists of two mutually coupled resonances with the expected intensity ratio of 2:1.[79] The solid-state structure of the *cis* isomer shows a pseudo-chair-shaped P$_3$N$_3$ ring with planar geometry at the nitrogen atoms (Figure 11.16a);[80] the trimer (DmpNPCl)$_3$ (Dmp = 2,6-dimethylphenyl) has been structurally characterised as the *trans* isomer (Figure 11.16b).[81] The treatment of (EtNPCl)$_3$ with phenols or trifluoroethanol produces the corresponding triaryloxy or trifluoroethoxy derivatives [EtNP(OR)]$_3$ as a mixture of *cis* and *trans* isomers.[82]

Cyclophosphazanes undergo interesting ring transformations under the influence of heat or Lewis acids. For example, distillation of the trimer (EtNPCl)$_3$ under reduced pressure produces the corresponding dimer (EtNPCl)$_2$ quantitatively; this process is reversed upon heating the dimer at 130 °C.[80]

The interconversion of dimeric and trimeric cyclophosphazanes has been investigated in the context of an alkene/cyclobutane analogy.[83a,b] Such studies may provide insight into the generation of phosphazane polymers (–RP=NR′–)$_n$ by a ring-opening process. The presence of the very bulky Mes* group on nitrogen favours the formation of monomers Mes*N=PX (X = Cl, OTf; OTf = CF$_3$SO$_3^-$) or the dimer (Mes*N=POTf)$_2$.[83b] By contrast, moderately bulky aryl substituents destabilise the dimer with respect to the trimer. Thus, treatment of the cyclic dimers (XPNR)$_2$ (X = Cl, Br; R=Dipp, Dmp) with gallium trihalides followed by 4-(dimethylamino)pyridine generates the corresponding trimers.[81] This oligomerisation process involves formation of the ionic compound [(DippN)$_3$P$_3$Cl$_2$][GaCl$_4$] as an intermediate (Figure 11.17a). However, the P$_2$N$_2$ ring is retained when extremely bulky aryl groups are installed on the nitrogen atoms of cyclodiphosphazanes. Thus the reaction of ClP(μ-NTer)$_2$PCl (Ter = 2, 6-Mes$_2$C$_6$H$_3$) with gallium trichloride produces the red monocation [P(μ-NTer)$_2$PCl]$^+$ in almost quantitative yield (Figure 11.17b).[83c] The mean P–N bond distance involving the two-coordinated phosphorus atoms in both the P$_2$N$_2$ and P$_3$N$_3$ monocations are substantially shorter (by *ca* 0.1 Å) than the other P–N bonds.[81,83c] Dicationic complexes of cyclodiphosphazanes

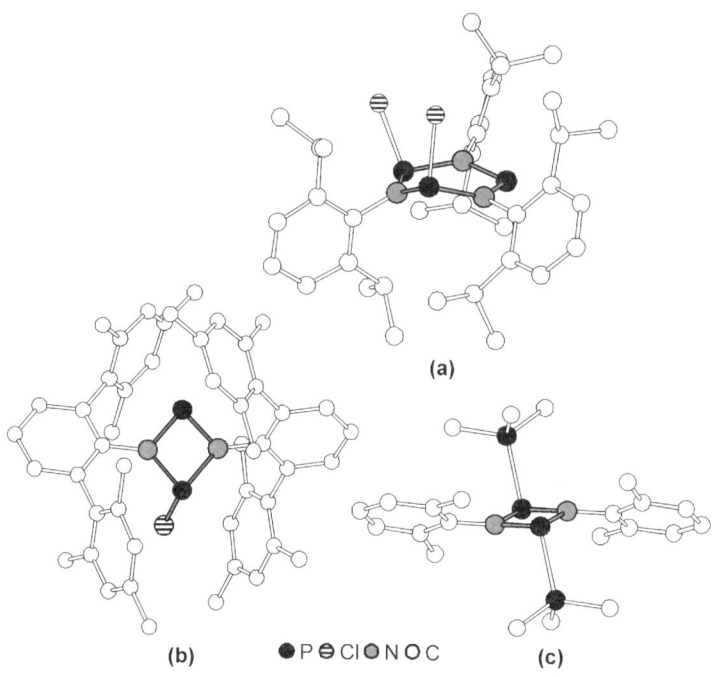

Figure 11.17 Molecular structures of the monocations (a) [(DippN)$_3$P$_3$Cl$_2$]$^+$ and (b) [P(μ-NTer)$_2$PCl]$^+$ and (c) the dication [DmpNP(PMe$_3$)]$_2^{2+}$.

[DmpNP(L)]$_2$[OTf]$_2$ (L = PMe$_3$, 4-dimethylaminopyridine) have also been structurally characterised (Figure 11.17c).[83d]

When the substituents on both the nitrogen and phosphorus atoms are small alkyl groups, eight-membered rings (cyclotetraphosphazanes) may be prepared *via* cyclocondensation. Thus, the tetramers cyclo-(MeNPR)$_4$ (R = Me, Et) are obtained in good yields from alkyldichlorophosphines and heptamethyldisilazane [eqn (11.23)].[84] The P$_4$N$_4$ ring in cyclo-(MeNPR)$_4$ adopts a crown conformation, *cf.* cyclo-S$_8$.

$$4 \text{ MePCl}_2 + 4 \text{ (Me}_3\text{Si)}_2\text{NMe} \xrightarrow{-8 \text{ Me}_3\text{SiCl}} \text{cyclo-(MeNPMe)}_4 \qquad (11.23)$$

The presence of two phosphorus(III) centres in cyclodiphosphazanes offers the possibility for the coordination of one or two metal centres and, in the latter case, the *cis* configuration of the exocyclic substituents has been exploited for the construction of metalated macrocycles.[85–87] Some representative examples of the coordination behaviour of these cyclic *P*-donor Lewis bases are illustrated in Figure 11.18.

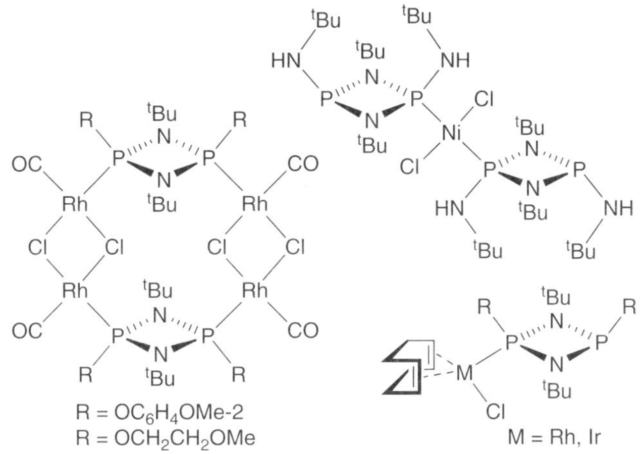

Figure 11.18 Coordination complexes of cyclodiphosph(III)azanes.

The *cis* arrangement of the exocyclic substituents in the bisamido cyclodiphosphazene $^tBu(H)NP(\mu-N^tBu)_2PN(H)^tBu$ has been exploited for the preparation of a wide range of main group and transition metal complexes.[73b] For example, group 4 complexes are prepared by either (a) salt metathesis of the dilithio reagent $^tBu(Li)NP(\mu-N^tBu)_2PN(Li)^tBu$, with MCl_4 (M = Zr, Hf), or (b) transamination reactions of $^tBu(H)NP(\mu-N^tBu)_2PN(H)^tBu$ with $M(NMe_2)_4$ (M = Ti, Zr, Hf).[88,89] A characteristic structural feature of these complexes is the involvement of one nitrogen of the P_2N_2 ring in a weak, intramolecular interaction with the metal centre, as exemplified by the hafnium complex **11.10**.[88b] The catalytic activity of these group 4 complexes towards ethene polymerisation has been investigated.[89,90] They have a tendency to be deactivated by Lewis acid coordination to the phosphorus lone pair of the P_2N_2 ring unless these donor centres are sterically protected by bulky groups on nitrogen.[90] The structures of nickel(II) complexes, *e.g.* **11.11**, in which three-membered rings are formed through coordination to P(III) and nitrogen atoms in preference to a strain-free N,N'-chelated metallacycle, are indicative of this behaviour.[91]

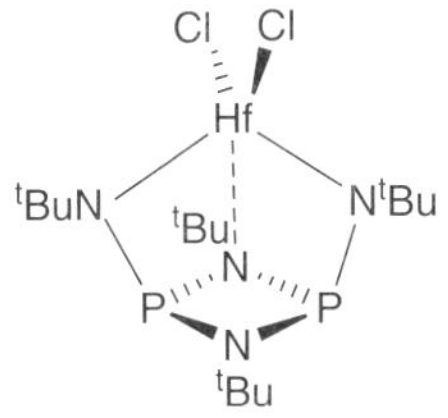

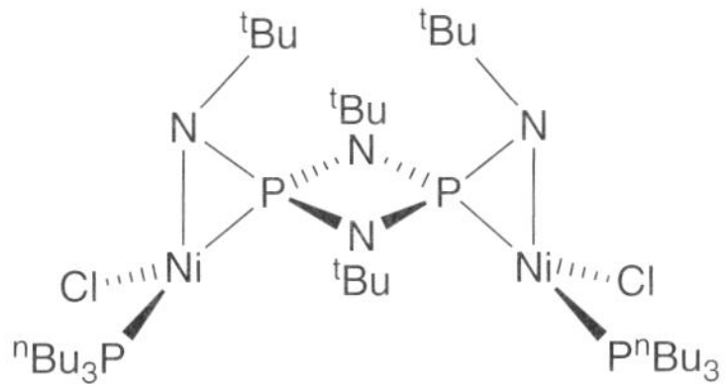

There is an emerging chemistry of macrocyclic systems in which P_2N_2 rings are connected by difunctional bridging groups such as NH, NR and 1,2- or 1,

4-$X_2C_6H_4$ (X = NH, O).[92] A simple example is the formation of the compounds $P_4(N^tBu)_6$ (**11.12**)[93] in which two N^tBu groups bridge two P_2N_2 rings. The formation of the tetrameric macrocycle [{P(μ-N^tBu)}$_2$(μ-NH)]$_4$ and the corresponding pentamer is discussed in Section 6.4 [eqn (6.4) and Figure 6.7]. These macrocycles act as anion acceptors.[94] The use of 1,5-diaminonaphthalene or 1,4-phenylene as spacer groups produces macrocycles **11.13** and **11.14**, which have been compared to calixarenes.[95] The presence of *endo*-NH groups and the flexibility of these macrocycles indicated by NMR studies suggest that they may be exploited for the encapsulation of anions.

11.4.2 Cyclophosphazenes

Cyclophosphazenes $(R_2PN)_n$ (R=Cl) are unsaturated inorganic ring systems that have been known since the early 19th century;[96a] a cyclic structure was proposed for the trimer $(Cl_2PN)_3$ in 1895.[96b] These heterocycles are sometimes referred to as phosphonitriles, because of the resemblance of the monomeric unit $R_2P\equiv N$ to organic nitriles $RC\equiv N$. The standard synthesis of cyclophosphazenes produces the dichloro derivatives $(Cl_2PN)_n$, which can be converted to a large variety of other derivatives by nucleophilic substitution reactions. Early interest in these classic inorganic heterocycles focused on fundamental aspects, especially (a) molecular and electronic structures as a function of ring size and (b) the factors governing substitution patterns in reactions with nucleophiles. A very important property of cyclophosphazenes is their susceptibility to ring-opening polymerisation. The potential applications of the resulting polymers has been a major incentive for recent work and these inorganic macromolecules have been accorded detailed treatment in three recent books.[97] Significant developments have also occurred in the use of the unsaturated P_3N_3 ring system as a scaffold on which to create multimetallic or supramolecular systems through reactions at the peripheral ligands.[98]

Various aspects of the chemistry of cyclophosphazenes have been discussed in earlier chapters, *viz.* synthesis (see Scheme 2.1), electronic structure (see Section 4.1.2.2 and Figure 4.2), macrocyclic ring systems (see Section 6.2.2 and Figure 6.4) and ligand behaviour (see **7.20** and **7.21** in Section 7.1.3). Phosphazene polymers are considered in Section 11.4.3. In this section, emphasis will be placed on new developments in cyclophosphazene chemistry that are not covered elsewhere in this book.

Although the majority of cyclophosphazenes are six- or eight-membered ring systems, the cyclodiphosphazene $[(^{i}Pr_2N)_2PN]_2$ is obtained in 42% yield by irradiation of the azide $(^{i}Pr_2N)_2PN_3$ in toluene at $-40\,^{\circ}C$ [eqn (11.24)].[99a] The four-membered ring in this cyclodiphosphazene is planar and the endocyclic and exocyclic P–N bond lengths are all close to 1.65 Å, indicating a strongly delocalised structure. Two other cyclodiphosphazenes, $[Ter(N_3)PN]_2$ $(Ter = 2,6-Mes_2C_6H_3)$[99b] and $[Cp^*(Mes^*NH)PN]_2$,[99c] have been structurally characterised.

$$(11.24)$$

The unsaturated four-membered P_2N_2 ring is also stabilised by the formation of diadducts with Lewis acids, which are surprisingly formed in the reaction of Ph_2PCl with bis(trimethylsilyl)sulfur diimide with $GaCl_3$ or $AlCl_3$ [eqn (11.25)].[100]

$$(11.25)$$

The standard synthesis of hexachlorocyclotriphosphazene $(Cl_2PN)_3$ involves the reaction of phosphorus pentachloride with ammonium chloride in boiling 1,1,2,2-tetrachloroethane.[101a] Alternatively, the reaction of PCl_5 with $N(SiMe_3)_3$ in hot methylene dichloride produces the cyclic trimer in 70% yield together with small amounts of larger cyclic oligomers $(NPCl_2)_n$ [$n = 4$ (4%), $n = 5$ (3%), $n = 6$ (5%)].[101b]

The complete replacement of the chlorine substituents in cyclotriphosphazenes $(Cl_2PN)_3$ by nucleophilic reagents, *e.g.* sodium alkoxides or aryloxides, primary or secondary amines, leads to a variety of other cyclotriphosphazenes, as illustrated in Scheme 11.5. Of particular interest is the synthesis of the hexaazido derivative

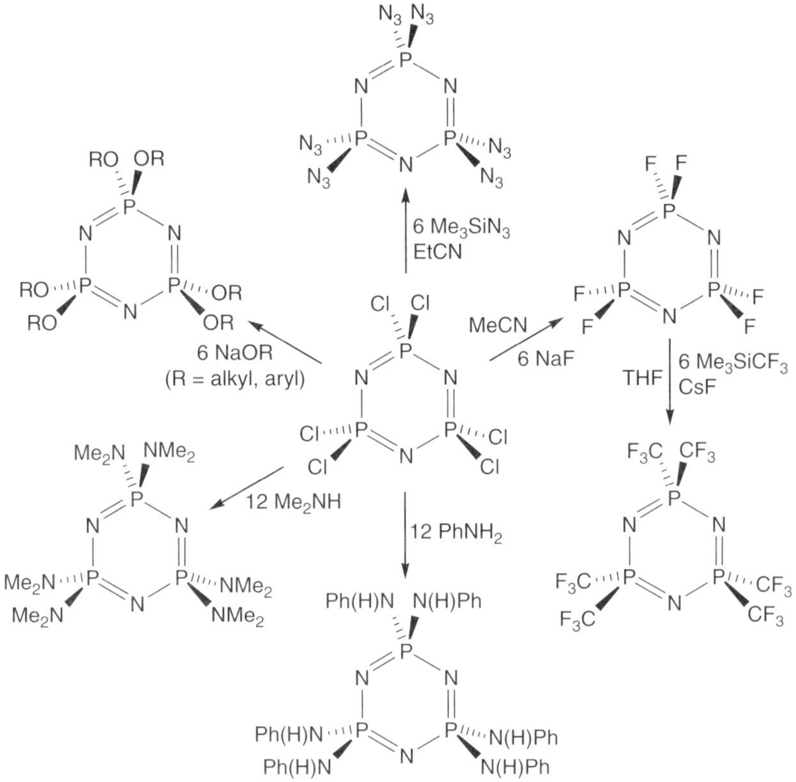

Scheme 11.5 Synthesis of cyclotriphosphazenes from $(NPX_2)_3$ (X = Cl, F).

from $(Cl_2PN)_3$ and trimethylsilyl azide.[102] Although this highly nitrogen-rich compound (with a molecular formula of P_3N_{21}!) is extremely shock sensitive, it is reported to have a decomposition temperature of 220 °C. The completely trifluoromethylated derivative $[NP(CF_3)_2]_3$ is obtained in 90% yield by treatment of $(F_2PN)_3$ with Me_3SiCF_3 in the presence of a catalytic amount of caesium fluoride in THF.[103]

The reactions of $(Cl_2PN)_3$ with Grignard reagents are not a good source of permethylated or perphenylated derivatives because side reactions occur, e.g. reductive coupling to give the P–P-bonded bicyclophosphazene $(Cl_5P_3N_3)_2$. As an alternative, condensation reactions involving the elimination of a trimethylsilyl halide from an acyclic precursor have been developed for the synthesis of fully substituted alkyl- or arylcyclophosphazenes [eqn (11.26)].[104] Side reactions are less problematic in reactions of cyclophosphazenes with organolithium reagents, especially if the fluorinated derivatives are used. For example, the reaction of $(F_2PN)_4$ with methyllithium in diethyl ether at −20 °C gives the octamethyl derivative $(Me_2PN)_4$ in 63% yield.[105] However, methylation of the trimer $(F_2PN)_3$ beyond gem-$N_3P_3F_4Me_2$ is accompanied by cleavage of the P_3N_3 ring and alkyl halide elimination.

$$Me_3SiN=P(R)(R)-Br \xrightarrow[-Me_3SiBr]{\Delta} \text{(cyclic trimer)} + \text{(cyclic tetramer)}$$

(11.26)

A large number of cyclophosphazenes have been structurally characterised.[97b] The halogenated six-membered rings $(X_2PN)_3$ (X = Cl, F) are planar, whereas other derivatives are somewhat distorted towards chair (X = Me, Ph) or boat (X = NMe$_2$) structures. The structure of trifluoromethyl derivative $[(CF_3)_2PN]_3$ has been determined in both the gas phase[106] and the solid state.[103] The metrical parameters of the planar hexagonal P_3N_3 ring are identical, within experimental error, in the two phases. In general, the P–N bond lengths in cyclotriphosphazenes $(X_2PN)_3$ fall in the range 1.57–1.60 Å, cf. 1.76 Å for a P–N single bond; the shortest bonds are observed for the most electronegative substituents (X = F). The endocyclic bond angles at both phosphorus and nitrogen are close to 120°. Eight-membered rings $(X_2PN)_4$ adopt non-planar, boat or saddle conformations. The fluorinated derivative $(F_2PN)_4$ is an interesting example; it undergoes a phase transition at −74 °C during which the saddle conformation of the P_4N_4 ring is converted to a pseudo-planar form (Figure 11.19a).[107] The addition of a fluoride ion to $(F_2PN)_4$, using the reagent $[(Me_2N)_3S][Me_3SiF_2]$, produces the $[P_4N_4F_9]^-$ anion in which the P_4N_4 ring has a boat conformation (Figure 11.19b).[108a] In contrast to the thermal stability of $[P_4N_4F_9]^-$, the six-membered ring $[P_3N_3F_7]^-$ decomposes above −40 °C to give $[P_3N_3F_5(NPF_2)_2NPF_5]^{2-}$ with a ring-opened side chain.[108b]

The tetrahedral geometry at the phosphorus centres in cyclophosphazenes gives rise to the possible formation of a variety of isomers in nucleophilic substitution reactions. For example, the dimethylated derivative $N_3P_3F_4Me_2$ can exist as the geminal isomer **11.15**, and also the *cis* or *trans* non-geminal isomers, **11.16** and **11.17**. In a similar manner, there are five possible isomers of the eight-membered ring $N_4P_4F_6Me_2$, since the two methyl groups in the non-geminal isomers may occupy vicinal (nearest neighbour) or antipodal positions. All five isomers are formed in the reaction of $(F_2PN)_4$ with methyllithum.[105]

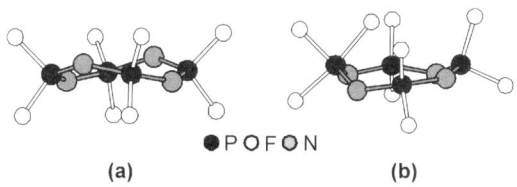

Figure 11.19 Molecular structures of (a) $(F_2PN)_4$ and (b) $[P_4N_4F_9]^-$ $[(Me_2N)_3S^+$ cation].

11.15 gem / 11.16 cis / 11.17 trans

In an interesting adaptation of the cyclocondensation synthesis depicted in eqn (11.26), both the *cis* and *trans* isomers of non-geminally substituted [(Ph)(Me)PN]$_3$ have been prepared and separated by using column chromatography on silica gel [eqn (11.27)].[109]

$$\text{Me}-\underset{\underset{\text{OCH}_2\text{CF}_3}{|}}{\overset{\overset{\text{NSiMe}_3}{\|}}{P}}-\text{Ph} \xrightarrow[-\text{Me}_3\text{SiOCH}_2\text{CF}_3]{\text{CF}_3\text{CH}_2\text{OH}} \text{cis} + \text{cis-trans}$$

(11.27)

The thermolysis of cyclotriphosphazenes (NPX$_2$)$_3$ produces phosphazene polymers when ionisable substituents are attached to phosphorus (*e.g.* X = halogen). However, alkyl or aryl substituents favour ring transformation processes, even when some halogen substituents are present along with the alkyl groups.[110] For example, the trimer (NPMe$_2$)$_3$ and the tetramer (NPMe$_2$)$_4$ participate in a ring–ring equilibration above 200 °C.[110a] This transformation has been utilised to generate the unsymmetrically substituted tetramer [NPMe(Ph)]$_4$ by prolonged heating of *trans*-[NPMe(Ph)]$_3$ at 250 °C. All four non-geminal isomers were isolated by column chromatography and structurally characterised by X-ray crystallography (Figure 11.20).[111]

The interaction of cyclotriphosphazenes, especially (NPCl$_2$)$_3$, with Lewis acids is of particular interest in connection with the mechanism of the ring-opening polymerisation process, which is postulated to occur *via* the intermediate formation of the cyclic cation [N$_3$P$_3$Cl$_5$]$^+$ (see Section 11.4.3). However, there is no evidence for the formation of this cation even in reactions with very potent halide-abstracting reagents. Instead, the six-membered ring in (NPCl$_2$)$_3$ behaves as a weak Lewis base in the 1:1 *N*-bonded monoadducts with the strong Lewis acids MCl$_3$ (M = Ga, Al) (Scheme 11.6).[112a] A similar coordination mode is observed in the 1:2 silver complex [Ag(N$_3$P$_3$Cl$_6$)$_2$]$^+$ (see Section 7.1.2 and Figure 7.5).[112b] Cationic *N*-bonded adducts are also formed in the interaction of (NPCl$_2$)$_3$ with protic, alkyl and silylium reagents using a very weakly coordinating carborane anion (Scheme 11.6).[113] The P$_3$N$_3$ ring in all these *N*-bonded adducts is distorted towards a chair conformation and the P–N bonds involving the three-coordinate nitrogen atom are elongated.

The remarkable, highly charged hexacation [P$_3$N$_3$]$^{6+}$, which is formed in the reaction of (NPCl$_2$)$_3$ with 4-(dimethylamino)pyridine (DMAP) in superheated

Figure 11.20 The four non-geminal isomers of [NPMe(Ph)]$_4$.

chloroform [eqn (11.28)], is stabilised by coordination of two strongly electron-donating DMAP ligands to each phosphorus centre.[114] Calculations of the charge distribution in [P$_3$N$_3$(DMAP)$_6$]$^{6+}$ indicate that the charge is shared equally between the P$_3$N$_3$ ring and the DMAP ligands. The P$_3$N$_3$ ring in the hexacation is a planar hexagon with average P–N bond lengths of 1.56 Å.

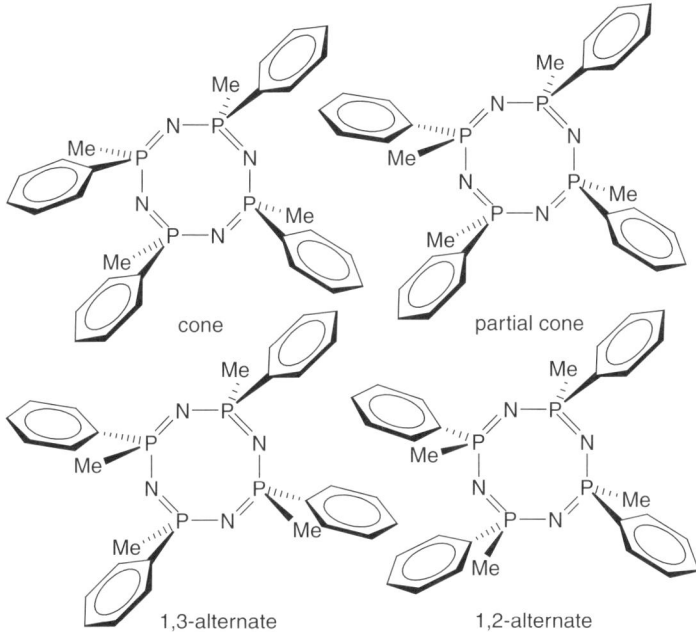

(11.28)

The Lewis basicity of cyclophosphazenes is enhanced by replacement of the chlorine substituents by electron-donating groups, *e.g.* Me or Me$_2$N, as illustrated by the formation of complexes of (NPMe$_2$)$_n$ ($n = 6, 8$) with transition metal

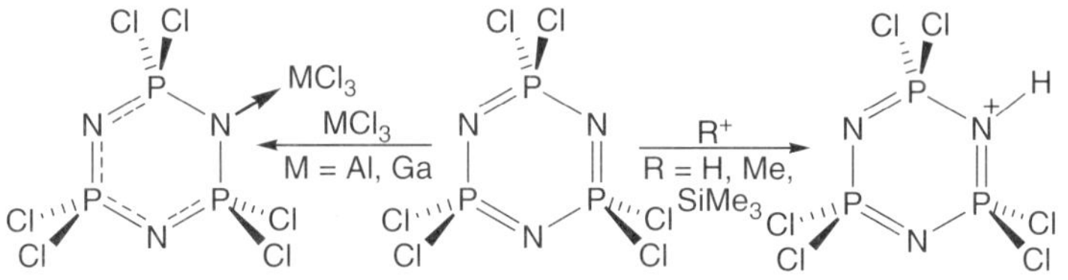

Scheme 11.6 Formation of N-bonded adducts of $(NPCl_2)_3$.

halides in which the inorganic ring system behaves as a multidentate ligand with endocyclic nitrogen donor centres (see **7.20** and **7.21** in Section 7.1.3). Cyclophosphazenes with exocyclic amino or other nitrogen-containing groups, *e.g.* pyrazolyl, hydrazino, have a very extensive coordination chemistry in which the inorganic ring provides a stable platform for multi-site ligands.[98] Primary amino-substituted cyclophosphazenes afford a particularly interesting illustration of this behaviour. For example, the trimers $N_3P_3(NHR)_6$ (R = Cy, Ph) can be completely deprotonated by *n*-butyllithium to give the hexaanions $[N_3P_3(NR)_6]^{6-}$ (**11.18**).[115] Both the deprotonation process and the reprotonation of $[N_3P_3(NR)_6]^{6-}$ to give the corresponding trianions $[N_3P_3(NHR)_3(NR)_3]^{3-}$ take place stereospecifically with the formation of the non-geminal *cis* isomer.[116] In a similar manner, the asymmetrically substituted cyclotetraphosphazene *trans*-$N_4P_4Ph_4(NHCy)_4$ can be deprotonated to give the tetraanion *trans*-$[N_4P_4Ph_4(NCy)_4]^{4-}$ (**11.19**).[117] The organic substituents in $Li_6[N_3P_3(NR)_6]$ confer solubility in non-polar aprotic organic solvents. The similarity of the P–N bond lengths in the hexaanions **11.18** indicates that the high negative charge is stabilised through delocalisation over the endocyclic and exocyclic nitrogen atoms, and also by ion-pairing with the Li^+ counterions.

The binary anion $[P_3N_9]^{12-}$, which is found in the solid state in the ternary compound $LiPN_3$, has a chair-shaped P_3N_3 ring (**11.20**),[118] *cf.* the isoelectronic cyclotrisilicate $[Si_3O_9]^{6-}$ (see **10.43** in Section 10.5).

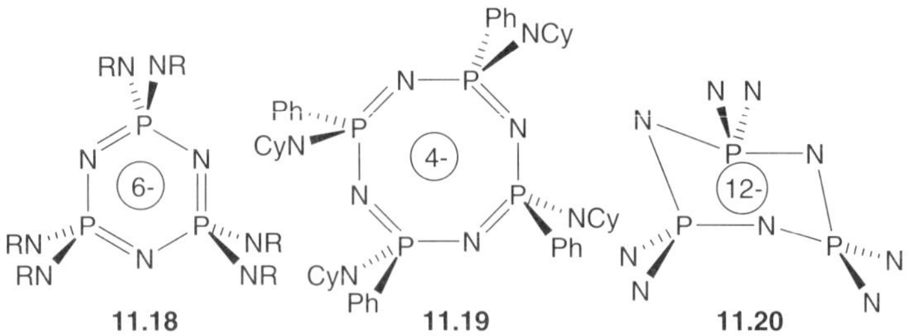

Inorganic heterocycles in which other p-block elements are inserted into a phosphazene ring are accessible *via* a variety of synthetic approaches. The primary interest in these hybrid ring systems is as potential precursors of linear polymers *via* ring-opening polymerisation (Section 11.4.3). Consequently, recent work has focused on halogenated derivatives. For example, the perhalogenated boratophosphazene N(PCl$_2$NMe)$_2$BCl$_2$ is prepared by reaction of [Cl$_3$P=N=PCl$_3$][BCl$_4$] with two equivalents of [MeNH$_3$]Cl (Scheme 11.7).[119] The boron atom in N(PCl$_2$NMe)$_2$BCl$_2$ is *ca* 0.39 Å out of the best plane of the other five ring atoms and one of the B–Cl bonds is significantly longer than the other; the hybrid borazine–phosphazene cation [N(PCl$_2$NMe)$_2$BCl]$^+$ is readily obtained by treatment of the boratophosphazene with chloride ion acceptors, MCl$_3$ (M = Al, Ga) (Scheme 11.7). The BN$_3$P$_2$ ring in this cation is almost planar and the B–N bond lengths of 1.43 Å are typical of those found in borazines, *cf.* 1.53 Å in the boratophosphazene.

The boratophosphazene N(PCl$_2$NMe)$_2$BCl$_2$ undergoes an unusual skeletal substitution reaction with the silver salts Ag[MF$_6$] (M = As, Sb) to give heterophosphazenes e.g. **11.21** in which the BCl$_2$ unit is replaced by an MF$_4$ group to give a P$_2$N$_3$M(V) ring with a boat conformation (Scheme 11.8).[120] In contrast, the reaction of N(PCl$_2$NMe)$_2$BCl$_2$ with Ag[BF$_4$] results in metathetical

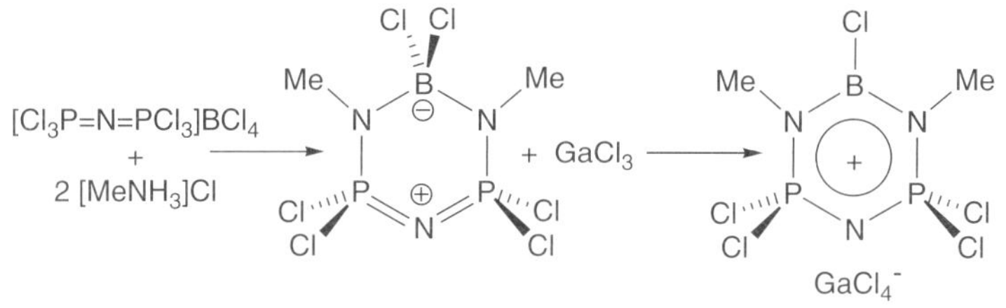

Scheme 11.7 Synthesis of N(PCl$_2$NMe)$_2$BCl$_2$ and [N(PCl$_2$NMe)$_2$BCl]$^+$.

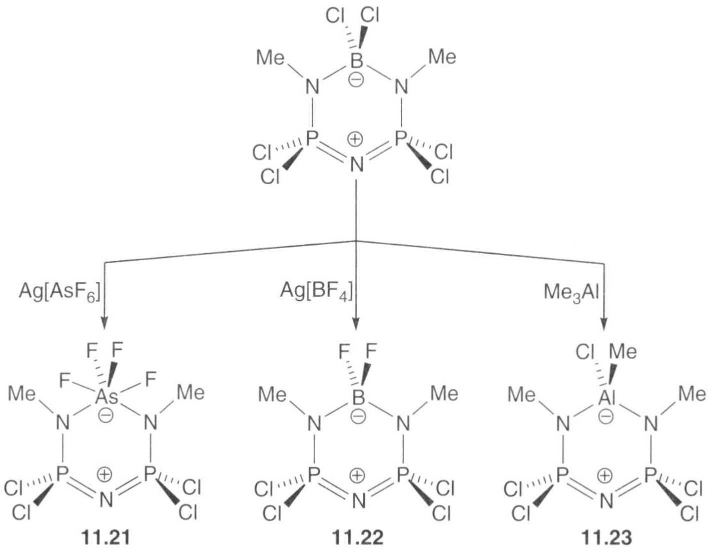

Scheme 11.8 Skeletal substitution reactions of N(PCl$_2$NMe)$_2$BCl$_2$.

replacement of the chlorides at the BCl_2 centre by fluoride to give **11.22**. Skeletal substitution of BCl_2 by an Al(Me)Cl group occurs in the reaction of $N(PCl_2NMe)_2BCl_2$ with $AlMe_3$ to produce **11.23**.[121] In this case, the skeletal substitution can be reversed by treatment of the product with $Ag[BF_4]$.

Cyclic phosphazenes containing three- or four-coordinate sulfur are prepared by several routes. The reaction of Ph_2PCl with S_4N_4 in a 3:1 molar ratio in boiling acetonitrile produces the hybrid six-membered ring $(Ph_2PN)_2(NSCl)$ (**11.24**) in good yield.[122] The sulfur atom in this six-membered ring lies ca 0.3 Å out of the P_2N_3 plane and the long S–Cl bond indicates partial ionic character. The introduction of electron-donating substituents on the sulfur centre renders the P_2N_3S ring susceptible to ring-opening processes that result in ring expansion. Thus, the treatment of $(Ph_2PN)_2(NSCl)$ with secondary amines produces the corresponding dialkylamino derivatives $(Ph_2PN)_2(NSNR_2)$, which undergo dimerisation to give the 12-membered rings $(Ph_2PN)_4(NSNR_2)_2$ (**11.25**) in acetonitrile solution or upon gentle heating.[123] Reduction of $(Ph_2PN)_2(NSCl)$ with $SbPh_3$ also results in ring expansion to give the 12-membered ring $(Ph_2PN)_4(NS)_2$ (**11.26**), which exhibits a weak transannular S···S interaction (2.39 Å).[124] The latter dimerisation process occurs *via* the intermediate formation of the cyclic radical $[(Ph_2PN)_2(NS)]^{\bullet}$ identified by EPR spectroscopy (see Figure 3.11 in Section 3.6).[125]

11.24   **11.25**   **11.26**

The perhalogenated thiophosphazene $(Cl_2PN)_2(NSCl)$ is obtained as a moisture-sensitive colourless liquid by the [3+3] cyclocondensation of bis(trimethylsilyl)sulfur diimide with $[Cl_3P=N=PCl_3]Cl$ [eqn (11.29)].[126] The six-membered ring in the corresponding cation $[(Cl_2PN)_2(NS)]^+$ is planar.[127] The cyclothiophosphazene $(Cl_2PN)_2(NSCl)$ polymerises at a much lower temperature (ca 90 °C) than the cyclotriphosphazene $(NPCl_2)_3$ (Section 11.4.3).

$$[Cl_3P=N=PCl_3]Cl + Me_3SiN=S=NSiMe_3 \xrightarrow[-2\ Me_3SiCl]{CCl_4}$$

(11.29)

Hybrid phosphazene–sulfanuric ring systems of the type $(NPCl_2)_{3-n}[NS(O)Cl]_n$ ($n=1, 2$) have been known since the early 1960s.[128] The presence of

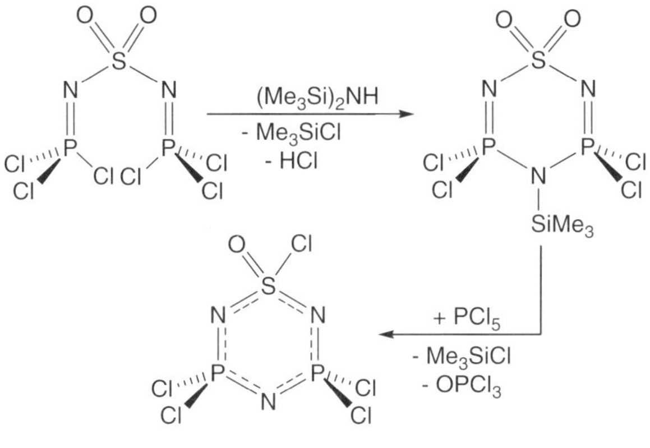

Group 15: Rings and Polymers

Scheme 11.9 Synthesis of (NPCl$_2$)$_2$[NS(O)Cl] *via* a [5 + 1] cycloaddition.

a four-coordinate sulfur(VI) centre renders these heterocycles more stable towards moisture than those with a three-coordinate sulfur(IV) and, consequently, the formation of polymers by ring-opening polymerisation has been studied in detail (Section 11.4.3). The best synthesis of the cyclic precursor (NPCl$_2$)$_2$[NS(O)Cl] involves the [5 + 1] cyclocondensation reaction of the bis(phosphazo)sulfone Cl$_3$P=N–SO$_2$–N=PCl$_3$ with hexamethyldisilazane (Scheme 11.9).[129] Interestingly, the thermal ring-opening polymerisation of (NPCl$_2$)$_2$[NS(O)Cl] produces small amounts of 12- and 24-membered macrocyclic oligomers (see **6.16** and **6.17** in Section 6.2.2), in addition to the polymer.[130]

11.4.3 Polyphosphazenes

Polyphosphazenes possess a phosphorus–nitrogen backbone. The first example, poly(dichlorophosphazene), (PCl$_2$=N)$_n$, was prepared in an insoluble, cross-linked form at the end of the 19th century by the thermal ROP of the cyclic trimer (Cl$_2$PN)$_3$.[131] However, this material with a network structure, referred to as 'inorganic rubber', remained only a chemical curiosity for over 60 years due its intractability and hydrolytic instability. Key developments concerning the phosphazene polymer system took place in the mid-1960s when it was shown that if the ROP of pure (Cl$_2$PN)$_3$ is carried out carefully, uncross-linked poly(dichlorophosphazene) (PCl$_2$=N)$_n$ is formed that is soluble in organic solvents. This permitted replacement of the halogen substituents by reaction with nucleophiles in solution to yield hydrolytically stable polyorganophosphazenes (Scheme 11.10).[97a,132]

The mechanism of the thermal ROP of (Cl$_2$PN)$_3$ has been proposed to involve a cationic mechanism (see Scheme 8.2 in Section 8.1.2.2). The unique reaction sequence involving ROP followed by nucleophilic substitution with oxygen- or nitrogen-based nucleophiles (generally alkoxides, aryloxides or primary amines) permits a diverse range of polyorganophosphazenes to be

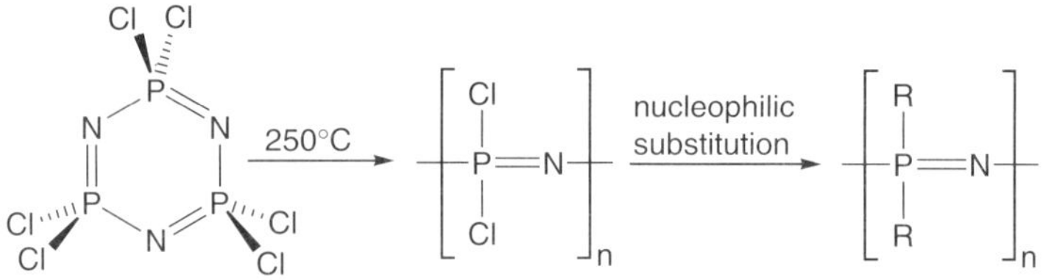

Scheme 11.10 Synthesis of polyphosphazenes *via* ROP and nucleophilic substitution.

prepared and allows the properties to be tuned and specific characteristics to be introduced.[97a]

The phosphazene backbone also possesses unusual features that lead to a range of potential applications for these easily processed materials (Figure 11.21).[97a,97c,133] For example, it is extremely flexible and polyalkoxyphosphazenes such as the *n*-butoxy derivative $[P(O^nBu)_2=N]_n$ possess glass transition temperatures T_g of below –100 °C.[97a] Furthermore, the P–N main chain is thermally and oxidatively stable, optically transparent from 220 nm to the near-infrared region and it imparts flame-retardant properties.

Some of the most useful polyphosphazenes are fluoroalkoxy derivatives and amorphous copolymers (**11.27**) that are practicable as flame-retardant, hydrocarbon solvent- and oil-resistant elastomers, which have found aerospace and automotive applications. Polymers such as the amorphous comb polymer poly[bis(methoxyethoxyethoxy)phosphazene] (**11.28**) weakly coordinate Li$^+$ ions and are of substantial interest as components of polymeric electrolytes in battery technology.[97a,97c,133,134] Polyphosphazenes are also of interest as biomedical materials and bioinert, bioactive, membrane-forming and bioerodable materials and hydrogels have been prepared.[97a,97c,133,134]

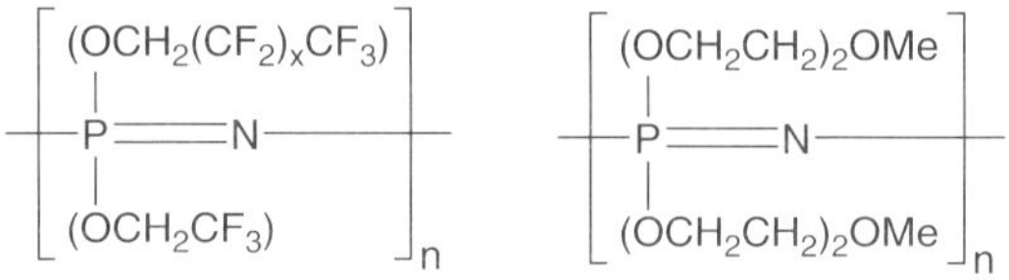

11.27   **11.28**

The reaction of poly(dichlorophosphazene) with organometallic reagents such as Grignard or organolithium reagents generally leads to chain cleavage in addition to substitution and does not provide a satisfactory route to polymers with only alkyl and aryl side groups bound by direct P–C bonds.[135] In the early 1980s, a solution to this problem was provided by the discovery of a condensation route to poly(alkyl/arylphosphazenes) from acyclic precursors of the type $(RO)R_2P=NSiMe_3$ [eqn (11.30)] [(cf. synthesis of cyclo-$(Me_2PN)_n$ ($n = 3, 4$) shown in eqn (11.26)].[136] The polymerisation is in fact a chain growth reaction that allows access to high molecular weight polyphosphazenes such as poly(dimethylphosphazene) $(PMe_2=N)_n$, poly(methylphenylphosphazene) $(PMePh=N)_n$ and

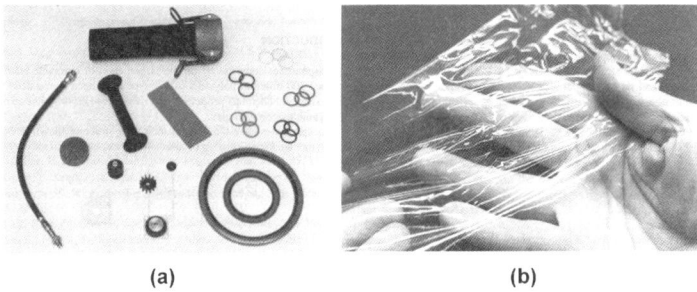

Figure 11.21 Polyphosphazenes: (a) processed into gaskets, O-rings and flexible hoses and (b) solution-cast thin films. Reproduced with permission from J. E. Mark, H. R. Allcock and R. West, *Inorganic Polymers*, 2nd edn., Oxford University Press, Toronto, 2005.[97c]

many others.[135,137]

$$R-\underset{\underset{OR}{|}}{\overset{R}{|}}P=N-SiMe_3 \xrightarrow[-Me_3SiOR]{190\ °C} \left[\begin{array}{c} R \\ | \\ P=N \\ | \\ R \end{array} \right]_n$$

(R = alkyl or aryl)

(11.30)

Although several polyphosphazenes have been commercialised, recent work has focused on the development of improved or alternative, cheaper and more convenient methods for making these fascinating materials. The thermal ROP route requires the synthesis and careful purification of the cyclic trimer $(Cl_2PN)_3$ and the use of elevated temperatures (*ca* 250 °C). Control of molecular weight is difficult and cross-linking can take place at high conversion, which can limit the yield. The process can be catalysed by Lewis acids such as BCl_3 to reduce the temperature required to *ca* 200 °C.[97a,134] Moreover, use of the silylcarborane reagent $[Et_3Si][CH_6B_{11}Br_6]$ allows ROP to proceed at ambient temperature.[138]

Condensation reactions to polyphosphazenes have been developed that provide an alternative, direct route to fluoroalkoxyphosphazene polymers and aryl derivatives [eqns (11.31) and (11.32)].[139,140] The development of condensation routes to poly(dichlorophosphazene) has also been reported; for example, a promising route that operates at 200 °C has been described [eqn (11.33)].[141]

$$RO-\underset{\underset{OR}{|}}{\overset{OR}{|}}P=N-SiMe_3 \xrightarrow[-Me_3SiOR]{[Bu_4N]F,\ 110°C} \left[\begin{array}{c} OR \\ | \\ P=N \\ | \\ OR \end{array} \right]_n$$

(11.31)

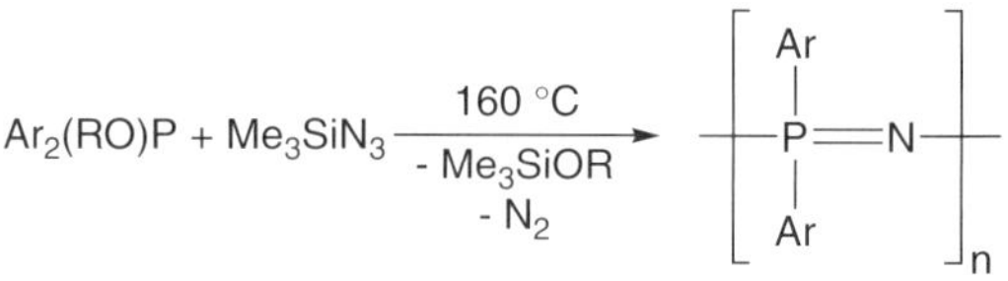

$$Ar_2(RO)P + Me_3SiN_3 \xrightarrow[\substack{-Me_3SiOR \\ -N_2}]{160\ °C} \left[\begin{array}{c} Ar \\ | \\ -P=N- \\ | \\ Ar \end{array} \right]_n \quad (11.32)$$

$$Cl_3P=N-P(O)Cl_2 \xrightarrow[-P(O)Cl_3]{200\ °C} \left[\begin{array}{c} Cl \\ | \\ -P=N- \\ | \\ Cl \end{array} \right]_n \quad (11.33)$$

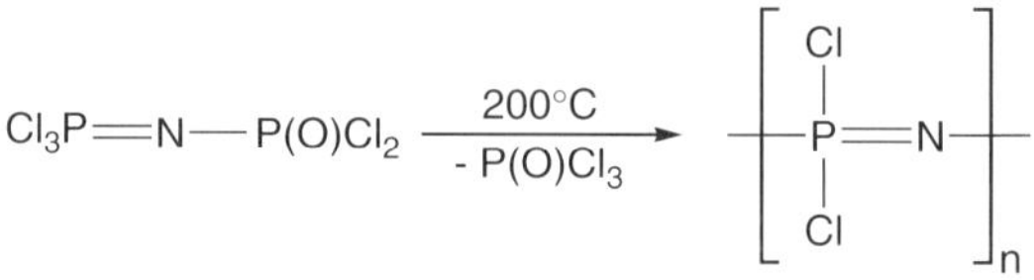

A synthesis of polydichlorophosphazene that operates at room temperature and allows molecular weight control has been developed starting from the trichlorophosphoranimine $Cl_3P=NSiMe_3$ [eqn (11.34)].[142]

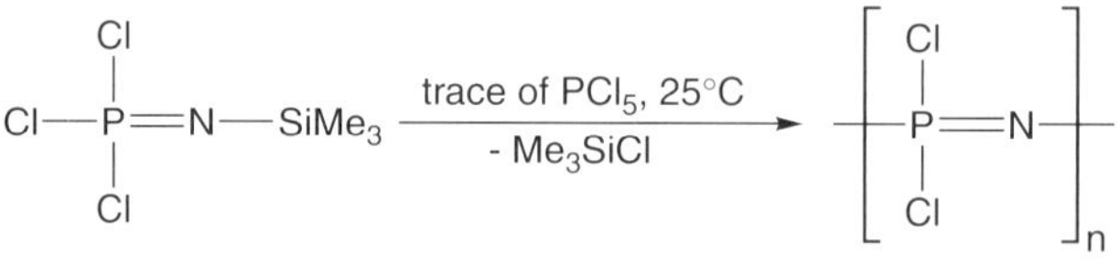

$$\begin{array}{c} Cl \\ | \\ Cl-P=N-SiMe_3 \\ | \\ Cl \end{array} \xrightarrow[-Me_3SiCl]{\text{trace of } PCl_5,\ 25\ °C} \left[\begin{array}{c} Cl \\ | \\ -P=N- \\ | \\ Cl \end{array} \right]_n \quad (11.34)$$

The phosphoranimine monomer is readily prepared in two steps from PCl_3 in a one-pot procedure [eqn (11.35)].[143]

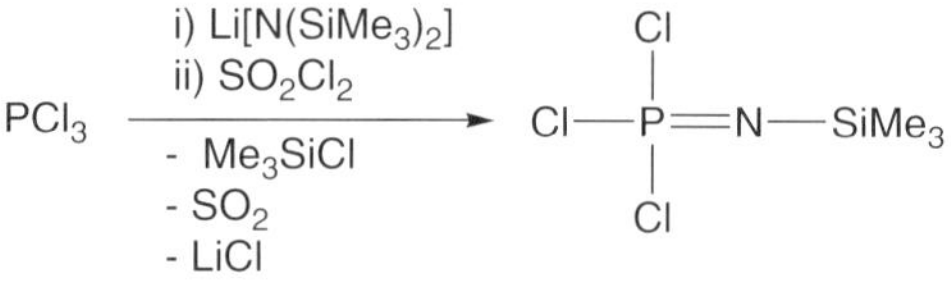

$$PCl_3 \xrightarrow[\substack{-Me_3SiCl \\ -SO_2 \\ -LiCl}]{\substack{\text{i) } Li[N(SiMe_3)_2] \\ \text{ii) } SO_2Cl_2}} \begin{array}{c} Cl \\ | \\ Cl-P=N-SiMe_3 \\ | \\ Cl \end{array} \quad (11.35)$$

The polymerisation of $Cl_3P=NSiMe_3$ initiated by PCl_5 is believed to proceed *via* a cationic mechanism (Scheme 11.11).[144] When the reaction is performed in solution, the process is termed 'living', as no significant chain transfer or chain

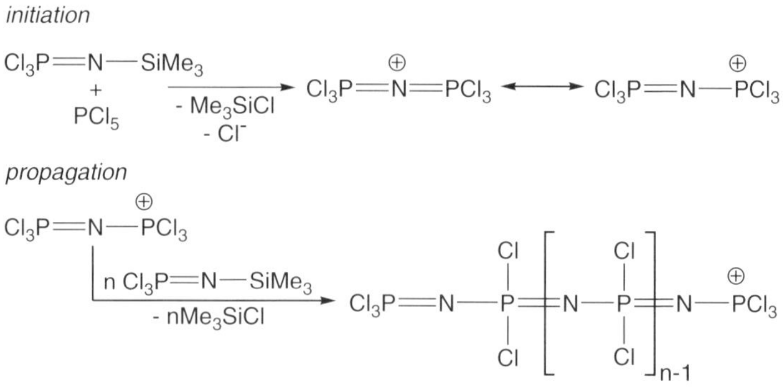

initiation

$$Cl_3P=N-SiMe_3 + PCl_5 \xrightarrow[-Cl^-]{-Me_3SiCl} Cl_3P\overset{\oplus}{=}N=PCl_3 \longleftrightarrow Cl_3P=N\overset{\oplus}{-}PCl_3$$

propagation

$$Cl_3P=N\overset{\oplus}{-}PCl_3 \xrightarrow[-nMe_3SiCl]{n\ Cl_3P=N-SiMe_3} Cl_3P=N-\left[\begin{array}{c} Cl \\ | \\ P=N \\ | \\ Cl \end{array} \right]_{n-1} \begin{array}{c} Cl \\ | \\ P\overset{\oplus}{=}N-PCl_3 \\ | \\ Cl \end{array}$$

Scheme 11.11 Proposed mechanism for the polymerisation of $Cl_3P=NSiMe_3$.

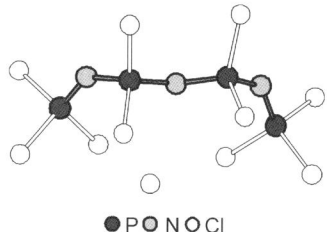

● P ○ N ○ Cl

Figure 11.22 Molecular structure of $[Cl_3P=N-(PCl_2=N)_2-PCl_3]^+$ Cl^-. Reproduced with permission from E. Rivard et al., Inorg. Chem., 2004, **43**, 2765.[146] Copyright (2004) American Chemical Society.

termination reactions are detected. Sequential addition of a different monomer to the active chain end allows the formation of block copolymers.[145]

The stoichiometric reactions of $Cl_3P=NSiMe_3$ with $[Cl_3P=N=PCl_3]^+$, which model the propagation step in the condensation polymerisation of the former species, yield discrete oligomers rather than high molecular weight polymers. For example, a seven-membered chain **11.29** results from a reaction in a 1:2 molar ratio [eqn (11.36)] and this species represents a useful conformational model for high molecular weight poly(dichlorophosphazene) (Figure 11.22).[146]

$$[Cl_3P=N=PCl_3]^+Cl^- \xrightarrow[-2Me_3SiCl]{2Cl_3P=NSiMe_3} [Cl_3P=N-(PCl_2=N)_2-PCl_3]^+Cl^-$$

11.29

(11.36)

Closely related to polyphosphazenes is the class of polymers known as polyheterophosphazenes, where one or more of the P atoms per repeat unit is substituted by an atom of a heteroelement. The first well-characterised example of such materials involved carbon as the replacement for phosphorus; the resulting macromolecules, polycarbophosphazenes, were prepared via ROP, but at a dramatically lower temperature than for $(NPCl_2)_3$ [eqn (11.37)].[147] Subsequently, polymers with three-coordinate sulfur(IV) and four-coordinate sulfur(VI) centres were obtained and these materials were termed polythiophosphazenes[148] and polythionylphosphazenes,[149] respectively. The latter polymers [eqn (11.38)] are much more stable than the sulfur(IV) analogues after halogen replacement and several have been explored as matrices for gas sensors as a consequence of their high permeability.[150]

(11.37)

11.5 ARSENIC-, ANTIMONY- AND BISMUTH–NITROGEN RINGS

11.5.1 Cyclopnictazanes

The chemistry of cyclopnictazanes $(RENR')_n$ (E = As, Sb, Bi) has developed more slowly than that of their phosphorus analogues (Section 11.4.1), for two main reasons.[151] First, the lability of E–N bonds (E = As, Sb, Bi) poses challenges for the synthesis of these inorganic rings, especially for the heavier pnictogens. Second, in contrast to phosphorus (^{31}P, $I=\frac{1}{2}$, 100%), none of the heavier pnictogens possess a suitable NMR nucleus ($I=\frac{1}{2}$) that could be used for monitoring the progress of reactions or providing insights into ring transformations that occur in solution.

The most important cyclodipnictazanes are the chloro derivatives [ClE(μ-NR)]$_2$ (E = As, Sb), which are prepared by cyclocondensation of primary amines with pnictogen halides.[151] Cyclodiars(III)azanes of the type [ClAs(μ-NR)]$_2$ (**11.30**) were first obtained in 1960 by reaction of a primary amine with AsCl$_3$ in a 2:1 molar ratio.[152] More recently, this method has been applied to the synthesis of cyclodiarsazanes with bulky aryl groups on nitrogen, e.g. R = 2,6-^{i}Pr$_2$C$_6$H$_3$, 2,6-Me$_2$C$_6$H$_3$, in low yields.[153] The trimer [ClAs(μ-NMe)]$_3$ (**11.31**) is formed, in addition to the dimer, in the reaction of methylamine with AsCl$_3$.[154] A rare example of a cyclotetrapnict(III)azane [Cp*As(μ-NH)]$_4$ is generated by treatment of Cp*AsCl$_2$ with ammonia in diethyl ether (**11.32**).[155] The cyclodistib(III)azane [ClSb(μ-N^{t}Bu)]$_2$ is prepared by reaction of SbCl$_3$ with *tert*-butylamine in the presence of triethylamine.[156] However, the bismuth analogue cannot be obtained by this route. The cyclotristib(III)azane [ClSb(μ-NDmp)]$_3$ (Dmp = 2,6-Me$_2$C$_6$H$_3$) has also been characterised.[157]

The reactions of lithium amides with pnictogen halides have also been used in the synthesis of heavy cyclopnict(III)azanes. For example, the combination of one equivalent of LiNH^{t}Bu and two equivalents of DippNHLi in reactions with ECl$_3$ produces the series of bis(amido)dipnict(III)azanes [DippN(H)E (μ-NDipp)]$_2$ (E = As, Sb, Bi) (**11.33**).[158] The cyclodipnict(III)azanes [PhE(μ-N^{t}Bu)]$_2$ (E = Sb, Bi) (**11.34**) are formed from the reaction of PhECl$_2$ and two equivalents of LiNH^{t}Bu, presumably *via* condensation of PhECl$_2$ with the acyclic intermediates PhE(NH^{t}Bu)$_2$.[159] A similar condensation process occurs between SbCl$_3$ and LiNHPh to give the macrocycle [{Sb(μ-NPh)}$_2$(μ-NPh)]$_6$ in which six Sb$_2$N$_2$ rings are bridged by phenylimido groups (see Figure 6.9 in Section 6.3).[160] The transamination reaction of primary amines, *e.g.* CyNH$_2$, with E(NMe$_2$)$_3$ is a very versatile method for the generation of an E$_2$N$_2$ ring in the form of bis(dimethylamino) derivatives [Me$_2$NE(μ-NCy)]$_2$ (**11.35**) that is applicable to all three heavy pnictogens (E = As, Sb, Bi).[161,162]

Cyclodipnict(III)azanes may exist as *cis* or *trans* isomers, *cf.* cyclodiphosph(III)azanes (Section 11.4.1). The solid-state structures of the dichloro derivatives [ClE(μ-N^{t}Bu)]$_2$ (E = As,[163] Sb[164]) reveal a *cis* configuration; however, the derivatives [XSb(μ-N^{t}Bu)]$_2$ (X = N$_3$, O^{t}Bu), prepared by nucleophilic substitution reactions, are isolated as *trans* isomers.[156] In general, there is a trend towards the preferential formation of *trans* isomers in the solid state for the heavier pnictogens.[151] In solution, however, NMR data indicate the existence of *cis* and *trans* isomers for some derivatives. In the case of the bisamido derivative [DippN(H)Sb(μ-N^{t}Bu)]$_2$, both the *cis* and *trans* isomers have been structurally characterised (Figure 11.23).[164]

Monomers of the type RN=ER' (E = As, Sb) exhibit an increasing tendency to dimerise compared with their phosphorus analogues (E = P, Section 11.4.1), as manifested by the derivatives MesN*=EX (X = Cl, OTf).[83b] Although the arsenic compound (E = As, X = Cl) is isolated as the dimer, it changes colour from yellow to red upon heating, suggesting monomer formation. This dissociation is utilised in the synthesis of arsatetrazoles *via in situ* generation of

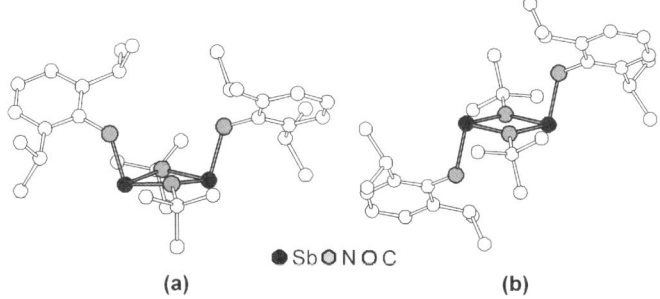

Figure 11.23 Molecular structures of *cis* and *trans* isomers of [DippN(H)Sb(μ-N^{t}Bu)]$_2$.

Mes*N=AsCl in dichloromethane (Section 11.1.2 and Scheme 11.1).[9b] The antimony derivative (Mes*N=SbOTf)$_2$ exists as a dimer that does not dissociate, indicating that the larger Sb$_2$N$_2$ ring is better able to accommodate the steric strain imposed by bulky Mes* groups than an As$_2$N$_2$ ring.

11.5.2 Cycloarsazenes

Examples of heavy pnictogen analogues of unsaturated P–N ring systems (Section 11.4.2) are limited to cycloarsazenes for which only six- and eight-membered rings are known. The following synthetic approaches have been used for the synthesis of cycloarsazenes: (a) thermal decomposition of diphenylarsenic(III) azide Ph$_2$AsN$_3$ or the phenylarsenic(V) azide PhAsCl$_3$N$_3$ produces the tetramer cyclo-(Ph$_2$AsN)$_4$ and the trimer [Ph(Cl)AsN]$_3$,[165] respectively; (b) treatment of Ph$_2$AsCl$_3$ with ammonia or Ph$_2$AsCl with ammonia–chloramine gives a mixture of the cyclic trimer and tetramer (Ph$_2$AsN)$_n$ (n = 3, 4);[166] (c) thermolysis of the dimeric arsenic(V) compound [(CF$_3$)$_2$As(Cl)NSiMe$_3$]$_2$ generates a mixture of cyclic oligomers [(CF$_3$)$_2$AsN]$_n$ (n = 3, 4) [eqn (11.39)];[167] and (d) the reaction of arsenic acid H$_3$AsO$_4$ with (Me$_3$Si)$_2$NH yields the trimer [(Me$_3$SiO)$_2$AsN]$_3$.[168]

(11.39)

The six-membered rings in cyclo-(R$_2$AsN)$_3$ are slightly puckered with As–N distances in the narrow ranges 1.74–1.77 Å (R = Ph)[166] and 1.73–1.75 Å (R = OSiMe$_3$),[168] *cf.* a single-bond value of 1.87 Å and an As–N bond length of 1.83 Å in the cyclotriars(III)azane (ClAsNMe)$_3$.[154] The eight-membered ring cyclo-(Ph$_2$AsN)$_4$ is isostructural with cyclo-(Ph$_2$PN)$_4$ and the conformation of the As$_4$N$_4$ ring is intermediate between ideal boat (S_4) and ideal saddle (D_{2d});[169] the local symmetry in the trifluoromethyl derivative cyclo-[(CF$_3$)$_2$AsN]$_4$ is close to S_4.[167]

11.6 PNICTOGEN–OXYGEN RINGS

11.6.1 Phosphorus–Oxygen Rings

Binary oxides of phosphorus, arsenic and antimony are polycyclic compounds, *e.g.* E$_4$O$_6$ (E = P, As, Sb), with an adamantane-like structure. Monocyclic

Group 15: Rings and Polymers 253

structures are exhibited by the cyclometaphosphates $(PO_3^-)_n$, which form an extensive homologous series of saturated ring systems; the trimer ($n=3$) and tetramer ($n=4$) are the most common. The structures of the octameric ($n=8$) and decameric ($n=10$) macrocycles are depicted in Figure 6.3 (see Section 6.2.1).

The six-membered ring $[P_3O_9]^{3-}$ forms a large number of transition metal complexes in which the cyclotriphosphate acts as a tripodal (O,O',O'') ligand, e.g. [^{n}Bu$_4$N][HB(pz)$_3$Fe(P$_3$O$_9$)] (pz = 1-pyrazolyl) and [Ph$_3$PNPPh$_3$]$_2$[PtMe$_3$(P$_3$O$_9$)] (**11.36**).[170] The greater versatility of the cyclotetraphosphate $[P_4O_{12}]^{4-}$ ligand is manifested in the formation of dinuclear or tetranuclear complexes with rhodium.[171] The P$_4$O$_4$ ring changes its conformation from chair in the dinuclear complex (**11.37**) to saddle in the tetranuclear complex.

11.36   **11.37**

Although they are less common than cyclic metaphosphates, phosphorus–oxygen rings containing trivalent phosphorus of the type cyclo-(RPO)$_n$ (R = R′$_2$N, ArO; $n=2, 3$) have been characterised. The first cyclic trimer cyclo-(^{i}Pr$_2$NPO)$_3$ was obtained in 1971 by the reaction of the iminophosphine ^{i}Pr$_2$NP=N^{t}Bu with sulfur dioxide [eqn (11.40)].[172] An alternative route to the P$^{III}_3$O$_3$ ring system is the hydrolysis of dichlorophosphites ArOPCl$_2$ containing bulky aryl groups [eqn (11.41)].[173a] The P$_3$O$_3$ rings in these two heterocycles have distorted boat conformations. Upon sublimation at 220 °C, the six-membered ring in cyclo- [(2,6-^{t}Bu$_2$-4-MeC$_6$H$_2$O)PO]$_3$ undergoes an interesting ring contraction to give a rhombic P$_2$O$_2$ ring with the bulky aryloxy substituents in a *cis* configuration.[173b]

^{i}Pr$_2$N-P=N^{t}Bu + O=S=O $\xrightarrow{-^t\text{BuNSO}}$ (11.40)

ArOPCl$_2$ + H$_2$O $\xrightarrow[-2\,[\text{Et}_3\text{NH}]\text{Cl}]{2\,\text{Et}_3\text{N}}$ (11.41)

Larger P_nO_n rings ($n = 4$–6) are assembled *via* transformations that occur at a metal centre. For example, reaction of bis(diisopropylamino)phosphine oxide (^{i}Pr$_2$N)$_2$P(O)H with molybdenum hexacarbonyl generates the dinuclear complex (CO)$_4$Mo{cyclo-(^{i}Pr$_2$NPO)$_4$}Mo(CO)$_4$ in which each metal centre is P,P'-chelated by the P_4O_4 ring, which is in a boat conformation [eqn (11.42)].[174] The dinuclear complex Cr$_2$(CO)$_7${cyclo-(^{i}Pr$_2$NPO)$_5$} contains a P_5O_5 ring coordinated through the five phosphorus(III) centres to one *cis*-Cr(CO)$_4$ and one *fac*-Cr(CO)$_3$ unit,[175] while the P_6O_6 ring in the related complex Cr$_2$(CO)$_6$[(DMP)PO]$_6$ (DMP = 2,6-dimethylpiperidino) acts as a hexadentate ligand towards two *fac*-Cr(CO)$_3$ units.[176]

$$2\ Mo(CO)_6 + 4\ (^iPr_2N)_2P(O)H \xrightarrow[-4\ ^iPr_2NH]{-4\ CO}$$

(11.42)

11.6.2 Arsenic–, Antimony– and Bismuth–Oxygen Rings

Cycloarsoxanes (RAsO)$_n$ were discovered 140 years ago and their oligomeric nature in solution was established in the 1930s.[177] Subsequent ^{1}H NMR studies showed that the trimer ($n = 3$) and tetramer ($n = 4$) predominate in solution for alkyl derivatives (R = Me, Et), although smaller amounts of the dimer ($n = 2$) and pentamer ($n = 5$) are also present.[178]

Several methods are available for the synthesis of cycloarsoxanes: (a) hydrolysis of organoarsenic(III) dihalides RAsX$_2$ (X = Cl, Br, I), (b) reduction of arsonic acids RAsO(OH)$_2$ and (c) oxidation of primary arsines RAsH$_2$ or cyclopolyarsines (RAs)$_n$.[177] The tetramer cyclo-(MeAsO)$_4$ exhibits the boat–chair conformation,[179a] whereas the mesityl derivative cyclo-(MesAsO)$_4$ adopts the crown conformation (Figure 11.24).[179b] The As–O bond distances are in the range 1.77–1.82 Å, typical for single bonds.

Like their phosphorus(III) analogues, cycloarsoxanes (RAsO)$_n$ undergo metal-mediated ring expansion to produce flexible macrocyclic ligands ($n = 5, 6, 8$) that

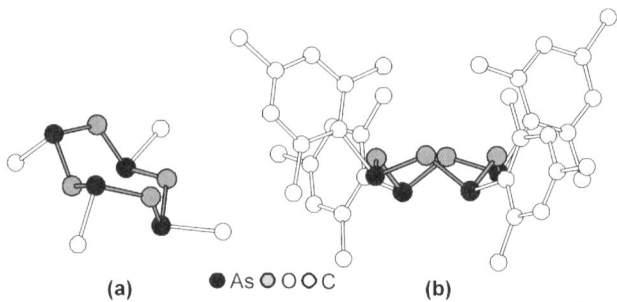

Figure 11.24 (a) Boat–chair and (b) crown conformations of the As$_4$O$_4$ ring in (RAsO)$_4$ (R = Me, Mes).

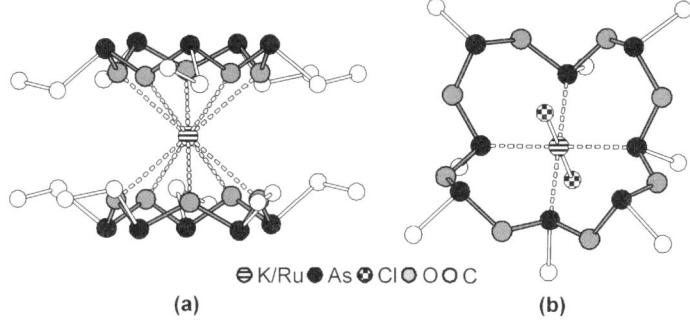

Figure 11.25 Molecular structures of (a) [K{cyclo-(EtAsO)$_5$}$_2$] (SCN$^-$ salt) and (b) [RuCl$_2${cyclo-(MeAsO)$_8$}].

coordinate to metal centres through the arsenic(III) donor sites. For example, the ligands (MeAsO)$_n$ ($n = 4$, 5) form sandwich complexes with alkali metals. Similarly to the behaviour of crown ethers, an increase in the radius of the alkali metal favours coordination of larger macrocycles in these complexes, *e.g.* [K{cyclo-(EtAsO)$_5$}$_2$]SCN (Figure 11.25a).[180a] The cyclic octamer (MeAsO)$_8$ forms the 1:1 complexes [MCl$_2${cyclo-(MeAsO)$_8$}] (M = Ru, Os) in which the 16-membered As$_8$O$_8$ ring acts as a tetradentate ligand and adopts a double crown conformation (Figure 11.25b).[180b] The tetrameric cycloarsoxanes (RAsO)$_4$ (R = Me, Et) can also serve as bridging ligands in the formation of a wide variety of coordination polymers with copper halides.[177,181]

Both antimony and bismuth form four-membered rings cyclo-(REO)$_2$ (E = Sb, Bi) when the substituent on the pnictogen is a very bulky aryl group.[182] These E$_2$O$_2$ systems are formed rapidly in solution by air oxidation of the corresponding distibene or dibismuthene [eqn (11.43)]. Remarkably, they are also produced slowly *in the solid state* as colourless crystals when the green distibene or purple dibismuthene crystals are exposed to air.[183] This transformation is reminiscent of the formation of E$_2$O$_2$ rings (E = Si, Ge) by oxidation of disilylene or digermylene precursors (see Schemes 10.19 and 10.22).

$$\underset{\text{Tbt}}{\overset{\text{Tbt}}{E=E}} \xrightarrow[23°C]{O_2 \text{ (air)}} \underset{\text{Tbt}}{\overset{\text{Tbt}}{E\underset{O}{\overset{O}{\diamond}}E}} \qquad \text{Tbt} = \begin{array}{c} Me_3Si \\ Me_3Si \end{array} \begin{array}{c} SiMe_3 \\ SiMe_3 \\ SiMe_3 \\ SiMe_3 \end{array}$$

(11.43)

11.7 PNICTOGEN–CHALCOGEN RINGS

11.7.1 Binary Phosphorus–Sulfur and –Selenium Rings

Binary phosphorus–sulfur and –selenium compounds have polycyclic structures.[184a] The most common examples are P$_4$E$_3$ and P$_4$E$_{10}$ (E = S, Se). The former is comprised of a P_4 tetrahedron in which three of the edges are bridged by chalcogen atoms; the latter have adamantane-like structures in which all six edges of the P_4 tetrahedron are bridged by chalcogen atoms and there are four terminal P=E bonds, *cf.* P_4O_{10}. In the case of sulfur, the entire series P_4S_n ($n = 3–10$) is known and isomers exist for several of the binary systems as a result of different combinations of bridging and terminal sulfur atoms.[184b,c] Interestingly, the oligomeric phosphorus selenides P_2Se_5 and P_4Se_{10} are distinct species; the former has a bicyclic structure (**11.38**).[185] Binary phosphorus–tellurium rings have not been structurally characterised

Sulfur-rich binary P–S anions of the type $[S_2P(\mu-S_n)_2PS_2]^{2-}$ form monocyclic structures with sulfide or disulfide bridging units ($n = 1, 2$) (**11.39**).[186a] Another type of cyclic P–S anion with the general formula $[(PS_2)_n]^{n-}$ is formed in the reaction of white phosphorus with polysulfides in a non-aqueous medium; these anions have homocyclic structures with two exocyclic sulfur atoms attached to each phosphorus atom, *e.g.* $[P_4S_8]^{4-}$ (**11.40**).[186b]

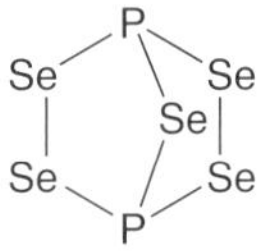

11.38 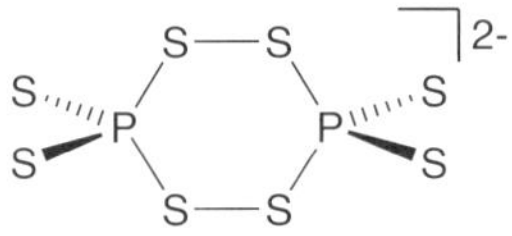   **11.39** 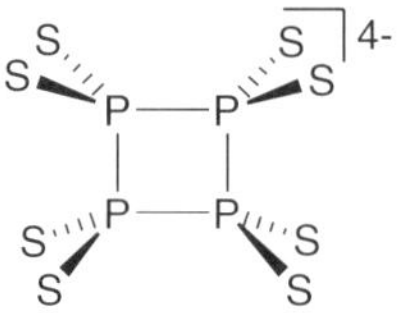   **11.40**

11.7.2 Organophosphorus–Sulfur Rings

A variety of organophosphorus–sulfur heterocycles including three-, four- five- six- and eight-membered ring systems are known. In some cases these rings are related by ring transformation processes. Three-membered rings of the type cyclo-$(RP)_2S$ are prepared by a number of different routes, including (a) [2 + 1] cyclo-condensation of Cl(R)P–P(R)Cl and $(Me_3Sn)_2S$ (R = ^tBu),[187] (b) oxidation of a diphosphene RP=PR with sulfur [R = N(SiMe$_3$)$_2$,[188] 2,4,6-$^tBu_3C_6H_2$,[189a] Fc[189b] and Bbt[189c]] or (c) reaction of RP(S)Cl$_2$ with magnesium (R = 2,4,6-$^tBu_3C_6H_2$).[190] Intriguingly, the formation of the supermesityl derivative by method (b) involves the initial formation of an acyclic monosulfide that is converted to the isomeric P_2S ring upon mild heating or UV irradiation (Scheme 11.12).[189] The C_2-symmetric P_2S ring in this derivative has endocyclic bond angles of 64.6 and 57.7° at the sulfur and phosphorus atoms, respectively.[190]

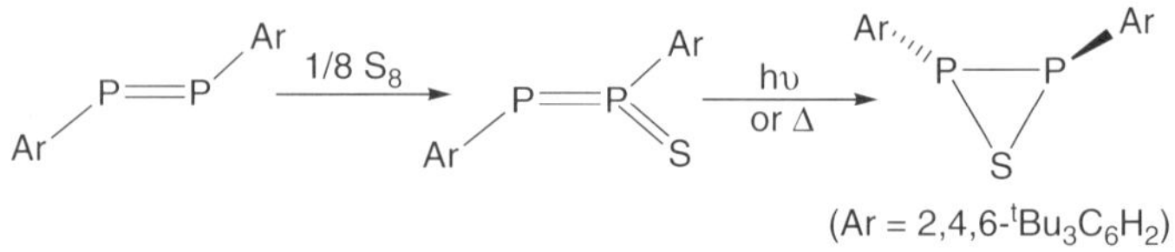

(Ar = 2,4,6-$^tBu_3C_6H_2$)

Scheme 11.12 Formation of P_2S ring *via* the acyclic isomer.

Four-membered rings of the type $R(S)P(\mu-S)_2P(S)R$ are the most important organophosphorus–sulfur heterocycles because of their widespread use as reagents in organic synthesis, *e.g.* as thionation reagents.[191] The aryl derivative ($R = 4\text{-MeOC}_6H_4$), commonly known as Lawesson's reagent, is prepared by heating a mixture of P_4S_{10} with anisole in a high-boiling solvent.[192] A soluble form of Lawesson's reagent ($R = 3\text{-}^tBu\text{-}4\text{-MeOC}_6H_3$)[193a] and the ferrocenyl derivative $Fc(S)P(\mu-S)_2P(S)Fc$[193b] are obtained in a similar manner. The reaction of the latter reagent with dicyclohexylcarbodiimide generates a four-membered P_2NS ring [eqn (11.44)].[194]

$$(11.44)$$

Dicationic P_2S_2 rings have also been characterised.[195] The diphosphonium systems $[\{(R_2N)_2P(\mu-S)\}_2][AlCl_4]_2$ (**11.41**, $R = Me$, Et) are obtained from the reaction of $(R_2N)_2P(S)Cl$ and aluminium trichloride in the absence of a solvent; the selenium analogue ($R = Et$) is prepared in a similar manner.[195a] A similar dicationic P_2S_2 ring system **11.42** is generated upon oxidation of the two-coordinate phosphorus centre in the zwitterionic complex $Cl_2Al(\mu\text{-NSiMe}_3)_2P$ with elemental sulfur.[195b]

11.41   **11.42**

The nature of the aryl substituent attached to phosphorus has a subtle influence on the ring size of the organophosphorus–sulfur heterocycle obtained from the reactions of $ArPCl_2$ ($Ar = aryl$) and a source of sulfide. The six-membered ring cyclo-$(Mes^*PS)_3$ is formed from treatment of Mes^*PCl_2 with lithium sulfide; the P_3S_3 ring adopts a chair conformation with the sulfur atoms 1.0 Å out of the plane of the three phosphorus atoms (Figure 11.26a).[196] In contrast, the phenyl derivative, which is obtained from the reaction of $PhPCl_2$ with $(Me_3Sn)_2S$ in CS_2, is comprised of a five-membered P_3S_2 ring with an exocyclic sulfur atom attached to one of the phosphorus centres (Figure 11.26b).[197a] Finally, the eight-membered ring cyclo-$(MesPS)_4$ is generated in high yield from $MesPCl_2$ and $(Me_3Si)_2S$; the P_4S_4 ring adopts a crown conformation (Figure 11.26c),[197b] *cf.* cyclo-S_8. The cyclic tetramer $(MesPS)_4$ undergoes a ring contraction at 35–40 °C over a period of 3 days to give the corresponding four-membered ring $Mes(S)P(\mu-S)_2P(S)Mes$.

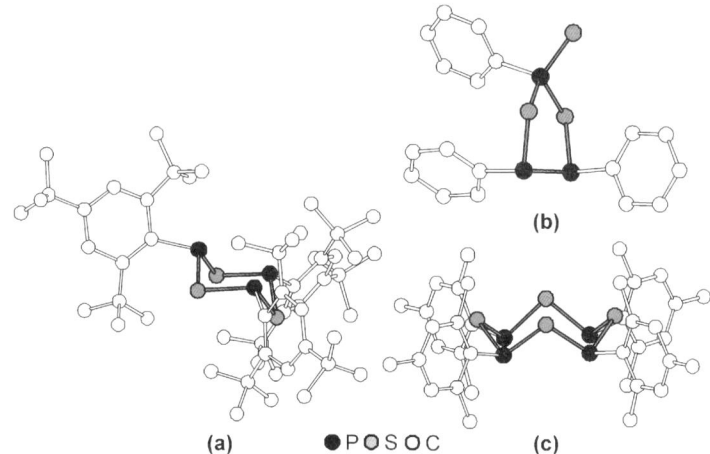

Figure 11.26 Molecular structures of five-, six- and eight-membered organophosphorus–sulfur heterocycles: (a) (Mes*PS)$_3$; (b) Ph$_3$P$_3$(μ-S)$_2$(=S); (c) (MesPS)$_4$.

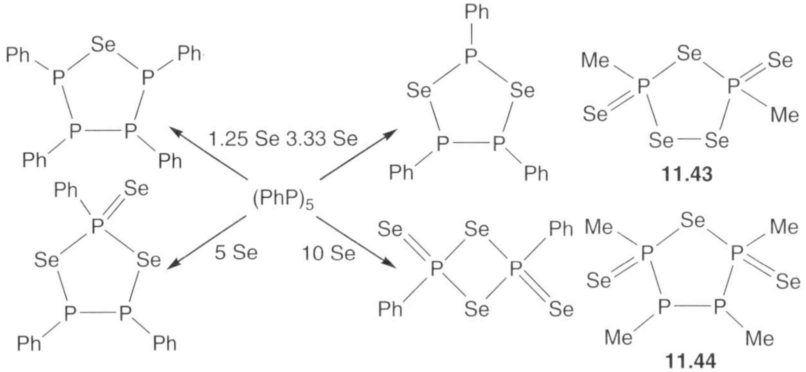

Scheme 11.13 Formation and structures of organophosphorus–selenium rings.

11.7.3 Organophosphorus–Selenium Rings

Several organophosphorus–selenium heterocycles are also known, some of which have no sulfur analogues.[198] As illustrated in Scheme 11.13, the reaction of cyclo-(PhP)$_5$ with selenium produces a variety of these ring systems depending on the P–Se stoichiometry.[198] The combination of ^{31}P and ^{77}Se NMR spectroscopy facilitates the identification of intermediates in the selenation of the more reactive methylated pentamer cyclo-(MeP)$_5$. The initial oxidation of one phosphorus centre is followed by replacement of two of the MeP groups in the five-membered ring by Se atoms and, subsequently, ring contraction to give

Me(Se)P(μ-Se)$_2$P(Se)Me with a central P$_2$Se$_2$ ring.[199] When a large excess of selenium is used for the oxidation, the P$_2$Se$_3$ ring is formed in the form of Me(Se)P(μ-Se)(μ-SeSe)P(Se)Me (**11.43**).[200] The methylated derivative (MeP)$_4$(μ-Se)(=Se)$_2$ (**11.44**) with the selenium atoms of the two terminal P=Se groups on the same side of folded P$_4$Se ring has been structurally characterised.[201] Phosphorus–selenium rings are also formed in the cyclocondensation reactions of RPCl$_2$ with Se(SiMe$_3$)$_2$ or Na$_2$Se which, in the case of bulky groups such as ^{t}Bu or mesityl, give rise to cyclic trimers and tetramers cyclo-(RPSe)$_n$ ($n = 3, 4$). The eight-membered P$_4$Se$_4$ ring adopts a crown conformation, *cf.* cyclo-(MesPS)$_4$ (Figure 11.26c).[199]

In common with the sulfur systems, the most important organophosphorus–selenium heterocycle is the four-membered ring R(Se)P(μ-Se)$_2$P(Se)R. In the solid state, the R groups adopt a *trans* arrangement with respect to the P$_2$Se$_2$ ring (R = ^{t}Bu,[202a] Ph[202b]). Solid-state ^{31}P and ^{77}Se NMR spectroscopy has been used to confirm the structures of insoluble derivatives, and also that of the five-membered ring in **11.43**.[200] Similarly to the behaviour of Lawesson's reagent, the phenyl derivative Ph(Se)P(μ-Se)$_2$P(Se)Ph has been used for the transformation of carbonyl groups to selenocarbonyls.[203,204] This selenium-transfer agent, which has become known as 'Woollins reagent', is prepared on a large scale with high purity from the reaction of PhPCl$_2$ with sodium selenide (prepared *in situ* from sodium metal and selenium in liquid ammonia).[204] It has also been used for the synthesis of selenoamides from formamides or aryl nitriles.[205] Woollins reagent reacts with azobenzene in hot toluene to produce a four-membered P$_2$NSe ring [eqn (11.45)].[202b]

(11.45)

11.7.4 Organophosphorus–Tellurium Rings

Phosphorus–tellurium ring systems have been prepared by reactions of (a) RPCl$_2$ [R = ^{t}Bu,[206a] NR$'_2$ (R$'$ = ^{i}Pr, Cy, Ph)][206b] with a source of telluride, *e.g.* M$_2$Te (M = Li, Na), Te(SiMe$_3$)$_2$, or (b) ^{t}BuP(Cl)–(Cl)P^{t}Bu and Na$_2$Te.[206a] Structural characterisation is limited to ^{31}P and ^{125}Te NMR spectra[206] and, in the case of R=^{t}Bu,[206a] mass spectrometric data. In general, the P–Te ring systems cyclo-(RP)$_n$Te$_m$ are limited to heterocycles for which $n > m$. Method (b) produces the three-membered ring cyclo-(^{t}BuP)$_2$Te, which is purified by trap-to-trap distillation. A three-membered P$_2$Te ring has also been prepared by the telluridation of the diphosphene BbtP=PBbt (Bbt = 2, 6-bis[bis(trimethylsilyl)methyl]-4-[tris(trimethylsilyl)methyl]phenyl) and structurally characterised.[206c]

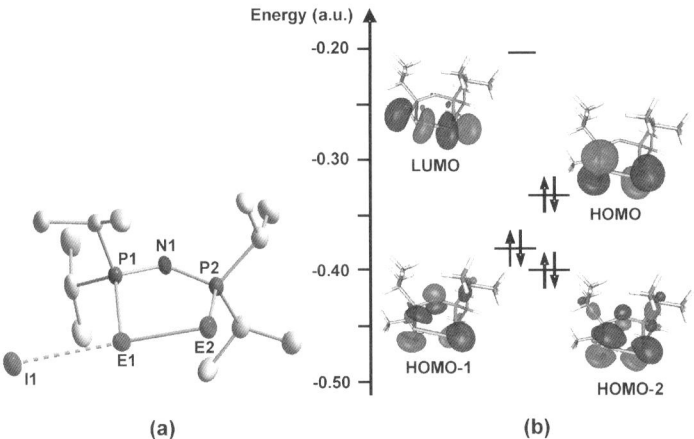

Figure 11.27 (a) Molecular structure of [N(P^{i}Pr$_2$E)$_2$]I; (b) frontier MOs of [N(P^{i}Pr$_2$Te)$_2$]$^+$. Reproduced with permission of the Royal Society of Chemistry from J. Konu et al., Chem. Commun., 2006, 1624.[207a]

A novel class of phosphorus–tellurium ring system, the five-membered cyclic cation [N(P^{i}Pr$_2$Te)$_2$]$^+$, is obtained as an iodide salt by the two-electron oxidation of the corresponding anion in the sodium salt [(TMEDA)Na(P^{i}Pr$_2$Te)$_2$]; the selenium analogue is obtained in a similar manner (Figure 11.27a).[207] The five-membered rings [NP$_2$E$_2$]$^+$ (E = Se, Te) are formally six π-electron systems. The π-bond order is low, however, since the bonding effect of the E–E bonding orbital (HOMO-2) is essentially cancelled by the double occupation of the E–E π* orbital (HOMO); the third occupied π orbital (HOMO-1) is primarily a non-bonding nitrogen-centred orbital. As an example, the frontier orbitals of the tellurium derivative [N(P^{i}Pr$_2$Te)$_2$]$^+$ are illustrated in Figure 11.27b. The iodide salts of the cyclic [N(P^{i}Pr$_2$E)$_2$]$^+$ cations exhibit strong E···I interactions (Figure 11.27a) and long chalcogen–chalcogen bonds as a result of the donation of electron density from a lone pair on the iodide ion into the σ* (E–E) orbital (LUMO) of the cation (Figure 11.27b).[207] The elongation of the E–E bonds is not observed in the ion-separated salts obtained by replacement of the iodide counterion by [SbF$_6$]$^-$.[207b]

11.7.5 Arsenic–, Antimony– and Bismuth–Chalcogen Rings

11.7.5.1 Binary Anions and Cations. The heavy pnictogens in combination with chalcogens form a large variety of anions.[208] Although many of these binary species form oligomers or polymers with layered or framework structures, there are a number of examples of discrete anions with cyclic structures that are formed with large counterions such as tetralkylammonium or tetraarylphosphonium cations.[208] Some examples are [As$_3$Se$_6$]$^{3-}$ (**11.45**),[209] [Sb$_4$S$_8$]$^{4-}$ (**11.46**),[210] [AsSe$_8$]$^-$ (**11.47**)[208] and [As$_2$S$_6$]$^{2-}$ (**11.48**);[211] the last two reflect the propensity of chalcogens to catenate in the formation of these binary anionic species.

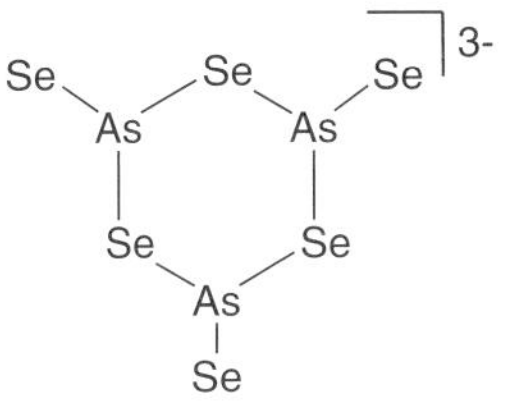

Group 15: Rings and Polymers 261

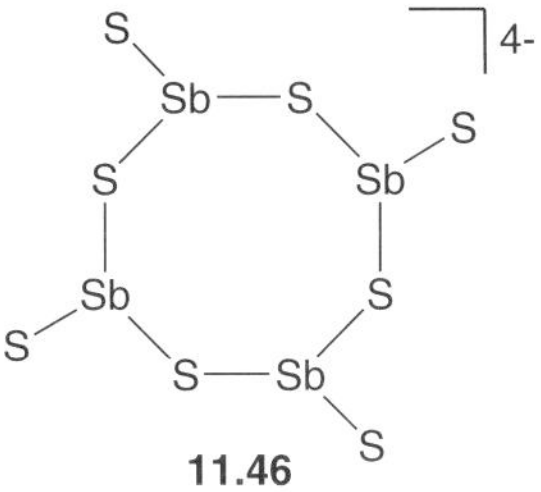

11.45 [As₃Se₆]³⁻ ring structure
11.46 [Sb₃S₇]⁴⁻... (structures shown)
11.47
11.48 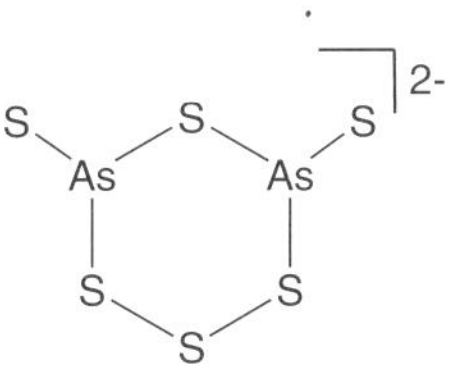

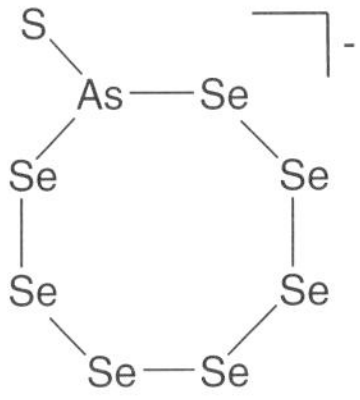

Mixed group 15–group 16 polycations are rare and tend to form clusters. For example, the arsenic–chalcogen cations $[As_3E_4]^+$ (E = S, Se)[212] have polycyclic structures similar to those of the well-known isovalent species P_4S_3 and P_7^{3-}; the bismuth–tellurium cation $[Bi_4Te_4]^{4+}$ adopts a slightly distorted cubic structure.[213] An exception is the antimony–tellurium cation cyclo-$[Sb_2Te_2]^+$, which is obtained as the $AlCl_4^-$ salt from the reaction of tellurium, antimony, antimony trichloride, sodium chloride and aluminium trichloride in a sealed tube at 130 °C.[214] The cyclo-$[Sb_2Te_2]^+$ cations form four-membered rings that are connected by Sb–Te and Sb–Sb bonds to give a polymeric strand with alternating planar and slightly bent Sb_2Te_2 rings (Figure 11.28).

11.7.5.2 Organic Derivatives. The most widely studied ring systems involving linkages between the heavy group 15 and 16 elements are the organocycloarsathianes cyclo-$(RAsS)_n$; the cyclic trimers (n = 3) and tetramers (n = 4) predominate in solution.[177] Several methods are used for the synthesis of these inorganic heterocycles: (a) treatment of $RAsX_2$ (X = Cl, Br) with H_2S (or another source of sulfide), (b) oxidation of $RAsH_2$ with sulfur or (c) reaction of $RAsH_2$ with $SOCl_2$ or PhNSO.[177] In addition, cycloarsoxanes $(RAsO)_n$ can be converted to their sulfur analogues by treatment with H_2S.[215] The trimer cyclo-$(MeAsS)_3$ (**11.49**) adopts a chair conformation in the solid state, whereas the tetramers cyclo-$(RAsS)_4$ (**11.50**, R = Me, Et, ᵗBu, Ph) have a crown-shaped structure, *cf.* cyclo-S_8.[177,216]

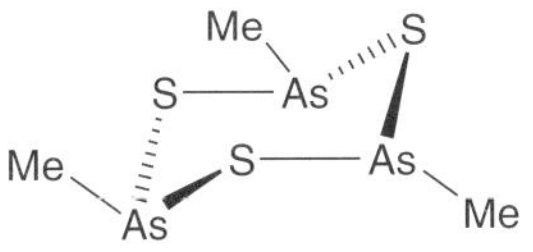

11.49
11.50

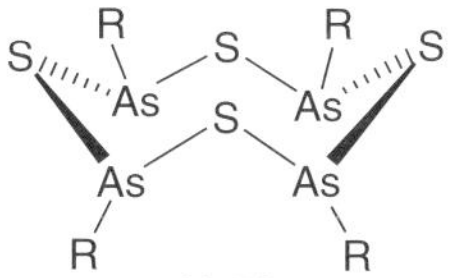

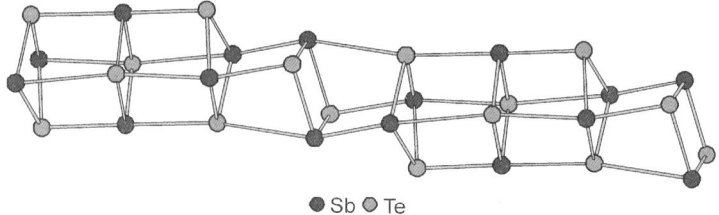

Figure 11.28 A section of the polymeric strand $(Sb_2Te_2^+)_n$ $(AlCl_4^-$ salt).

Three- and five-membered rings (As_2S and As_2S_3) have also been structurally characterised for the arsenic–sulfur system. The diarsathiiran cyclo-$(RAs)_2S$ [R = C(SiMe_3)_3] is prepared by the addition of sulfur to the diarsene RAs=AsR.[217] The non-planar five-membered ring cyclo-$(PhAs)_2S_3$ is obtained, in addition to cyclo-$(PhAsS)_4$, from treatment of phenylarsenic acid with aqueous ammonia and hydrogen sulfide.[218]

The coordination chemistry of cyclo-$(RAsS)_n$ ligands is potentially very extensive, since both As and S may serve as soft donor atoms towards transition and coinage metals in low oxidation states. For example, in the silver(I) complex [Ag{cyclo-$(EtAsS)_4$}_2][CF_3SO_3] the two cyclic ligands are S,S'-chelated to the copper centre (Figure 11.29a),[215] whereas three As atoms of the cycloarsathiane occupy facial sites in the octahedral ruthenium(II) complex [$RuCl_2${cyclo-$(EtAsS)_4$}(PPh_3)] (Figure 11.29b).[219] The photolysis of cyclo-$(MeAsS)_n$ ($n = 3, 4$) with metal carbonyls $M(CO)_6$ (M = Cr, W) in THF results in ring expansion to give complexes containing pentameric ($n = 5$) and hexameric ($n = 6$) ligands, e.g. [Cr{cyclo-$(MeAsS)_5$}($CO)_3$] and [W{cyclo-$(MeAsS)_6$}($CO)_3$].[220] In the chromium complex the ligand binds to the metal centre via two arsenic and one sulfur atom (Figure 11.29c).

In addition to ring expansion, metal-mediated reactions of cycloarsathianes may also result in As–S bond cleavage with the formation of anionic linear or macrocyclic ligands at elevated temperatures.[177] For example, the reactions of cyclo-$(EtAsS)_n$ ($n = 3, 4$) with MCl_3 (M = Ru, Os) in superheated toluene at temperatures above 100 °C produce the dianionic 14-membered $[Et_4As_6S_{10}]^{2-}$ (**11.51**) and 16-membered $[Et_6As_8S_{10}]^{2-}$ (**11.52**) rings as octahedral ruthenium(II) and osmium(II) complexes, respectively.[219] In both of these complexes the anionic As–S ligand binds to the metal through four arsenic atoms and the two exocyclic sulfur atoms that bear the dinegative charge.

11.51   **11.52**

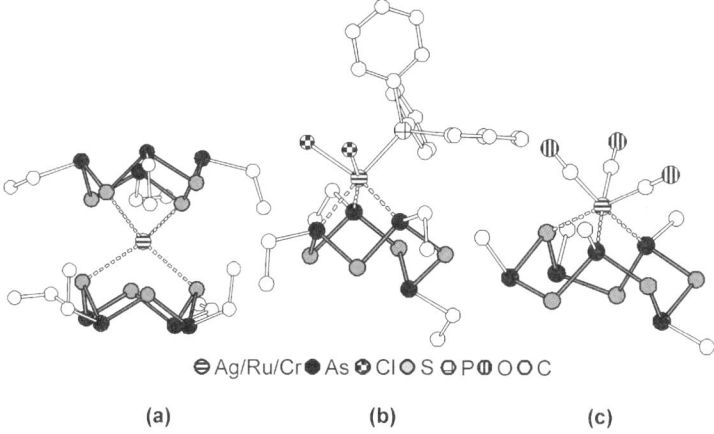

Figure 11.29 Coordination modes of cyclo-(EtAsS)$_4$ and cyclo-(MeAsS)$_5$.

Ring systems involving arsenic, antimony or bismuth linked to selenium or tellurium have not been widely investigated. The compounds cyclo-(PhAs)$_2$Se$_3$ and cyclo-(MeAsSe)$_3$, obtained by heating cyclo-(PhAs)$_6$ and cyclo-(MeAs)$_5$, respectively, with selenium, have only been characterised spectroscopically; they are thought to have cyclic structures similar to their sulfur analogues.[221] The six-membered rings cyclo-(RESe)$_3$ are especially important as a source of the first examples of stable distibenes and dibismuthenes RE=ER (E = Sb, Bi; R = bulky aryl group).[222,223] They are prepared by the reaction of RECl$_2$ {R = 2,4,6-tris[bis(trimethylsilyl)methyl]phenyl (Tbt)} with lithium selenide in THF. Subsequent deselenation of cyclo-(TbtESe)$_3$ with a phosphine reagent generates the deep-purple TbtBi=BiTbt or green TbtSb=SbTbt (Scheme 11.14).

The reactions of the dipnictenes TbtE=ETbt with a suitable source of chalcogen, e.g. selenium or ^{n}Bu$_3$P=Te, produces three-membered E$_2$Se or E$_2$Te rings; however, elemental sulfur generates four-membered Sb$_2$S$_2$ and

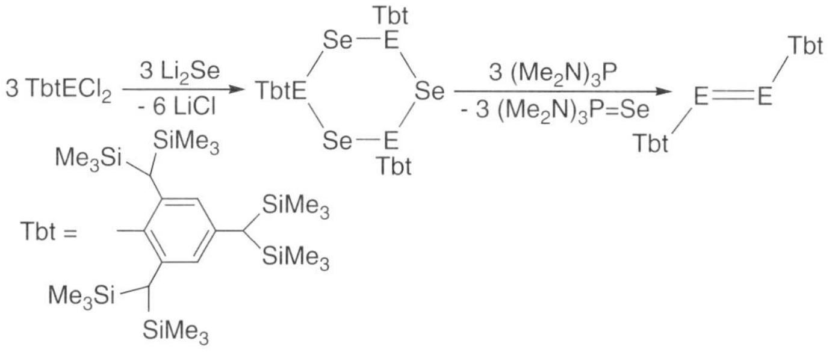

Scheme 11.14 Synthesis of dipnictenes from E$_3$Se$_3$ (E=Sb, Bi) ring systems.

five-membered E_2S_3 rings, in addition to the Sb_2S ring.[224] The treatment of $RSbCl_2$ (R = $(Me_3Si)_2CH$) with sodium chalcogenides in liquid ammonia yields oligomers of the type cyclo-$(RSbE)_n$ (E = Se, Te), which, on the basis on NMR spectra, exist predominantly as the trimer in solution.[225]

REFERENCES

1. R. Huisgen and I. Ugi, *Chem. Ber.*, 1957, **90**, 2914.
2. (a) J. D. Wallis and J. D. Dunitz, *J. Chem. Soc., Chem. Commun.*, 1983, 910; (b) F. Biesemeier, U. Müller and W. Massa, *Z. Anorg. Allg. Chem.*, 2002, **628**, 1933.
3. R. Müller, J. D. Wallis and W. von Philipsborn, *Angew. Chem. Int. Ed. Engl.*, 1985, **24**, 513.
4. A. Hammerl and T. M. Klapötke, *Inorg. Chem.*, 2002, **41**, 906.
5. (a) V. Benin, P. Kaszynski and G. Radziszewski, *J. Org. Chem.*, 2002, **67**, 1354; (b) A. Vij, J. G. Pavlovich, W. W. Wilson, V. Vij and K. O. Christe, *Angew. Chem. Int. Ed.*, 2002, **41**, 3051.
6. A. Vij, W. W. Wilson, V. Vij, F. S. Tham, J. A. Sheey and K. O. Christe, *J. Am. Chem. Soc.*, 2001, **123**, 6308.
7. (a) S. Fau, K. J. Wilson and R. J. Bartlett, *J. Phys. Chem. A*, 2002, **106**, 4639; (b) D. A. Dixon, D. Feller, K. O. Christe, W. W. Wilson, A. Vij, V. Vij, H. D. B. Jenkins, R. M. Olson and M. S. Gordon, *J. Am. Chem. Soc.*, 2004, **126**, 834.
8. (a) R. N. Butler, in *Comprehensive Heterocyclic Chemistry*, ed. A. R. Katritzky, C. W. Rees and E. F. V. Scriven, Pergamon Press, Oxford, 1996, Vol. **4**; (b) M. Friedrich, J. C. Gálvez-Ruiz, T. M. Klapötke, P. Mayer, B. Weber and J. J. Weigand, *Inorg. Chem.*, 2005, **44**, 8044.
9. (a) A. Villinger, P. Mayer and A. Schulz, *Chem. Commun.*, 2006, 1236; (b) A. Schulz and A. Villinger, *Angew. Chem. Int. Ed.*, 2008, **47**, 603; (c) E. Niecke, M. Nieger and F. Reichert, *Angew. Chem. Int. Ed. Engl.*, 1988, **27**, 1715.
10. C. E. Housecroft and A. G. Sharpe, *Inorganic Chemistry*, 2nd edn., Pearson Education, Harlow, 2005, pp. 392–393 and 402.
11. O. J. Scherer, *Acc. Chem. Res.*, 1999, **32**, 751.
12. H. J. Breunig, N. Burford and R. Roesler, *Angew. Chem. Int. Ed.*, 2000, **39**, 4148.
13. (a) J. Bai, A. V. Virovets and M. Scheer, *Science*, 2003, **300**, 781; (b) M. Scheer, J. Bai, B. P. Johnson, R. Merkle, A. V. Virovets and C. E. Anson, *Eur. J. Inorg. Chem.*, 2005, 4023; (c) M. Scheer, L. J. Gregoriades, A. V. Virovets, W. Kunz, R. Neueder and I. Krossing, *Angew. Chem. Int. Ed.*, 2006, **45**, 5689; (d) L. J. Gregoriades, H. Krauss, J. Wachter, A. V. Virovets, M. Soerka and M. Scheer, *Angew. Chem. Int. Ed.*, 2006, **45**, 4189.
14. (a) M. Baudler and T. Etzbach, *Chem. Ber.*, 1991, **124**, 1159; (b) M. Baudler and T. Etzbach, *Angew. Chem. Int. Ed. Engl.*, 1991, **30**, 580.
15. E. Urnezius, W. W. Brennessel, C. J. Cramer, J. E. Ellis and P. v. R. Schleyer, *Science*, 2002, **295**, 832.

16. (a) A. Cisar and J. D. Corbett, *Inorg. Chem.*, 1977, **16**, 2482; (b) A. Cisar and J. D. Corbett, *Inorg. Chem.*, 1984, **23**, 770; (c) N. Korber and M. Reil, *Chem. Commun.*, 2002, 84; (d) F. Kraus, J. C. Aschenbrenner and N. Korber, *Angew. Chem. Int. Ed.*, 2003, **42**, 4030.
17. F. Kraus, T. Hanauer and N. Korber, *Inorg. Chem.*, 2006, **45**, 1117.
18. F. Kraus and N. Korber, *Chem. Eur. J.*, 2005, **11**, 5945.
19. O. J. Scherer, H. Sitzmann and G. Wolmershäuser, *Angew. Chem. Int. Ed. Engl.*, 1985, **24**, 351.
20. (a) W. Schmettow, A. Lipka and H. G. von Schnering, *Angew. Chem.. Int. Ed. Engl.*, 1974, **13**, 345; (b) H. G. von Schnering, T. Meyer, W. Hönle, W. Schmettow, U. Hinze, W. Bauhofer and G. Kliche, *Z. Anorg. Allg. Chem.*, 1987, **553**, 261.
21. (a) W. Hönle, G. Krogull, K. Peters and H. G. von Schnering, *Z. Kristallogr. New. Cryst. Struct.*, 1999, **214**, 17; (b) F. Kraus, T. Hanauer and N. Korber, *Angew. Chem. Int. Ed.*, 2005, **44**, 7200.
22. F. Kraus, J. Schmedt auf der Günne, B. F. DiSalle and N. Korber, *Chem. Commun.*, 2006, 218.
23. N. Korber and F. Richter, *Angew. Chem. Int. Ed. Engl.*, 1997, **36**, 1512.
24. M. Reil and N. Korber, *Z. Anorg. Allg. Chem.*, 2007, **633**, 1599.
25. (a) H. G. von Schnering, J. Wolf, D. Weber, R. Ramirez and T. Meyer, *Angew. Chem. Int. Ed. Engl.*, 1986, **25**, 353; (b) B. W. Eichhorn, S. P. Mattamana, D. R. Gardner and J. C. Fettinger, *Inorg. Chem.*, 1998, **120**, 9708.
26. M. Baudler, *Angew. Chem. Int Ed. Engl.*, 1982, **21**, 492.
27. K. B. Dillon, F. Mathey and J. F. Nixon, *Phosphorus: the Carbon Copy*, Wiley, New York, 1997.
28. M. Baudler and K. Glinka, *Chem. Rev.*, 1993, **93**, 1623.
29. M. C. Fermin and D. W. Stephan, *J. Am. Chem. Soc.*, 1995, **117**, 12645.
30. D. Dou, E. N. Duesler and R. T. Paine, *Inorg. Chem.*, 1999, **38**, 788.
31. (a) J. J. Daly, *J. Chem. Soc.*, 1966, 428; (b) J. J. Daly, *J. Chem. Soc.*, 1964, 6147.
32. W. Weigand, A. W. Cordes and P. N. Swepston, *Acta Crystallogr., Sect. B*, 1981, **37**, 1631.
33. J. Hahn, M. Baudler, C. Krüger and Y.-H. Tsay, *Z. Naturforsch., Teil B*, 1982, **37**, 797.
34. H.-G. Schäfer, W. W. Schoeller, J. Niemann, W. Haug, T. Dabisch and E. Niecke, *J. Am. Chem. Soc.*, 1986, **108**, 7481.
35. (a) H.-G. Ang, S.-G. Ang and Q. Zhang, *J. Chem. Soc. Dalton Trans.*, 1996, 2773; (b) S. J. Geier and D. W. Stephan, *Chem. Commun.*, 2008, 2779.
36. R. Wolf, M. Finger, C. Limburg, A. C. Willis, S. B. Wild and E. Hey-Hawkins, *Dalton Trans.*, 2006, 831.
37. C. A. Dyker and N. Burford, *Chem. Asian J.*, 2008, **3**, 28.
38. (a) N. Burford, C. A. Dyker and A. Decken, *Angew. Chem. Int. Ed.*, 2005, **44**, 2364; (b) N. Burford, C. A. Dyker, M. Lumsden and A. Decken, *Angew. Chem. Int. Ed.*, 2005, **44**, 6196.

39. (a) J. J. Weigand, N. Burford, M. Lumsden and A. Decken, *Angew. Chem. Int. Ed.*, 2006, **45**, 6733; (b) J. J. Weigand, N. Burford and A. Decken, *Eur. J. Inorg. Chem.*, 2008, 4343.
40. S. D. Riegel, N. Burford, M. D. Lumsden and A. Decken, *Chem. Commun.*, 2007, 4668.
41. M. Baudler and B. Makowka, *Z. Anorg. Allg. Chem.*, 1985, **528**, 7.
42. A. Schmidpeter and G. Burget, *Phosphorus Sulfur Silicon*, 1985, **22**, 323.
43. (a) G. Fritz and K. Stoll, *Z. Anorg. Allg. Chem.*, 1986, **538**, 78; (b) G. Fritz, K. Blastoch, K. Stoll, T. Vaahs, D. Hanke and H. W. Schneider, *Phosphorus Sulfur Silicon*, 1987, **30**, 385.
44. A. Schisler, P. Lönnecke, U. Huniar, R. Ahlrichs and E. Hey-Hawkins, *Angew. Chem. Int. Ed.*, 2001, **40**, 4217.
45. A. Schisler, P. Lönnecke, T. Gelbrich and E. Hey-Hawkins, *Dalton Trans.*, 2004, 2895.
46. (a) A. Schisler, P. Lönnecke and E. Hey-Hawkins, *Inorg. Chem.*, 2005, **44**, 461; (b) S. Gómez-Ruiz, A. Schisler, P. Lönnecke and E. Hey-Hawkins, *Chem. Eur. J.*, 2007, **13**, 7974.
47. S. Gómez-Ruiz, R. Wolf, S. Bauer, H. Bittig, A. Schisler, P. Lönnecke and E. Hey-Hawkins, *Chem. Eur. J.*, 2008, **14**, 4511.
48. P. R. Hofmann and K. G. Caulton, *J. Am. Chem. Soc.*, 1975, **97**, 6270.
49. M. Baudler and D. Koch, *Z. Anorg. Allg. Chem.*, 1976, **425**, 227.
50. D. Stein, A. Dransfeld, M. Flock, H. Rüegger and H. Grützmacher, *Eur. J. Inorg. Chem.*, 2006, 4157.
51. (a) J. Geier, H. Rüegger, M. Wörle and H. Grützmacher, *Angew. Chem. Int. Ed.*, 2003, **42**, 3951; (b) J. Geier, J. Harmer and H. Grützmacher, *Angew. Chem. Int. Ed.*, 2004, **43**, 4093.
52. R. Wolf, A. Schisler, P. Lönnecke, C. Jones and E. Hey-Hawkins, *Eur. J. Inorg. Chem.*, 2004, 3277.
53. R. Wolf and E. Hey-Hawkins, *Z. Anorg. Allg. Chem.*, 2006, **632**, 727.
54. I. Haiduc and D. B. Sowerby, in *The Chemistry of Inorganic Homo- and Heterocycles*, ed. I. Haiduc and D. B. Sowerby, Academic Press, London, 1987, Vol. **2**, pp. 701–711.
55. M. Baudler and P. Bachmann, *Angew. Chem. Int. Ed. Engl.*, 1981, **20**, 123.
56. A.-J. Dimaio and A. L. Rheingold, *Chem. Rev.*, 1990, **90**, 169.
57. A. L. Reingold, M. L. Fountain and A.-J. Dimaio, *J. Am. Chem. Soc.*, 1987, **109**, 141.
58. J. Grobe, A. Karst, B. Krebs, M. Läge and E.-U. Würthwein, *Z. Anorg. Allg. Chem.*, 2006, **632**, 599.
59. N. C. Lloyd, H. W. Morgan, B. K. Nicholson and R. S. Ronimus, *Angew. Chem. Int. Ed.*, 2005, **44**, 941.
60. (a) H. J. Breunig and R. Roesler, *Chem. Soc. Rev.*, 2000, **29**, 403; (b) L. Balács and H. J. Breunig, *Coord. Chem. Rev.*, 2004, **248**, 603; (c) H. J. Breunig, *Z. Anorg. Allg. Chem.*, 2005, **631**, 621.
61. K. Issleib, B. Hamann and L. Schmidt, *Z. Anorg. Allg. Chem.*, 1965, **339**, 289.
62. P. P. Power, *J. Chem. Soc. Dalton. Trans.*, 1998, 2939.

63. H. J. Breunig, R. Roesler and E. Lork, *Angew. Chem. Int. Ed. Engl.*, 1997, **36**, 2237.
64. H. J. Breunig, R. Roesler and E. Lork, *Organometallics*, 1998, **17**, 5594.
65. (a) R. Waterman and T. D. Tilley, *Angew. Chem. Int. Ed.*, 2006, **45**, 2926; (b) L. Dostál, R. Jambor, A. Růžička and J. Holeèek, *Organometallics*, 2008, **27**, 2169.
66. H. J. Breunig, R. Roesler and E. Lork, *Angew. Chem. Int. Ed.*, 1998, **37**, 3175.
67. (a) G. Lintl and W. Köstler, *Z. Anorg. Allg. Chem.*, 2002, **628**, 63; (b) G. Lintl, W. Köstler and H. Pritzkow, *Eur. J. Inorg. Chem.*, 2002, 2642.
68. L. Balács, H. J. Breunig and E. Lork, *Angew. Chem. Int. Ed.*, 2002, **41**, 2309.
69. G. Balács, L. Balács, H. J. Breunig and E. Lork, *Organometallics*, 2003, **22**, 2919.
70. (a) H. J. Breunig, R. Roesler and E. Lork, *Angew. Chem. Int. Ed. Engl.*, 1997, **36**, 2819; (b) G. Balács, H. J. Breunig and E. Lork, *Z. Anorg. Allg. Chem.*, 2003, **629**, 1937.
71. H. Althaus, H. J. Breunig, J. Probst, R. Roesler and E. Lork, *J. Organomet. Chem.*, 1999, **585**, 285.
72. (a) A. L. Rheingold, J. E. Lewis and J. M. Bellama, *Inorg. Chem.*, 1973, **12**, 2845; (b) J. J. Daly and F. Sanz, *Helv. Chim. Acta*, 1970, **53**, 1879; (c) P. Choudhury, M. F. El-Shazly, C. Spring and A. L. Rheingold, *Inorg. Chem.*, 1979, **18**, 543.
73. (a) R. Keat, in *The Chemistry of Inorganic Homo- and Heterocycles*, ed. I. Haiduc and D. B. Sowerby, Academic Press, London, 1987, Vol. 2, pp. 467–500; (b) L. Stahl, *Coord. Chem. Rev.*, 2000, **210**, 203.
74. (a) N. Burford, T. S. Cameron, K. D. Conroy, B. Ellis, C. L. B. Macdonald, R. Ovans, A. D. Phillips, P. J. Ragogna and D. Walsh, *Can. J. Chem.*, 2002, **80**, 1404; (b) E. Niecke and D. Gudat, *Angew. Chem. Int. Ed. Engl.*, 1991, **30**, 217.
75. N. Burford, C. A. Dyker, A. D. Phillips, H. A. Spinney, A. Decken, R. McDonald, P. J. Ragogna and A. L. Rheingold, *Inorg. Chem.*, 2004, **43**, 7502.
76. (a) K. W. Muir and J. F. Nixon, *Chem. Commun.*, 1971, 1405; (b) K. W. Muir, *J. Chem. Soc., Dalton Trans.*, 1975, 259.
77. (a) V. S. Reddy, A. J. Elias and A. Vij, *J. Chem. Soc., Dalton Trans.*, 1997, 2167; (b) I. Schranz, L. Stahl and R. J. Staples, *Inorg. Chem.*, 1998, **37**, 1493.
78. W. Zeiss and K. Barlos, *Z. Naturforsch., Teil B*, 1979, **34**, 423.
79. D. A. Harvey, R. Keat and D. S. Rycroft, *J. Chem. Soc., Dalton Trans.*, 1983, 425.
80. F. Garcia, R. A. Kowenicki, L. Riera and D. S. Wright, *Dalton Trans.*, 2005, 2495.
81. N. Burford, K. D. Conroy, J. C. Landry, P. J. Ragogna, M. J. Ferguson and R. McDonald, *Inorg. Chem.*, 2004, **43**, 8245.
82. R. Muragavel, S. S. Krishnamurthy, J. Chandrasekhar and M. Nethaji, *Inorg. Chem.*, 1993, **32**, 5447.

83. (a) N. Burford, T. S. Cameron, K. D. Conroy, B. Ellis, M. Lumsden, C. L. B. MacDonald, R. McDonald, A. D. Phillips, P. J. Ragogna, R. W. Schurko, D. Walsh and R. E. Wasylishen, *J. Am. Chem. Soc.*, 2002, **124**, 14012; (b) N. Burford, T. S. Cameron, C. L. B. MacDonald, R. McDonald, K. N. Robertson, R. W. Schurko and D. Walsh, *Inorg. Chem.*, 2005, **44**, 8058; (c) R. J. Davidson, J. J. Weigand, N. Burford, T. S. Cameron, A. Decken and U. Werner-Zwanziger, *Chem. Commun.*, 2007, 4671; (d) D. Michalik, A. Schulz, A. Villinger and N. Weding, *Angew. Chem. Int. Ed.*, 2008, **47**, 6465.
84. W. Zeiss, W. Schwarz and H. Hess, *Angew. Chem. Int. Ed. Engl.*, 1977, **16**, 407.
85. M. S. Balakrishna, V. S. Reddy, S. S. Krishnamurthy, J. F. Nixon and J. C. T. R. B. S. Laurent, *Coord. Chem. Rev.*, 1994, **129**, 1.
86. P. Chandrasekharan, J. T. Mague and M. S. Balakrishna, *Inorg. Chem.*, 2005, **44**, 7925.
87. P. Chandrasekharan, J. T. Mague and M. S. Balakrishna, *Organometallics*, 2005, **24**, 3780.
88. (a) L. Grocholl, L. Stahl and R. J. Staples, *Chem. Commun.*, 1997, 1465; (b) D. F. Moser, C. J. Carrow, L. Stahl and R. J. Staples, *J. Chem. Soc. Dalton Trans.*, 2001, 1246.
89. K. V. Axenov, I. Kilpeläinen, M. Klinga, M. Leskelä and T. Repo, *Organometallics*, 2006, **25**, 463.
90. K. V. Axenov, M. Klinga, M. Leskelä, V. Kotoy and T. Repo, *Eur. J. Inorg. Chem.*, 2004, 4702.
91. G. R. Lief, C. J. Carrow, L. Stahl and R. J. Staples, *Chem. Commun.*, 2002, 1562.
92. F. Garcia, J. M. Goodman, R. A. Kowenicki, I. Kozu, M. McPartlin. M. A. Silva, L. Riera, A. D. Woods and D. S. Wright, *Chem. Eur. J.*, 2004, **10**, 6066.
93. J. K. Brask, T. Chivers, M. L. Krahn and M. Parvez, *Inorg. Chem.*, 1999, **38**, 290.
94. A. Bashall, A. D. Bond, E. L. Doyle, F. Garcia, S. Kidd, G. T. Lawson, M. C. Parry, M. McPartlin, A. D. Woods and D. S. Wright, *Chem. Eur. J.*, 2002, **8**, 3377.
95. (a) F. Dodds, F. Garcia, R. A. Kowenicki, M. McPartlin, A. Steiner and D. S. Wright, *Chem. Commun.*, 2005, 3733; (b) F. Dodds, F. Garcia, R. A. Kowenicki, S. P. Parsons, M. McPartlin and D. S. Wright, *Dalton Trans.*, 2006, 4235.
96. (a) J. Liebig, *Justus Liebigs Ann. Chem.*, 1834, **11**, 139; (b) H. N. Stokes, *J. Am. Chem. Soc.*, 1895, **17**, 275.
97. (a) H. R. Allcock, *Chemistry and Applications of Polyphosphazenes*, Wiley-Interscience, New York, 2003; (b) V. Chandrasekhar, *Inorganic and Organometallic Polymers*, Springer-Verlag, Berlin, 2005; (c) J. E. Mark, H. R. Allcock and R. West, *Inorganic Polymers*, 2nd edn., Oxford University Press, Toronto, 2005.

98. (a) V. Chandrasekhar, P. Thilagar and B. M. Pandian, *Coord. Chem. Rev.*, 2007, **251**, 1045; (b) V. Chandrasekhar and S. Nagendran, *Chem. Soc. Rev.*, 2001, **30**, 193.
99. (a) A. Baceiredo, G. Bertrand, J.-P. Majoral, G. Sicard, J. Jaud and J. Galy, *J. Am. Chem. Soc.*, 1984, **106**, 6088; (b) R. Wehmschulte, M. A. Khan and S. I. Hossain, *Inorg. Chem.*, 2001, **40**, 2756; (c) J. Tireé, D. Gudat, M. Nieger and E. Niecke, *Angew. Chem. Int. Ed.*, 2001, **40**, 3025.
100. C. Hubrich, D. Michalik, A. Schulz and A. Villinger, *Z. Anorg. Allg. Chem.*, 2008, **634**, 1403.
101. (a) J. Novosad and M. Alberti, in *Inorganic Experiments*, 2nd edn., ed. J. D. Woollins, Wiley-VCH, Weinheim, 2003, p. 176; (b) H. R. Allcock, C. A. Crane, C. T. Morrissey and M. A. Olshavsky, *Inorg. Chem.*, 1999, **38**, 280.
102. M. Göbel, K. Karaghiosoff and T. M. Klapötke, *Angew. Chem. Int. Ed.*, 2006, **45**, 6037.
103. R. P. Singh, A. Vij, R. L. Kirchmeier and J. M. Shreeve, *Inorg. Chem.*, 2000, **39**, 375.
104. R. H. Neilson and P. Wisian-Neilson, *Inorg. Chem.*, 1980, **19**, 1875.
105. T. N. Ranganathan, S. M. Todd and N. L. Paddock, *Inorg. Chem.*, 1973, **12**, 316.
106. F. Trautner, R. P. Singh, R. L. Kirchmeier, J. M. Shreeve and H. Oberhammer, *Inorg. Chem.*, 2000, **39**, 5398.
107. A. J. Elias, B. Twamley, R. Haist, H. Oberhammer, G. Henkel, B. Krebs, E. Lork, R. Mews and J. M. Shreeve, *J. Am. Chem. Soc.*, 2001, **123**, 10299.
108. (a) E. Lork, D. Bohler and R. Mews, *Angew. Chem. Int. Ed. Engl.*, 1995, **34**, 2696; (b) E. Lork, P. G. Watson and R. Mews, *J. Chem. Soc. Chem. Commun.*, 1995, 1717.
109. P. Wisian-Neilson, R. S. Johnson, H. Zhang, J.-H. Jung, R. H. Neilson, J. Ji, W. H. Watson and M. Krawiec, *Inorg. Chem.*, 2002, **41**, 4775.
110. (a) H. R. Allcock and D. B. Patterson, *Inorg. Chem.*, 1977, **16**, 197; (b) H. R. Allcock, G. S. McDonnell and J. L. Desorcie, *Inorg. Chem.*1990, **29**, 3839.
111. J. H. Jung, J. C. Pomery, H. Zhang and P. Wisian-Neilson, *J. Am. Chem. Soc.*, 2003, **50**, 15537.
112. (a) A. J. Heston, M. J. Panzer, W. J. Youngs and C. A. Tessier, *Inorg. Chem.*, 2005, **44**, 6518; (b) M. Gonsior, S. Antonijevic and I. Krossing, *Chem. Eur. J.*, 2006, **12**, 1997.
113. Y. Zhang, F. S. Tham and C. A. Reed, *Inorg. Chem.*, 2006, **45**, 10446.
114. R. Boomishankar, J. Ledger, J.-B. Guilbaud, N. L. Campbell, J. Bacsa, R. Bonar-Law, Y. Z. Khimyak and A. Steiner, *Chem. Commun.*, 2007, 5152.
115. A. Steiner and D. S. Wright, *Angew. Chem. Int. Ed. Engl.*, 1996, **35**, 936.
116. G. T. Lawson, F. Rivals, M. Tascher, C. Jacob, J. F. Bickley and A. Steiner, *Chem. Commun.*, 2000, 341.

117. A. Steiner and D. S. Wright, *Chem. Commun.*, 1997, 283.
118. W. Schnick, *Angew. Chem. Int. Ed. Engl.*, 1993, **32**, 806.
119. D. P. Gates, R. Ziembinski, A. L Rheingold, B. S. Haggerty and I. Manners, *Angew. Chem. Int. Ed. Engl.*, 1994, **33**, 2277.
120. (a) D. P. Gates, L. M. Liable-Sands, G. P. A. Yap, A. L. Rheingold and I. Manners, *J. Am. Chem. Soc.*, 1997, **119**, 1125; (b) D. P. Gates, A. R. McWilliams, R. Ziembinski, L. M. Liable-Sands, I. A. Guzei, G. P. A. Yap, A. L. Rheingold and I. Manners, *Chem. Eur. J.*, 1998, **4**, 1489.
121. (a) A. R. McWilliams, E. Rivard, A. J. Lough and I. Manners, *Chem. Commun.*, 2002, 1102; (b) E. Rivard, P. J. Ragogna, A. R. McWilliams, A. J. Lough and I. Manners, *Inorg. Chem.*, 2005, **44**, 6788.
122. T. Chivers, M. N. S. Rao and J. F. Richardson, *J. Chem. Soc., Chem. Commun.*, 1982, 982.
123. (a) T. Chivers and M. N. S. Rao, *Inorg. Chem.*, 1984, **23**, 3605; (b) T. Chivers, M. N. S. Rao and J. F. Richardson, *J. Chem. Soc., Chem. Commun.*, 1983, 702.
124. T. Chivers, M. N. S. Rao and J. F. Richardson, *J. Chem. Soc., Chem. Commun.*, 1983, 186.
125. R. T. Oakley, *J. Chem. Soc., Chem. Commun.*, 1986, 596.
126. J. A. Dodge, H. R. Allcock, G. Renner and O. Nuyken, *J. Am. Chem. Soc.*, 1990, **112**, 1268.
127. S. Pohl, O. Petersen and H. W. Roesky, *Chem. Ber.*, 1979, **112**, 1545.
128. J. C. van de Grampel, *Coord. Chem. Rev.*, 1992, **112**, 247.
129. D. Suzuki, H. Akagi and K. Matsumura, *Synth. Commun.*, 1983, 369.
130. Y. Ni, A. J. Lough, A. L. Rheingold and I. Manners, *Angew. Chem. Int. Ed. Engl.*, 1995, **34**, 998.
131. H. N. Stokes, *J. Am. Chem. Soc.*, 1897, **19**, 782.
132. H. R. Allcock and R. L. Kugel, *J. Am. Chem. Soc.*, 1965, **87**, 4216.
133. H. R. Allcock, *Adv. Mater.*, 1994, **6**, 106.
134. R. de Jaeger and M. Gleria (eds), *Phosphazenes: a Worldwide Insight*, Nova Science, Hauppauge, NY, 2002.
135. R. H. Neilson and P. Wisian-Neilson, *Chem. Rev.*, 1988, **88**, 541.
136. P. Wisian-Neilson and R. Neilson, *J. Am. Chem. Soc.*, 1980, **102**, 2848.
137. P. Wisian-Neilson, Poly(alkyl/arylphosphazenes) and their derivatives, in *Phosphazenes: a Worldwide Insight*, ed. R. de Jaeger and M. Gleria, Nova Science, Hauppauge, NY, 2002, Chapter 5.
138. Y. Zhang, K. Huynh, I. Manners and C. A. Reed, *Chem. Commun.*, 2008, 494.
139. R. A. Montague and K. Matyjaszewski, *J. Am. Chem. Soc.*, 1990, **112**, 6721.
140. K. Matyjaszewski, U. Franz, R. A. Montague and M. L. White, *Polymer*, 1994, **23**, 5005.
141. G. D'Halluin, R. de Jaeger, J. P. Chambrette and P. Potin, *Macromolecules*, 1992, **25**, 1254.
142. C. H. Honeyman, I. Manners, C. T. Morrissey and H. R. Allcock, *J. Am. Chem. Soc.*, 1995, **117**, 7035.

143. B. Wang, E. Rivard and I. Manners, *Inorg. Chem.*, 2002, **41**, 1690.
144. H. R. Allcock, C. A. Crane, C. T. Morrissey, J. M. Nelson, S. D. Reeves, C. H. Honeyman and I. Manners, *Macromolecules*, 1996, **29**, 7740.
145. J. M. Nelson, A. P. Primrose, T. J. Hartie, H. R. Allcock and I. Manners, *Macromolecules*, 1998, **31**, 947.
146. E. Rivard, A. J. Lough and I. Manners, *Inorg. Chem.*, 2004, **43**, 2765.
147. I. Manners, G. Renner, H. R. Allcock and O. Nuyken, *J. Am. Chem. Soc.*, 1989, **111**, 5478.
148. J. A. Dodge, I. Manners, G. Renner, H. R. Allcock and O. Nuyken, *J. Am. Chem. Soc.*, 1990, **112**, 1268.
149. M. Liang and I. Manners, *J. Am. Chem. Soc.*, 1991, **113**, 4044.
150. Z. Wang, A. R. McWilliams, C. E. B Evans, X. Lu, S. Chung, M. A. Winnik and I. Manners, *Adv. Funct. Mater.*, 2002, **12**, 415.
151. M. S. Balakrishna, D. J. Eisler and T. Chivers, *Chem. Soc. Rev.*, 2007, **36**, 650.
152. G. A. Olah and A. A. Oswald, *Can. J. Chem.*, 1960, **38**, 1428.
153. N. Burford, J. C. Landry, M. J. Ferguson and R. McDonald, *Inorg. Chem.*, 2005, **44**, 5897.
154. J. Weiss and W. Eisenhuth, *Z. Naturforsch., Teil B*, 1967, **22**, 454.
155. E. V. Avtomonov, K. Megges, X. Li, J. Lorberth, S. Wocadlo, W. Massa, K. Harms, A. V. Churakov and J. A. K. Howard, *J. Organomet. Chem.*, 1997, **544**, 79.
156. D. C. Haagenson and L. Stahl, *Inorg. Chem.*, 2001, **40**, 4491.
157. N. Burford, E. Edelstein, J. C. Landry, M. J. Ferguson and R. McDonald, *Chem. Commun.*, 2005, 5074.
158. N. Burford, T. S. Cameron, K.-C. Lam, D. J. LeBlanc, C. L. B. McDonald, A. D. Philips, A. L. Rheingold, L. Stark and D. Walsh, *Can. J. Chem.*, 2001, **79**, 342.
159. G. G. Briand, T. Chivers and M. Parvez, *Can. J. Chem.*, 2003, **81**, 169.
160. R. Bryant, S. C. James, J. C. Jeffery, N. C. Norman, A. G. Orpen and U. Weckenmann, *J. Chem. Soc., Dalton Trans.*, 2000, 4007.
161. M. A. Beswick and D. S. Wright, *Coord. Chem. Rev.*, 1998, **176**, 373.
162. A. J. Edwards, M. A. Paver, M. A. Rennie, P. R. Raithby, C. A. Russell and D. S. Wright, *J. Chem. Soc., Dalton Trans.*, 1994, 2963.
163. R. Bohra, H. W. Roesky, M. Noltemeyer and G. M. Sheldrick, *Acta Crystallogr., Sect. C*, 1984, **40**, 1150.
164. D. J. Eisler and T. Chivers, *Inorg. Chem.*, 2006, **45**, 10734.
165. (a) W. T. Reichle, *Tetrahedron Lett.*, 1962, **2**, 51; (b) V. Krieg and J. Weidlein, *Angew. Chem. Int. Ed. Engl.*, 1971, **7**, 516.
166. L. K. Krannich, U. Thewalt, W. J. Cook, S. R. Jain and H. H. Sisler, *Inorg. Chem.*, 1973, **12**, 2304.
167. R. Bohra, H. W. Roesky, J. Lucas, M. Noltemeyer and G. M. Sheldrick, *J. Chem. Soc., Dalton Trans.*, 1983, 1011.
168. M. Baier, P. Bissinger and H. Schmidbaur, *Chem. Ber.*, 1993, **126**, 351.
169. M. J. Begley, D. B. Sowerby and R. J. Tillott, *J. Chem. Soc., Dalton Trans.*, 1974, 2527.

170. (a) K.-N. Han, D. Whang, H.-J. Lee, Y. Do and K. Kim, *Inorg. Chem.*, 1993, **32**, 2597; (b) S. Kamimura, S. Kuwata, M. Iwasaki and Y. Ishii, *Dalton Trans.*, 2003, 2666.
171. S. Kamimura, S. Kuwata, M. Iwasaki and Y. Ishii, *Inorg. Chem.*, 2004, **43**, 399.
172. E. Niecke, H. Zorn, B. Krebs and G. Henkel, *Angew. Chem. Int. Ed. Engl.*, 1980, **19**, 709.
173. (a) D. W. Chasar, J. P. Fackler, A. M. Mazany, R. A. Komoroski and W. J. Kroenke, *J. Am. Chem. Soc.*, 1986, **108**, 5956; (b) D. W. Chasar, J. P. Fackler, R. A. Komoroski, W. J. Kroenke and A. M. Mazany, *J. Am. Chem. Soc.*, 1987, **109**, 5690.
174. E. H. Wong, M. M. Turnbull, K. D. Hutchinson, C. Valdez, E. J. Gabe, F. L. Lee and Y. Le Page, *J. Am Chem. Soc.*, 1988, **110**, 8422.
175. E. H. Wong, X. Sun, E. J. Gabe, F. L. Lee and J.-P. Charland, *Organometallics*, 1991, **10**, 3010.
176. X. Sun, E. H. Wong, M. M. Turnbull, A. L. Rheingold, B. E. Waltermire and R. L. Ostrander, *Organometallics*, 1995, **14**, 83.
177. W. S. Sheldrick and I. M. Müller, *Coord. Chem. Rev.*, 1999, **182**, 125.
178. (a) M. Durand and J.-P. Laurent, *J. Organomet. Chem.*, 1974, **77**, 225; (b) W. S. Sheldrick and T. Häusler, *Z. Naturforsch., Teil B*, 1993, **48**, 1069.
179. (a) A.-J. Dimaio and A. L. Rheingold, *Organometallics*, 1991, **10**, 3764; (b) A. M. Arif, A. H. Cowley and M. Pakulski, *J. Chem. Soc., Chem. Commun.*, 1987, 165.
180. (a) W. S. Sheldrick and T. Häusler, *Z. Anorg. Allg. Chem.*, 1993, **619**, 1984; (b) I. M. Müller and W. S. Sheldrick, *Eur. J. Inorg. Chem.*, 1998, 1999.
181. (a) I. M. Müller and W. S. Sheldrick, *Z. Anorg. Allg. Chem.*, 1999, **625**, 443; (b) M. Heller, O. Teichert and W. S. Sheldrick, *Z. Anorg. Allg. Chem.*, 2005, **631**, 709.
182. N. Tokitoh, *J. Organomet. Chem.*, 2000, **611**, 217.
183. (a) N. Tokitoh, Y. Arai, R. Okazaki and S. Nagase, *Science*, 1997, **277**, 78; (b) N. Tokitoh, Y. Arai, T. Sasamori, R. Okazaki, S. Nagase, H. Uekusa and Y. Ohashi, *J. Am. Chem. Soc.*, 1998, **120**, 433.
184. (a) C. E. Housecroft and A. G. Sharpe, *Inorganic Chemistry*, 2nd edn., Pearson Education, Harlow, 2005, pp. 426–427; (b) M. E. Jason, T. Ngo and S. Rahman, *Inorg. Chem.*, 1997, **36**, 2633; (c) M. E. Jason, *Inorg. Chem.*, 1997, **36**, 2641.
185. R. Blachnik, P. Lönnecke, K. Boldt and B. Engelen, *Acta Crystallogr. Sect. C*, 1994, **50**, 659.
186. (a) P. C. Minshall and G. M. Sheldrick, *Acta Crystallogr., Sect. B*, 1978, **34**, 1378; (b) H. Falius, W. Krause and W. S. Sheldrick, *Angew. Chem. Int. Ed. Engl.*, 1981, **20**, 103.
187. M. Baudler, H. Suchomel, G. Fürstenburg and U. Schings, *Angew. Chem. Int. Ed. Engl.*, 1981, **20**, 1044.
188. E. Niecke and R. Rüger, *Angew. Chem. Int. Ed. Engl.*, 1983, **22**, 155.

189. (a) M. Yoshifuji, K. Shibayama, N. Inamoto, K. Hirotsu and T. Higuchi, *J. Chem. Soc., Chem. Commun.*, 1983, 862; (b) N. Nagahora, T. Sasamori and N. Tokitoh, *Heteroat. Chem.*, 2008, **19**, 443; (c) T. Sasamori and N. Tokitoh, *Dalton Trans.*, 2008, 1395.
190. M. Yoshifuji, K. Ando, K. Shibayama, N. Inamoto, K. Hiotsu and T. Higuchi, *Angew. Chem. Int. Ed. Engl.*, 1983, **22**, 418.
191. (a) R. A. Cherkasov, G. A. Kutyrex and A. N. Pudovik, *Tetrahedron*, 1985, **41**, 2567; (b) M. C. Cava and M. I. Levinson, *Tetrahedron*, 1985, **41**, 5061.
192. H. Z. Lecher, R. A. Greenwood, K. C. Whitehouse and T. H. Chao, *J. Am. Chem. Soc.*, 1952, **74**, 4933.
193. (a) M. St. J. Foreman, A. M. Z. Slawin and J. D. Woollins, *Heteroat. Chem.*, 1999, **10**, 651; (b) M. St. J. Foreman, A. M. Z. Slawin and J. D. Woollins, *J. Chem. Soc., Dalton Trans.*, 1996, 3653.
194. M. St. J. Foreman, A. M. Z. Slawin and J. D. Woollins, *Chem. Commun.*, 1997, 1269.
195. (a) N. Burford, R. E. v. H. Spence and R. D. Rogers, *J. Chem. Soc., Dalton Trans.*, 1990, 3611; (b) N. Burford, P. Losier, S. Mason, P. K. Bakshi and T. S. Cameron, *Inorg. Chem.*, 1994, **33**, 5613.
196. B. Cetinkaya, P. B. Hitchcock, M. F. Lappert, A. J. Thorne and H. Goldwhite, *J. Chem. Soc., Chem. Commun.*, 1982, 691.
197. (a) C. Lensch, W. Clegg and G. M. Sheldrick, *J. Chem. Soc., Dalton Trans.*, 1984, 723; (b) C. Lensch and G. M. Sheldrick, *J. Chem. Soc., Dalton Trans.*, 1984, 2855.
198. (a) G. Hua and J. D. Woollins, *Angew. Chem. Int. Ed.*, 2009, **48**, 1368; (b) P. Bhattacharyya, A. M Z. Slawin and J. D. Woollins, *Chem. Eur. J.*, 2002, **8**, 2705.
199. K. Karaghiosoff, K. Eckstein and R. Motzer, *Phosphorus Sulfur Silicon*, 1994, **93–94**, 185.
200. G. Grossman, G. Ohms, K. Krüger, K. Karaghiosoff, K. Eckstein, J. Hahn, A. Hopp, O. Malkina and P. Hrobarik, *Z. Anorg. Allg. Chem.*, 2001, **627**, 1269.
201. S. W. Hall, M. J. Pilkington, A. M. Z. Slawin, D. J. Williams and J. D. Woollins, *Polyhedron*, 1991, **10**, 261.
202. (a) J. T. Shore, W. T. Pennington, M. C. Noble and A. W. Cordes, *Phosphorus Sulfur Silicon*, 1988, **39**, 153; (b) P. Bhattacharyya, A. M Z. Slawin and J. D. Woollins, *J. Chem. Soc., Dalton Trans.*, 2001, 300.
203. I. Baxter, A. F. Hill, J. M. Malget, A. J. P. White and D. J. Williams, *Chem. Commun.*, 1997, 2049.
204. I. P. Gray, P. Bhattacharyya, A. M Z. Slawin and J. D. Woollins, *Chem. Eur. J.*, 2005, **11**, 6221.
205. (a) J. Bethke, K. Karaghiosoff and L. Wessjohann, *Tetrahedron Lett.*, 2003, **44**, 6911; (b) G. Hua, Y. Li, A. M. Z. Slawin and J. D. Woollins, *Org. Lett.*, 2006, **8**, 5251.
206. (a) W.-W. du Mont and T. Severengiz, *Z. Anorg. Allg. Chem.*, 1993, **619**, 1083; (b) S. Grimm, K. Karaghiasoff, P. Mayer and D. Ross, *Phosphorus*

Sulfur Silicon, 2001, **169**, 51; (c) T. Sasamori, E. Mieda, A. Tsurusaki, N. Nagahora and N. Tokitoh, *Phosphorus Sulfur Silicon*, 2008, **183**, 998.
207. (a) J. Konu, T. Chivers and H. M. Tuononen, *Chem. Commun.*, 2006, 1624; (b) J. Konu, T. Chivers and H. M. Tuononen, *Inorg. Chem.*, 2006, **45**, 10678.
208. W. Sheldrick and M. Wachhold, *Coord. Chem. Rev.*, 1998, **176**, 211.
209. W. S. Sheldrick and J. Kaub, *Z. Naturforsch., Teil B*, 1985, **40**, 1020.
210. W. Bensch, C. Näther and R. Stähler, *Chem. Commun.*, 2001, 477.
211. K. Wendel and U. Müller, *Z. Anorg. Allg. Chem.*, 1995, **621**, 979.
212. B. H. Christian, R. J. Gillespie and J. F. Sawyer, *Inorg. Chem.*, 1981, **20**, 3410.
213. J. Beck, M. Dolg and S. Schlüter, *Angew. Chem. Int. Ed.*, 2001, **40**, 2287.
214. J. Beck and S. Schlüter, *Z. Anorg. Allg. Chem.*, 2005, **631**, 569.
215. T. Häusler and W. S. Sheldrick, *Z. Naturforsch., Teil B*, 1994, **49**, 1215.
216. A.-J. Dimaio and A. L. Rheingold, *Inorg. Chem.*, 1990, **29**, 798.
217. C. Couret, J. Escudié, Y. Madaule, H. Ranaivonjatovo and J.-G. Wolf, *Tetrahedron Lett.*, 1983, **24**, 2769.
218. A. W. Cordes, P. D. Gwinup and M. C. Malmstrom, *Inorg. Chem.*, 1972, **11**, 836.
219. T. Häusler and W. S. Sheldrick, *Chem. Ber.*, 1996, **129**, 131.
220. O. M. Kekia and A. L. Rheingold, *Organometallics*, 1998, **17**, 726.
221. D. Herrman, K. Klostermann and A. Franke, *Z. Anorg. Allg. Chem.*, 1977, **431**, 164.
222. N. Tokitoh, Y. Arai, T. Sasamori, R. Okazaki, S. Nagase, H. Uekesa and Y. Ohashi, *J. Am. Chem. Soc.*, 1998, **120**, 433.
223. N. Tokitoh, Y. Arai, R. Okazaki and S. Nagase, *Science*, 1997, **277**, 78.
224. T. Sasamori and N. Tokitoh, *Dalton Trans.*, 2008, 1395.
225. M. A. Mohammed, K. H. Ebert and H. J. Breunig, *Z. Naturforsch., Teil B*, 1996, **51**, 149.

CHAPTER 12
Group 16: Rings and Polymers

Ring systems of the group 16 elements reflect the strong tendency of the chalcogens to catenate, as illustrated by the formation of an extensive series of homocyclic rings cyclo-S_n ($n = 6-26$) (Section 12.1.1).[1] Although the smaller rings have been characterised for selenium ($n = 6-8$), homocyclic tellurium rings have only been stabilised by trapping in solid-state materials. As a result of the relatively weak S–S bond, especially in the smaller allotropes cyclo-S_6 and cyclo-S_7, the formation of polymeric sulfur *via* ring-opening polymerisation occurs at relatively mild temperatures (Section 12.5). Although polymeric sulfur does not have any useful properties, the photoreceptive properties of grey polymeric selenium are responsible for the use of this chalcogen (or As_2Se_3) in photocopiers. The only stable allotrope of tellurium has a helical chain structure (Section 12.5).

A fascinating aspect of the chemistry of homocyclic chalcogen rings is the facility with which cationic systems are formed from non-metallic elements (Section 12.3). These highly electrophilic ring systems are stabilised in the solid state by combination with weakly nucleophilic counteranions. The chalcogens also form a wide variety of homonuclear dianions which, in the case of sulfur, are invariably linear, unbranched species *catena*-S_x^{2-} ($x = 2-8$) (Section 12.4). In contrast, the dianions of the heavier chalcogens (Se and Te) form bicyclic and spirocyclic structures, in addition to linear arrangements. The formation of bicyclic or polycyclic structures, in preference to monocyclic systems, is also a noteworthy trend for the numerous cations formed by selenium and tellurium.

With one exception, heterocyclic systems involving a chalcogen and another group 13, 14 or 15 element have been discussed in Chapters 9, 10 and 11, respectively. An intriguing aspect of the chemistry of these inorganic heterocycles is their use as precursors to compounds containing terminal bonds between p-block elements of these groups and a chalcogen (see Sections 9.6.2 and 10.7.2). The exception is the combination of a chalcogen and nitrogen in a

---

Inorganic Rings and Polymers of the p-Block Elements: From Fundamentals to Applications
By Tristram Chivers and Ian Manners
© Tristram Chivers and Ian Manners 2009
Published by the Royal Society of Chemistry, www.rsc.org

ring system. A wide variety of chalcogen–nitrogen heterocycles, both saturated and unsaturated, are known. The saturated systems can be viewed as derivatives of chalcogen homocycles in which one or more chalcogen atoms are replaced by an imido group (NR; R = H, alkyl or aryl) (Section 12.6.1). Unsaturated sulfur–nitrogen rings involving two-coordinate sulfur form a fascinating series of neutral, cationic and anionic heterocycles, which have been described as 'electron-rich aromatics' (see Section 4.1.2.3). An accurate, detailed description of the electronic structures of these inorganic rings continues to pose challenges to theoretical chemists. The ability of sulfur to adopt a number of different formal oxidation states (+2, +4 or +6) in which the sulfur centres are two-, three- or four- coordinate is illustrated by the different classes of ring systems formed by sulfur and nitrogen (Section 12.6.2). The remarkable properties of poly(sulfur nitride), $(SN)_x$, in which the alternation of two non-metals in a linear chain gives rise to a conducting material, are another noteworthy feature of sulfur–nitrogen chemistry (Section 12.7).

12.1 NEUTRAL SULFUR, SELENIUM AND TELLURIUM RINGS

12.1.1 Homoatomic Rings

The synthesis, identification, X-ray structures and coordination chemistry of cyclic sulfur allotropes have been discussed in earlier parts of this book, so only a brief review with appropriate cross-references will be given here; more emphasis will be accorded to selenium and tellurium analogues. The existence of an extensive series of sulfur homocycles cyclo-S_n ($n = 6$–22) is demonstrated convincingly by high-performance liquid chromatography on the mixture of cyclic allotropes formed from the reactions of SCl_2 with potassium iodide in CS_2 (see Figure 3.3 in Section 3.2). This reaction, which also produces iodine, presumably involves the decomposition of diodosulfanes S_nI_2 via homolytic cleavage of the weak S–I bonds. A similar series of cyclic sulfur allotropes cyclo-S_n ($n = 6$–26), together with polymeric sulfur, are also formed when elemental sulfur is heated above its melting point (Section 12.5).[1b] Although mass spectrometry has been used to identify individual sulfur allotropes, e.g. cyclo-S_{12},[2] their thermal sensitivity makes the use of this technique difficult, especially with high-energy electron ionisation (see Section 3.3). A more informative method of characterisation is vibrational spectroscopy, since the number of peaks observed in the vibrational spectra depends on the molecular size and also on the symmetry of the molecule.[3] Raman spectroscopy is the method of choice for this application, because the readily polarisable S–S vibrations exhibit strong Raman intensities (see Figure 3.8 in Section 3.5).

The synthesis of specific sulfur allotropes is achieved by reactions of titanocene pentasulfide Cp_2TiS_5 with dichlorosulfanes S_nCl_2 ($n = 2, 4, 6$) or sulfuryl chloride SO_2Cl_2 [see eqns (2.1) and (6.1)].[4] The X-ray structures of many of these cyclic sulfur allotropes have been determined (see Figure 6.2 in Section 6.1). In general, they adopt puckered molecular conformations, e.g. the crown-shaped cyclo-S_8, which minimise repulsions between non-bonding electron pairs on neighbouring sulfur atoms. A notable exception is cyclo-S_7 in which the sulfur atoms in an S–S–S–S- unit are in a planar arrangement with an elongated central S–S bond of 2.18 Å, cf. 2.06 Å in cyclo-S_8 (see Section 3.1 and Figure 3.2).[5]

There are three crystalline forms of cyclo-Se_8. Monoclinic α- and β-selenium are obtained by crystallising amorphous selenium from CS_2,[6a,b] whereas monoclinic γ-selenium is formed upon decomposition of $Se_4(NC_5H_{10})_2$ in CS_2.[6c] In all three forms of monoclinic selenium the eight-membered ring is crown-shaped (D_{4d} symmetry),[6a–c] cf. cyclo-S_8. The mean Se–Se bond lengths of α- and β-selenium, which are ca 2.33 Å at room temperature, become significantly shorter (by ca 0.03 Å) at −150 °C.[6d,e] Cyclo-Se_6 adopts a chair conformation in the solid state,[7] cf. cyclo-S_6.

Cyclic selenium molecules may be stabilised by encapsulation in composite materials (adducts) that are formed under solvothermal/hydrothermal conditions or *via* solid-state reactions. For example, heating a mixture of $PdBr_2$ and selenium at 150 °C for 10 days produces the compound $PdBr_2 \cdot$ cyclo-Se_6 in which two cyclo-Se_6 molecules are *trans*-coordinated to $PdBr_2$; these monomeric units are connected to give a one-dimensional chain through the opposite selenium atoms of the Se_6 rings (Figure 12.1a).[8] Other examples of adducts involving neutral cyclic selenium molecules that are entrapped in a metal halide matrix include $(AgI)_2$ cyclo-Se_6,[9] $K_3PSe_4 \cdot$ 2cyclo-Se_6[10] and $Rb_2[Pd(Se_4)_2] \cdot$ cyclo-Se_8.[11] The [Rb(cyclo-Se_8)]$^+$ cations in the last complex form a polymeric chain in which the Rb^+ cation is sandwiched between tetradentate, crown-shaped Se_8 ligands (Figure 12.1b). In the non-stoichiometric molybdenum heteropolychalcogenide $[\{(NH_4)_2[Mo_3S_{11.72}Se_{1.28}]\}_2(Se_{12})]$[12] the cyclo-$Se_{12}$ molecule exhibits a structure similar to that of cyclo-S_{12} (see Figure 6.2).[2b]

Although the crystal structure of cyclo-Se_7 is unknown,[13] vibrational spectra indicate a conformation similar to that of cyclo-S_7.[5] This ring conformation is adopted in the structures of the polyselenides [Na(12-crown-4)]$_2$[Se_8]$\cdot$(cyclo-Se_6, cyclo-Se_7)[14a] (Figure 12.2) and [Et_4N]$_2$[Se_5]$\cdot$1/2cyclo-$Se_6 \cdot$ cyclo-Se_7[14b] in which

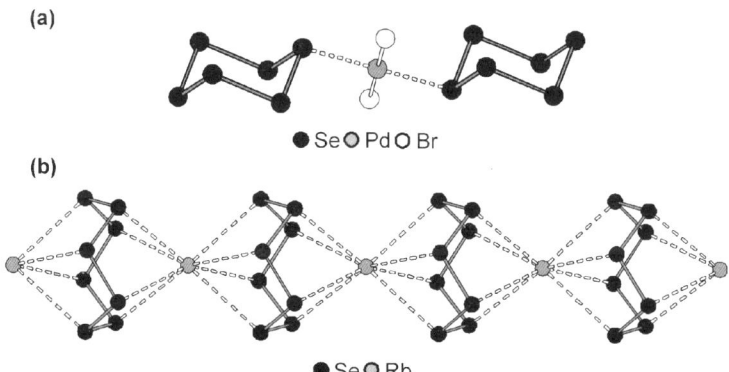

Figure 12.1 (a) Coordination of cyclo-Se_6 to palladium in $PdBr_2 \cdot Se_6$ and (b) the [Rb(cyclo-Se_8)]$^+$ chain in $Rb_2[Pd(Se_4)_2] \cdot Se_8$.

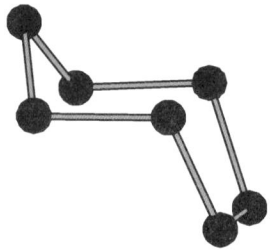

Figure 12.2 Conformation of cyclo-Se$_7$ in [Na(12-crown-4)]$_2$[Se$_8$] · (Se$_6$,Se$_7$).

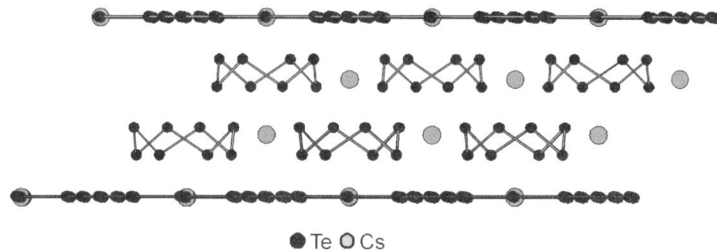

● Te ○ Cs

Figure 12.3 Cyclo-Te$_8$ in the Cs$_3$Te$_{22}$ lattice.

cyclo-Se$_7$ molecules are interlayered with cyclo-Se$_6$. The chair conformation of cyclo-Se$_7$ is maintained in the η^2-Se,Se' coordination complex [Re$_2$I$_2$(CO)$_6$(Se$_7$)] (see Figure 7.4 in Section 7.1.1).

NMR spectroscopy (^{77}Se, $I = 1/2$, 7%) is a powerful technique for identifying cyclic selenium molecules, especially the heteroatomic ring systems that contain sulfur or tellurium in addition to selenium, for which several isomers are possible for most compositions (Section 12.1.2). Solutions of monoclinic selenium in CS$_2$ have been shown by high-performance liquid chromatography to form an equilibrium mixture of cyclo-Se$_8$, cyclo-Se$_7$ and cyclo-Se$_6$.[15] The ^{77}Se NMR spectra of such solutions show two singlets that are attributable to cyclo-Se$_8$ and cyclo-Se$_6$ with relative intensities that correspond to a molar ratio of ca 6:1.[16] No resonance is observed for cyclo-Se$_7$, presumably as a result of the fluxional behaviour (pseudorotation) of the seven-membered ring (Section 12.1.2).

Elemental tellurium occurs in only one crystalline form, which consists of helical chains of tellurium atoms (Section 12.5); no isolated cyclic tellurium allotropes have been structurally characterised. However, both six- and eight-membered homocyclic tellurium rings are stabilised in solid-state materials. For example, the compound Re$_6$Te$_{16}$Cl$_6$ is comprised of [Re$_6$Te$_8$]$^{2+}$ clusters joined by cyclo-Te$_6$ and [TeCl$_3$]$^-$ ligands.[17] The cyclo-Te$_6$ molecules in the complex (AgI)$_2$ · Te$_6$ are embedded in a matrix of silver iodide.[9] The cyclo-Te$_6$ ligand in both of these complexes adopts a chair conformation. Cyclo-Te$_8$ is found in the caesium polytellurides Cs$_3$Te$_{22}$ and Cs$_4$Te$_{28}$ (Section 12.4).[18] In the former complex crown-shaped Te$_8$ rings exist between layered polytelluride anions (Figure 12.3).[18a]

Although cyclo-Te$_8$ has not been isolated, this tellurium homocycle has been tentatively identified by the observation of a ^{125}Te NMR resonance in the 820–870 ppm region for (a) the product of the decomposition of thermally unstable tellurium halides Te$_2$X$_2$ (X = Cl, Br),[19] (b) the cyclocondensation product of the reaction of two equivalents of Te$_2$Cl$_2$ with Cp$'_2$Ti(μ-Te$_2$)$_2$TiCp$'_2$ (Cp$'$ = MeC$_5$H$_4$)[19] and (c) a molten S–Se–Te mixture at 145 °C.[20]

12.1.2 Heteroatomic Rings

In principle, one or more chalcogen atoms in a homocyclic chalcogen ring can be replaced by a different chalcogen, leading to a large number of cyclic heteroatomic systems. In fact, there are 30 possible structural isomers for the eight-membered selenium sulfides cyclo-Se$_n$S$_{8-n}$ ($n = 1$–7).[1a] As a specific example, the replacement of two sulfur atoms in cyclo-S$_8$ by selenium atoms gives rise to four structural isomers, viz. cyclo-1,2-, -1,3- -1,4- and -1,5-Se$_2$S$_6$ (**12.1**–**12.4**). In addition, several of the heteroatomic chalcogen rings, e.g. cyclo-1,4-Se$_2$S$_6$ (**12.4**), are chiral and exhibit optical isomerism.

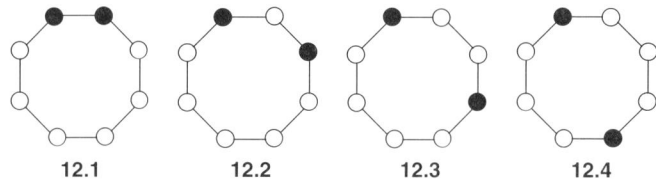

12.1   12.2   12.3   12.4

^{77}Se NMR spectroscopy is useful for the identification of the individual selenium sulfides present in a mixture. NMR studies show that the main components of sulfur–selenium melts containing up to 45 mol% selenium are cyclo-S$_7$Se and cyclo-1,2-Se$_2$S$_6$ (**12.1**).[21] A comparison of the spectra of natural-abundance samples (^{77}Se, 7.6%) with those of samples enriched (by >90%) with the ^{77}Se isotope is informative because the latter spectra provide values of ^{77}Se–^{77}Se coupling constants that are often not apparent in the natural-abundance spectra owing to the low intensity of the satellite peaks.[1a] For example, the isomer cyclo-1,2,3-Se$_3$S$_5$ was identified by the observation that the two singlets (relative intensity 2:1) observed in the natural-abundance NMR spectrum are replaced by an equally intense doublet and 1:2:1 triplet, with $J(Se_A, Se_B) = 34$ Hz, in the NMR spectrum of ^{77}Se-enriched samples.[22]

The synthesis of individual heteroatomic chalcogen rings containing six, seven or eight atoms is achieved by applying cyclocondensation methodology to various combinations of titanocene polychalcogenides and chalcogen halides, cf. the preparation of specific sulfur homocycles (Sections 2.1 and 12.1.1). The syntheses of the six-membered ring cyclo-Se$_5$S[23] and the seven-membered rings cyclo-1,2-E$_2$S$_5$ (E = Se,[24] Te[19]) serve as illustrative examples of this approach [eqns (12.1) and (12.2)].

$$Cp_2TiSe_5 + SCl_2 \rightarrow cyclo\text{-}Se_5S + Cp_2TiCl_2 \qquad (12.1)$$

$$Cp_2TiS_5 + E_2Cl_2 \rightarrow \text{cyclo-}E_2S_5 + Cp_2TiCl_2 \qquad (12.2)$$

$$(E = Se, Te)$$

The preparation of the symmetrical eight-membered ring cyclo-1,2,5,6-Se$_4$S$_4$ (**12.5**)[25] involves a further elaboration of the method [eqn (12.3)]. These syntheses are typically conducted at room temperature in CS$_2$.

$$Cp_2Ti\begin{pmatrix}Se-Se\\Se-Se\end{pmatrix}TiCp_2 + 2\,S_2Cl_2 \longrightarrow \underset{\textbf{12.5}}{\text{cyclo-Se}_4\text{S}_4} + 2\,Cp_2TiCl_2 \qquad (12.3)$$

Cyclic selenium sulfides are obtained as orange or red solids that are stable for short periods in the solid state at 20 °C. X-ray structural determinations indicate that they adopt the same ring conformations as their homocyclic analogues, *i.e.* cyclo-E$_6$ (E = S, Se; chair), cyclo-S$_7$ (chair with one long S–S bond) and cyclo-E$_8$ (crown, E = S, Se) (Section 12.1.1). However, crystallographic disorder between the positions of the sulfur and selenium atoms prevents the determination of accurate structural parameters in many cases. In solution, seven-membered rings, *e.g.* cyclo-1,2,3,4,5-Se$_5$S$_2$, undergo facile pseudorotation, as demonstrated by ^{77}Se NMR studies of ^{77}Se-enriched samples.[26]

12.1.3 Ring Transformations

The phenomenon of ring transformations, involving either contraction or expansion (or both simultaneously), is a pervasive feature of the behaviour of cyclic chalcogen molecules. Such interconversions with a change in ring size occur for cyclic sulfur allotropes under the influence of heat[1b] or photolysis.[27] These processes involve the input of significant amounts of energy. However, the interconversion of homoatomic or heteroatomic chalcogen ring systems occurs readily under ambient conditions in the presence of trace amounts impurities such as nucleophiles, *e.g.* traces of water present in glassware or strong electrophiles.[28]

Several mechanisms have been proposed for the intriguing interconversions of sulfur (or selenium) rings. These include the formation of (i) radicals by homolytic S–S bond cleavage, (ii) thiosulfoxides of the type S$_n$=S *via* ring contraction (an intramolecular process) or (iii) spirocyclic sulfuranes (or selenanes) *via* an intermolecular process. A fourth alternative (iv) invokes nucleophilic displacement reactions. Generic examples of mechanisms (ii)–(iv) for homoatomic sulfur or selenium rings are depicted in Scheme 12.1.

Since the generation of radicals from cyclo-S$_7$ or -S$_8$ requires an activation energy of more than 29 kcal mol^{-1}, the radical mechanism may account for the thermally or photochemically initiated processes, but it cannot prevail for the transformations that occur at ambient temperature. Hypervalent thiosulfoxides

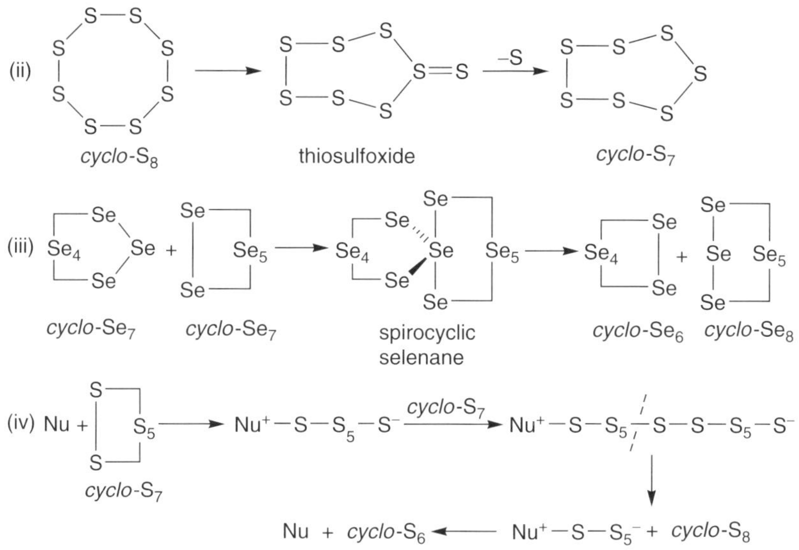

Scheme 12.1 Possible mechanisms for the interconversion of chalcogen rings (Nu = nucleophile).

are known as thermodynamically stable species only when the highly electronegative groups are attached to one of the sulfur centres, e.g. $F_2S=S$. Consistently, the high activation energies calculated for the conversion of disulfanes into thiosulfoxides, i.e. $R-S-S-R \rightarrow R_2S=S$ (R = H, Me, Cl),[28,29] preclude mechanism (ii) as a reasonable alternative for the interconversion of sulfur homocycles. Similarly, calculations on the formation of a hypervalent spirocyclic intermediate (sulfurane or selenane) via insertion of a chalcogen atom into a chalcogen-chalcogen bond [mechanism (iii)] carried out on model systems show that this process is highly endothermic for sulfur.[30]

Most of the early observations on ring transformations were made for homocyclic sulfur allotropes, but similar interconversions occur for both homoatomic selenium rings and heteroatomic selenium sulfides. An Se–Se bond is significantly weaker than an S–S bond and the formation of stable spirocyclic selenium structures is observed for certain polyselenides, e.g. Se_{11}^{2-} (Section 12.4.2). In this context, mechanism (iii) is a possible alternative for ring transformations involving selenium sulfides, e.g. the conversion of the seven-membered ring cyclo-1,2-Se_2S_5 into the eight-membered ring cyclo-1,2,3-Se_3S_5 and the six-membered ring SeS_5,[24] and for the analogous interconversion starting from cyclo-1,2,3,4,5-Se_5S_2.[26b] The formation of hypervalent intermediates is not necessary, however, to explain the observed product distribution in these ring transformations. As indicated by the example shown in Scheme 12.1 [mechanism (iv)], ring opening promoted by a nucleophile followed by chain growth and subsequent cyclisation reactions can account for the experimental observations.

12.2 OXIDISED CHALCOGEN HOMOCYCLES

12.2.1 With Exocyclic Oxygen Substituents

The oxidation of chalcogen homocycles may occur with either retention or a change of ring size. Depending on the nature of the oxidising agent, the products may contain one or more exocyclic substituents or be salts of homoatomic cyclic cations (Section 12.3.1). The reagent of choice for the preparation of homocyclic sulfur oxides is trifluoroperoxyacetic acid, CF_3CO_3H,[31] which has been used to synthesise the monoxides S_6O,[32] S_7O,[33] S_8O,[34] S_9O[35] and $S_{10}O$[35] from the corresponding cyclic sulfur allotropes [eqn (12.4)]. Interestingly, the use of an excess of CF_3CO_3H for the oxidation of cyclo-S_8 results in ring contraction to give the dioxide S_7O_2.[36]

$$\text{cyclo-}S_n + CF_3CO_3H \rightarrow S_nO + CF_3CO_2H \tag{12.4}$$

$$(n = 6 - 10)$$

The homocyclic sulfur oxides form dark-yellow crystals that decompose with the elimination of sulfur dioxide at room temperature. The monoxide S_8O exhibits the same crown conformation as cyclo-S_8 (Figure 12.4). The exocyclic oxygen atom adopts an axial position and induces a significant perturbation of the S–S bond lengths; those involving the three-coordinate sulfur atom are long (ca 2.20 Å) whereas the adjacent S–S bonds are short (ca 2.00 Å). These two structural features are attributed to a $\pi^*(SO)-\pi^*(SS)$ interaction between two singly occupied π^* orbitals.[37]

The monoxide S_8O forms the O-bonded adduct $S_8O \cdot SbCl_5$ upon treatment with antimony pentachloride in CS_2; adduct formation results in pyramidal inversion at the three-coordinate sulfur atom.[38] In contrast, the treatment of S_6O with antimony pentachloride induces dimerisation to give the 2:1 adduct of $SbCl_5$ with the cyclic dioxide $S_{12}O_2$.[39] The chair form of the two S_6O constituents of this dimerisation process are evident in the conformation of the 12-membered S_{12} ring in the adduct.

12.2.2 With Exocyclic Halogen Substituents

Sulfur or selenium homocycles containing one (or more) exocyclic halogen substituent are prepared by the oxidation of the element with arsenic or

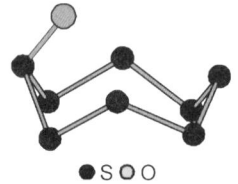

Figure 12.4 Molecular structure of S_8O.

Group 16: Rings and Polymers

antimony pentafluoride *in the presence of a halogen* using liquid sulfur dioxide as solvent.[40] For example, the hexafluoroarsenate salts of the $[S_7X]^+$ (X = I, Br)[41] and $[Se_6I_n]^{n+}$ (n = 1, 2)[42,43] cations are obtained by using this methodology [eqns (12.5) and (12.6)].

$$14/8 S_8 + X_2 + 3AsF_5 \rightarrow 2[S_7X][AsF_6] + AsF_3 \qquad (12.5)$$

$$(X = I, Br)$$

$$12/n Se + I_2 + 3AsF_5 \rightarrow 2/n[Se_6I_n][AsF_6]_n + AsF_3 \qquad (12.6)$$

$$(n = 1, 2)$$

The structures of the cations $[S_7X]^+$ (X = I, Br) show bond length alternations similar to those found in the isovalent neutral molecule S_7O (Figure 12.5). The bond elongations are most pronounced near the source of the perturbation, *i.e.* the three-coordinate sulfur centre. The S–S bond length of *ca* 2.39 Å in $[S_7I]^+$ is especially long, indicative of a bond order of *ca* 0.4.[40]

The structure of the salt $\{[Se_6I][AsF_6]\}_n$ incorporates polymeric strands of $[Se_6I]_n^+$ cations (Figure 12.6).[42] The chair-shaped hexaselenium rings are joined to neighbouring Se_6 rings by two weak 1,4-axial Se···I contacts. The polymeric structure of this selenium–iodine cation provides an interesting contrast to the monomeric structures of $[S_7I]^+$ salts.

A comparison of the structures of the dications $[Se_6I_2]^{2+}$ and $[Ph_2Se_6]^{2+}$ is informative (Figure 12.7). The diphenyl derivative is obtained in high yield by the reaction of $[Se_4][AsF_6]_2$ with diphenyl diselenide in liquid sulfur dioxide.[44] In both derivatives, the two exocyclic substituents are in the 1,4-positions of the six-membered ring. In the diiodo compound the Se_6 ring is in a chair conformation

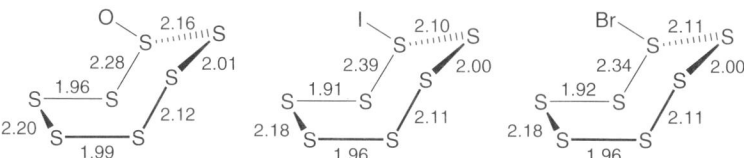

Figure 12.5 Comparison of S–S bond lengths in S_7O and $[S_7X]^+$ (X = I, Br) (SbF_6^- salts).

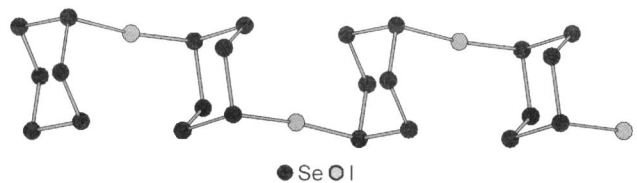

Figure 12.6 The polymeric $[Se_6I]_n^+$ cations in the AsF_6^- salt.

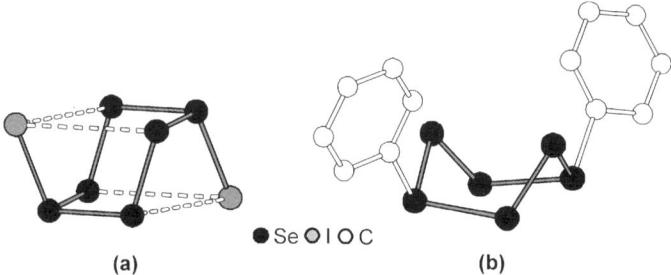

Figure 12.7 Structures of the cyclic dications (a) $[Se_6I_2]^{2+}$ and (b) $[Ph_2Se_6]^{2+}$ (AsF_6^- salts).

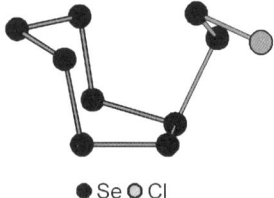

Figure 12.8 Structure of the $[Se_7Se_2Cl]^+$ cation in $[Se_9Cl][SbCl_6]$.

and the axially substituted iodine substituents complete a distorted cubic structure *via* weak Se···I contacts.[43] The mean Se–Se bond lengths involving the three-coordinate selenium centres are *ca* 0.25 Å longer than the other Se–Se bond distances. In contrast, the Se_6 ring in the 1,4-diphenyl derivative is in a boat conformation with phenyl groups at opposite corners of the bottom of the boat. The variation in Se–Se bond lengths in $[Ph_2Se_6]^{2+}$ (2.38–2.45 Å) is much smaller than that in $[Se_6I_2]^{2+}$. The cationic silver complex $[Ag_2(Se_6)]^{2+}$ has a distorted cubic structure similar to that of $[Se_6I_2]^{2+}$, *i.e.* an Se_6 ring in the chair conformation is bicapped by two Ag^+ ions.[43b]

A seven-membered ring is present in the monocation $[Se_9Cl]^+$, which is obtained as the $[SbCl_6]^-$ salt by treatment of selenium with $[NO][SbCl_6]$ in sulfur dioxide.[45] The Se_7 ring in $[Se_9Cl]^+$ is in the chair conformation with an SeSeCl substituent attached to one of the selenium atoms (Figure 12.8). The Se–Se bond lengths vary from 2.27 to 2.43 Å.

12.3 CATIONIC CHALCOGEN RINGS

In contrast to the heavier chalcogens, oxygen does not form homoatomic ring systems. The formation of the monocation O_2^+ removes an electron from one of the singly occupied π^* orbitals of neutral O_2 resulting in an increase in the π-bond order of the diatomic species. The oxidation of molecular Cl_2 with $O_2[SbF_6]$ in anhydrous HF produces the fascinating salt $[Cl_2O_2][SbF_6]$, which is black in the solid state and violet in solution.[46] The $[Cl_2O_2]^+$ cation has a

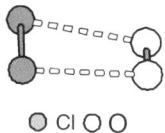

Figure 12.9 The trapezoidal $[Cl_2O_2]^+$ cation (SbF_6^- salt).

trapezoidal structure (Figure 12.9) with O–O, Cl–Cl and Cl–O distances of 1.19, 1.91 and 2.42 Å, respectively. The bonding in the cation involves a $\pi^*-\pi^*$ interaction in which *ca* 0.6 electrons are transferred from Cl_2^+ to the O_2 fragment.

12.3.1 Homocyclic Sulfur and Selenium Cations

The heavy chalcogens form an extensive series of cationic homoatomic ring systems, which exhibit fascinating, chalcogen-dependent structural trends. Investigations of this class of inorganic compounds have provided significant insights into our understanding of (a) the reaction conditions necessary for the synthesis of highly electrophilic species of main group elements, (b) the importance of lattice energy considerations for the stabilisation of such species in the solid state and (c) the use of non-classical bonding concepts to describe unusual structures.

For historical reasons, this section will begin with a discussion of homoatomic sulfur cations (the first structural information on a sulfur cation, $S_8[AsF_6]_2$, appeared in 1971).[47] In addition, the sulfur cations provide a benchmark for comparison of the syntheses, structures and bonding in analogous selenium and tellurium systems. The preparation of the highly electrophilic sulfur cations requires the use of strong oxidising agents in weakly basic and oxidation-resistant solvents.[48] The most effective reagents are arsenic or antimony pentafluoride with liquid sulfur dioxide as the reaction medium. By using appropriate variations in the stoichiometry, the cations S_4^{2+}, S_8^{2+} and S_{19}^{2+} are obtained as $[MF_6]^-$ (M = As, Sb) salts *via* this methodology [eqns (12.7)–(12.9)]; a trace of halogen is added to facilitate the formation of the most highly oxidised species cyclo-S_4^{2+}.[49]

$$4/8 S_8 + 3AsF_5 \rightarrow S_4[AsF_6]_2 + AsF_3 \quad (12.7)$$

$$S_8 + 3AsF_5 \rightarrow S_8[AsF_6]_2 + AsF_3 \quad (12.8)$$

$$19/8 S_8 + 3AsF_5 \rightarrow S_{19}[AsF_6]_2 + AsF_3 \quad (12.9)$$

The structures of S_4^{2+} (ref. 50a), S_8^{2+} (ref. 50b) and S_{19}^{2+} (ref. 51) are depicted in Figure 12.10; they are the only homoatomic sulfur cations that have been characterised in the solid state. The cyclo-S_4^{2+} cation forms an undistorted square-planar arrangement. The framework of the eight-membered S_8^{2+} cation can be related to that of cyclo-S_8. The removal of two electrons converts the crown-shaped ring (D_{4d}) into an *exo–endo* conformation with approximately C_s symmetry and three transannular contacts that are <3.0 Å. An analysis of the

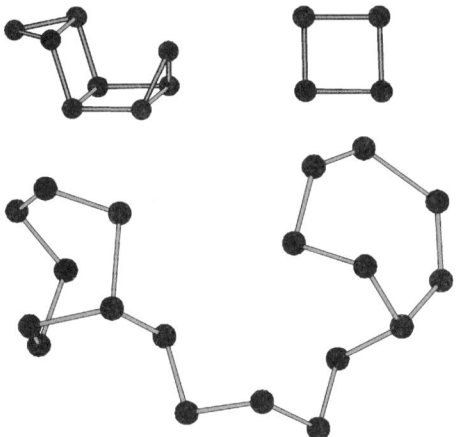

Figure 12.10 Structures of the cations S_4^{2+} and S_8^{2+} (AsF$_6^-$ salts) and S_{19}^{2+} (SbF$_6^-$ salt).

contacts between these cations and the AsF$_6^-$ counterions indicates that the positive charge is delocalised almost equally over the entire ring system in both cases.[49,50] The S_{19}^{2+} cation consists of two seven-membered (S$_7$) rings linked by a pentasulfur (S$_5$) chain. In common with the structures of the related [S$_7$X]$^+$ (X = I, Br) cations (Figure 12.5), the mean S–S bond lengths involving the three-coordinate sulfur centres in S_{19}^{2+} are significantly longer than the other S–S bonds, although the effect is not as pronounced as in the halogen-substituted cations.

The development of an understanding of the bonding in cyclic sulfur cations such as cyclo-S_4^{2+} and S_8^{2+} has revealed the presence of non-classical 3pπ–3pπ, π^*–π^* and 3p$^2 \rightarrow$ 3σ^* interactions that delocalise the positive charges expected from classical electron-precise structures.[48] The square-planar cyclo-S_4^{2+} cation and the heavier congeners cyclo-E_4^{2+} (E = Se, Te) can be viewed as six π-electron species. These inorganic ring systems are isovalent with the cyclobutadienide dianion [C$_4$H$_4$]$^{2-}$. As indicated by the π molecular orbital diagram for cyclo-S_4^{2+} in Figure 12.11, there is one occupied π-bonding molecular orbital delocalised over all four sulfur atoms giving an overall S–S bond order of 1.25 (1σ + 0.25π), consistent with the S–S bond lengths, which are *ca* 0.05 Å shorter than the single bond value.

The bonding description for the S_8^{2+} cation must take account of the three transannular contacts in the range 2.86–3.00 Å. This structural feature is attributed to the weakly bonding π^*–π^* interaction of the partially occupied 3p orbitals of the six central sulfur atoms of the eight-membered ring (Figure 12.12a).[50] An additional interaction that contributes to the delocalisation of the positive charge onto the *exo* and *endo* sulfur atoms of S_8^{2+} involves donation of electron density from the lone pair orbitals of those sulfur atoms into the empty σ^* orbitals of the vicinal S–S bonds (3p$^2 \rightarrow$ 3σ^* hyperconjugation) as depicted in Figure 12.12b.

Group 16: Rings and Polymers 287

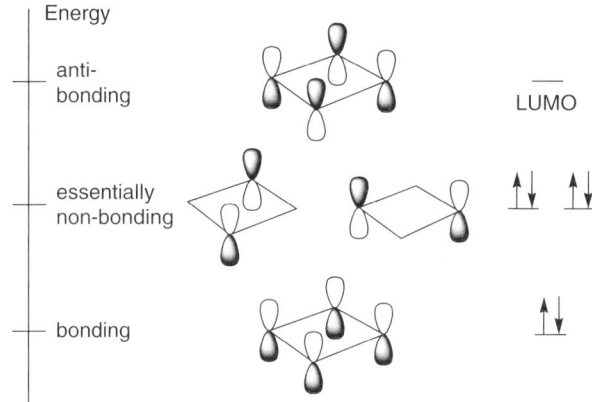

Figure 12.11 π-Molecular orbitals in cyclo-S_4^{2+}.

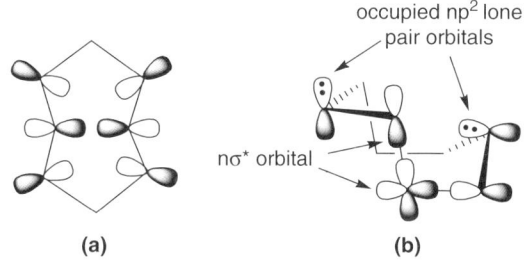

Figure 12.12 (a) π*–π* (top view) and (b) $3p^2 \rightarrow 3\sigma^*$ (side view) interactions in S_8^{2+}.

The cyclo-S_4^{2+} and -S_8^{2+} cations are stabilised in the solid state with respect to their possible dissociation products, *i.e.* the radical monocations $S_2^{+\bullet}$ and $S_4^{+\bullet}$ (or a mixture of $S_3^{+\bullet}$ and $S_5^{+\bullet}$), respectively, by lattice energy effects.[49,50] For example, the lattice stabilisation enthalpies of cyclo-E_4^{2+} relative to $E_2^{+\bullet}$ ions {*i.e.* in $E_4[AsF_6]_2$ (s) with respect to two molecules of $E_2[AsF_6]$ (s)} are estimated to be 52, 69 and 87 kcal mol^{-1} for M = S, Se, Te, respectively, on the basis of thermochemical calculations (see Scheme 12.2).[49] In a similar manner, although the gas-phase dissociation of S_8^{2+} into $S_3^{+\bullet}$ and $S_5^{+\bullet}$ is favourable by -32.5 kcal mol^{-1}, solid $S_8[AsF_6]_2$ (s) is more stable than the salts of the monocations.[50]

In solution, however, lattice-stabilisation effects no longer apply and the dissociation of homocyclic sulfur dications occurs readily. The formation of the cation radical $S_5^{+\bullet}$ in solutions of sulfur in 65% oleum has been firmly established by EPR studies of samples that were 92% enriched with ^{33}S ($I = 3/2$).[52] The 16-line spectrum shown in Figure 12.13 is consistent with a cyclic structure in which the unpaired electron interacts with five equivalent ^{33}S nuclei: $[(2 \times 5 \times 3/2) + 1 = 16]$.

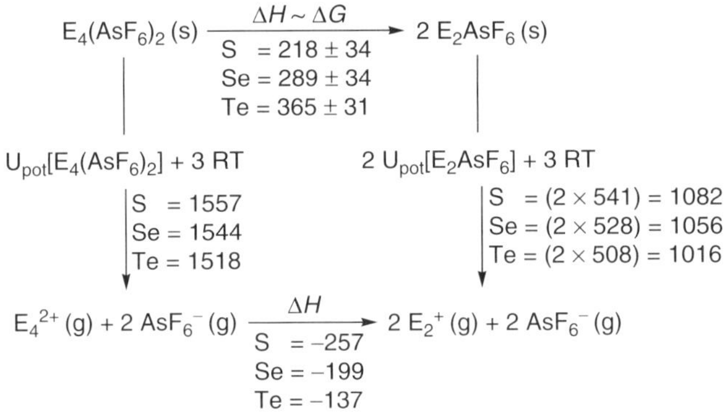

Scheme 12.2 Thermochemical cycle for formation of $E_2[AsF_6]$ salts from $E_4[AsF_6]_2$ (numerical values in kJ mol^{-1}; E = S, Se, Te).

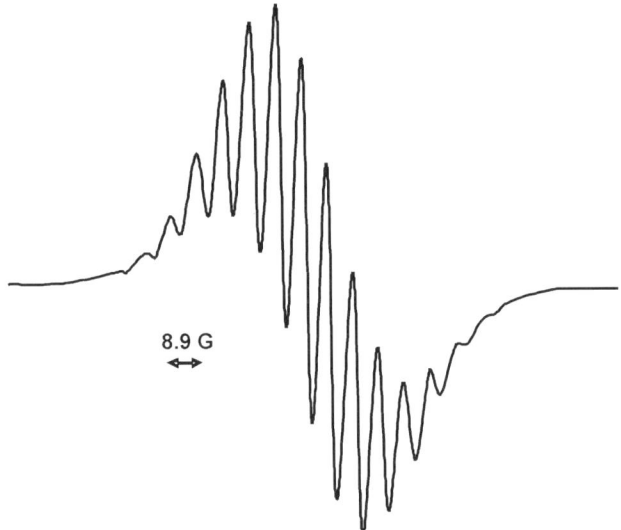

Figure 12.13 ESR spectrum of 92% enriched sulfur in 65% oleum. Reproduced with permission from H. S. Low and R. A. Beaudet, *J. Am. Chem. Soc.*, 1976, **98**, 3849.[52] Copyright (1976) American Chemical Society.

The radical cation $S_5^{+\bullet}$ is also formed by the dissociation of S_8^{2+} in solvents such as fluorosulfuric acid and liquid sulfur dioxide. The second product of this dissociation has been identified as cyclo-S_6^{2+}, *i.e.* the dimer of the radical monocation $S_3^{+\bullet}$ [eqn (12.10)], by a combination of computational and spectroscopic investigations.[53] However, stable salts of cyclo-S_6^{2+} have not been isolated in the solid state, presumably because disproportionation into the corresponding cyclo-S_4^{2+} and -S_8^{2+} salts is exothermic by 1.4–4.1 kcal mol^{-1} [eqn (12.11)]. Calculations show that the most stable form of cyclo-S_6^{2+} is a flattened six-membered ring in a chair conformation with the positive charges delocalised over all six atoms. Despite the non-planarity of this conformation, an analysis of the molecular orbitals shows a clear distinction between the σ and π MOs in cyclo-S_6^{2+}, which may therefore be described as a 10 π-electron system, cf. $[S_3N_3]^-$ (see Figure 4.3 in Section 4.1.2.3). The blue colour of solutions of S_8^{2+} is attributed to the π* → π* transition of cyclo-S_6^{2+}.[53]

$$S_8^{2+} \rightleftharpoons S_5^{+\bullet} + \tfrac{1}{2} S_6^{2+} \qquad (12.10)$$

$$2S_6[AsF_6]_2(s) \rightarrow S_8[AsF_6]_2(s) + S_4[AsF_6]_2(s) \qquad (12.11)$$

Selenium cations exhibit similarities, but also some significant differences, compared with their sulfur analogues. The structurally characterised homoatomic selenium cations include Se_4^{2+},[54] Se_8^{2+},[55] Se_{10}^{2+},[56] and Se_{17}^{2+}.[57] Although these cations are obtained as MF_6^- (M = As, Sb) salts by methods similar to those employed for the sulfur cations [eqns (12.7)–(12.9)], other synthetic approaches are also successful for their preparation. For example, the reaction of selenium with a molten mixture of $SeCl_4$ and $AlCl_3$ at 250 °C produces $Se_8[AlCl_4]$.[55] Transition metal halides with the metal in a high oxidation state may also serve as selective, mild oxidants towards selenium. Thus, the reaction of selenium with tungsten hexachloride at 350 °C produces the Se_{17}^{2+} dication as the $[WCl_6]^{2-}$ salt;[57] the formation of this unique selenium polycation, together with Se_8^{2+}, was first indicated by ^{77}Se NMR studies of the disproportionation solutions of Se_{10}^{2+} in liquid sulfur dioxide.[58]

The selenium cations cyclo-Se_4^{2+} and Se_8^{2+} have structures similar to those of the sulfur analogues (Figure 12.10).[54,55] The chiral Se_{10}^{2+} cation has a tricyclic structure in which a boat-shaped six-membered Se_6 ring is bridged in the 1,4-positions by an Se_4 chain (Figure 12.14a).[56] The framework of the Se_{17}^{2+} cation is similar to that of S_{19}^{2+} (Figure 12.10), except that the bridge between the two seven-membered rings is comprised of three rather than five chalcogen atoms (Figure 12.14b).[57] The monocyclic $[Se_7Se_2Cl]^+$ cation (Figure 12.8) may be formed by halogen cleavage of one of Se–Se bonds in the bridging Se_3 unit of Se_{17}^{2+}.[47a]

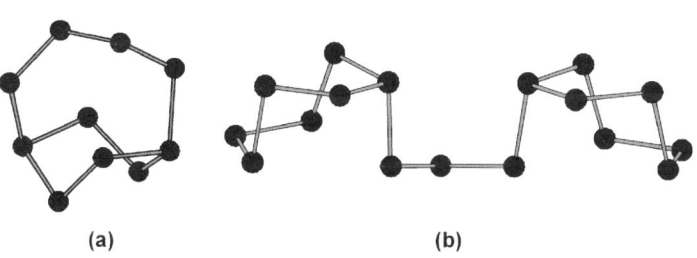

Figure 12.14 Structures of (a) Se_{10}^{2+} (SbF_6^{2-} salt) and (b) Se_{17}^{2+} (WCl_6^{2-} salt).

12.3.2 Homocyclic Tellurium Cations

Tellurium forms the most extensive series of homoatomic chalcogen cations. The list of structurally characterised tellurium polycations includes Te_4^{2+}, Te_6^{2+}, Te_6^{4+}, Te_8^{2+} and Te_8^{4+}, and also the polymeric species $(Te_4^{2+})_n$, $(Te_6^{2+})_n$, $(Te_7^{2+})_n$, $(Te_8^{2+})_n$ and $(Te_{10}^{2+})_n$.[59] The cyclo-Te_4^{2+} cation is produced by using methodology similar to that employed for the synthesis of cyclo-S_4^{2+} [eqn (12.7)]. For example, the oxidation of tellurium with AsF_5 in liquid sulfur dioxide generates $Te_4[AsF_6]_2$.[60] More common approaches to tellurium cations, however, are (a) the direct oxidation of tellurium with a transition metal halide or (b) the treatment of a mixture of tellurium and tellurium tetrachloride with a transition metal (oxo)halide. The latter method involves the *in situ* generation of a chalcogen subhalide, *e.g.* Te_3Cl_2, which has a polymeric structure based on a helical chain of tellurium atoms. In both of these synthetic approaches, the stoichiometry of the reagents can be adjusted to favour the synthesis of a specific tellurium cation.

The tellurium cations exhibit significant structural differences compared with those of sulfur and selenium as a result of the more metallic character of tellurium. The square-planar structure of cyclo-Te_4^{2+} cation, *e.g.* in $Te_4[AlCl_4]_2$ and $Te_4[Al_2Cl_7]_2$,[61] resembles that of the lighter chalcogen analogues. The mean Te–Te bond distance of 2.66 Å is shorter than the single-bond value by *ca* 0.06 Å. However, in the salt $(Te_4)(Te_{10})[Bi_4Cl_{16}]$, prepared by heating a Te–$TeCl_4$–$BiCl_3$ mixture to 170 °C, the cyclo-Te_4^{2+} cation exhibits a polymeric structure with Te–Te bond lengths of 2.75 and 2.81 Å in the ring and inter-ring Te–Te contacts of 2.98 Å (Figure 12.15).[62]

The six-atom systems Te_6^{2+} and Te_6^{4+} exhibit non-classical structures that differ from those established for related mixed chalcogen cations (Section 12.3.2). The tetracation Te_6^{4+} was first prepared as $Te_6[AsF_6]_4$ in 1979 by reaction of tellurium with a large excess of AsF_5 in liquid SO_2,[63] and the dication Te_6^{2+} was originally obtained as the salt $Te_6[WOCl_4]_2$ by oxidation of elemental tellurium with $WOCl_4$.[64] The Te_6^{4+} cation has the structure of an elongated prism with six short Te–Te bonds of 2.67 Å forming the triangular faces and three Te–Te bonds of *ca* 3.12 Å, which link the two triangular faces (Figure 12.16a). The hypothetical Te_6^{6+} hexacation is expected to have a regular prismane structure with all six Te atoms carrying a formal positive charge. The addition of two electrons causes the lengthening of all three bonds parallel to the C_3 axis to give the elongated ground-state structure of Te_6^{4+}.[65] The

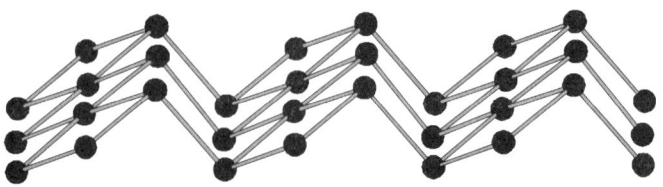

Figure 12.15 Section of the polymeric $(Te_4^{2+})_n$ cationic chain in $(Te_4)(Te_{10})[Bi_4Cl_{16}]_2$.

Group 16: Rings and Polymers 291

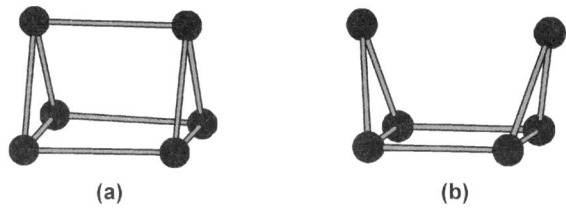

Figure 12.16 Structures of the cations (a) Te_6^{4+} (AsF_6^- salt) and (b) Te_6^{2+} ($WOCl_4^-$ salt)

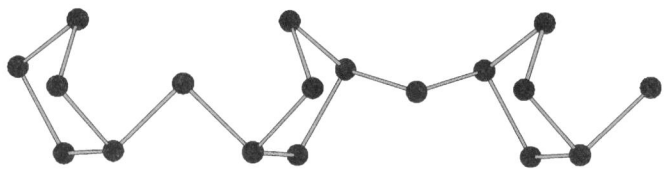

Figure 12.17 A section of the polymeric $(Te_6^{2+})_n$ cationic chain in $Te_6[HfCl_6]$.

addition of a further two electrons formally generates the dication Te_6^{2+}, which, in contrast to the classical structure with localised bonds (Section 12.3.2) and the flattened chair conformation calculated for cyclo-S_6^{2+},[53] forms a prism in which one edge is fully opened and two edges are elongated with Te–Te distances of 3.21 and 3.38 Å (Figure 12.16b).[64,66]

The polymeric hexatellurium dication $(Te_6^{2+})_n$ in the salts $Te_6[MCl_6]$ (M = Zr, Hf) is comprised of puckered five-membered rings connected by a single tellurium atom in a polymeric chain (Figure 12.17).[67] The unique Te–Te bond involving the two-coordinate tellurium atoms is short (2.69 Å) relative to those involving the three-coordinate tellurium atoms (2.77–2.82 Å).

The seven-atom polymeric tellurium cation $(Te_7^{2+})_n$ is unique among chalcogen polycations. It has been structurally characterised in two different forms. In the compound $Te_7[AsF_6]_2$, obtained unexpectedly from the reaction of $Te_7[AsF_6]_2$ with iron pentacarbonyl in liquid SO_2,[68] it is comprised of six-membered rings in the chair conformation linked by a single tellurium atom (Figure 12.18a). In distinct contrast, the polymeric $(Te_7^{2+})_n$ cation in $Te_7[WOBr_5]_2$ incorporates a square-planar tellurium centre in a spirocyclic arrangement with Te–Te bond lengths that vary from 2.74 to 2.97 Å (Figure 12.18b).[69] A central, hypervalent tellurium atom is also a common feature in tellurium polyanions (Section 12.4.3).

The eight-atom systems provide a compelling illustration of the structural diversity of tellurium polycations. There are examples of (a) structural isomers of the monomeric dication Te_8^{2+}, (b) a polymeric $(Te_8^{2+})_n$ cation and (c) a tetracation Te_8^{4+}. Unlike cyclo-S_8^{2+} and -Se_8^{2+}, the Te_8^{2+} dication cannot be made by oxidising tellurium with MF_5 (M = As or Sb) in liquid SO_2. However, the reactions of (a) the tellurium subhalide Te_3Cl_2 with $ReCl_4$ at 200 °C and (b) tellurium with WCl_6 at 180 °C produce the salts $Te_8[MCl_6]_n$ (n = 1, M = Re;

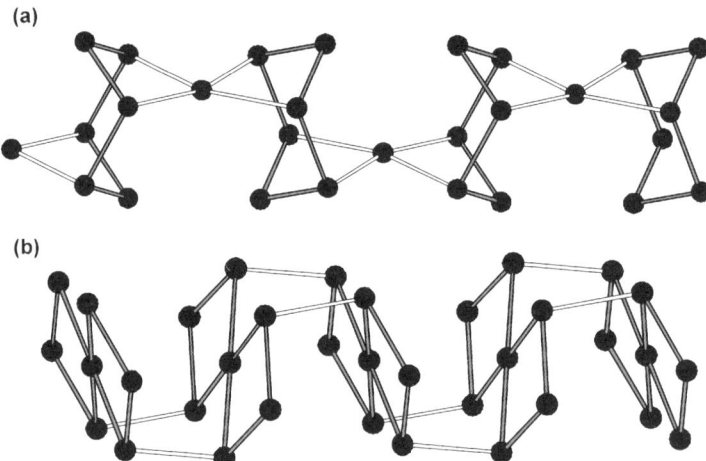

Figure 12.18 Section of the polymeric $(Te_7^{2+})_n$ cationic chains in (a) $Te_7[AsF_6]_2$ and (b) $Te_7[WOBr_5]_2$.

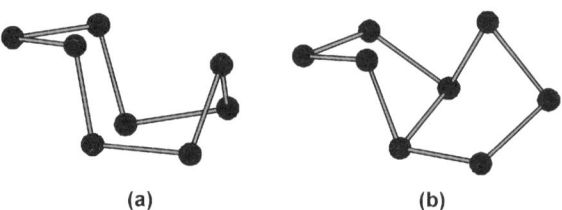

Figure 12.19 Structures of the Te_8^{2+} dication in (a) $Te_8[ReCl_6]$ and (b) $Te_8[WCl_6]_2$.

$n = 2$, M = W).[70,71] The Te_8^{2+} dication in $Te_8[ReCl_6]$[70] exhibits the *exo*, *endo* conformation of the eight-membered ring with a transannular Te–Te separation of 3.15 Å (Figure 12.19a) that is also found in cyclo-S_8^{2+} and -Se_8^{2+}, whereas the same cation in $Te_8[WCl_6]_2$ has a more pronounced bicyclic structure comprised of two five-membered rings each in an envelope conformation and a shorter transannular Te–Te bond (2.99 Å) (Figure 12.19b). The polymeric cation $(Te_8^{2+})_n$ in $Te_8[U_2Br_{10}]$[72a] and $Te_8[Bi_4Cl_{14}]$[72b] has a structure that is closely related to that of $(Te_7^{2+})_n$ cation in $Te_7[AsF_6]_2$ (Figure 12.18a), except that ditelluro (–Te–Te–) units rather than single tellurium atoms bridge the Te_6 rings. The six-membered tellurium rings have a boat conformation in $Te_8[U_2Br_{10}]$,[72a] whereas they adopt a chair conformation in $Te_8[Bi_4Cl_{14}]$.[72b]

The tetracation Te_8^{4+} is formed as $Te_8[VOCl_4]_2$ from the treatment of tellurium with $VOCl_3$ at 200 °C.[73] This reaction involves the intermediate formation of $VOCl_2$ and Te_3Cl_2 by a redox process. The Te_8^{4+} cation is comprised of two Te_4^{2+} cations linked by two Te–Te interactions of 3.01 Å. The mean Te–Te bond length in the four-membered Te_4 rings is 2.77 Å, indicating single bonds,

Group 16: Rings and Polymers

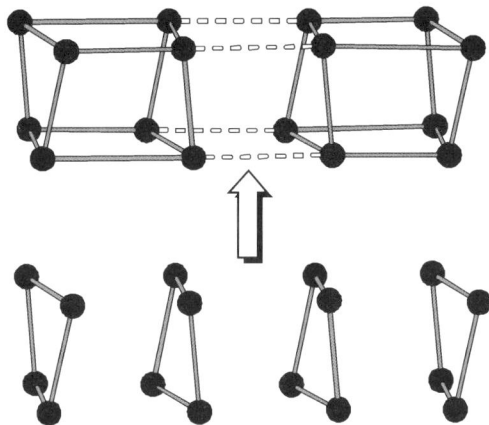

Figure 12.20 Structure of the Te_8^{4+} cation in $Te_8[VOCl_4]_2$.

i.e. loss of ring delocalisation. The Te_8^{4+} cation may also be viewed as a cube with two open edges; the Te···Te separation on these edges is 3.85 Å. The structural data for the tetracation are consistent with a classical bonding description in which the three-coordinate tellurium atoms each bear a positive charge. There are weak Te···Te interactions (3.59 Å) between neighbouring Te_8^{4+} cations in the solid state. Thus, the Te_8^{4+} cation is formally derived from a stack of equidistant cyclic Te_4^{2+} cations, which undergo Peirls distortion upon pairing (Figure 12.20).

12.3.3 Heteroatomic Chalcogen Cations

In a similar manner to neutral homoatomic chalcogen rings (Section 12.1.2), the replacement of one or more of the chalcogen atoms in a homoatomic cation by another chalcogen potentially gives rise to a large number of heteroatomic cations that differ in composition for a given ring size. Additionally, structural isomers are possible for some compositions. As a simple example, there are four conceivable cations **12.6**–**12.9** for the four atom Te–Se system.

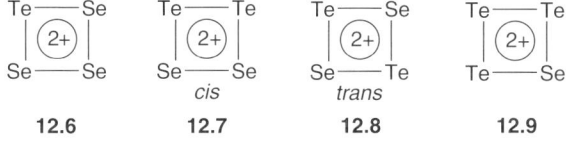

Several approaches have been used for the synthesis of heteroatomic chalcogen cations, including (a) direct oxidation of a mixture of two chalcogens with oleum, MF_5 (M = As, Sb) or transition metal halides, (b) reaction of a chalcogen with a homoatomic cation of another chalcogen and (c) comproportionation of the homoatomic cations of two different chalcogens.[74–81]

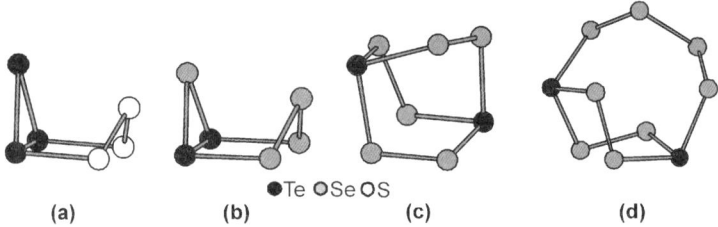

Figure 12.21 Structures of (a) $[Te_3S_3]^{2+}$, (b) $[Te_2Se_4]^{2+}$, (c) $[Te_2Se_6]^{2+}$ and (d) $[Te_2Se_8]^{2+}$.

The Se–Te system has been studied most extensively because both elements have spin-$\frac{1}{2}$ isotopes and information about the species present in solution may be obtained by both ^{77}Se and ^{125}Te NMR studies. For example, all members of the series cyclo-$[Te_3Se]^{2+}$ (**12.9**) *cis*- and *trans*-cyclo-$[Te_2Se_2]^{2+}$ (**12.7** and **12.8**) and cyclo-$[TeSe_3]^{2+}$ (**12.6**) have been detected in the ^{125}Te NMR spectrum of a mixture of tellurium and selenium in oleum.[76] The ^{125}Te resonances move progressively downfield with the replacement of tellurium in cyclo-Te_4^{2+} by the more electronegative atoms. The structure of *trans*-$[Te_2Se_2]^{2+}$ (**12.8**) in the solid state exhibits an Se–Te bond length that is the mean of the bond lengths found for the homoatomic cations cyclo-Te_4^{2+} and cyclo-Se_4^{2+} in various salts, consistent with delocalisation and significant double bond character in the four-membered *trans*-$[Te_2Se_2]^{2+}$ ring.[78]

The structures of the heteroatomic six-atom systems $[Te_3S_3]^{2+}$ (Figure 12.21a)[75] and $[Te_2Se_4]^{2+}$ (Figure 12.21b)[80] show an important difference when compared with that of Te_6^{2+}. In both cases, two edges of the hexagonal prism are opened, *cf.* one in Te_6^{2+} (Figure 12.16b), and there are two three-coordinate tellurium atoms that formally bear a positive charge. Therefore, in contrast to Te_6^{2+}, the structures of these heteroatomic cations may be described by localised two-electron, two-centre bonds. Other examples of this classical bonding description for heteroatomic chalcogen cations are the eight-atom system $[Te_2Se_6]^{2+}$ (Figure 12.21c) and the 10-atom system $[Te_2Se_8]^{2+}$ (Figure 12.21d).[81] In $[Te_2Se_6]^{2+}$, the two three-coordinate tellurium atoms are bridged by three –Se–Se– units to give an architecture that differs significantly from that of Se_8^{2+}. In contrast, the structure of $[Te_2Se_8]^{2+}$ resembles that of the homoatomic cation Se_{10}^{2+} (Figure 12.15), with the two Te atoms occupying the three-coordinate positions.

12.4 ANIONIC CHALCOGEN RINGS AND CHAINS

12.4.1 *Catena*-Sulfur Anions

The propensity of sulfur, selenium and tellurium to catenate is illustrated by the formation of an extensive series of polyanions for all three chalcogens. The structures of these polyanions exhibit interesting trends within the series in which the ability of tellurium and, to a lesser extent, selenium to adopt hypervalent bonding arrangements is evident, *cf.* tellurium polycations (Section 12.3.1). Tellurium exhibits a more diverse range of polyanions with respect to both structure and charge than the other members of the series.

Sulfur forms a series of homoatomic dianions *catena*-S_x^{2-} ($x=2$–8), which, without exception, have unbranched chain structures in the solid state.[82] The electrochemical reduction of cyclo-S_8 in aprotic solvents occurs *via* an initial two-electron process to produce *catena*-S_8^{2-}.[83] In solution, *catena*-S_8^{2-} and other long-chain polysulfides, *e.g. catena*-S_6^{2-} and *catena*-S_7^{2-}, dissociate *via* an entropy-driven process to give radical anions $S_x^{-\bullet}$ ($x=2$–4), including the ubiquitous trisulfur radical anion ($x=3$).[84] This intensely blue species is the chromophore in the mineral lapis lazuli, which is used in the manufacture of jewellery.

The long-chain polysulfides are stabilised in the solid state by combination with a large counteranion, *e.g.* $[R_4N]^+$ (R = alkyl), $[Ph_4P]^+$, $[(Ph_3P)_2N]^+$ (PPN^+).[82] Typically they are prepared by the reaction of cyclo-S_8 with a nucleophile in the presence of one of these cations. For example, the reaction of [PPN]SH with cyclo-S_8 in ethanol yields the heptasulfide [PPN][*catena*-S_7]·2EtOH.[85] Orange–red crystals of the octasulfide $[Et_3NH]_2$[*catena*-S_8] are formed on treatment of a solution of cyclo-S_8 in formamide with triethylamine and hydrogen sulfide [eqn (12.12)].[86]

$$\tfrac{7}{8}S_8 + 2NEt_3 + H_2S \rightarrow [Et_3NH]_2S_8 \qquad (12.12)$$

The S–S bond lengths in *catena*-polysulfides fall within the range 2.03–2.07 Å, typical of single bonds; the negative charge is primarily located on the two terminal sulfur atoms. The SSS bond angles of 106–111° are close to the tetrahedral values, whereas the –SSSS– torsion angles of 65–98° reflect the influence of the two lone pairs on each of the internal sulfur atoms. The structures of long-chain polysulfides can be viewed as part of the helical chain of polymeric sulfur (Section 12.5). However, the structural motif (order of the signs of the torsion angles) may vary as a result of packing effects in the ionic compounds. For right- and left-handed helices the order of the signs is + + + and − − −, respectively. The structures of the *catena*-S_7^{2-} and *catena*-S_8^{2-} dianions are shown in Figure 12.22.[85,86]

Alkali metal polysulfides are prepared by the direct reaction of an alkali metal with cyclo-S_8. Alternatively, *n*-butyllithium may be used to generate

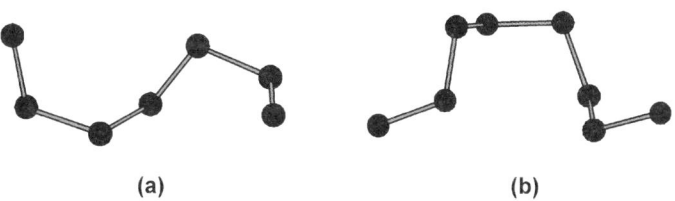

Figure 12.22 Structures of (a) S_7^{2-} (PPN^+ salt) and (b) S_8^{2-} [$(Et_3N)_3H^+$ salt].

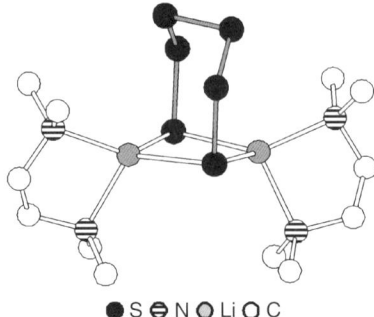

Figure 12.23 Molecular structure of [Li(TMEDA)]$_2$S$_6$.

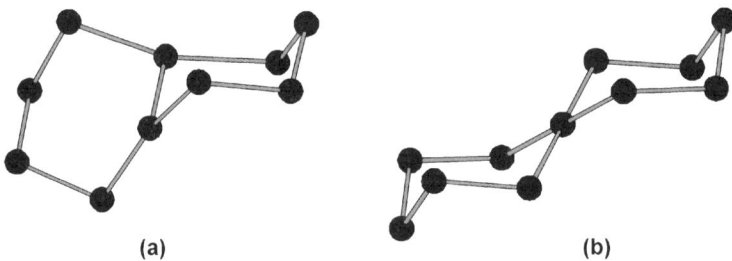

Figure 12.24 Structures of (a) Se$_{10}$$^{2-}$ (PPN$^+$ salt) (b) Se$_{11}$$^{2-}$ (Ph$_4$P$^+$ salt).

lithium derivatives. For example, the reaction of cyclo-S$_8$ with ^{n}BuLi in toluene containing tetramethylethylenediamine (TMEDA) produces the orange hexasulfide [Li(TMEDA)]$_2$S$_6$.[87] In the solid state, the two TMEDA-solvated Li$^+$ cations bridge the terminal sulfur atoms of the S$_6$$^{2-}$ chain to give two LiS$_6$ rings (Figure 12.23). In donor solvents, *e.g.* acetone, [Li(TMEDA)]$_2$S$_6$ readily dissociates to give the blue trisulfur radical anion. The fascinating structure of the radical monoanion S$_6$$^{-\bullet}$ is illustrated in Figure 5.2 and the bonding is compared with that of cyclo-S$_6$ in Figure 5.3 (see Section 5.3).

12.4.2 *Catena-* and Spirocyclic Selenium Anions

Selenium forms a series of *catena* dianions Se$_x$$^{2-}$ similar to those described for sulfur, but the chain length is limited to six atoms ($x=6$) in simple salts. However, the *catena*-Se$_8$$^{2-}$ dianion has been structurally characterised in the black complex [Na(12-crown-4)$^+$]$_2$[Se$_8$$^{2-}$]·(cyclo-Se$_6$,cyclo-Se$_7$).[14a] In addition, selenium forms several higher polyselenides ($x > 8$), in which one or more of the selenium atoms is three- or four-coordinate. The structures of the hypervalent species Se$_{10}$$^{2-}$,[88] Se$_{11}$$^{2-}$,[89] and Se$_{16}$$^{4-}$ [90] are especially interesting. The structure of the Se$_{10}$$^{2-}$ dianion in the black PPN$^+$ salt is bicyclic with both six-membered rings in chair conformations (Figure 12.24a). The three-coordinate selenium atoms are in a pseudo-trigonal-pyramidal arrangement involving three selenium neighbours and two lone pairs. The transannular Se–Se bond occupies an equatorial position with a length of 2.46 Å whereas the two axial Se–Se bond distances are 2.57 and 2.76 Å. The Se_{11}^{2-} dianion in the red–brown $[Ph_4P]^+$ salt has a spirocyclic structure in which a central Se^{2+} cation is chelated by two pentaselenide Se_5^{2-} dianions, resulting in a square-planar geometry with two lone pairs completing the pseudo-octahedron (Figure 12.24b). The Se–Se bonds involving the central Se atom are substantially longer (2.57 Å) than the other Se–Se bonds (2.32–2.33 Å). The heteroatomic dianion $TeSe_{10}^{2-}$ has a similar spirocyclic arrangement with the tellurium atom in the central position.[91] The Se_{16}^{4-} dianion in the caesium salt is comprised of an Se_6 ring and two Se_5^{2-} chains.[90]

12.4.3 Homoatomic Tellurium Anions

Like their cationic counterparts (Section 12.3.2), the structural diversity of polyanions of the chalcogens increases down the series S, Se, Te.[92] Tellurium forms a remarkable variety of polyanions, including some in which the charge is >-2, owing to the ability of the heaviest chalcogen to adopt non-classical bonding structures, i.e. those in which the octet rule is violated. The ability of tellurium chains to associate even though they both carry a negative charge is illustrated by the simple examples of the linear Te_3^{4-} and T-shaped Te_4^{4-} anions shown in Figure 12.25.

Some examples of the unique behaviour of tellurium polyanions are found in the compounds $RbTe_6$,[93] MTe_3 (M = Li, Na),[94] $[Cr(en)_3]Te_6$,[95] $[K(15\text{-crown-}5)_2]_2Te_8$,[96] $[Et_4N]_2Te_{12}$[97] and the caesium salts Cs_2Te_{13},[18b] Cs_4Te_{28}[18b] and Cs_3Te_{22}.[18a]

A comparison of the structures of the series Te_6^{n-} ($n = 1, 2, 3$) is informative. The monoanion Te_6^- in $RbTe_6$, is comprised of two-dimensional $^2_\infty[Te_6^-]$ sheets separated by Rb^+ cations. The sheets are made up of rows of Te_6 rings (in the chair conformation with Te–Te bond lengths of 2.78 to 3.20 Å) that are bonded to other rows of Te_6 rings through four of the Te atoms [d(Te–Te) = 3.21 Å].[93] In addition, the Te_6 rings within a row interact through weak Te$\cdots$Te contacts of

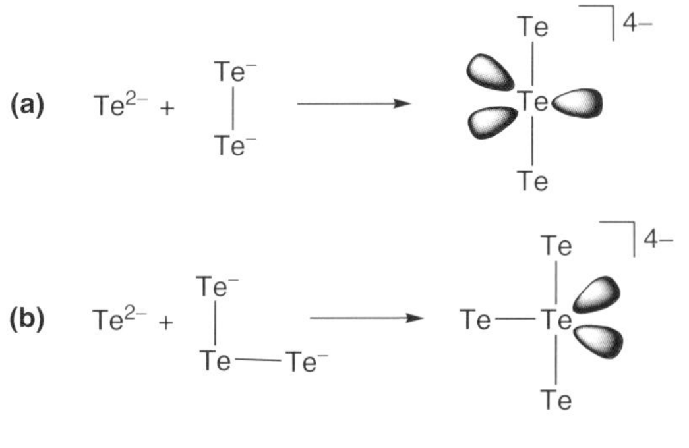

Figure 12.25 Formation and structures of (a) Te_3^{4-} and (b) Te_4^{4-}.

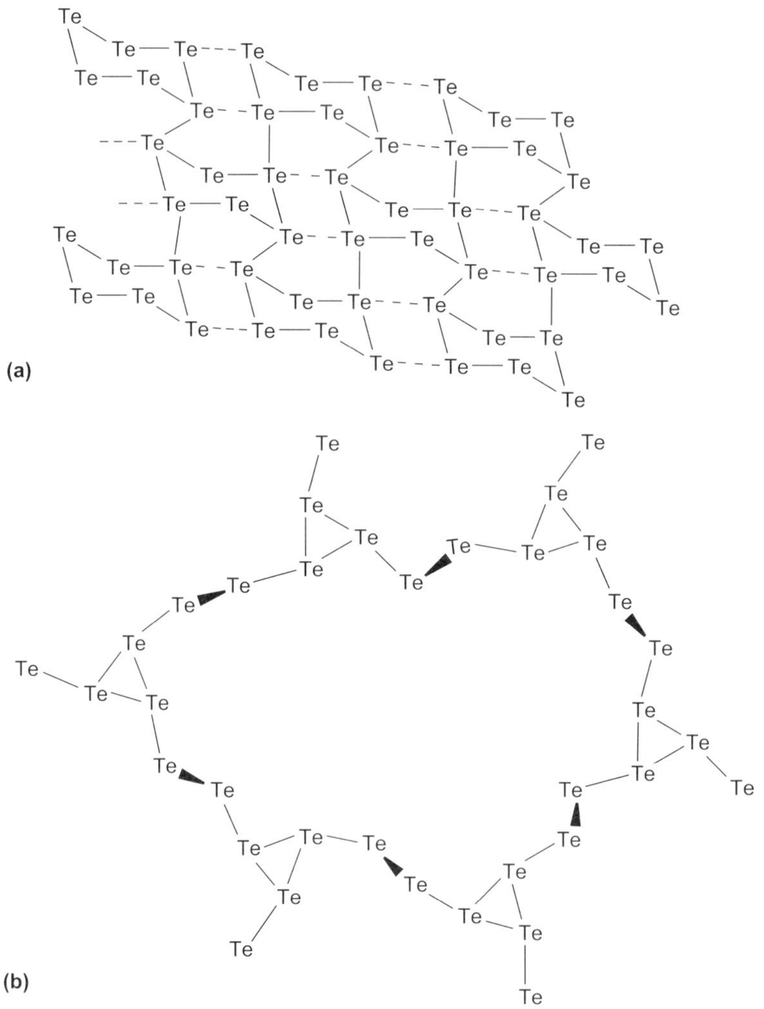

Figure 12.26 Structures of (a) $^2_\infty[Te_6^-]$ in RbTe$_6$ and (b) Te$_6^{3-}$ in [Cr(en)$_3$]Te$_6$.

3.45 Å (Figure 12.26a). The dianion Te$_6^{2-}$ found in MTe$_3$ (M = Li, Na) is arranged in infinite helical chains $^1_\infty[Te_6^{2-}]$ separated by channels of alkali metal cations. The Te–Te distances within the six-atom chains are 2.86–3.02 Å (M = Li) or 2.77–2.98 Å (M = Na), while the interchain Te···Te linkages are 3.14–3.32 and 3.16 Å, respectively.[94] The polymeric trianion Te$_6^{3-}$ in [Cr(en)$_3$][Te$_6$] forms a unique three-dimensional framework in which the [Cr(en)$_3$]$^{3+}$ cations occupy the cavity of a porous structure.[95] The Te$_6^{3-}$ anion contains a central Te$_3$ ring with one Te atom attached to each member of the ring (Figure 12.26b).[95] The mean Te–Te separation within the three-membered Te$_3$ ring is long (3.14 Å), whereas the distance between ring and exocyclic Te atoms is only 2.92 Å.

Group 16: Rings and Polymers

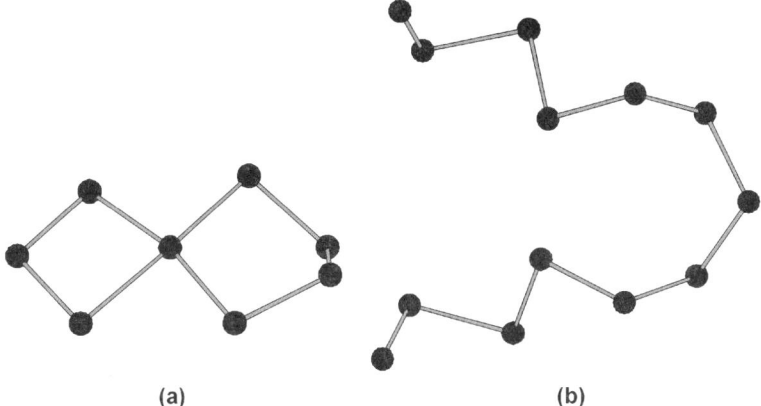

Figure 12.27 Structures of (a) Te_8^{2-} in $[K(15\text{-crown-}5)_2]_2Te_8$ and (b) Te_{13}^{2-} in Cs_2Te_{13}.

In contrast to the acyclic structures of *catena*-E_8^{2-} (E = S, Se), the Te_8^{2-} dianion in $[K(15\text{-crown-}5)_2]_2Te_8$ has a spirocyclic framework in which a central Te^{2+} cation is chelated by Te_3^{2-} and Te_4^{2-} anions (Figure 12.27a).[96] The Te–Te distances in the planar Te_3^{2-} and envelope-shaped Te_4^{2-} rings are *ca* 0.3 Å shorter than the mean distances to the central Te atom. The longest isolated tellurium anionic chain is found in Cs_2Te_{13}, in which the Te–Te bond distances in the 13-atom chain range from 2.75 to 2.91 Å (Figure 12.27b).[18b] These chains are connected through Te···Te interactions (3.18–3.26 Å) to create a ladder of Te_4 rectangles and Te_6 rings.

The structure of the remarkable Cs_3Te_{22} lattice, in which cyclo-Te_8 rings are embedded between layered polytelluride anions, is shown in Figure 12.3.[18a] The network structure of Cs_4Te_{28} also contains neutral cyclo-Te_8 molecules.[18b] All three tellurium-rich caesium polytellurides Cs_2Te_{13}, Cs_3Te_{22} and Cs_4Te_{28} are obtained from the solvothermal reaction of caesium carbonate with As_2Te_3 in superheated methanol at 160–195 °C.[18] Inspection of the structure of Cs_2Te_{13} (Figure 12.27b), which is formed at lower temperatures, suggests that the Te_8 ring in Cs_3Te_{22} and Cs_4Te_{28} is produced from the incomplete crown-shaped Te_7 unit in the Te_{13}^{2-} chain.

12.4.4 Heteroatomic Selenium–Tellurium Anions

Mixed Te–Se polychalcogenide anions are formed by reactions between polytellurides and polyselenides in DMF in the presence of tetraalkylammonium or tetraphenylphosphonium cations.[98] The $[Te_3Se_6]^{2-}$ dianion in $[NEt_4]_2[Te_3Se_6]$ forms a one-dimensional infinite chain of Te_3Se_5 eight-membered rings bridged by selenium atoms. The Te_3Se_5 rings have an open-book conformation and alternate in their orientation (Figure 12.28a). The $Te_3Se_7^{2-}$ dianion in $[NEt_4]_2[Te_3Se_7]$ also forms one-dimensional chains. However, this anion has a bicyclic structure of five- and six-membered rings with a central Se–Se single

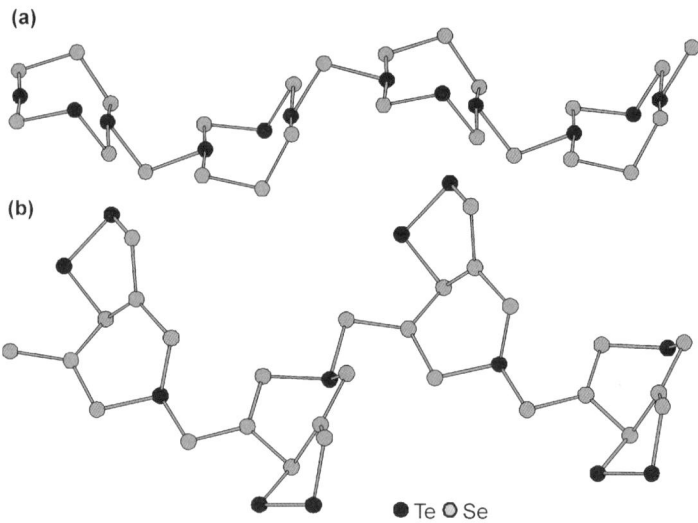

Figure 12.28 Sections of the infinite chains of mixed Se–Te anions in (a) [NEt$_4$]$_2$[Te$_3$Se$_6$] and (b) [NEt$_4$]$_2$[Te$_3$Se$_7$].

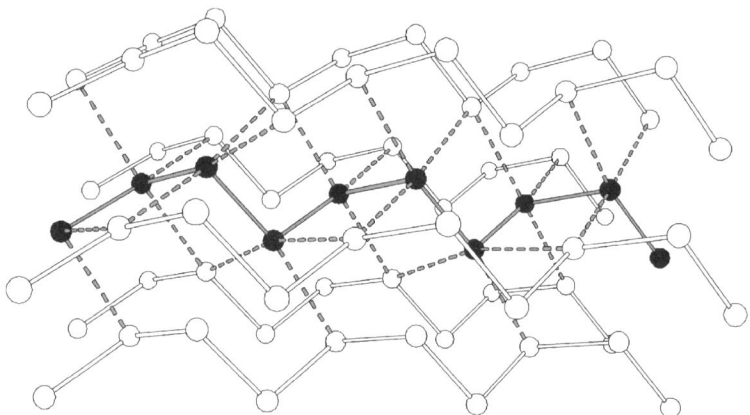

Figure 12.29 The helical chain structure of hexagonal tellurium.

bond (2.37 Å). The five-membered ring has a Te–Te bond (2.71 Å) and the bridging Se atom is bonded to a Te atom of the six-membered ring and an Se atom of the six-membered ring of the next repeating unit (Figure 12.28b).

12.5 POLYMERIC SULFUR, SELENIUM AND TELLURIUM

All three heavy chalcogens form polymers with an infinite helical chain structure; however, the conditions under which they are produced differ significantly. Poly-*catena*-sulfur S_∞ exists in a variety of forms, all of which revert slowly to cyclo-S_8 at room temperature.[99] The best known is plastic sulfur, which is obtained by quenching molten sulfur after it has been heated to *ca* 400 °C (see Section 5.4.3). The S–S bond distances in the helical chains of fibrous sulfur are 2.07 Å, while the S–S–S bond angle is 106.0° and the S–S–S–S dihedral angle is 85.3°.

The most thermodynamically stable form of selenium is grey selenium, which is a photoconductor. It is generated by warming other modifications (Section 12.1.1) or by slowly cooling the molten element. The structure consist of helical chains with d(Se–Se) = 2.37 Å and an Se–Se–Se bond angle of 103.1°.[100] Similarly to fibrous sulfur, there is a repeat unit every three atoms. In contrast to the other chalcogens, tellurium has only one crystalline form; it is composed of helical chains similar to those in hexagonal selenium (Figure 12.29).[101] The Te–Te distance is 2.84 Å. The closest interatomic distances between the chains in polymeric selenium and tellurium are similar at 3.44 and 3.50 Å, respectively.

Homogeneous crystalline alloys of selenium and tellurium, Se_xTe_{1-x}, of various compositions (x = 0.1–0.9) are prepared by the co-reduction of solutions of Se(IV) and Te(IV) compounds, *e.g.* dialkylselenites and tetraalkoxytelluranes (or glycol solutions of SeO_2 and TeO_2), with hydrazine.[102]

12.6 SULFUR–NITROGEN HETEROCYCLES

Sulfur and nitrogen forms a vast number of ring systems in which the sulfur atoms may be in oxidation states ranging from +2 (or lower) to +6 and the nitrogen atoms can be either two- or three-coordinate.[103] These fascinating heterocycles are conveniently divided into saturated systems, in which the N atoms are three-coordinate, and unsaturated systems involving two-coordinate nitrogen. The category of unsaturated sulfur–nitrogen heterocycles may be further subdivided according to the coordination number of the sulfur atoms.

12.6.1 Saturated Sulfur–Nitrogen Rings: Cyclic Sulfur Imides

Cyclic sulfur imides are saturated ring systems involving two-coordinate sulfur atoms and three-coordinate nitrogen atoms. Their structures can be drawn without double bonds, but metrical parameters supported by electron density measurements indicate some involvement of the nitrogen lone-pair electrons in π-bonding to the adjacent sulfur atoms. The best known examples are eight-membered rings in which one or more of the sulfur atoms in cyclo-S_8 are replaced by an imido (NH) group, *e.g.* cyclo-S_7NH and the diimide isomers cyclo-1,3-, 1,4- and 1,5-$S_6(NH)_2$ (see **2.1–2.4** in Section 2.1). The tetraimide cyclo-$S_4N_4H_4$ is well known and two triimides cyclo-1,3,5- and 1,3,6-$S_5(NH)_3$ have also been characterised. Saturated S–N ring systems involving adjacent nitrogen atoms are relatively rare.

The standard preparation of these cyclic sulfur imides, with the exception of cyclo-$S_4N_4H_4$, involves the reaction of S_2Cl_2 with gaseous ammonia in a polar solvent, *e.g.* DMF, at −10 °C followed by hydrolysis with cold, dilute HCl (see Section 2.1).[104] This method gives mainly cyclo-S_7NH, but the imides with higher nitrogen content can be separated by chromatography of CS_2 solutions on silica gel or by high-performance liquid chromatography.[105] The tetraimide cyclo-$S_4N_4H_4$ is prepared in good yields by the reduction of S_4N_4 with methanolic $SnCl_2 \cdot 2H_2O$.

The cyclic sulfur imides all adopt the crown conformation of cyclo-S_8 with S–S bond lengths in the range 2.04–2.06 Å, typical of single bonds. However, the S–N bond distances (1.66–1.68 Å) are significantly shorter than a S–N single bond and the geometry around nitrogen is almost planar, indicating three-centre π-bonding in the S–N(H)–S units.[106] This conclusion is supported by experimental electron deformation density measurements for cyclo-S_7NH and cyclo-$S_4N_4H_4$.[107] In addition, the values of the NMR coupling constant $^1J(^{15}N,^1H)$ for all cyclic sulfur imides fall within the narrow range 93–96 Hz, consistent with sp^2-hybridised nitrogen atoms.[108]

The cyclic sulfur imides are weak Brønsted acids that are readily deprotonated by strong bases. The treatment of S_7NH with tetra-*n*-butylammonium hydroxide in diethyl ether at low temperatures produces the thermally unstable yellow $[S_7N]^-$ anion, which decomposes at room temperature to give the deep-blue $[S_4N]^-$ anion,[109a] which has an acyclic structure (Scheme 12.3a).[109b] Similarly, the deprotonation of the tetraimide cyclo-$S_4N_4H_4$ results in ring opening to give the acyclic $[S_2N_2H]^-$ anion (Scheme 12.3b).[110]

The sulfur centre in cyclic sulfur imides is oxidised to an S=O functionality by using methodology similar to that described for cyclo-S_8 (Section 12.2.1). Thus, the treatment of S_7NH with trifluoroperoxyacetic acid produces $S_7NH(O)$.[111] Stronger oxidising agents induce ring opening, as illustrated by the formation of $[NS_2]^+$ upon oxidation of S_7NH with $SbCl_5$ in liquid SO_2.[112] This reaction provided the first synthesis of this sulfur analogue of $[NO_2]^+$.

Two types of behaviour are observed in the interaction of cyclic sulfur imides with metal centres: (a) adduct formation and (b) oxidative addition. The formation of S-bonded adducts, *e.g.* $(S_4N_4H_4)M(CO)_5$ (M = Cr, W) and the sandwich complex $(S_4N_4H_4)_2 \cdot AgClO_4$ (see Figure 7.2), is discussed in Section 7.1.1. Oxidative addition involving insertion of the metal into an S–S bond of the cyclic sulfur imide S_7NR (R = H, Me) occurs with $Cp_2Ti(CO)_2$, as illustrated in Scheme 12.4.[113]

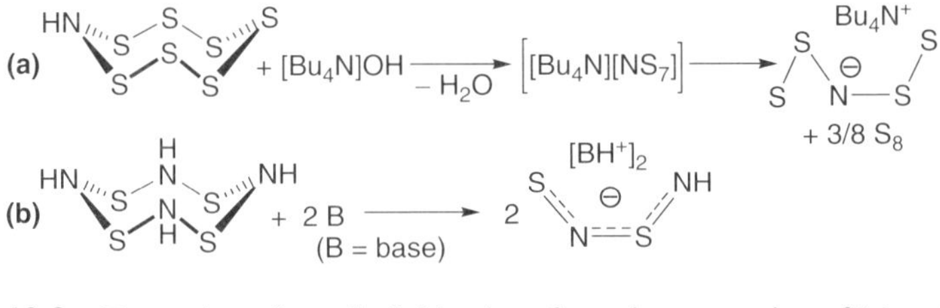

Scheme 12.3 Formation of acyclic S–N anions from deprotonation of (a) cyclo-S_7NH and (b) cyclo-$S_4N_4H_4$.

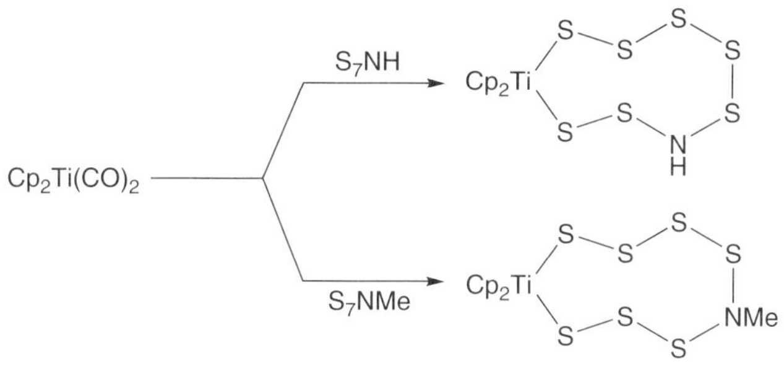

Scheme 12.4 Oxidative addition of (a) S_7NH and (b) S_7NMe to $Cp_2Ti(CO)_2$.

Like the cyclic sulfur allotropes, cyclic sulfur imides with ring sizes either smaller or larger than eight are thermally much less stable than the eight-membered rings; however, synthetic methods for these unsaturated sulfur–nitrogen heterocycles have been developed and several of them have been structurally characterised (**12.10–12.13**). The diimides 1,4-$S_4(NR)_2$ (**12.10**, R = Et, Bz, Cy, CH_2CH_2Ph) are prepared in low yields from the cyclo-condensation of the appropriate primary amine with S_2Cl_2 in diethyl ether under high dilution conditions.[114a] These six-membered rings adopt a chair conformation with S–N bond lengths that are ca 0.05 Å longer than those in eight-membered cyclic sulfur imides and the bond angles at nitrogen reveal a pyramidal (sp^3) geometry. The nitrogen-rich six-membered ring $S_2(NR)_4$ (**12.11**, R = CO_2Et), prepared from the hydrazine RNHNHR and S_2Cl_2 in the presence of Et_3N, also assumes a chair conformation.[114b] In contrast to the behaviour of S_7NMe (Scheme 12.4), the reaction of 1,4-$S_4(NR)_2$ (R = Me, Cy) with $Cp_2Ti(CO)_2$ results in replacement of one of the NR groups by the metal to give the six-membered rings Cp_2TiS_4NR, which maintain a chair conformation.[115a] The chair-shaped six-membered ring RNS_5 (**12.12**, R = 2,6-dimesityl-4-methylphenyl) is isolated in low yield when the acyclic N-thiosulfinylaniline RNSS is passed through silica gel.[115b]

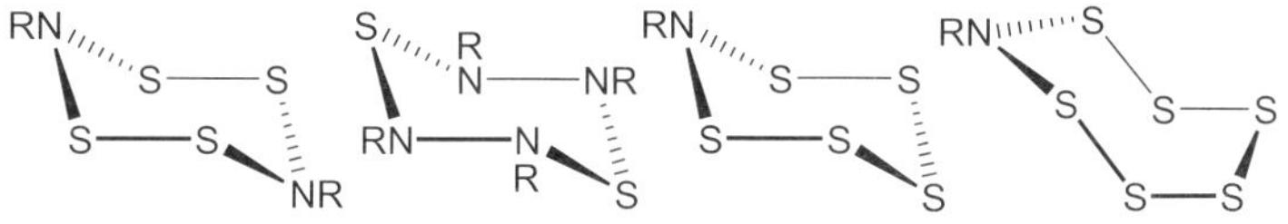

12.10   12.11   12.12   12.13

The cyclic sulfur imides with a single NR functionality, cyclo-S_nNR (**12.12**, n = 5; **12.13**, n = 6)[115a] and cyclo-S_nNH (n = 8, 9, 11)[116] are obtained by a methodology similar to that used for the preparation of unstable sulfur allotropes, e.g. cyclo-S_9 and -S_{10} (Section 12.1.1). The synthesis of the nine- and 10-membered rings cyclo-S_nNH (n = 8, 9) is shown in eqn (6.2) and their structures are depicted as **6.7** and **6.8** in Section 6.2.1.

12.6.2 Unsaturated Sulfur–Nitrogen Rings

Unsaturated sulfur–nitrogen rings are comprised of two-coordinate nitrogen atoms and sulfur atoms that are either two-, three- or four-coordinate. As discussed in Section 4.1.2.3, investigations of planar unsaturated S–N heterocycles that contain two-coordinate sulfur have enhanced our understanding of the electronic structures of π-electron-rich systems. This class of compound, specifically the four-membered ring S_2N_2, is also important in the generation of the unique polymer poly(sulfur nitride), $(SN)_x$ (see Sections 5.4.3 and 12.7.1). Unsaturated cyclic S–N halides, *e.g.* the six-membered ring $(NSCl)_3$, are versatile reagents for the synthesis of other S–N compounds and they have also been used for some organic transformations.

12.6.2.1 Two-coordinate Sulfur. The series of planar, unsaturated cyclic S–N heterocycles includes cations and anions, and also neutral species: cyclo-S_2N_2, cyclo-$[S_2N_3]^+$, cyclo-$[S_3N_2]^{2+}$, cyclo-$[S_3N_3]^-$, cyclo-$[S_4N_3]^+$, cyclo-$[S_4N_4]^{2+}$ and cyclo-$[S_5N_5]^+$ (see structures **4.3**–**4.8**). As discussed in Section 4.1.2.3, these inorganic rings all formally conform to the Hückel $(4n+2)$ π-electron rule for aromaticity. It is important to note, however, that they are all π-electron rich, *i.e.* the number of π-electrons exceeds the number of atoms in the ring. The excess of π-electrons is accommodated in antibonding (π*) orbitals or, in the case of S_2N_2, in π orbitals that are primarily non-bonding. This feature of their electronic structures has important consequences for the properties and reactivity of these S–N ring systems, as will be evident in the following discussion of specific examples.

The four-membered ring cyclo-S_2N_2 is prepared by the thermolysis of other S–N compounds. The potentially explosive cage compound S_4N_4 is most frequently used for this purpose using silver wool as a catalyst at 220 °C;[117] the S–N heterocycles $[S_4N_3]Cl$[118a] and $Ph_3AsNS_3N_3$[118b] are alternative sources of cyclo-S_2N_2 (Scheme 12.5). Disulfur dinitride, cyclo-S_2N_2, forms large, colourless crystals which detonate upon heating. It has a regular square-planar structure with S–N bond distances (1.65 Å) intermediate between single- and double-bond values.[117] The four-membered ring forms *N*-bonded adducts with

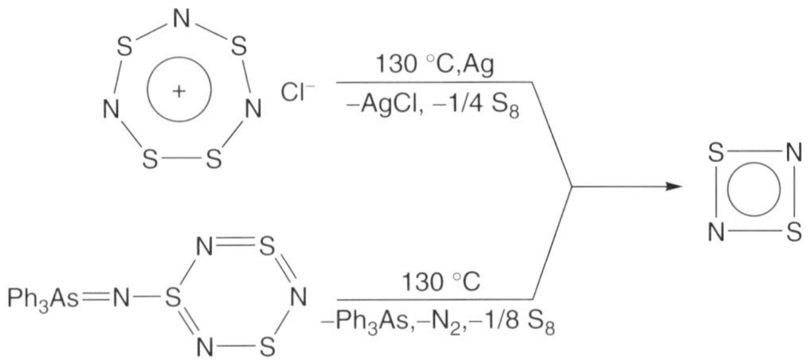

Scheme 12.5 Synthesis of S_2N_2 *via* thermolysis of S–N heterocycles.

Lewis acids, e.g. cyclo-$S_2N_2 \cdot 2AlBr_3$, and with a variety of transition metal halides.[119] In practice, insertion of metal fragments into an S–N bond of the S_2N_2 ring to give cyclometallathiazenes of the type $L_nMS_2N_2$[120] is preferred over the predicted formation of π-complexes, e.g. $(\eta^4\text{-}S_2N_2)M(CO)_3$ (M = Cr, Mo, W), for this six π-electron system.[121]

The five-membered cyclo-$[S_2N_3]^+$ cation is the only nitrogen-rich member of this series of binary S–N ring systems. It is obtained in low yield as the thermally stable salt $[S_2N_3]_2[Hg_2Cl_6]$ from the reaction of $(NSCl)_3$ with $HgCl_2$ in dichloromethane.[122] The bond lengths in the planar S_2N_3 ring indicate delocalised π-bonding that is attenuated across the S–S bond.

A variety of cyclic S–N cations are prepared by the treatment of S_4N_4 with various oxidising agents (Scheme 12.6). The sulfur-rich radical cation cyclo-$[S_3N_2]^{+\bullet}$ is obtained by treatment of S_4N_4 with AsF_5 or HSO_3F.[123] In solution it exhibits a characteristic five-line (1:2:3:2:1) EPR signal.[124] In the solid state, these five-membered rings dimerise with the formation of two weak intermolecular S···S contacts (π*–π* interactions) (see **4.11** in Section 4.2.2). The calculated barrier to dissociation of the corresponding dication cyclo-$[S_3N_2]^{2+}$ into $[NS]^+$ and $[NS_2]^+$ in the gas phase is 10.9 kcal mol^{-1}.[125a] However, lattice-stabilisation effects allow the isolation of $[MF_6]^-$ salts (M = As, Sb) of this six π-electron system in the solid state from the cycloaddition of $[NS]^+$ and $[NS_2]^+$ in liquid sulfur dioxide [eqn (12.13)].[125b] The lattice energy of the 1:2 salt $[S_3N_2][AsF_6]_2$ is calculated to be 357 kcal mol^{-1} compared with the combined lattice energies of 255 kcal mol^{-1} for the two 1:1 salts $[NS][AsF_6]$ and $[NS_2][AsF_6]$.

$$[NS][AsF_6] + [NS_2][AsF_6] \rightarrow [S_3N_2][AsF_6]_2 \quad (12.13)$$

The use of an excess of Lewis acid, such as $SbCl_5$ or AsF_5 or the oxidising agent peroxydisulfuryl difluoride $S_2O_6F_2$, converts S_4N_4 to the corresponding dication cyclo-$[S_4N_4]^{2+}$ without a change in ring size (Scheme 12.6).[126] The insertion reaction of the $[NS]^+$ cation with S_4N_4 produces salts of the cation cyclo-$[S_5N_5]^+$.[127] This planar 10-membered ring usually adopts the shape of azulene in the solid state.

The seven-membered cyclo-$[S_4N_3]^+$ cation was one of the first binary S–N heterocycles to be structurally characterised. The chloride salt is obtained as a yellow solid by the reaction of $[S_3N_2Cl]Cl$ with S_2Cl_2 in carbon tetrachloride

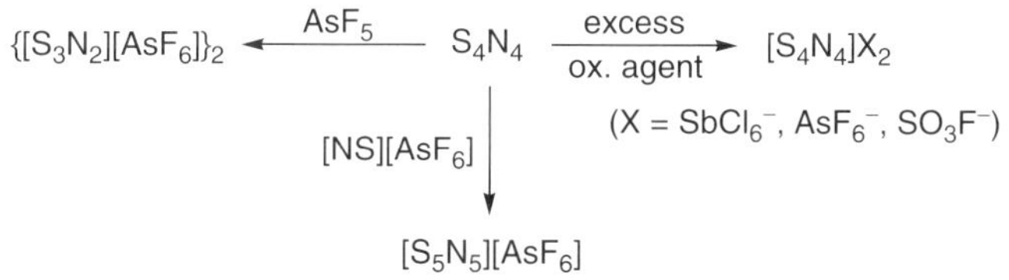

$\{[S_3N_2][AsF_6]\}_2 \xleftarrow{AsF_5} S_4N_4 \xrightarrow[\text{ox. agent}]{\text{excess}} [S_4N_4]X_2$

$[NS][AsF_6] \downarrow \quad (X = SbCl_6^-, AsF_6^-, SO_3F^-)$

$[S_5N_5][AsF_6]$

Scheme 12.6 Synthesis of cyclic S–N cations via oxidation of S_4N_4.

(see Scheme 2.3 in Section 2.1). A deformation density study indicates charge transfer of *ca* 0.75 electrons between the Cl⁻ ion and the two sulfur atoms of the S–S bond.[128]

The most electron-rich binary S–N species is the 10 π-electron, six-membered anion cyclo-[S_3N_3]⁻. The best preparation of this cyclic anion involves the reaction of S_4N_4 with an azide of a large cation, *e.g.* [(Ph₃P)₂N]⁺ [eqn (12.14)].[129] The large cation is necessary to stabilise the anion, because alkali metal salts of cyclo-[S_3N_3]⁻ are explosive in the solid state. However, the controlled thermolysis of yellow cyclo-[S_3N_3]⁻ salts in acetonitrile results in ring opening to give the deep-blue, acyclic anion *catena*-[SSNSS]⁻.[109b] The cobaltocenium salt [Cp₂Co][S_3N_3] is obtained from the redox reaction of S_4N_4 and cobaltocene.[130]

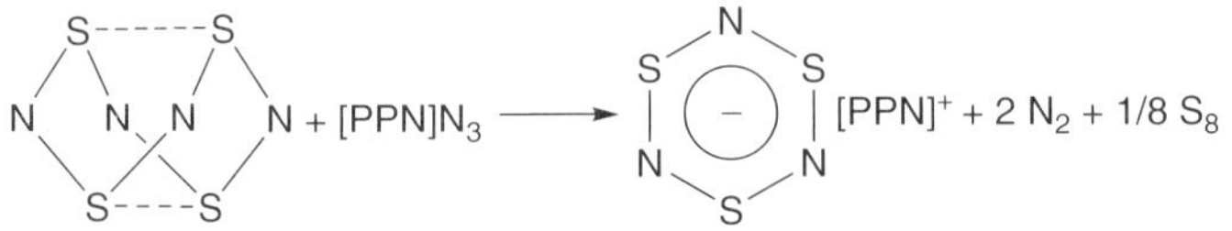

S_4N_4 + [PPN]N₃ ⟶ [S_3N_3]⁻ [PPN]⁺ + 2 N₂ + 1/8 S₈

(12.14)

The controlled air oxidation of yellow cyclo-[S_3N_3]⁻ salts allows the installation of one or two oxygen atoms on one of the sulfur atoms to give red [S_3N_3O]⁻ and, subsequently, purple [$S_3N_3O_2$]⁻ ions (Scheme 12.7); the progress of these transformations has been monitored by ¹⁵N NMR spectroscopy by using ¹⁵N-enriched cyclo-[S_3N_3]⁻.[131]

12.6.2.2 Three-coordinate Sulfur. The most important cyclic sulfur–nitrogen halides are the five-membered cation [S_3N_2Cl]⁺ and the neutral six-membered ring (NSCl)₃, both of which are used for the synthesis of other S–N heterocycles. The dark-orange salt [S_3N_2Cl]Cl is conveniently prepared by refluxing S_2Cl_2 with dry, finely ground ammonium chloride.[132] As illustrated in Scheme 2.3 (see Section 2.1), the five-membered ring is converted to (NSCl)₃ or [S_4N_3]Cl by the action of Cl₂ (preferably in the form of SO₂Cl₂) or an excess of S_2Cl_2, respectively. The sulfoxide S_3N_2O is obtained by treatment of [S_3N_2Cl]Cl with warm, anhydrous formic acid in dichloromethane [eqn (12.15)].[133] The five-membered rings in the [S_3N_2Cl]⁺ cation and the isoelectronic neutral molecule

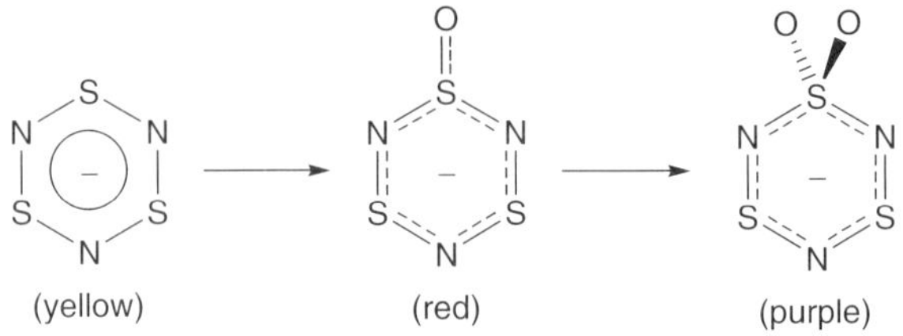

(yellow) ⟶ (red) ⟶ (purple)

Scheme 12.7 Air oxidation of [S_3N_3]⁻ to [S_3N_3O]⁻ and [$S_3N_3O_2$]⁻.

S_3N_2O are slightly puckered; the S–S bond length in the oxide is considerably greater than that in the cation (2.26 vs 2.18 Å).

(12.15)

Pale-yellow $(NSCl)_3$ may be recrystallised from carbon tetrachloride without decomposition when the temperature is kept below 50 °C. In the solid state, the six-membered ring adopts a chair conformation with all three chlorine atoms in axial positions and equal S–N bond lengths (ca 1.60 Å). This arrangement is stabilised by the anomeric effect (delocalisation of the nitrogen lone pair into an S–Cl σ^* orbital).[134]

An indication of the versatility of $(NSCl)_3$ as a reagent is given by the examples shown in Scheme 12.8. Some reactions take place with the retention of the six-membered ring. For example, the fluoride $(NSF)_3$ is obtained by halogen exchange between $(NSCl)_3$ and AgF_2 in carbon tetrachloride[135] and alkoxy derivatives $(NSOR)_3$ are prepared by metathesis with sodium alkoxides.[136] However, the six-membered ring readily dissociates to the monomer NSCl in solution,[137] hence it serves as a facile source of the $[NS]^+$ cation, e.g. by reaction with $Ag[AsF_6]$.[138] Complexes of the N-bonded NSCl (chlorothionitrene) ligand with transition metals are prepared by treatment of $(NSCl)_3$ with a variety of transition metal halides under mild conditions.[139] As another example of the

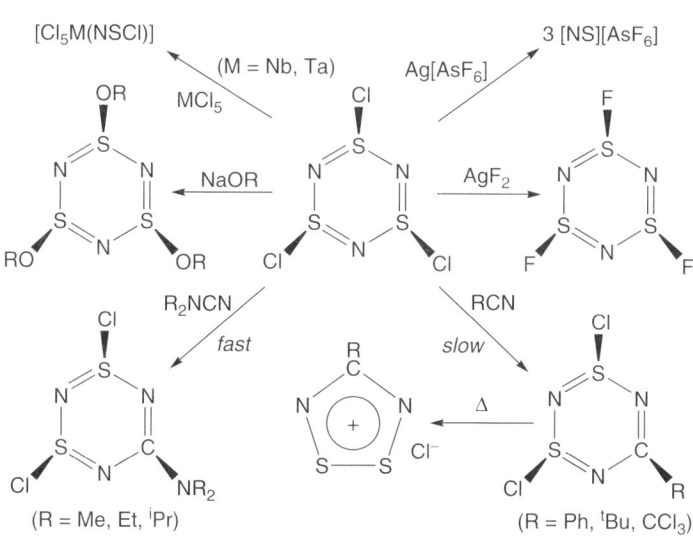

Scheme 12.8 Reactions of $(NSCl)_3$.

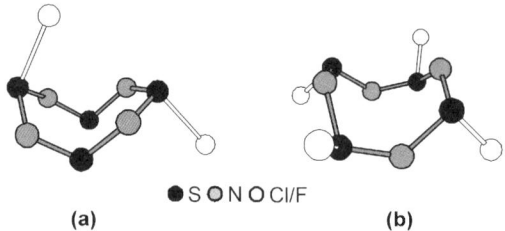

Figure 12.30 Eight-membered ring conformations in (a) 1,5-$S_4N_4Cl_2$ and (b) $(NSF)_4$.

versatility of $(NSCl)_3$, carbon atoms may be introduced into a sulfur–nitrogen ring by reactions with nitriles upon warming.[140] The S_3N_3 ring in the trifluoride $(NSF)_3$ is more robust than that in $(NSCl)_3$, as demonstrated by formation of the salts $[N_3S_3F_2][MF_6]$ from the reaction of $(NSF)_3$ with MF_5 (M = As, Sb).[135]

The sulfur–nitrogen halide $(NSCl)_3$ is also a useful reagent in organic chemistry.[141] For example, alkenes or alkynes react readily with $(NSCl)_3$ to produce 1,2,5-thiadiazoles[142a] and the reaction with N-alkylpyrroles gives bis-1,2,5-thiadiazoles.[142b]

Halogenated eight-membered S_4N_4 rings are thermally less stable than the corresponding six-membered rings. Dihalogenated derivatives 1,5-$S_4N_4X_2$ (X = Cl, F) are obtained by oxidative addition of one equivalent of X_2 to S_4N_4 under mild conditions.[143a,b] The structure of 1,5-$S_4N_4Cl_2$ consists of a folded eight-membered ring [$d(S \cdots S) = 2.45$ Å] with the exocyclic substituents in the *exo,endo* positions (Figure 12.30a).[144] The fluorination of S_4N_4 with an excess of AgF_2 produces the tetrafluoride $(NSF)_4$,[135] which is a boat-shaped eight-membered ring with alternating short (1.54 Å) and long (1.66 Å) S–N bonds as a result of Jahn–Teller distortion; the fluorine substituents occupy alternate axial and equatorial positions (Figure 12.30b).[145] In contrast, the tetrachloro derivative $(NSCl)_4$ is thermally unstable with respect to the formation of the six-membered ring $(NSCl)_3$. Ring contraction also occurs upon treatment of $(NSF)_4$ with AsF_5 to give $[S_3N_3F_2][AsF_6]$.[135]

12.6.2.3 Four-coordinate Sulfur. Sulfur–nitrogen heterocycles that are oligomers of the monomer –N=S(O)X– (X = Cl, F) are known as sulfanuric halides. Unlike the isoelectronic cyclophosphazenes $(NPCl_2)_n$, for which an extensive series of cyclic oligomers is known (see Section 11.4.2), only six-membered rings (n = 3) are well characterised for $[NS(O)X]_n$. The sulfanuric halides are colourless solids (X = Cl) or liquids (X = F). The chloride is prepared by the treatment of thionyl chloride with sodium azide in acetonitrile at -30 °C [eqn (12.16)];[146] however, this procedure is potentially hazardous. Fluorination of $[NS(O)Cl]_3$ with SbF_3 produces the corresponding fluoride.[147]

$$SOCl_2 + NaN_3 \rightarrow 1/3[NS(O)Cl]_3 + NaCl + N_2 \quad (12.16)$$

Two isomers are known for each of the sulfanuric halides. The six-membered ring in the isomer α-$[NS(O)Cl]_3$ adopts a chair conformation with equal S–N

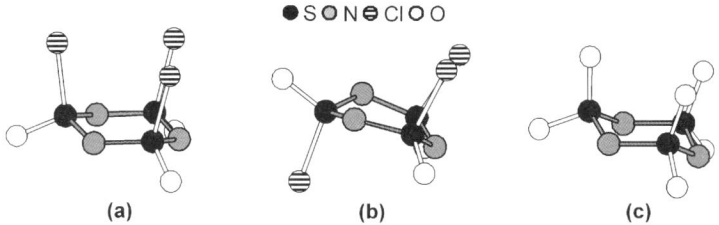

Figure 12.31 Molecular structures of (a) α-[NS(O)Cl]$_3$, (b) β-[NS(O)Cl]$_3$ and (c) [N$_3$S$_3$O$_6$]$^{3-}$.

bond lengths (1.57 Å). The three chlorine atoms are in axial positions on the same side of the ring in the α-isomer (Figure 12.31a),[148a] whereas the β-isomer has two axial and one equatorial chlorine atoms (Figure 12.31b).[148b] In contrast to cyclophosphazenes, the S$_3$N$_3$ ring in sulfanuric chloride is readily degraded by nucleophiles. However, the fluoride [NS(O)F]$_3$ is more robust towards nucleophilic reagents. For example, all three fluorine atoms can be replaced by primary or secondary amines and mono- or diphenyl derivatives are formed upon treatment with phenyllithium.[149]

The six-membered ring [N$_3$S$_3$O$_6$]$^{3-}$ is formally isoelectronic with sulfanuric fluoride [NS(O)F]$_3$; it is the trimer of the [NSO$_2$]$^-$ anion, isoelectronic with SO$_3$ (Section 12.9). The trianion [N$_3$S$_3$O$_6$]$^{3-}$ is obtained, as the ammonium salt, by the reaction of sulfamide SO$_2$(NH$_2$)$_2$ with sulfuryl chloride followed by treatment with ammonia gas; the mean S–N bond length in the chair-shaped six-membered ring is 1.60 Å (Figure 12.31c).[150a] The selenium analogue [N$_3$Se$_3$O$_6$]$^{3-}$ adopts a similar conformation in the tripotassium salt.[150b]

12.7 SULFUR–NITROGEN CHAINS AND POLYMERS

12.7.1 Poly(Sulfur Nitride)

The exciting discovery of the metal-like properties and superconducting behaviour of the non-metallic polymer poly(sulfur nitride) (or polythiazyl), (SN)$_x$, in 1973 sparked much activity in sulfur–nitrogen chemistry.[151] This interest continues as a result of the prediction that molecular chains incorporating thiazyl units could serve as molecular wires in the development of nanoscale technology.[152]

Poly(sulfur nitride) (or polythiazyl) is prepared by the topochemical, solid-state polymerisation of S$_2$N$_2$ at 0 °C (Figure 5.9).[153] A time-resolved X-ray diffraction study showed that this process is non-diffusive and produces monoclinic β-(SN)$_x$ (90%) and orthorhombic α-(SN)$_x$ (10%).[154a] Recent work has shown that the polymer can be incorporated into the channels of a zeolite by carrying out the polymerisation in a framework with the appropriate dimensions to act as a host for S$_2$N$_2$ and (SN)$_x$.[154b] Poly(sulfur nitride) is a shiny metallic solid consisting of highly oriented parallel fibres. It is an almost planar *cis,trans*-polymer with alternating S–N bond lengths of *ca* 1.63 and

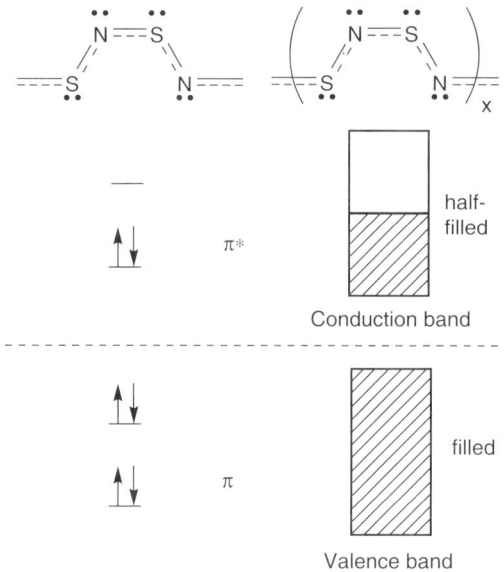

Figure 12.32 Molecular orbital description of the bonding in acyclic S_2N_2 and $(SN)_x$.

1.59 Å, intermediate between single- and double-bond values; the bond angles at sulfur and nitrogen are ca 106 and 120°, respectively.[153a] There are weak interactions between the polymer chains $[d(S \cdots S) = 3.48$ Å$]$ that play an important role in preventing a substantial Peierls distortion in the polymer chain.

Poly(sulfur nitride) behaves as a highly anisotropic conductor at room temperature. The conductivity along the polymer chains is similar to that of mercury and about 50 times greater than that perpendicular to the chain. The conductivity increases by three orders of magnitude upon cooling to 4 K and $(SN)_x$ becomes a superconductor at 0.3 K. The metallic behaviour can be understood from inspection of the molecular description of the π-bonding in the polymer chain (Figure 12.34). The repeating unit is considered to be the hypothetical acyclic S_2N_2 molecule, which is a six π-electron species. One of the antibonding π-orbitals is occupied by two electrons and the other is unoccupied. The $(SN)_x$ polymer is made up of an infinite number of S_2N_2 units linked in a planar arrangement. In the molecular orbital description of $(SN)_x$, this gives rise to a completely filled valence band and a half-filled conducting band (Figure 12.32). Thus the electronic structure is reminiscent of that of alkali metals, for which the half-filled conducting band is the result of the overlap of an infinite number of singly occupied ns orbitals, e.g. $2s^1$ for Li. The electrons in the partially filled band in $(SN)_x$ are free to move under the influence of an applied potential difference and conduction occurs along the polymer chain. The increase in conductivity with decreasing temperature is characteristic of a metallic conductor.

Although $(SN)_x$ does not react with water or acidic solutions, the sensitivity of this inorganic polymer to alkaline reagents and to oxidation imposes limitations on applications. Nevertheless, several applications have been considered for poly(sulfur nitride). For example, $(SN)_x$ can be used to increase the efficiency of GaAs solar cells by up to 35%.[151b] The polymer has also been investigated for possible use as an electrode material, since $(SN)_x$ interacts more strongly than metal electrodes with metal ions.[155a] An intriguing potential application in forensic science involves the formation of $(SN)_x$ upon exposure of S_2N_2 vapour to fingerprints.[155b]

Partial bromination of $(SN)_x$ with bromine vapour yields the blue–back polymer $(SNBr_{0.4})_x$, which has a room temperature conductivity higher than that of $(SN)_x$. The bromine is present as tribromide (Br_3^-) anions and Br_2 molecules intercalated between partially oxidised polymer chains.[156]

12.7.2 Sulfur–Nitrogen Chains

A variety of chain compounds in which thiazyl fragments are capped by organic groups are known. Since acyclic structures favour an even number of π-electrons, most of the known S–N chains contain an even number of nitrogen atoms and are neutral, e.g. $^tBuNSNSNSN^tBu$ (**12.14**)[157] and 4-MeC$_6$H$_4$SNSNSNSNSC$_6$H$_4$Me-4 (**12.15**);[158] however, cationic systems with an odd number of nitrogen atoms are also known, e.g. [4-MeC$_6$H$_4$SNSNSNSC$_6$H$_4$Me-4]$^+$ (**12.16**).[159] In contrast to $(SN)_x$, there is a distinct alternation of short and long sulfur–nitrogen bonds in the oligomeric chains, consistent with more localised [–S–N=S=N–] structures. The colours of acyclic sulfur–nitrogen compounds are dependent on chain length; shorter chains are bright yellow or orange, whereas the longer chains (more than six heteroatoms) produce deep-green, blue or purple solutions and exhibit a metallic lustre in the solid state. This trend is consistent with a narrower HOMO–LUMO energy gap and, hence, lower energy electronic transitions as the chain length increases.

12.7.3 Sulfanuric Polymers

The sulfanuric polymers, poly(oxathiazenes) $[-N=S(O)R-]_n$, represent another interesting class of S–N backbone polymer. Unlike the six-membered ring $(NPCl_2)_3$, the cyclic sulfanuric trimer $[N=S(O)Cl]_3$ does not undergo ring-opening polymerisation on heating; it decomposes exothermically above 250 °C. Consequently, a condensation route was developed for the synthesis of high molecular weight sulfanuric polymers [eqn (12.17)].[160]

$$Me_3SiN=S(R)(O)-OR' \xrightarrow[-Me_3SiOR']{\substack{120-127\,°C \\ (R'=Ph,\,CH_2CF_3)}} [-N=S(R)(O)-]_n$$

(R = Me, Et, Ph) (12.17)

The introduction of organic substituents and the change in oxidation state of sulfur results in properties for $[-N=S(O)R-]_n$ that are very different from those of $(SN)_x$. The sulfanuric polymers are thermoplastics with glass transition temperatures (T_g) in the range 30–85 °C, which are considerably higher than those of the better known polyphosphazenes $(NPX_2)_n$ (see Section 8.2.3, Table 8.1). The higher T_gs are attributed to a combination of increased intermolecular interactions as a result of the polar S=O groups and also greater side group–main chain interactions due to the smaller NSN bond angle [calculated value is 103°, cf. ca 120° for NPN in $(NPX_2)_n$]. Polyoxathiazenes appear to be very polar materials as a consequence of the presence of the S=O groups. For example, $[MeS(O)=N]_n$ is soluble in DMF, DMSO, hot water and concentrated H_2SO_4.[160]

12.8 SELENIUM– AND TELLURIUM–NITROGEN RINGS

The chemistry of selenium– and tellurium–nitrogen heterocycles has been much slower to develop than that of the corresponding sulfur systems. In part, this is the result of the hazardous nature of unsaturated binary nitrides of the heavier chalcogens. For example, Se_4N_4 is much more susceptible to explosion than S_4N_4. Nevertheless, several methods for the synthesis of Se_4N_4 are available.[161] No binary tellurium nitrides have been structurally characterised, but the composition Te_3N_4 has been established. It is likely that this involves a μ_3-nitrido functionality,[162] as established in the adduct $Te_6N_8 \cdot 2TeCl_4$.[163]

12.8.1 Cyclic Selenium and Tellurium Imides

There are very few similarities between the cyclic selenium imides and their sulfur analogues (Section 12.6.1). The selenium imides that have been structurally characterised include examples of five-, six-, eight- and 15-membered rings: cyclo-$Se_3(NR)_2$ (**12.17**, R = ^{t}Bu, Ad),[164] cyclo-$Se_3(N^tBu)_3$ (**12.18**),[165] cyclo-$Se_6(N^tBu)_2$ (**12.19**)[166] and cyclo-$Se_9(N^tBu)_6$ (**12.20**).[166] All of these derivatives have bulky alkyl substituents on the nitrogen atoms; the parent selenium imides (R = H) are unknown.

12.17   12.18   12.19

Group 16: Rings and Polymers 313

12.20 — 15-membered Se-N ring with ^{t}Bu groups

12.21 — ClSe(N^{t}Bu)Se(N^{t}Bu)SeCl-type acyclic structure

The cyclocondensation reaction of *tert*-butylamine with SeCl$_2$ in THF produces five- and six-membered rings, **12.17** (R = ^{t}Bu) and **12.18**, in addition to smaller amounts of larger rings, **12.19** and **12.20**;[164b] the latter were originally prepared *via* reaction of LiN(^{t}Bu)SiMe$_3$ with Se$_2$Cl$_2$ or SeOCl$_2$.[166] The composition of the mixture of Se–N compounds obtained from the ^{t}BuNH$_2$–SeCl$_2$ reaction depends markedly on the stoichiometry. When a 2:3 molar ratio is used, acyclic imidoselenium(II) halides, *e.g.* ClSeN(^{t}Bu)SeN(^{t}Bu)SeCl (**12.21**), are the major products.[165] This bifunctional compound probably serves as a building block for the formation of ring systems, *e.g.* cyclo-Se$_3$(N^{t}Bu)$_3$ (**12.18**). A better preparation of **12.18** involves the slow decomposition of the selenium(IV) diimide ^{t}BuN=Se=N^{t}Bu in toluene at 20 °C.[164b]

^{77}Se NMR spectroscopy is an excellent tool for analysing complex mixtures of cyclic selenium imides, as illustrated in Figure 3.7. The five- and 15-membered rings, **12.17** and **12.20**, each show two resonances with a relative intensity of 2:1, corresponding to the diselenido (–SeSe–) and monoselenido (–Se–) bridging units, respectively. The eight-membered ring **12.19** also exhibits two resonances attributable to the central selenium atoms in the N*Se*Se and Se*Se*Se units. The characteristic chemical shifts for the different selenium environments in cyclic selenium imides show a marked upfield shift as the electronegativity of the neighbouring atoms decreases: δ 1400–1625 (N*Se*N), 1100–1200 (N*Se*Se) and 500–600 (Se*Se*Se).[164b]

The six-membered ring cyclo-Se$_3$(N^{t}Bu)$_3$ adopts a chair conformation, *cf.* cyclo-Se$_6$, and the eight-membered ring cyclo-Se$_6$(N^{t}Bu)$_2$ is crown-shaped, *cf.* cyclo-Se$_8$. These cyclic selenium imides exhibit typical Se–N and Se–Se single bond lengths. However, both the Se–N and Se–Se bonds are significantly elongated in the puckered five-membered ring cyclo-Se$_3$(NAd)$_2$ as result of ring strain.[164a]

Although selenium(IV) diimides RN=Se=NR (R = adamantyl) and sulfinylamines RN=S=O are monomeric in the solid state, the seleninylamine OSe(μ-N^{t}Bu)$_2$SeO crystallises as the *cis* dimer (Figure 12.33).[164c] Calculations reveal that the cyclodimerisation of RNSeO ([2 + 2] cycloaddition) is energetically favoured.[164a]

The only known cyclic tellurium(II) imide is the six-membered ring cyclo-Te$_3$(N^{t}Bu)$_3$, which has a chair conformation with single Te–N bond lengths (Figure 12.34a).[167] Significantly, tellurium(IV) diimides, *e.g.* ^{t}BuNTe(μ-N^{t}Bu)$_2$TeN^{t}Bu, adopt dimeric structures with a central Te$_2$N$_2$ ring (Figure 12.34b),[167]

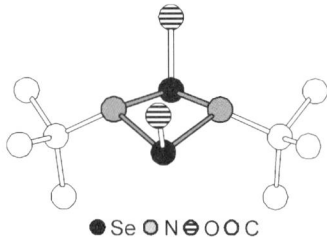

Figure 12.33 The four-membered Se_2N_2 ring in $OSe(\mu-N^tBu)_2SeO$.

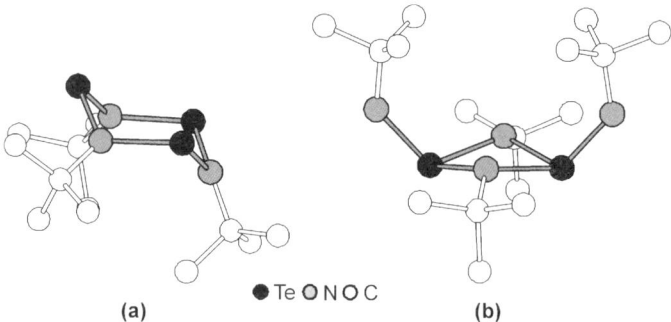

Figure 12.34 Molecular structures of (a) cyclo-$Te_3(N^tBu)_3$ and (b) $^tBuNTe(\mu-N^tBu)_2TeN^tBu$.

whereas sulfur(IV) and selenium(IV) diimides are monomeric. A four-membered Te_2N_2 ring is also the central feature of the related derivatives of tellurium(IV) imides, $(^tBuO)_2Te(\mu-N^tBu)_2Te(O^tBu)_2$ and $Cl_2Te(\mu-N^tBu)_2TeCl_2$.[168] The increasing tendency for dimer formation to occur for the heavier chalcogen(IV) diimides RN=E=NR (and related derivatives) reflects the expected trend to lower π-bond energies for chalcogen–nitrogen (np–2p)π-bonds along the series S ($n = 3$), Se ($n = 4$), Te ($n = 5$).[169]

12.8.2 Unsaturated Systems

12.8.2.1 Binary Selenium–Nitrogen Rings. There are only a few selenium analogues of the unsaturated sulfur–nitrogen compounds described in Section 12.6.2. Recent *ab initio* calculations for cyclo-E_2N_2 (E = S, Se) and cyclo-$SSeN_2$ indicate that these four-membered rings can be described as two π-electron aromatic systems with 6–8% singlet diradical character.[170] Although cyclo-Se_2N_2 has not been structurally characterised as a free species, it forms 1:2 adducts with main group or transition metal halides (**12.22** and **12.23**). These adducts are produced by reactions of Se_4N_4 with either $AlBr_3$ or $[PPh_4]_2[Pd_2X_6]$ (X = Cl, Br); the mean Se–N bond length in the four-membered ring is *ca* 1.79 Å.[171]

Addition of the thio crown ether [14]aneS$_4$ to **12.23** (X = Br) produces a reactive intermediate, which forms a platinum adduct of cyclo-Se$_2$N$_2$ upon treatment with [PtCl$_2$(PMe$_2$Ph)]$_2$ (see Scheme 7.2).[172] This intermediate has been tentatively identified as Se$_2$N$_2$ by IR spectroscopy. The mixed chalcogen nitride 1,5-Se$_2$S$_2$N$_4$, which is a potential precursor of cyclo-SeSN$_2$, has a cage structure similar to that of the binary nitrides E$_4$N$_4$ (E = S, Se).[173] However, no adducts of cyclo-SeSN$_2$ have been structurally characterised.

The five-membered cations cyclo-[Se$_3$N$_2$]$^{+\cdot}$ and cyclo-[Se$_3$N$_2$]$^{2+}$ are prepared by oxidation of Se$_4$N$_4$ with AsF$_5$.[174] In the solid state the radical cation cyclo-[Se$_3$N$_2$]$^{+\cdot}$ forms a dimer *via* weak Se$\cdots$Se contacts of 3.12–3.15 Å (**12.24**). In contrast to the sulfur analogue, the dication [Se$_3$N$_2$]$^{2+}$ (**12.25**) does not dissociate significantly into [SeN]$^+$ and [Se$_2$N]$^+$ in solution. However, the ^{77}Se NMR spectrum of **12.25** shows only one resonance (rather than the expected two resonances), even at low temperatures, indicative of a rapid exchange process. The mixed chalcogen radical cyclo-[Se$_2$SN$_2$]$^{+\cdot}$ is formed by reaction of Se$_4$[AsF$_6$]$_2$ with S$_4$N$_4$ in liquid SO$_2$ and further oxidation with AsF$_5$ produces the corresponding dication cyclo-[Se$_2$SN$_2$]$^{2+}$.[175] All the possible mixed chalcogen radicals cyclo-[Se$_{3-n}$S$_n$N$_2$]$^{+\cdot}$ ($n = 0$–3) have been characterised in solution by their EPR spectra.[176]

The mixed chalcogen system [(Se$_2$SN$_2$)Cl]$_2$ is formed, in addition to 1,5-Se$_2$S$_2$N$_4$, from the reaction of [(Me$_3$Si)$_2$N]S with SeCl$_4$ in dioxane.[177] The crystal structure contains two planar [Se$_2$SN$_2$]$^{+\cdot}$ rings linked by Se$\cdots$Se interactions (3.07 Å). As indicated in Figure 12.35, the chloride (Cl$^-$) counterions show close contacts to the Se atoms of the cyclo-[Se$_2$SN$_2$]$_2^{2+}$ dication *and* to the sulfur atom of the neighbouring [Se$_2$SN$_2$]$_2^{2+}$ dication. These strong intermolecular interactions account for the insolubility of this and related chalcogen–nitrogen halides in organic solvents.

12.8.2.2 Selenium– and Tellurium–Nitrogen Halides. The best characterised selenium–nitrogen halide is the cation [Se$_3$N$_2$Cl]$^+$ (**12.26**), which forms a

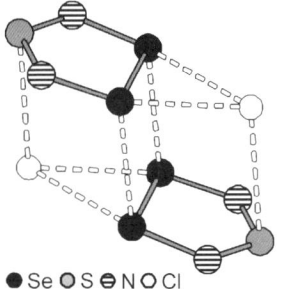

Figure 12.35 Interionic close contacts in [(Se$_2$SN$_2$)Cl]$_2$.

slightly puckered five-membered ring with a chlorine atom attached to one of the Se centres.[178] This cation is obtained as the red [GaCl$_4$]$^-$ salt from the reaction of the acyclic cation [N(SeCl)$_2$]$^+$ with triphenylantimony in methylene dichloride.[178a] The explosive and insoluble compound Se$_3$N$_2$Cl$_2$, which is prepared by the reaction of Se$_2$Cl$_2$ with trimethylsilyl azide in CH$_2$Cl$_2$ [eqn (12.18)], also contains the [Se$_3$N$_2$Cl]$^+$ cation.[179]

$$3Se_2Cl_2 + 2Me_3SiN_3 \rightarrow 2Se_3N_2Cl_2 + 2Me_3SiCl + N_2 \qquad (12.18)$$

There are no tellurium analogues of these selenium–nitrogen halides or the binary selenium–nitrogen cations. The different bonding properties of tellurium are reflected in the formation of cyclic tellurium–nitrogen halides that have no counterparts among the lighter chalcogens. For example, the dication [Te$_4$N$_2$Cl$_8$]$^{2+}$ (**12.27**) is obtained from the reaction of TeCl$_4$ with N(SiMe$_3$)$_3$ in acetonitrile.[180] This cyclic dication is a dimer of the hypothetical tellurium(IV) imide [Cl$_3$Te–N=TeCl]$^+$, cf. ^{t}BuNTe(μ-N^{t}Bu)$_2$TeN^{t}Bu (Figure 12.34b). In boiling toluene, the reaction of TeCl$_4$ with N(SiMe$_3$)$_3$ produces a more complex tellurium–nitrogen chloride that contains the tricyclic cation [Te$_5$N$_3$Cl$_{10}$]$^+$ (**12.28**),[181] which embodies a μ$_3$-nitrido functionality, as proposed for the binary tellurium nitride Te$_3$N$_4$.[162,163]

The reaction of Me$_3$SiN=S=NSiMe$_3$ with TeCl$_4$ is a fruitful source of tellurium–nitrogen chlorides that also contain sulfur, as depicted by structures **12.29**–**12.33**.[182] The structures of these mixed chalcogen systems provide an opportunity for a direct comparison of tellurium–nitrogen and sulfur–nitrogen bonding in the same molecule. In general, they consistently form sulfur–nitrogen double bonds (in –N=S=N– units) and Te–N single bonds, a clear indication that nitrogen forms stronger π-bonds with sulfur than with tellurium. The tricyclic compound **12.31** provides another example of the μ_3-nitrido functionality favoured by tellurium.[183] Interestingly, both Cl substituents in the five-membered ring $Cl_2Te_2SN_2$ (**12.29**) are covalently bonded to tellurium,[184] cf. $[S_3N_2Cl]Cl$ in which a chloride ion is present. The eight-membered $[Te_2S_2N_4]^{2+}$ dication (**12.32**) forms a folded bicyclic structure with a transannular Te–Te bond length of 2.88 Å,[184] whereas the corresponding binary system cyclo-$[S_4N_4]^{2+}$ is a planar delocalised ring.

12.9 SULFUR–, SELENIUM– AND TELLURIUM–OXYGEN RINGS AND POLYMERS

Binary oxides of sulfur, selenium and tellurium form a range of structures including monomers, dimers and one- or three-dimensional polymers. The preference of the heavier chalcogens for singly bonded structures over those involving double bonds is attributable to the fact that π-bond energies decrease relative to σ-bond energies down the series S, Se, Te. This trend is manifested by the structures of chalcogen dioxides EO_2 (E = S, Se, Te). Sulfur dioxide is a monomeric gaseous species with a bent (C_{2v}) structure and two S=O double bonds. In contrast, both selenium and tellurium dioxides are white solids with polymeric structures. The former consists of chains in which the selenium centres are in trigonal pyramidal environments with selenium–oxygen bond lengths of ca 1.81 and 1.63 Å, consistent with the presence of single and double bonds, respectively, in this one-dimensional polymer (Figure 12.36).[164a,185] In

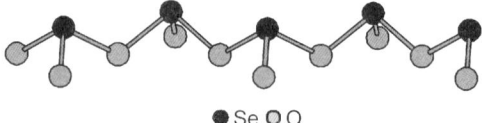

Figure 12.36 Structure of $(SeO_2)_\infty$.

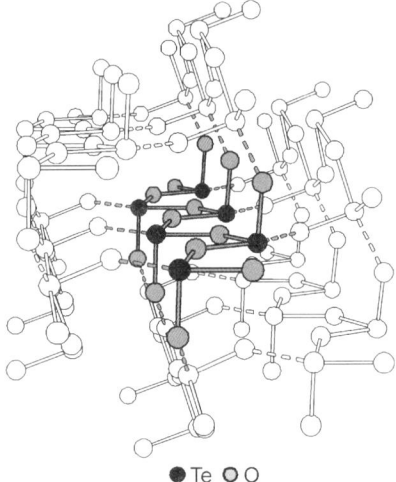

Figure 12.37 Structure of $(\gamma\text{-}TeO_2)_\infty$.

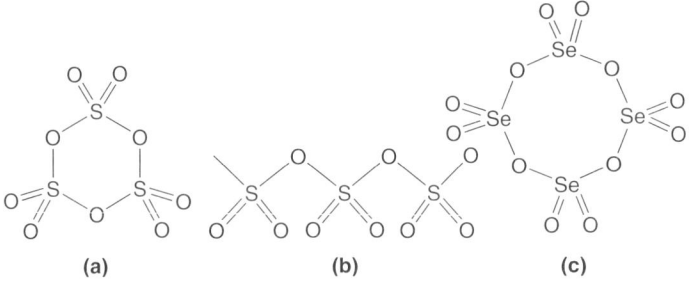

Figure 12.38 Structures of (a) γ-SO$_3$, (b) β-SO$_3$ and (c) (SeO$_3$)$_4$.

the gas phase, selenium dioxide forms the dimer OSe(μ-O)$_2$SeO with a central Se$_2$O$_2$ ring.[186]

Tellurium dioxide exists in three crystallographic forms: tetragonal, colourless α-TeO$_2$, yellow, orthorhombic β-TeO$_2$, which occurs naturally in the mineral tellurite, and γ-TeO$_2$. The coordination geometry around tellurium in all three forms of TeO$_2$ is a pseudo-trigonal bipyramid with a lone pair of electrons occupying an equatorial position. The structure of γ-TeO$_2$ (Figure 12.37) illustrates the three-dimensional network that is formed.[187] The Te–O bond lengths are close to single-bond values.

Solid sulfur trioxide exists in two well-defined modifications. Orthorhombic γ-SO$_3$ consists of cyclic trimers (Figure 12.38a),[188] which are converted by traces of water to the one-dimensional polymeric structure of monoclinic β-SO$_3$ (Figure 12.38b).[189] In contrast, selenium trioxide forms a cyclic tetramer (Figure 12.38c).[190]

REFERENCES

1. (a) R. S. Laitinen, P. Pekonen and R. J. Suontamo, *Coord. Chem. Rev.*, 1994, **130**, 1; (b) R. Steudel, H.-J. Mäusle, D. Rosenbauer, H. Möckel and T. Freyholdt, *Angew. Chem. Int. Ed. Engl.*, 1981, **20**, 394.
2. (a) M. Schmidt and E. Wilhelm, *Angew. Chem. Int. Ed. Engl.*, 1966, **5**, 964; (b) J. Buchler, *Angew. Chem. Int. Ed. Engl.*, 1966, **5**, 965; (c) A. Kutoglu and E. Hellner, *Angew. Chem. Int. Ed. Engl.*, 1966, **5**, 965.
3. B. Eckert and R. Steudel, *Top. Curr. Chem.*, 2003, **31**, 231.
4. (a) M. Schmidt, B. Block, H. D. Block, H. Köpf and E. Wilhelm, *Angew. Chem. Int. Ed. Engl.*, 1968, **7**, 632; (b) T. Sandow, J. Steidel and R. Steudel, *Angew. Chem. Int. Ed. Engl.*, 1982, **21**, 794; (c) R. Steudel and R. Strauss, *J. Chem. Soc., Dalton Trans.*, 1984, 1775.
5. R. Steudel, R. Reinhardt and F. Schuster, *Angew. Chem. Int. Ed. Engl.*, 1977, **16**, 715.
6. (a) P. Cherin and P. Unger, *Acta Crystallogr., Sect. B*, 1972, **28**, 313; (b) R. E. Marsh, L. Pauling and J. D. McCullough, *Acta Crystallogr.*, 1953, **6**, 71; (c) O. Foss and V. Janickis, *J. Chem. Soc., Dalton Trans.*, 1980, 624; (d) A. Maaninen, J. Konu, R. S. Laitinen, T. Chivers, G. Schatte, J. Pietikäinen and M. Ahlgren, *Inorg. Chem.*, 2001, **40**, 3539; (e) T. Maaninen, J. Konu and R. S. Laitinen, *Acta Crystallogr., Sect. E*, 2004, **60**, o2235.
7. Y. Miyamoto, *Jpn. J. Appl. Phys.*, 1980, **19**, 1813.
8. H.-J. Deiseroth, C. Reiner, M. Schlosser, X. Wang, H. Ajaz and L. Kienle, *Inorg. Chem.*, 2007, **46**, 8418.
9. H.-J. Deiseroth, M. Wagener and E. Neumann, *Eur. J. Inorg. Chem.*, 2004, 4755.
10. C. A. Dickerson, M. J. Fisher, R. E. Sykora, T. E. Albrecht-Schmitt and J. A. Cody, *Inorg. Chem.*, 2002, **41**, 640.
11. M. Wachhold and M. G. Kanatzidis, *J. Am. Chem. Soc.*, 1999, **121**, 4189.
12. R. A. Stevens, C. C. Raymond and P. K. Dorhout, *Angew. Chem. Int. Ed. Engl.*, 1995, **34**, 2509.
13. K. Nigata and Y. Miyamoto, *Jpn. J. Appl. Phys.*, 1984, **23**, 704.
14. (a) R. Staffel, U. Müller, A. Ahle and K. Dehnicke, *Z. Naturforsch., Teil B*, 1991, **46**, 1287; (b) J. Dietz, U. Müller, V. Müller and K. Dehnicke, *Z. Naturforsch., Teil B*, 1991, **46**, 1293.

15. R. Steudel and E.-M. Strauss, *Z. Naturforsch., Teil B*, 1981, **36**, 1085.
16. R. S. Laitinen and T. A. Pakkanen, *J. Chem. Soc., Chem Commun.*, 1986, 1381.
17. Y. V. Mironov, M. A. Pell and J. A. Ibers, *Angew. Chem. Int. Ed. Engl.*, 1996, **35**, 2854.
18. (a) W. S. Sheldrick and M. Wachold, *Angew. Chem. Int. Ed. Engl.*, 1995, **34**, 450; (b) W. S. Sheldrick and M. Wachold, *Chem. Commun.*, 1996, 607.
19. J. Pietikäinen and R. S. Laitinen, *Chem. Commun.*, 1998, 2381.
20. T. Chivers, R. S. Laitinen, K. J. Schmidt and J. Taavitsainen, *Inorg. Chem.*, 1993, **32**, 337.
21. T. Chivers, R. S. Laitinen and K. J. Schmidt, *Can. J. Chem.*, 1992, **70**, 719.
22. (a) R. S. Laitinen and T. A. Pakkanen, *J. Chem. Soc., Chem. Commun.*, 1986, 1381; (b) R. S. Laitinen and T. A. Pakkanen, *Inorg. Chem.*, 1987, **26**, 2598.
23. R. Steudel, M. Papavassiliou, E.-M. Strauss and R. S. Laitinen, *Angew. Chem. Int. Ed. Engl.*, 1986, **25**, 99.
24. R. Steudel and E.-M. Strauss, *Angew. Chem. Int. Ed. Engl.*, 1984, **23**, 362.
25. D. M. Giolando, M. Papavassiliou, J. Pickardt, T. B. Rauchfuss and R. Steudel, *Inorg. Chem.*, 1988, **27**, 2596.
26. (a) P. Pekonen, Y. Hiltunen, R. S. Laitinen and T. A. Pakkanen, *Inorg. Chem.*, 1990, **29**, 2770; (b) P. Pekonen, Y. Hiltunen, R. S. Laitinen and T. A. Pakkanen, *Inorg. Chem.*, 1991, **30**, 3679.
27. E.-M. Strauss and R. Steudel, *Z. Naturforsch., Teil B*, 1987, **42**, 682.
28. R. Steudel, Y. Steudel and K. Maiskiewicz, *Chem. Eur. J.*, 2001, **7**, 3281.
29. R. Steudel, Y. Drozdova, K. Maiskiewicz, R. H. Hertwig and W. Koch, *J. Am. Chem. Soc.*, 1997, **119**, 1990.
30. R. S. Laitinen, T. A. Pakkanen and R. Steudel, *J. Am. Chem. Soc.*, 1987, **109**, 710.
31. (a) R. Steudel, *Comments Inorg. Chem.*, 1982, **1**, 313; (b) R. Steudel and T. Sandow, *Inorg. Synth.*, 1982, **21**, 172.
32. R. Steudel and J. Steidel, *Angew. Chem. Int. Ed. Engl.*, 1978, **17**, 134.
33. (a) R. Steudel and T. Sandow, *Angew. Chem. Int. Ed. Engl.*, 1976, **15**, 772; (b) R. Steudel, R. Reinhardt and T. Sandow, *Angew. Chem. Int. Ed. Engl.*, 1977, **16**, 716.
34. (a) R. Steudel and J. Latte, *Angew. Chem. Int. Ed. Engl.*, 1974, **13**, 603; (b) P. Luger, H. Bradaczek, R. Steudel and M. Rebsch, *Chem. Ber.*, 1976, **109**, 180.
35. R. Steudel, T. Sandow and J. Steidel, *Z. Naturforsch., Teil B*, 1985, **40**, 594.
36. R. Steudel and T. Sandow, *Angew. Chem. Int. Ed. Engl.*, 1978, **17**, 611.
37. M. W. Wang, Y. Steudel and R. Steudel, *Chem. Eur. J.*, 2007, **13**, 502.
38. R. Steudel, T. Sandow and J. Steidel, *J. Chem. Soc., Chem. Commun.*, 1980, 180.
39. R. Steudel, J. Steidel and J. Pickard, *Angew. Chem. Int. Ed. Engl.*, 1980, **19**, 325.

Group 16: Rings and Polymers 321

40. T. Klapötke and J. Passmore, *Acc. Chem. Res.*, 1989, **22**, 234.
41. (a) J. Passmore, G. Sutherland, P. Taylor, T. K. Whidden and P. S. White, *Inorg. Chem.*, 1981, **20**, 3839; (b) J. Passmore, G. Sutherland, T. K. Whidden, P. S. White and C.-M. Wong, *Can. J. Chem.*, 1985, **63**, 1209.
42. W. A. Shantha Nandana, J. Passmore and P. S. White, *J. Chem. Soc., Chem. Commun.*, 1983, 526.
43. (a) J. Passmore, P. S. White and C.-M. Wong, *J. Chem. Soc., Chem. Commun.*, 1985, 1178; (b) D. Aris, J. Beck, A. Decken, I. Dionne, I. Krossing, J. Passmore, E. Rivard, F. Steden and X. Wang, *Phosphorus Sulfur Silicon*, 2004, **179**, 859.
44. R. Faggiani, R. J. Gillespie and J. W. Kolis, *J. Chem. Soc., Chem. Commun.*, 1987, 592.
45. R. Faggiani, R. J. Gillespie, J. W. Kolis and K. C. Malhotra, *J. Chem. Soc., Chem. Commun.*, 1987, 591.
46. T. Drews, W. Koch and K. Seppelt, *J. Am. Chem. Soc.*, 1999, **121**, 4379.
47. (a) R. J. Gillespie, J. Passmore, P. K. Ummat and O. C. Vaidya, *Inorg. Chem.*, 1971, **10**, 1327; (b) C. G. Davies, R. J. Gillespie, J. J. Park and J. Passmore, *Inorg. Chem.*, 1971, **10**, 2781.
48. (a) I. Krossing, *Top. Curr. Chem.*, 2003, **230**, 135; (b) S. Brownridge, I. Krossing, J. Passmore, H. D. B. Jenkins and H. K. Roobottom, *Coord. Chem. Rev.*, 2000, **197**, 397.
49. M. P. Murchie, J. Passmore, G. W. Sutherland and R. Kapoor, *Dalton Trans.*, 1992, 503.
50. (a) T. S. Cameron, H. K. Roobottom, I. Dionne, J. Passmore and H. D. B. Jenkins, *Inorg. Chem.*, 2000, **39**, 2042; (b) T. S. Cameron, R. J. Deeth, I. Dionne, H. D. B. Jenkins, I. Krossing, J. Passmore and H. K. Roobottom, *Inorg. Chem.*, 2000, **39**, 5614.
51. R. Faggiani, R. J. Gillespie, J. F. Sawyer and J. E. Vekris, *Acta Crystallogr., Sect. C*, 1989, **45**, 1847.
52. H. S. Low and R. A. Beaudet, *J. Am. Chem. Soc.*, 1976, **98**, 3849.
53. I. Krossing and J. Passmore, *Inorg. Chem.*, 2004, **43**, 1000.
54. I. D. Brown, D. B. Crump and R. J. Gillespie, *Inorg. Chem.*, 1971, **10**, 2319.
55. R. K. McMullan, D. J. Prince and J. D. Corbett, *Inorg. Chem.*, 1971, **10**, 1749.
56. R. C. Burns, W. C. Chan, R. J. Gillespie, W. C. Luk, J. F. Sawyer and D. R. Slim, *Inorg. Chem.*, 1980, **19**, 1432.
57. J. Beck and J. Wetterau, *Inorg. Chem.*, 1995, **34**, 6202.
58. R. C. Burns, M. J. Collins, R. J. Gillespie and G. J. Schrobilgen, *Inorg. Chem.*, 1986, **25**, 4465.
59. J. Beck, *Coord. Chem. Rev.*, 1997, **163**, 55.
60. J. Beck, F. Steden, A. Reich and H. Fölsing, *Z. Anorg. Allg. Chem.*, 2003, **629**, 1073.
61. T. W. Couch, D. Lokken and J. D. Corbett, *Inorg. Chem.*, 1972, **11**, 357.
62. J. Beck, A. Fischer and A. Stankowski, *Z. Anorg. Allg. Chem.*, 2002, **628**, 2542.

63. R. C. Burns, R. J. Gillespie, W.-C. Luk and D. R. Slim, *Inorg. Chem.*, 1979, **18**, 3086.
64. J. Beck, *Chem. Ber.*, 1995, **128**, 23.
65. P. D. Lyne, D. M. P. Mingos and T. Ziegler, *J. Chem. Soc., Dalton Trans.*, 1992, 2743.
66. J. Beck and G. Bock, *Z. Anorg. Allg. Chem.*, 1996, **622**, 823.
67. A. Baumann and J. Beck, *Z. Anorg. Allg. Chem.*, 2004, **630**, 2078.
68. G. W. Drake, G. L. Schimek and J. W. Kolis, *Inorg. Chem.*, 1996, **35**, 1740.
69. J. Beck, *Angew. Chem. Int. Ed. Engl.*, 1991, **30**, 1128.
70. J. Beck and K. Muller-Buschbaum, *Z. Anorg. Allg. Chem.*, 1997, **623**, 409.
71. J. Beck, *Angew. Chem. Int. Ed. Engl.*, 1990, **29**, 293.
72. (a) J. Beck and A. Fischer, *Z. Anorg. Allg. Chem.*, 2002, **628**, 369; (b) J. Beck and A. Stankowski, *Z. Naturforsch., Teil B*, 2001, **56**, 453.
73. J. Beck and G. Bock, *Angew. Chem. Int. Ed. Engl.*, 1995, **34**, 2559.
74. P. Boldrini, I. D. Brown, R. J. Gillespie, P. R. Ireland, W. Luk, D. R. Slim and J. E. Vekris, *Inorg. Chem.*, 1976, **15**, 765.
75. R. J. Gillespie, W. Luk, E. Maharajh and D. R. Slim, *Inorg. Chem.*, 1977, **16**, 892.
76. C. R. Lassigne and E. J. Wells, *J. Chem. Soc., Chem. Commun.*, 1978, 956.
77. G. J. Schrobilgen, R. C. Burns and P. Granger, *J. Chem. Soc., Chem. Commun.*, 1978, 957.
78. P. Boldrini, I. D. Brown, M. J. Collins, R. J. Gillespie, E. Maharajh, D. R. Slim and J. F. Sawyer, *Inorg. Chem.*, 1985, **24**, 4302.
79. R. Faggiani, R. J. Gillespie and J. E. Vekris, *J. Chem. Soc., Chem. Commun.*, 1988, 902.
80. R. C. Burns, M. J. Collins, S. J. Eicher, R. J. Gillespie and J. F. Sawyer, *Inorg. Chem.*, 1988, **27**, 1807.
81. M. J. Collins, R. J. Gillespie and J. F. Sawyer, *Inorg. Chem.*, 1987, **26**, 1476.
82. R. Steudel, *Top. Curr. Chem.*, 2003, **231**, 127.
83. E. Levillain, F. Gaillard, P. Leghie, A. Demortier and J. P. Lelieur, *J. Electroanal. Chem.*, 1997, **420**, 167.
84. T. Chivers, *Nature*, 1974, **252**, 32.
85. T. Chivers, F. Edelmann, J. F. Richardson and K. J. Schmidt, *Can. J. Chem.*, 1986, **64**, 1509.
86. A. Schliephake, H. Falius, H. Buchkremer-Hermanns and P. Böttcher, *Z. Naturforsch., Teil B*, 1988, **43**, 21.
87. A. J. Banister, D. Barr, A. T. Brooker, W. Clegg, M. J. Cunnington, M. J. Doyle, S. R. Drake, W. R. Gill, K. Manning, P. R. Raithby, R. Snaith, K. Wade and D. S. Wright, *J. Chem. Soc., Chem. Commun.*, 1990, 105.
88. D. Fenske, G. Kräuter and K. Dehnicke, *Angew. Chem. Int. Ed. Engl.*, 1990, **29**, 390.
89. M. Kantzidis and S.-P. Huang, *Inorg. Chem.*, 1989, **28**, 4667.
90. W. S. Sheldrick and H. G. Braunbach, *Z. Naturforsch., Teil B*, 1989, **44**, 1397.

91. (a) R. Zagler and B. Eisenmann, *Z. Naturforsch., Teil B*, 1991, **46**, 5943; (b) S.-P. Huang, S. Dhingra and M. G. Kanatzidis, *Polyhedron*, 1992, **11**, 1869.
92. (a) M. G. Kanatzidis, *Angew. Chem. Int. Ed. Engl.*, 1995, **34**, 2109; (b) D. M. Smith and J. A. Ibers, *Coord. Chem. Rev.*, 2000, **200–202**, 187.
93. W. S. Sheldrick and B. Schaaf, *Z. Naturforsch., Teil B*, 1994, **49**, 993.
94. (a) D. Y. Valentine, O. B. Cavin and H. L. Yakel, *Acta Crystallogr., Sect. B*, 1977, **33**, 1389; (b) P. Bottcher and R. Keller, *Z. Anorg. Allg. Chem.*, 1986, **542**, 144.
95. C. Reisner and W. Tremel, *J. Chem. Soc., Chem. Commun.*, 1997, 387.
96. B. Schreiner, K. Dehnicke, K. Maczek and D. Fenske, *Z. Anorg. Allg. Chem.*, 1993, **619**, 1414.
97. C. J. Warren, R. C. Haushalter and A. B. Bocarsly, *J. Alloys Compd.*, 1996, **233**, 23.
98. P. Sekar and J. A. Ibers, *Inorg. Chem.*, 2004, **43**, 5436.
99. N. N. Greenwood and A. Earnshaw, *Chemistry of the Elements*, Butterworth-Heinemann, Oxford, 1997, pp. 659–660.
100. R. Keller, W. B. Holzapfel and H. Schulz, *Phys. Rev. B*, 1977, **16**, 4404.
101. C. Adenic, V. Langer and O. Lindquist, *Acta Crystallogr.*, 1989, **45**, 941.
102. T. W. Smith, S. D. Smith and S. S. Badesha, *J. Am. Chem. Soc.*, 1984, **106**, 7247.
103. T. Chivers, *A Guide to Chalcogen–Nitrogen Chemistry*, World Scientific, Singapore, 2005.
104. H. G. Heal and J. Kane, *Inorg. Synth.*, 1968, **11**, 184.
105. R. Steudel and F. Rose, *J. Chromatogr.*, 1981, **216**, 399.
106. H.-J. Hecht, R. Reinhardt, R. Steudel and H. Bradaczek, *Z. Anorg. Allg. Chem.*, 1976, **426**, 43.
107. (a) C.-C. Wang, Y.-Y. Hong, C.-H. Ueng and Y. Wang, *J. Chem. Soc., Dalton Trans.*, 1992, 3331; (b) D. Gregson, G. Klebe and H. Fuess, *J. Am. Chem. Soc.*, 1988, **110**, 8488.
108. T. Chivers, M. Edwards, D. D. McIntyre, K. J. Schmidt and H. J. Vogel, *Magn. Reson. Chem.*, 1992, **30**, 177.
109. (a) T. Chivers and I. Drummond, *Inorg. Chem.*, 1974, **13**, 122; (b) T. Chivers, W. G. Laidlaw, R. T. Oakley and M. Trsic, *J. Am. Chem. Soc.*, 1980, **102**, 5773.
110. T. Chivers and K. J. Schmidt, *J. Chem. Soc., Chem. Commun.*, 1990, 1342.
111. R. Steudel and F. Rose, *Z. Naturforsch., Teil B*, 1978, **30**, 122.
112. R. Faggiani, R. J. Gillespie, C. J. L. Lock and J. D. Tyrer, *Inorg. Chem.*, 1978, **17**, 2975.
113. K. Bergemann, M. Kustos, P. Krüger and R. Steudel, *Angew. Chem. Int. Ed. Engl.*, 1995, **34**, 1330.
114. (a) R. Jones, D. J. Williams and J. D. Woollins, *Angew. Chem. Int. Ed. Engl.*, 1985, **24**, 760; (b) J. Novosad, D. J. Williams and J. D. Woollins, *Z. Anorg. Allg. Chem.*, 1994, **620**, 495.
115. (a) R. Steudel, O. Schumann, J. Buschmann and P. Luger, *Angew. Chem. Int. Ed.*, 1998, **37**, 492; (b) S. Sasaki, H. Hatsushiba and M. Yoshifuji, *Chem. Commun.*, 1998, 2221.

116. R. Steudel, K. Bergemann, J. Buschmann and P. Luger, *Angew. Chem. Int. Ed.*, 1998, **35**, 2537.
117. C. M. Mikulski, P. J. Russo, M. S. Soran, A. G. MacDiarmid, A. F. Garito and A. J. Heeger, *J. Am. Chem. Soc.*, 1975, **97**, 6358.
118. (a) A. J. Banister and Z. V. Hauptman, *J. Chem. Soc., Dalton Trans.*, 1980, 731; (b) T. Chivers, A. W. Cordes, R. T. Oakley and P. N. Swepston, *Inorg. Chem.*, 1981, **20**, 2376.
119. K. Dehnicke and U. Müller, *Transition Met. Chem.*, 1985, **10**, 261.
120. M. Bénard, *New J. Chem.*, 1986, **10**, 529.
121. T. Chivers and F. Edelmann, *Polyhedron*, 1986, **5**, 1661.
122. S. Herler, P. Mayer, H. Nöth, A. Schulz, M. Suter and M. Vogt, *Angew. Chem. Int. Ed.*, 2001, **40**, 3173.
123. R. J. Gillespie, J. P. Kent and J. F. Sawyer, *Inorg. Chem.*, 1981, **20**, 3784.
124. K. M. Johnson, K. F. Preston and L. H. Sutcliffe, *Mag. Reson. Chem.*, 1988, **26**, 1015.
125. (a) F. Grein, *Can. J. Chem.*, 1993, **71**, 335; (b) W. V. F. Brooks, T. S. Cameron, S. Parsons, J. Passmore and M. J. Schriver, *Inorg. Chem.*, 1994, **33**, 6230.
126. R. J. Gillespie, J. P. Kent, J. F. Sawyer, D. R. Slim and J. D. Tyrer, *Inorg. Chem.*, 1981, **20**, 3799.
127. A. J. Banister, Z. V. Hauptman and A. G. Kendrick, *J. Chem. Soc., Dalton Trans.*, 1987, 915.
128. H. Johansen, *J. Am. Chem. Soc.*, 1988, **110**, 5322.
129. J. Bojes, T. Chivers, W. G. Laidlaw and M. Trsic, *J. Am. Chem. Soc.*, 1979, **101**, 4517.
130. P. N. Jagg, P. F. Kelly, H. S. Rzepa, D. J. Williams, J. D. Woollins and W. Wylie, *J. Chem. Soc., Chem. Commun.*, 1991, 942.
131. T. Chivers, A. W. Cordes, R. T. Oakley and W. T. Pennington, *Inorg. Chem.*, 1983, **22**, 2429.
132. W. L. Jolly and K. D. Maguire, *Inorg. Synth.*, 1967, **9**, 102.
133. K. Tersago, V. Matuska, C. Van Alsenoy, A. M. Z. Slawin, J. D. Woollins and F. Blockhuys, *Dalton Trans.*, 2007, 4529.
134. E. Jaudas-Prezel, R. Maggiulli, R. Mews, H. Oberhammer and W.-D. Stohrer, *Chem. Ber.*, 1990, **123**, 2117.
135. O. Glemser and R. Mews, *Angew. Chem. Int. Ed. Engl.*, 1980, **19**, 883.
136. R. Jones, I. P. Parkin, D. J. Williams and J. D. Woollins, *Polyhedron*, 1987, **6**, 2161.
137. J. Passmore and M. Schriver, *Inorg. Chem.*, 1988, **27**, 2749.
138. A. Apblett, A. J. Banister, D. Biron, A. G. Kendrick, J. Passmore, M. Schriver and M. Stojanac, *Inorg. Chem.*, 1986, **25**, 4451.
139. K. Dehnicke and U. Muller, *Comments Inorg. Chem.*, 1985, **4**, 213.
140. A. Apblett and T. Chivers, *Inorg. Chem.*, 1989, **28**, 4544.
141. (a) C. W. Rees, *J. Heterocycl. Chem.*, 1992, **29**, 639; (b) T. Torroba, *J. Prakt. Chem.*, 1999, **341**, 99.

142. (a) X. G. Duan, X. L. Duan, C. W. Rees and T. Y. Yue, *J. Chem. Soc., Perkin Trans.*, 1997, 2597; (b) X. L. Duan and C. W. Rees, *Chem. Commun.*, 1997, 1493.
143. (a) I. Zborilova and P. Gebauer, *Z. Anorg. Allg. Chem.*, 1979, **448**, 5; (b) I. Ruppert, *J. Fluorine Chem.*, 1982, **20**, 241.
144. Z. Zak, *Acta Crystallogr., Sect. B*, 1981, **37**, 23.
145. (a) G. A. Wiegers and A. Vos, *Acta Crystallogr.*, 1963, **16**, 152; (b) M. H. Palmer, R. T. Oakley and N. P. C. Westwood, *Chem. Phys.*, 1989, **131**, 255.
146. H. Kluver and O. Glemser, *Z. Naturforsch., Teil B*, 1977, **32**, 1209.
147. T.-P. Lin, U. Klingebiel and O. Glemser, *Angew. Chem. Int. Ed. Engl.*, 1972, **11**, 1095.
148. (a) A. C. Hazell, G. Wiegers and A. Vos, *Acta Crystallogr.*, 1966, **20**, 186; (b) E. Lork, U. Behrens, G. Steinke and R. Mews, *Z. Naturforsch., Teil B*, 1994, **49**, 437.
149. H. Wagner, D. L. Wagner and O. Glemser, *Chem. Ber.*, 1977, **110**, 683.
150. (a) C. Leben and M. Jansen, *Z. Naturforsch., Teil B*, 1999, **54**, 757; (b) V. Kocman and J. Rucklidge, *Acta Crystallogr. Sect. B*, 1974, **30**, 6.
151. (a) For reviews, see (a) M. M. Labes, P. Love and L. F. Nichols, *Chem. Rev.*, 1979, **79**, 1; (b) A. J. Banister and I. B. Gorrell, *Adv. Mater.*, 1998, **10**, 1415.
152. J. M. Rawson and J. J. Longridge, *Chem. Soc. Rev.*, 1997, 53.
153. (a) C. M. Mikulski, P. J. Russo, M. S. Saran, A. G. MacDiarmid, A. F. Garito and A. J. Heeger, *J. Am. Chem. Soc.*, 1975, **97**, 6358; (b) M. J. Cohen, A. F. Garito, A. J. Heeger, A. G. MacDiarmid, C. M. Mikulski, M. S. Saran and J. Kleppinger, *J. Am. Chem. Soc.*, 1976, **98**, 3844.
154. (a) H. Müller, S. O. Svensson, J. Birch and Å. Kvick, *Inorg. Chem.*, 1997, **36**, 1488; (b) R. S. P. King, P. F. Kelly, S. E. Dann and R. J. Mortimer, *Chem. Commun.*, 2007, **1**, 4812.
155. (a) J. F. Rubison, T. D. Behymer and H. B. Mark Jr, *J. Am. Chem. Soc.*, 1982, **104**, 1224; (b) P. F. Kelly, R. S. P. King and R. J. Mortimer, *J. Chem. Soc., Chem. Commun.*, 2008, 6111.
156. J. W. Macklin, G. B. Street and W. D. Gill, *J. Chem. Phys.*, 1979, **70**, 2425.
157. W. Isenberg, R. Mews and G. M. Sheldrick, *Z. Anorg. Allg. Chem.*, 1985, **525**, 54.
158. G. Wolmerhäuser and P. R. Mann, *Z. Naturforsch., Teil B*, 1991, **46**, 315.
159. J. J. Mayerle, J. Kuyper and G. B. Street, *Inorg. Chem.*, 1978, **17**, 2610.
160. A. K. Roy, G. T. Burns, G. C. Lie and S. Grigoras, *J. Am Chem. Soc.*, 1993, **115**, 2604.
161. (a) V. C. Ginn, P. F. Kelly and J. D. Woollins, *J. Chem. Soc., Dalton Trans.*, 1992, 2129; (b) J. Siivari, T. Chivers and R. S. Laitinen, *Inorg. Chem.*, 1993, **32**, 1519.
162. T. Chivers, *J. Chem. Soc., Dalton Trans.*, 1996, 1185.

163. W. Mosa, C. Lau, M. Möhlen, B. Neumüller and K. Dehnicke, *Angew. Chem. Int. Ed.*, 1998, **37**, 2840.
164. (a) T. Maaninen, H. M. Tuononen, G. Schatte, R. Suontamo, J. Valkonen, R. S. Laitinen and T. Chivers, *Inorg. Chem.*, 2004, **43**, 2097; (b) T. Maaninen, T. Chivers, R. S. Laitinen, G. Schatte and M. Nissinen, *Inorg. Chem.*, 2000, **39**, 5341; (c) T. Maaninen, R. Laitinen and T. Chivers, *Chem. Commun.*, 2002, 1812.
165. T. Maaninen, T. Chivers, R. S. Laitinen, G. Schatte and E. Wegelius, *Chem. Commun.*, 2000, 759.
166. H. W. Roesky, K.-L. Weber and J. W. Bats, *Chem. Ber.*, 1984, **117**, 2686.
167. T. Chivers, X. Gao and M. Parvez, *J. Am. Chem. Soc.*, 1995, **117**, 2359.
168. T. Chivers, G. D. Enright, N. Sandblom, G. Schatte and M. Parvez, *Inorg. Chem.*, 1999, **38**, 5431.
169. N. Sandblom, T. Ziegler and T. Chivers, *Inorg. Chem.*, 1998, **37**, 354.
170. H. M. Tuononen, R. Suontamo, J. Valkonen, R. S. Laitinen and T. Chivers, *J. Phys. Chem. A*, 2005, **109**, 6309.
171. (a) P. F. Kelly and A. M. Z. Slawin, *J. Chem. Soc., Dalton Trans.*, 1996, 4029; (b) P. F. Kelly and A. M. Z. Slawin, *Angew. Chem. Int. Ed. Engl.*, 1995, **34**, 1758.
172. S. M. Aucott, D. Drennan, S. L. James, P. F. Kelly and A. M. Z. Slawin, *Chem. Commun.*, 2007, 3054.
173. A. Maaninen, R. S. Laitinen, T. Chivers and T. A. Pakkanen, *Inorg. Chem.*, 1999, **38**, 3450.
174. E. G. Awere, J. Passmore and P. S. White, *J. Chem. Soc., Dalton Trans.*, 1993, 299.
175. E. G. Awere, J. Passmore and P. S. White, *J. Chem. Soc., Dalton Trans.*, 1992, 1267.
176. E. G. Awere, J. Passmore, K. F. Preston and L. H. Sutcliffe, *Can. J. Chem.*, 1988, **66**, 1776.
177. A. Maaninen, J. Konu, R. S. Laitinen, T. Chivers, G. Schatte, J. Pietikäinen and M. Ahlgrén, *Inorg. Chem.*, 2001, **40**, 3539.
178. (a) R. Wollert, B. Neumüller and K. Dehnicke, *Z. Anorg. Allg. Chem.*, 1992, **616**, 191; (b) C. Lau, B. Neumüller and K. Dehnicke, *Z. Naturforsch., Teil B*, 1997, **52**, 543.
179. T. Chivers, J. Siivari and R. S. Laitinen, *Inorg. Chem.*, 1993, **32**, 4391.
180. J. Passmore, G. Schatte and T. S. Cameron, *J. Chem. Soc., Chem. Commun.*, 1995, 2311.
181. C. Lau, B. Neumüller and K. Dehnicke, *Z. Anorg. Allg. Chem.*, 1996, **622**, 739.
182. (a) A. Haas, *J. Organomet. Chem.*, 2002, **646**, 80; (b) A. Haas, *Adv. Heterocycl. Chem.*, 1998, **71**, 115.
183. H. W. Roesky, J. Münzenberg and M. Noltemeyer, *Angew. Chem. Int. Ed. Engl.*, 1990, **29**, 61.
184. (a) A. Haas and M. Pryka, *J. Chem. Soc., Chem. Commun.*, 1994, 391; (b) A. Haas and M. Pryka, *Chem. Ber.*, 1995, **128**, 11.
185. K. Stahl, J.-P. Legros and J. Galy, *Z. Kristallogr.*, 1992, **202**, 99.

186. G. A. Ozin and A. Vander Voet, *J. Mol. Struct.*, 1971, **10**, 173.
187. J. C. Champarnaud-Mesjard, S. Blanchadin, P. Thomas, A. Mirgorodsky, T. Merle-Mejean and B. Frit, *J. Phys. Chem. Solids*, 2000, **61**, 1499.
188. W. S. McDonald and D. W. J. Cruikshank, *Acta Crystallogr.*, 1967, **22**, 48.
189. R. Westrik and C. H. McGillavry, *Acta Crystallogr.*, 1954, **7**, 764.
190. F. C. Mijlhoff, *Acta Crystallogr.*, 1965, **18**, 795.

Subject Index

Notes: Page numbers in **bold** type indicate a more comprehensive coverage of a topic. Page numbers in *italic* refer to figures or schemes. Inorganic ring compounds are listed without the cyclo- prefix. For example, cycloborazines are listed as borazines.

alkane elimination reactions 15–16, 137
aluminium homocycles
   anions 114–15
   radicals 53
aluminium–chalcogen rings 148–50
aluminium–nitrogen rings 135–8
aluminium–oxygen rings 147
aluminium–pnictogen rings
   saturated 140–2
   unsaturated 138–40
alumoxanes 147
amine–borane adducts 15, *16*, 126
aminoboranes *see* borazanes
ammonia–borane 124, 126
anion radicals 34, 52, 56–7
annulenes 66–7
antimony homocycles
   cyclostibines 14, 84, 228–30
   as ligands 84, 214
   polyanions 215, 217
antimony–chalcogen rings 260–1, 263–4
antimony–nitrogen rings 250–2
   macrocycles 78
antimony–oxygen rings 255
aromaticity *see* delocalisation
arsatetrazoles 251–2
arsathianes 261–2
arsazanes 250–1

arsazenes 252
arsenic homocycles
   cycloarsines 12, 25, 84, 226–7
   as ligands 84, 93, 214, 215–17
   polyanions 214–17
arsenic–chalcogen rings 260–3
arsenic–nitrogen rings
   cycloarsazanes and pnictazanes 250–2
   cycloarsazenes 252
arsenic–oxygen rings 254–5
arsines 12, 25, 226–7
   as ligands 84
arsoxane rings 254–5

bicyclo[1.1.0]butanes 59–60
biradicals 52, 130–3
   biradicaloids 58, 59–62
   as reaction intermediates 62–4
   stable 58–9
bismuth homocycles
   cyclobismuthines 84, 229–30
   polyanions 215, 217
bismuth–chalcogen rings 260–1, 263
bismuth–nitrogen rings 250–2
bismuth–oxygen rings 255
bond angles 21–2
bond radii 21, *22*
bond-stretch isomers 131

Subject Index

bonding *see* electronic structure and bonding
borates 143
boratophosphazene rings 243
borazanes 15, 16, **124–5**
borazines 2, 15, **116–20**
 carbocation-containing 119
 as ceramic precursors 125–6
 metal complexes 93–4, 117–18
 structure and bonding 22–3, 40–1
 trichloroborazine synthesis 8
 UV spectroscopy 36
boric oxide 144
boron disulfide 71
boron halides 110–11
boron homocycles **110–13**
 polyboron ligands 113
boron nitride ceramics 125
boron sulfide 144
boron–chalcogen rings 144–7
boron–nitrogen polymers 126–8
boron–nitrogen rings 15–16, **116–28**
 applications 125–6
 saturated cycloborazanes 124–5
 unsaturated 116–24
  cycloborazines *see* borazines
  Dewar borazines 120–1
  hetero-substituted 122
boron–oxygen rings 22–3, 143–4
boron–phosphorus polymers/chains 134–5
boron–phosphorus rings *11*, 15, 25, **128–35**
 saturated 133–4
 unsaturated 128–9
  biradicals 62, 130–3
boron–selenium rings 144–7
boron–sulfur rings 144–7
boroxanes 143–4
boroxines 143
butadiene ring analogues 174, 177

carbophosphazene macrocycles 73
catena-selenium dianions 296
catena-sulfur anions 57, 63, **294–6**
cation radicals 34, 52, 55–6

ceramic precursors
 borazines 125–6
 boron–nitrogen polymers 126–8
 polysilanes 108
 polysilazanes 184
chain conformations 107
chalcogen heterocycles 279–80
 anions 3, 299–300
 cations 3, 293–4
 ring transformations 280–2
chalcogen homocycles 276–8
 anions 294–9
 cations 285–93
 with exocyclic oxygen/halogen substituents 282–4
chalcogen imides 69–70, 312–14
chalcogen ketone congeners 201
chalcogen–gallium rings 148–50
chalcogen–germanium rings 196–202
chalcogen-lead rings 196, 202
chalcogen–nitrogen rings 10, 276, 312–17
 see also sulfur–nitrogen
chalcogen–oxygen rings 317–19
chalcogen–silicon rings 179, 196–202
chalcogen–tin rings *11*, 196–202
characterisation methods
 inorganic polymers 102–8
 inorganic rings 20–37
chromatography 23–4
complexes *see* ligand chemistry
condensation polymerisation 99–100
condensation reactions 7–11
conducting polymers 10, 276, **309–11**
conjugated systems 39–40
coordination complexes *see* ligand chemistry
covalent bond radii 21, *22*
crown ethers/crown ether analogues 66–7, 74–7, 191, 216
crystallisation transition 106
cycloaddition reactions
 and frontier orbitals 48–50
 synthetic routes 16–18
cyclobutadiene analogues 44, 174, 177
cyclobutanediyls 59–60
cyclocondensation reactions 7–11

dehydrocoupling reactions 13–15
delocalisation/aromaticity 39–45
  and cyclosilanes 160
  lone pair aromaticity 215
  and polygermanes 169
  and polysilanes 167–8
deprotonation reactions 74–5
Dewar borazines 120–1
diarsazanes 250–1
dichalcogenadiazolyl radicals 21, 48, 54–5, 58–9
dichlorosulfane 7–8
Diels–Alder reactions 16
1,3-diphosphacyclobutane-2,4-diyls 61–2
diphosphazane complexes 234–5
diphosphenes 17, 91
disilenes 17, 162, 170
disjoint biradicals 58
dithiadiazolium cations 49
dithiadiazolyl radicals 54–5, 58–9
  EPR spectra 34, 35
dithionitryl cation 17, 49

electroluminescence 169
electron diffraction 22–3
electron paramagnetic resonance see EPR
electron-rich aromatics 43, 276, 304–6
electronic structure and bonding **39–51**
  frontier orbital considerations 45–50
  π-electron delocalisation 40–5
  σ-electron delocalisation 45
EPR spectroscopy 20, 33–6

frontier orbitals 45–50
fullerene analogues 78–80

gallium homocycles 114–15
gallium nitride semiconductor 142
gallium–chalcogen rings 148–50
gallium–nitrogen rings 135–8
gallium–oxygen rings 147
gallium–pnictogen rings
  saturated 140–3
  unsaturated 138–40
galloxanes 147
germanazanes 181

germanium homocycles
  cyclogermanes 12, 161, 162–3
  cyclogermenes 170–1
    radicals/ions 53, 173, 175–6
  polyanions 165–7
germanium–chalcogen rings 196–202
germanium–nitrogen rings 181–3
  biradicals 183
germanium–oxygen rings 192
germanium–phosphorus rings 186
germanium–silicon rings 162, 172–3, 176–7
germaphosphanes 186
germenyl radicals 175–6
germoxanes 192
glass transition 106
group 13 systems **110–56**
  aluminium/gallium/indium–chalcogen rings 148–50
  aluminium/gallium/indium–nitrogen rings
    saturated 138
    unsaturated 135–8
  aluminium/gallium/indium–pnictogen rings
    saturated 140–2
    semiconductor precursors 142–3
    unsaturated 138–40
  boron–nitrogen rings 116
    applications 125–6
    saturated rings 124–5
    unsaturated rings 116–24
  boron–oxygen rings 143–4
  boron–phosphorus rings
    saturated 133–4
    unsaturated 128–9
  biradicals 130–3
  boron–sulfur/seleniun rings 144–7
  homoatomic anionic rings 113–15
  polymers/chains
    boron–nitrogen 126–8
    boron–phosphorus 134–5
group 14 systems **158–210**
  homoatomic rings
    saturated 159–65
    unsaturated 170–9

polymers
  homoatomic 167–9
  silicon–nitrogen 184
  silicon–oxygen (silicones) 193–6
silicon/germanium/tin–nitrogen rings
  saturated 179
  unsaturated and biradicals 181–3
silicon/germaniun/tin/lead–chalcogen rings 192–202
silicon/germaniun/tin–oxygen rings 187–93
silicon/germaniun/tin–phosphorus rings 184–7
group 15 systems **211–74**
  arsenic/antimony/bismuth–nitrogen rings 250–2
  homoatomic rings
    cycloarsines/stilbenes/bismuthines 226–31
    cyclophosphines 217–26
    polyanions 214–17
    polynitrogen 212–13
  phosphorus–nitrogen rings
    saturated 231–6
    unsaturated 236–45
  pnictogen–chalcogen rings 255–64
  pnictogen–oxygen rings 253–5
  polymers
    ladder polymers 230–1
    polyphosphazenes 245–50
group 16 systems **275–327**
  anionic chalcogen rings and chains 294–300
  cationic chalcogen rings 284–94
  neutral sulfur/selenium/tellurium rings
    heteroatomic 279–80
    homoatomic 276–9
    ring transformations 280–1
  oxidised chalcogen homocycles 282–4
  polymers
    poly(oxathiazenes)/sulfanuric polymers 311–12
    poly(sulfur nitride) 309–11
    sulfur/selenium/tellurium chains 300–1
    sulfur/selenium/tellurium–oxygen chains 317–19
    sulfur–nitrogen chains 311

  selenium/tellurium–nitrogen rings
    saturated 312–14
    unsaturated 314–17
  selenium/tellurium–oxygen rings 317–19
  sulfur–nitrogen rings
    saturated 301–3
    unsaturated 304–9
  sulfur–oxygen rings 317–19

heavy ketones 201
historical background 2
host–guest chemistry **74–7**, 89–90
Hückel rule 39
hydrogen storage 15, 126
hyperfine coupling constant 34

iminoalanes 135, 137
iminoboranes 116, 121, 122
indium homocycles 114
indium–chalcogen rings 148–50
indium–nitrogen rings 136, 137–8
indium–pnictogen rings 140–1
  semiconductor precursors 142–3
infrared spectroscopy 31–3
inorganic heterocycles introduced 1–3
inorganic polymers *see* polymers
intermolecular interactions 21
inverse crown structures 74–7, 191
IR spectroscopy 31–3

Jahn–Teller distortion 39, 50

ketone congeners 109

ladder polymers 230–1
Landé splitting factor 34
lead homocycles
  cycloplumbanes 164–5
  polyanions 165–7
lead–chalcogen rings 196, 202
lead–sulfur rings 202
ligand chemistry **82–96**
  cyclobutadiene analogues 44, 174, 177
  cyclodiphosphazane complexes 234–5
  group 13 anionic rings 113–14
  π-complexes 90–4

σ-complexes
   heterocyclic ligands 86–8
   homocyclic ligands 82–6
   macrocyclic ligands 88–90

macrocycles **66–80**
   heterocyclic systems
      as ligands 88–90
      saturated 68–71, 191, 234–6
      unsaturated 71–4
   homocyclic systems 67–8
   host–guest chemistry 74–7
   supramolecular assemblies 77–80, 237
magnetic equivalence 30
mass spectrometry 24–5, 103, 104
melting temperature 105
molar extinction coefficients 36
molecular conductors 10, 276, **309–11**
molecular weight of polymers 103–4

negative hyperconjugation 42
neutral radicals 52, 53–5
nitrogen homocycles 212–13
nitrogen–arsenic rings 213
nitrogen–phosphorus rings 213
nitrogen–selenium systems *see* selenium–nitrogen
nitrogen–sulfur systems *see* sulfur–nitrogen
NMR spectroscopy 20, 25–33
   applications 27–30, 102
   and boron–phosphorus biradicals 131
   NMR properties of elements 25–7
   spin system notation 30–1
norbornadiene adducts 50

oxygen 284–5

paramagnetic rings 52
   anion radicals 56–7
   cation radicals 55–6
   EPR spectroscopy 33–6
   intermolecular interactions 21
   neutral radicals 53–5
pentazoles 212
permethylcyclosilanes 27

phosphaalkenes 99
phosphapolysilanes 184
phosphates 70, 253
phosphazanes 75–7, **231–6**
phosphazenes *see under* phosphorus–nitrogen rings
phosphazene–sulfanuric macrocycles 72
phosphines *see* phosphorus homocycles
phosphinoboranes 133–4
   chains and polymers 134–5
phosphinophosphonium ions 31
phosphorus homocycles (cyclophosphines) 12, 14, 39, 40, **217–26**
   anionic systems 12–13, 33, 224–6
   binary anion 242
   cationic systems 221–4
   as ligands 82–3, 91–3
   polyanions 213–17
   reactions of 219–21
phosphorus–nitrogen polymers (polyphosphazenes) 97, 245–50
phosphorus–nitrogen rings
   cyclophosphazanes 75–7, **231–6**
      macrocycles 234–6
   cyclophosphazenes 2, 8–9, **236–45**
      hexacation 240–1
      hexachloro derivative 237–9, *246*
      isomerisation 239–41, *241*
      as ligands 86–7, 90
      macrocyclic systems 71, *72*, 90, 241–2
      NMR spectroscopy 27–8
      and polymerisation 101
      structure and bonding 41–2
      synthesis 8–9
      UV–visible spectroscopy 36–7
phosphorus–nitrogen–boron rings 243–4
phosphorus–nitrogen–sulfur rings 23–4, 87–8, 94, 244
phosphorus–oxygen rings 252–4
phosphorus–selenium rings 30–1
   binary 255–6
   organophosphorus 258–9
phosphorus–sulfur rings 30–1
   binary 255–6
   organophosphorus 256–8
phosphorus–tellurium rings 259–60

Subject Index

photoconductivity 169
photosensitivity 170
π-complexes 90–4
π-electron delocalisation 40–5
   electron precision 41
   manifolds 44
π–π interactions 46–7
plumbanes 164–5
pnictazane rings 250–2
pnictogen anions 214–15
pnictogen–aluminium rings 138–42
pnictogen–chalcogen rings 255–64
pnictogen–gallium rings 140–3
pnictogen–indium rings 140–3
pnictogen–oxygen rings 252–5
poly (sulfur nitride) 276, **309–11**
polyaminoboranes 128
polyborazylenes 108, 126–7
polycarbophosphazenes 249–50
polycarbosilanes 108
poly(dimethylsiloxane) 100
   chain conformation 107, *108*
   thermal transitions 107
polydispersity index 103
polygermanes 169
polyheterophosphazenes 249–50
polyiminoboranes 128
polymers, inorganic **97–109**
   characterisation 102–3
      molecular weights/polydispersion 103–4
      thermal transitions 104–7
   conducting 10, 276, 309–11
   ladder 230–1
   potential synthetic routes 98–100
      ring-opening polymerisation 100–2
poly(oxathiazene) 311–12
polyphosphate anions 70
polyphosphate macrocycles 70
polyphosphazenes 29–30, 97, **245–50**
polyphosphide anions 215
polyphosphines 99
polyphosphinoboranes 134–5
polysilanes 2, 13, 39, 97, 158, **167–9**
   as ceramic precursors 108
polysilazanes 184

polysiloxanes (silicones) 2, 97, 101, 158, **193–6**
polystannanes 13, 169–70
poly(sulfur nitride) 64, 102, 276, **309–11**
polythionylphosphazene 97, 249–50
pyridine 1

radicals *see* paramagnetic rings
Raman spectrocopy 31–3
reductive coupling reactions 11–13
   and boron homocycles 111–12
   and cyclopolysilanes 159
ring tranformations 280–1
ring-opening polymerisation 100–2
   cyclotriphosphazene interactions 240–1
   and polysilanes 168

salt-elimination reactions 10–11
Salvarsan 13, 25, *26*, 227
sandwich complexes 84–5, 92–3, 113, 216
selenaboranes 146–7
selenium homocycles 275, 277–8
   catena- and spirocyclic anions 296–7
   cations 289
   with exocyclic halogen substituents 282–4
   as ligands 85, *86*
   ring transformations 281
selenium imides 69–70, **312–13**, *314*
selenium nitride ligands 87–8
selenium polymers 301
selenium sulfides 279–80, 281
selenium-nitrogen halides 315–16
selenium–nitrogen rings
   cations 56, 315
   as ligands 87–8
   NMR spectroscopy 28, *29*
   saturated (imides) 69–70, 312–13, *314*
   unsaturated
      binary rings 314–15
      halides 315–16
selenium–oxygen rings 317–19
selenium–silicon rings 196–202
selenium–tellurium anions 299–300
selenium–tellurium cations 293–4
semiconductor precursors 142–3

σ-complexes 82–90
σ-electron delocalisation 45
  cyclosilanes 160
  and polygermanes 169
  and polysilanes 167–8
silanes *see under* silicon homocycles
silaphosphanes 184–6
silazanes 179–81
silenes
  cyclosilenes 12, 170–2, 174–6, 178
  disilenes 17, 162, 170
silenyl radical 53–4
silicon carbide 169
silicon homocycles
  anion radical 57
  saturated (cyclosilanes) 2, 40, **159–62**
    as ligands 92
    macrocyclic systems 67–8
    NMR spectroscopy 27
    polyanions 165–7
    and polymerisation 101–2
    radical anions/cations 56, 160
    σ-electron delocalisation 45, 160
    synthesis 11–12
    UV–visible spectroscopy 36
  unsaturated
    bicyclic cyclopentasilene 178
    cyclotetrasilenes 174–6
    cyclotrisilenes 12, 170–2
    silacyclobutadienes 178–9
silicon nitride 181
silicones *see* polysiloxanes
silicon–carbon cation 173–4
silicon–chalcogen rings 179, **196–202**
silicon–germanium rings 162, 172–3, 173–4, 176–7
silicon–nitrogen polymers (polysilazanes) 184
silicon–nitrogen rings (cyclosilazanes) 179–81
silicon–oxygen polymers *see* polysiloxanes
silicon–oxygen rings (cyclosiloxanes) 70, 158, **187–9**
  incorporating heteroatoms 189–91
  as ligands 88–90
  ring-opening polymerisation 101–2

silicon–oxygen–nitrogen rings 191
silicon–phosphorus rings
  cyclophosphapolysilanes 184
  cyclosilaphosphanes 184–6
silicon–tin rings 165
Silly Putty 195–6
siloxanes *see* silicon–oxygen
spectroscopic analysis 20–37
spin system notation 30–1
spin-active nuclei 27
spirocyclic selenium anions 296–7
stannanes 11–12, 161, **162–5**
stannaphosphanes 186–7
stannazanes 181
stannenes 171
stannoxanes 192–3
stibines 14, 84, 228–30
structure and bonding *see* electronic structure
sufanuric halides 308–9
sulfanuric macrocycle systems 72, 74
sulfanuric polymers 311–12
sulfur homocycles 2, 67–8, 275, **276–7**
  catena-sulfur anions 57, 63, **294–6**
  cations 56, 283, **285–9**
  chromatographic separation 23, *24*
  with exocyclic oxygen/halogen substituents 282–4
  IR spectrosopy 32
  as ligands 84–6
  and mass spectrometry 25
  polymerisation 62–3
  ring tranformations 280–1
  synthetic methods 7–8
  torsional bond angles 21–2
  UV-visible spectroscopy 36
sulfur imides 9, **301–3**
  chromatographic separation 23
  macrocyclic systems 68–9
sulfur nitride ligands 87–8
sulfur nitride polymers 64, 102, 276, **309–11**
sulfur polymers 300–1
sulfur–nitrogen halides 9, 306–8
sulfur–nitrogen polymers and chains 64, 102, 276, **309–12**

Subject Index

sulfur–nitrogen rings 9–10, 40, **301–8**
  cations and anions 43–5, 56, 304–7
  cyclic sulfur imides 301–3
  electronic structure/aromaticity 43–5
  infrared spectra 33
  as ligands 87–8
  NMR spectra of ions 28, 29, 30
  π-π interactions 46, 47
  unsaturated
    two-coordinate sulfur 43–5, 304–6
    three-coordinate sulfur 306–8
    four-coordinate sulfur 308–9
  UV–visible spectroscopy 37
sulfur–oxygen rings 317–19
sulfur–selenium rings 279–80
sulfur–silicon rings 196–202
supramolecular assemblies 77–80
synthesis
  inorganic polymers 98–100
    ring-opening polymerisation 100–2
  inorganic rings 7
    alkane elimination 15–16
    cycloaddition 16–18
    cyclocondensation 7–11
    dehydrocoupling 13–15
    reductive coupling 11–13

tellurium homocycles 275, 278–9
  anions 297–9
  cations 290–3
tellurium imides 313–4
tellurium polymers 301
tellurium sulfides 279–80
tellurium–nitrogen rings
  saturated (imides) 313–14
  unsaturated (nitrogen halides) 315–17
tellurium–oxygen rings 317–19
tellurium–selenium anions 299–300
tellurium–selenium cations 293–4
tellurium–silicon rings 196–202
tellurium–sulfur cations 294

tetra-t-butylcyclopentaphosphanide anion 12–13, 224–5
tetrahydrofuran 1
tetrasilabicyclo[1.1.0]silanes 60–1
tetrasilabutadiene 92
tetrazoles 213
thermal transitions 104–7
thermogravimetric analysis 107–8
thiaboranes 146–7
thiatriazinyl radical 46–7
thiazyl macrocycle 73–4
thionitrosyl cation 17
thionylphosphazene 100
thiophene 1
thiophosphazenes 244–5
thiosulfoxides 280–1
tin dihydrides 13
tin homocycles
  cyclostannanes 161, 163–4
  cyclotristannenes 171
  polyanions 165–7
tin–chalcogen rings *11*, 196–202
tin–nitrogen rings 182, *183*
  saturated (cyclostannazanes) 181
  unsaturated 182–3
tin–oxygen rings 192–3
tin–phosphorus rings 186–7
torsional bond angles 21, *22*
trichloroborazines 8
trisilylated benzamidines 10

UV–visible spectroscopy 36–7

van der Waals radii 21, *22*
visible spectroscopy 36–7

Wade's rules 110
Wurtz coupling 168

X-ray diffraction 20, 21–3, 105

Zintl anions 165, 215
zwitterions 42